Communications
Receivers:
Principles and Design

Other McGraw-Hill Communications Books of Interest

Communications Receivers: Principles and Design

Ulrich L. Rohde

Jerry C. Whitaker

T. T. N. Bucher

Second Edition

McGraw-Hill

New York San Francisco Washington, D.C. Auckland Bogotá
Caracas Lisbon London Madrid Mexico City Milan
Montreal New Delhi San Juan Singapore
Sydney Tokyo Toronto

Library of Congress Cataloging-in-Publication Data

Rohde, Ulrich L.
    Communications receivers: principles and design / Ulrich L.
Rohde, Jerry Whitaker, T.T.N. Bucher.—2nd ed.
        p.   cm.
    Includes index.
    ISBN 0-07-053608-2 (hardcover)
    1. Radio—Receivers and reception—Design and construction.
I. Whitaker, Jerry.   II. Bucher, T. T. N. (T. T. Nelson)
III. Title.
TK6563.R57   1996
621.384'18—dc20

                                                96-31725
                                                CIP

McGraw-Hill

A Division of The McGraw-Hill Companies

1 2 3 4 5 6 7 8 9 0   BKP/BKP   9 0 1 0 9 8 7 6

ISBN 0-07-053608-2

*The sponsoring editor for this books was Steve Chapman, the editing
supervisor was Bernard Onken, and the production supervisor was
Pamela Pelton. It was set in Times Roman by Jerry Whitaker.*

Printed and bound by Quebecor / Book Press.

McGraw-Hill books are available at special quantity discounts to use
as premiums and sales promotions, or for use in corporate training programs. For more information, please write to the Director of Special
Sales, McGraw-Hill, 11 West 19th Street, New York, NY 10011. Or
contact your local bookstore.

This book is printed on recycled, acid-free paper containing a
minimum of 50% recycled, de-inked fiber.

Table Of Contents

Chapter 1: Basic Radio Considerations

Chapter 2: Radio Receiver Characteristics

Chapter 3: Receiver System Planning

Chapter 4: Antennas and Antenna Coupling

Chapter 5: Amplifiers and Gain Control

Chapter 6: Mixers

Chapter 7: Frequency Control and Local Oscillators

Chapter 8: Demodulation and Demodulators

Chapter 9: Other Receiver Circuits

Chapter 10: Receiver Design Trends

Preface

Powerful microprocessors and digital signal processing devices have reshaped communications technology within the last decade. Devices with 64-bit word lengths and clock frequencies of 200 MHz and more permit analog signals to be processed digitally, dramatically reducing—or even eliminating—circuit alignment and adjustment procedures. The application of digital techniques reaches from the frequency synthesizer to the demodulator. Curiously enough, as digital technology moves forward, one of the frontiers of development is the design and manufacture of hybrid analog-digital circuitry on a single chip. Clearly, there are still functions for which analog is superior.

Technologies have been developed to allow the generation of digital waveforms shaped arbitrarily within bandwidth constraints. It is thus possible to combine advanced waveforms and modulation schemes, spinoffs of MSK and PSK, to provide secure point-to-point communications at a fraction of the cost of specialized systems just a few years ago. These communications can be made both evasive and adaptive. Thus, the radio system has the ability to make a channel-quality analysis and to determine by link analysis the required transmitter power. As a result of *sounding* assigned channels, it is possible to avoid interference and determine the best available frequencies for reliable communications.

Many communications links exchange analog and digital voice signals as well as digital data. Redundancy in the message and forward error-correcting techniques improve reliability and are often essential to protect against interference and jamming. Spread-spectrum waveforms may be used to allow the application of correlation techniques for detection. This permits operation below the signal-to-noise-density ratio of conventional systems to reduce detectability. Conversely, the signal processing spreads the received interfering spectrum, so such systems are far more resistant to interference than other services.

As a result of these and other developments, radio receivers are in the process of a significant change in design. It is for this reason that we felt it was important to produce a new, second edition of *Communications Receivers: Principles and Design*. In this book you will find detailed information on the basics of receiver design, including hundreds of equations, illustrations, and tables. In writing the second edition, we kept in mind the need to retain receiver design fundamentals, while at the same time providing information on current and emerging technologies.

The major changes are digital signal processing and computer aided design using PC's. There is an interesting race going on between the commercial companies developing software radio and the companies building radios for the amateur radio community. The definition of software radio means that the analog stage that is being released moves as much as possible toward the antenna. A good example for this is the Rohde & Schwarz EK-895 and the Kenwood TS-870. Both use digital signal processing with slightly different implementation and both will probably show great enhancement as the computational power of the DSP increases. Another candidate for this technology is the R&S XK-2100 transceiver. The simultaneous availability of providing the selectivity in the IF and doing the DSP "clean-up" is at the limits of current processors. The next generation will have not only good selectivity with short delay, but also simultaneous auto notch, beat note removal, DSP noise blanker, and adjustable squelch. It can even consider adaptive bandwidth selection. In the area of frequency synthesis, new custom chips that have fractional division N will be available, which will provide better performance than the current chips on the market The previously-mentioned XK-2100 combines many of the relevant techniques, and those interested in receiver and transmitter technology may be interested in obtaining a service manual from R&S Munich.

It is no secret that the method of design is also changing. The widespread use of computer-aided design and simulation has significantly reduced the dependence of the engineer on manual calculations and on lengthy breadboard experimentation. In this book, a variety of software tools were used for the synthesis of circuits and their verification, in particular those provided by Communications Consulting Corp. of Upper Saddle River, NJ, and Compact Software, Inc., of Paterson, NJ.

The book includes 10 chapters:

Chapter 1 provides the reader with basic information on radio communications systems, radio transmission, noise, and modulation theory and techniques. The latter are extremely important because of their implications for digital implementation of modulators and demodulators. The chapter then describes radio receiver configurations, with some historical background, and provides the basis for possible overall configuration tradeoffs. The chapter closes with descriptions of typical radio receivers representative of modern designs for communications and surveillance, including a software radio (the EK-985).

Chapter 2 points out the important characteristics of radio receivers and describes techniques for their measurement. Included in this discussion are guidelines for digital system quality analysis.

Chapter 3 is devoted to receiver system planning so that the designer can make appropriate tradeoffs among conflicting performance characteristics. It includes sections on filter design for selectivity, on spurious response locations, and on tracking of variable-frequency circuits. The noise correlation matrix is also discussed.

Chapter 4 covers antennas and antenna matching, including active antennas. Active antennas can be extremely useful in some applications, but are frequently overlooked.

Chapter 5 surveys the design of amplifiers and gain-control techniques. Extensive data is provided on the analysis of gain-control loops and digital IF processing.

Chapter 6 provides a discussion of various mixer types. The many design considerations for these important circuits are discussed in detail. Extensive analysis of the Gilbert cell mixer is also included to demonstrate CAD design capabilities.

Chapter 7 deals with local oscillator frequency control. The accuracy and stability of the oscillators is one of the most important aspects of superheterodyne receiver design, and the treatment of oscillators could easily form a book by itself. In modern receivers, the required oscillator signals are produced or controlled by synthesizers. Detailed analysis of the nonlinear noise phenomenon in an oscillator is also presented.

Chapter 8 provides an overview of demodulation techniques and demodulator design. Standard analog demodulation technologies are covered, such as AM, SSB, and FM demodulators. Another section reviews common demodulators for digital signal modulation, including an introduction to DSP using modern integrated circuits.

Chapter 9 provides a description of some of the auxiliary and special circuits often required in radio receivers. These include noise limiters and blankers, squelch circuits, automatic frequency control, diversity combining, link-quality analysis, and the very important area of adaptive receiver processing for a combination of multiple antenna inputs and for signal equalization.

Chapter 10 considers modern receiver design trends. These include the increasing use of more and more digital signal processing, a trend that is extensively reflected in earlier chapters. Discussed are new concepts in circuit design for high-performance synthesizers, mixers, and switches, including a detailed analysis of DSP-based radios. The second trend, which has already spawned many books, is the use of spread-spectrum modulation. Some of the techniques and problems are touched upon, and references are given for more detailed information. The final trend discussed is the use of simulation in the design and evaluation of receivers. Overall system simulation can provide valuable guidance for receiver design, especially in the case of multipath media and nonlinear operations.

Although this second edition of *Communications Receivers* is nearly 20 percent larger than the first edition, it is difficult to select material on such a broad subject to fit within the confines of a book of this size. From the start, the effort has concentrated on single-channel communications receivers at frequencies where lumped circuit elements (or some simple transmission-line elements) would serve for RF and oscillator circuits. This has ruled out most receivers above 1 GHz, especially those for point-to-point radio relay and satellite communications relay. It has also made it necessary to treat some areas in less detail than we would prefer. However, throughout the book we have tried to provide references where more information can be found.

It is our sincere hope that *Communications Receivers: Principles and Design,* 2nd edition, will aid your efforts in improving communications systems around the world.

Ulrich L. Rohde, Ph.D., Sc.D.

Jerry C. Whitaker

Chapter 1

Basic Radio Considerations

1.1 Radio Communications Systems

The capability of radio waves to provide almost instantaneous distant communications without interconnecting wires has been a major factor in the explosive growth of communications during the twentieth century. The invention of the vacuum tube made radio a practical and affordable communications medium. The replacement of vacuum tubes by transistors and integrated circuits has allowed the development of a wealth of economical yet complex communications systems, which have become an integral part of our society.

In this book, we review the principles and design of modern single-channel radio receivers for frequencies below microwave. While it is possible to design a receiver to meet specified requirements without knowing the system in which it is to be used, such ignorance can prove time-consuming and costly when the inevitable need for design compromises arises. We strongly urge that the receiver designer take the time to understand thoroughly the system and the operational environment in which the receiver is to be used. Here we can outline only a few of the wide variety of systems and environments in which radio receivers may be used.

Figure 1.1 is a simplified block diagram of a communications system that allows the transfer of information between a *source* where information is generated and a *destination* that requires it. In the systems with which we are concerned, the transmission medium is radio, which is used when alternative media, such as light or electrical cable, are not technically feasible or are uneconomical. Figure 1.1 represents the simplest kind of communications system, where a single source transmits to a single destination. Such a system is often referred to as a *simplex* system. When two such links are used, the second sending information from the destination location to the source location, the system is referred to as *duplex*. Such a system may be used for two-way communication or, in some cases, simply to provide information on the quality of received information to the source. If only one transmitter may transmit at a time, the system is said to be *half-duplex*.

Figure 1.2 is a diagram representing the simplex and duplex circuits, where a single block T represents all of the information functions at the source end of the link and a single block R represents those at the destination end of the link. In this simple diagram, we encounter one of the problems which arise in communications systems—a definition of the boundaries between parts of the system. The blocks T and R, which might be thought of as transmitter and receiver, incorporate several

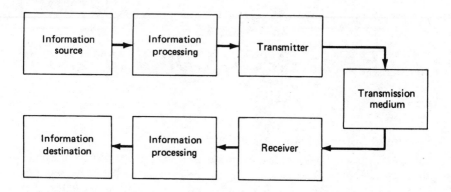

Figure 1.1 Simplified block diagram of a communications link.

functions that were portrayed separately in Figure 1.1. Later in the chapter we return to this question as it applies to receivers.

Many radio communications systems are much more complex than the simplex and duplex links shown in Figures 1.1 and 1.2. For example, a broadcast system has a star configuration in which one transmitter sends to many receivers. A data-collection network may be organized into a star where there are one receiver and many transmitters. These configurations are indicated in Figure 1.3. A consequence of a star system is that the peripheral elements, insofar as technically feasible, are made as simple as possible, and any necessary complexity is concentrated in the central element.

Examples of the transmitter-centered star are the familiar *amplitude-modulated* (AM), *frequency-modulated* (FM), and television broadcast systems. In these systems, high-power transmitters with large antenna configurations are employed at the transmitter, whereas most receivers use simple antennas and are themselves relatively simple. An example of the receiver-centered star is a weather-data-collection network, with many unattended measuring stations that send data at regular intervals to a central receiving site. Star networks can be configured using duplex rather than simplex links, if this proves desirable. Mobile radio networks have been configured largely in this manner, with the shorter-range mobile sets transmitting to a central radio relay located for wide coverage. Cellular radio systems incorporate

(a)

(b)

Figure 1.2 Simplified portrayal of communications links: (*a*) simplex link and (*b*) duplex link.

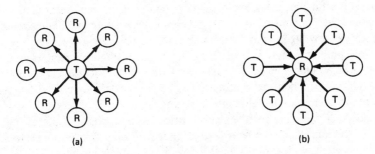

Figure 1.3 Star-type communications networks: (*a*) broadcast system and (*b*) data-collection network.

a number of low-power relay stations that provide contiguous coverage over a large area, communicating with low-power mobile units. The relays are interconnected by various means to a central switch. This system uses far less spectrum than conventional mobile systems because of the capability for reuse of frequencies in noncontiguous cells.

Packet radio transmission is another example of a duplex star network. Stations transmit at random times to a central computer terminal and receive responses sent from the computer. The communications consist of brief bursts of data, sent asynchronously and containing the necessary address information to be properly directed. The term *packet* network is applied to this scheme and related schemes using similar protocols. A packet system typically incorporates many radios, which can serve either as terminals or as relays, and uses a *flooding-type* transmission scheme.

The most complex system configuration occurs when there are many stations, each having both a transmitter and receiver, and where any station can transmit to one or more other stations simultaneously. In some networks only one station transmits at a time. One may be designated as a network controller to maintain a calling discipline. In other cases, it is necessary to design a system where more than one station can transmit simultaneously to one or more other stations.

In many radio communications systems, the range of transmissions, because of terrain or technology restrictions, is not adequate to bridge the gap between potential stations. In such a case, radio repeaters may be used to extend the range. The repeater comprises a receiving system connected to a transmitting system, so that a series of radio links may be established to achieve the required range. Prime examples are the multichannel radio relay system used by long-distance telephone companies and the satellite multichannel relay systems that are used extensively to distribute voice, video, and data signals over a wide geographic area. Satellite relay systems are essential where physical features of the earth (oceans, high mountains, and other physical restrictions) preclude direct surface relay.

On a link-for-link basis, radio relay systems tend to require a much higher investment than direct (wired) links, depending on the terrain being covered and the distances involved. To make them economically sound, it is common practice in the telecommunications industry to *multiplex* many single communications onto one radio relay link. Typically, hundreds of channels are sent over one link. The radio links connect between central offices in large population centers and gather the various users together through switching systems. The hundreds of trunks destined for a particular remote central office are multiplexed together into one wider-bandwidth channel and provided as input to the radio transmitter. At the other central office, the wide-band channel is demultiplexed into the individual channels and distributed appropriately by the switching system. Telephone and data common carriers are probably the largest users of such duplex radio transmission. The block diagram of Figure 1.4 shows the functions that must be performed in a radio relay system. At the receiving terminal, the radio signal is intercepted by an antenna, amplified and changed in frequency, demodulated, and demultiplexed so that it may be distributed to the individual users.

In addition to the simple communications use of radio receivers outlined previously, there are many special-purpose systems that also require radio receivers. While the principles of design are essentially the same, such receivers have peculiarities that have led to their own design specialties. For example, in receivers used for direction finding, the antenna systems have specified directional patterns. The receivers must accept one or more inputs and process them so that the output signal can indicate the direction from which the signal arrived. Older techniques include the use of loop antennas, crossed loops, *Adcock* antennas, and other specialized designs, and determine the direction from a pattern null. More modern systems use complex antennas, such as the *Wullenweber*. Others determine both direction and range from the delay differences found by cross-correlating signals from different antenna structures.

Radio ranging can be accomplished using radio receivers with either *cooperative* or *noncooperative* targets. Cooperative targets use a radio relay with known delay to return a signal to the transmitting location, which is also used for the receiver. Measurement of the round-trip delay (less the calibrated internal system delays) permits the range to be estimated very closely. Noncooperative ranging receivers are found in radar applications. In this case, reflections from high-power transmissions are used to determine delays. The strength of the return signal depends on a number of factors, including the transmission wavelength, target size, and target reflectivity. By using very narrow beam antennas and scanning the azimuth and elevation angles, radar systems are also capable of determining target direction. Radar receivers have the same basic principles as communications receivers, but they also have special requirements, depending upon the particular radar design.

Another area of specialized design is that of telemetry and control systems. Examples of such systems are found in almost all space vehicles. The telemetry channels return to earth data on temperatures, equipment conditions, fuel status,

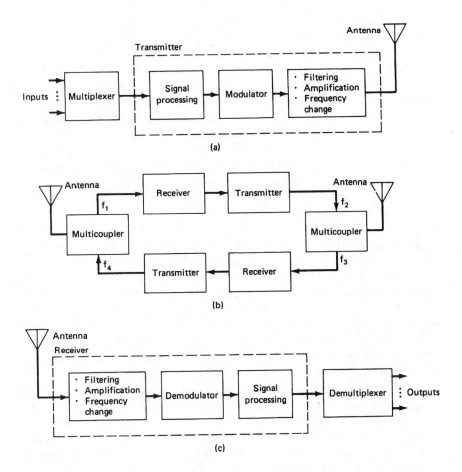

Figure 1.4 Block diagram of simplified radio relay functions: (*a*) terminal transmitter, (*b*) repeater (without drop or insert capabilities), and (*c*) terminal receiver.

and other important parameters, while the control channels allow remote operation of equipment modes and vehicle attitude, and the firing of rocket engines. The principal difference between these systems and conventional communications systems lies in the multiplexing and demultiplexing of a large number of analog and digital data signals for transmission over a single radio channel. The principal difference in receiver design resides in the multiplexing equipment and in designing the receiver to handle the composite signal with minimum crosstalk between channels.

Electronic countermeasure (ECM) systems, used primarily for military purposes, give rise to special receiver designs, both in the systems themselves and in their target communications systems. The objectives of countermeasure receivers is to detect activity of the target transmitters, to identify them from their electro-magnetic signatures, to locate their positions, and in some cases to demodulate their signals. Such receivers must have high detection sensitivity and the ability to demodulate a wide variety of signal types. Moreover, spectrum analysis capability and other analysis techniques are required for signature determination. Either the same receivers or separate receivers can be used for the radio-location function. To counter such actions, the communications circuit may use minimum power, direct its power toward its receiver in as narrow a beam as possible, and spread its spectrum in a manner such that the intercept receiver cannot despread it, thus decreasing the *signal-to-noise ratio* (SNR, also referred to as S/N) to render detection more difficult. This technique is referred to as *low probability of intercept* (LPI).

Some ECM systems are designed primarily for interception and analysis. In other cases, however, the purpose is to jam the communications receivers so as to disrupt communications. To this end, once the transmission of a target system has been detected, the ECM system transmits a strong signal on the same frequency, with a randomly controlled modulation that produces a spectrum similar to the communications sequence. Another alternative is to transmit a "spoofing" signal that is similar to the communications signal but contains false or out-of-date information. The *electronic countercountermeasure* (ECCM) against spoofing is good cryptographic security. The countermeasures against jamming are high-powered, narrow-beam, or adaptive-nulling receiver antenna systems, and a spread-spectrum system with secure control so that the jamming transmitter cannot emulate it. In this case, the communications receiver must be designed to correlate the received signal using the secure spread-spectrum control. Thus, the jammer power is spread over the transmission bandwidth, while the communication power is restored to the original signal bandwidth before spreading. This provides an improvement in signal-to-jamming ratio equal to the spreading multiple, which is referred to as the *processing gain.*

Special receivers are also designed for testing radio communications systems. In general, they follow the design principles of the communications receivers, but their design must be of even higher quality and accuracy because their purpose is to measure various performance aspects of the system under test. Test receivers usually may be calibrated accurately to determine frequency and input voltage stability. They also may be designed for use with special antennas for measuring the electromagnetic field strength from the system under test at a particular location. Because such receivers are normally used for testing a variety of communications links, they must be provided with an adjustable, accurately calibrated bandwidth and a number of different standard types of demodulators. The receivers may include or provide signals for the use of spectrum analyzers. In many regards, test receiver requirements are similar to those for intercept receivers. While test receivers are not treated separately in this book, many of our design examples are taken from test receiver design.

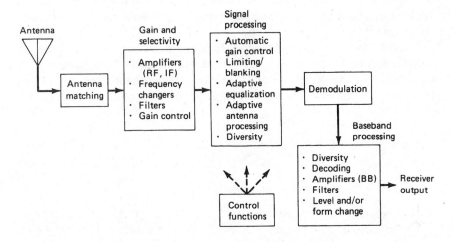

Figure 1.5 Block diagram of a communications receiver. (RF = radio frequency, IF = intermediate frequency, and BB = baseband.)

From this brief discussion of communications systems, we hope that the reader will gain some insight into the scope of receiver design, and the difficulty of isolating the treatment of the receiver design from the system. There are also difficulties in setting hard boundaries to the receiver within a given communications system. For the purposes of our book, we have decided to treat as the receiver that portion of the system that accepts input from the antenna and produces a demodulated output for further processing at the destination or possibly by a demultiplexer. We consider modulation and demodulation to be a part of the receiver, but we recognize that for data systems especially there is an ever-increasing volume of modems (*modulator-demodulators*) that are designed and packaged separately from the receiver. For convenience, Figure 1.5 shows a block diagram of the receiver as we have chosen to treat it in this book. It should be noted that signal processing may be accomplished both before and after modulation.

1.1.1 Radio Transmission and Noise

Light and X rays, like radio waves, are electromagnetic waves that may be attenuated, reflected, refracted, scattered, and diffracted by the changes in the media through which they propagate. In free space, the waves have electric and magnetic field components that are mutually perpendicular and lie in a plane transverse to the direction of propagation. In common with other electromagnetic waves, they travel with a velocity c of 299,793 km/s, a value that is conveniently rounded to 300,000 km/s for most calculations. In rationalized *meter, kilogram, and second* (MKS) units, the power flow across a surface is expressed in watts per square meter and is the product of the electric-field (volts per meter) and the magnetic-field (amperes per meter) strengths at the point over the surface of measurement.

A radio wave propagates spherically from its source, so that the total radiated power is distributed over the surface of a sphere with radius R (meters) equal to the distance between the transmitter and the point of measurement. The power density S (watts per square meter) at the point for a transmitted power P_t (watts) is:

$$S = \frac{G_t \times P_t}{4\,\pi \times R^2} \tag{1.1}$$

where G_t is the transmitting antenna gain in the direction of the measurement over a uniform distribution of power over the entire spherical surface. Thus, the gain of a hypothetical isotropic antenna is unity.

The power intercepted by the receiver antenna is equal to the power density multiplied by the effective area of the antenna. Antenna theory shows that this area is related to the antenna gain in the direction of the received signal by the expression:

$$Ae_r = \frac{G_r\,\lambda^2}{4\,\pi} \tag{1.2}$$

When Equations (1.1) and (1.2) are multiplied to obtain the received power, the result is:

$$\frac{P_r}{P_t} = \frac{G_r\,G_t\,\lambda^2}{16\,\pi^2\,R^2} \tag{1.3}$$

This is usually given as a loss L (in decibels), and the wavelength λ is generally replaced by velocity divided by frequency. When the frequency is measured in megahertz, the range in kilometers, and the gains in decibels, the loss becomes:

$$L = [32.4 + 20\log R + 20\log F] - G_t - G_r \equiv A_{fs} - G_t - G_R \tag{1.4}$$

A_{fs} is referred to as the loss in free space between isotropic antennas. Sometimes the loss is given between half-wave dipole antennas. The gain of such a dipole is 2.15 dB above isotropic, so that the constant in Equation (1.4) must be increased to 36.7 to obtain the loss between dipoles.

Because of the earth and its atmosphere, most terrestrial communications links cannot be considered free-space links. Additional losses occur in transmission. Moreover, the received signal field is accompanied by an inevitable noise field generated in the atmosphere or space, or by machinery. In addition, the receiver itself is a source of noise. Electrical noise limits the performance of radio communications by requiring a signal field sufficiently great to overcome its effects.

While the characteristics of transmission and noise are of general interest in receiver design, it is far more important to consider how these characteristics affect the design. The following sections summarize the nature of noise and transmission effects in frequency bands through SHF (30 GHz).

A. ELF and VLF (up to 30 kHz)

Transmission in the *extremely-low frequency* (ELF) and *very-low frequency* (VLF) range is primarily via surface wave with some of the higher-order waveguide modes introduced by the ionosphere appearing at the shorter ranges. Because transmission in these frequency bands is intended for long distances, the higher-order modes are normally unimportant. These frequencies also provide the only radio communications that can penetrate the oceans substantially. Because the transmission in saltwater has an attenuation that increases rapidly with increasing frequency, it may be necessary to design depth-sensitive equalizers for receivers intended for this service. At long ranges, the field strength of the signals is very stable, varying only a few decibels diurnally and seasonally, and being minimally affected by changes in solar activity. There is more variation at shorter ranges. Variation of the phase of the signal can be substantial during diurnal changes and especially during solar flares and magnetic storms. For most communications designs, these phase changes are of little importance. The noise at these low frequencies is very high and highly impulsive. This situation has given rise to the design of many noise-limiting or noise-canceling schemes, which find particular use in these receivers. Transmitting antennas must be very large to produce only moderate efficiency; however, the noise limitations permit the use of relatively short receiving antennas because receiver noise is negligible in comparison with atmospheric noise at the earth's surface. In the case of submarine reception, the high attenuation of the surface fields, both signal and noise, requires that more attention be given to receiving antenna efficiency and receiver sensitivity.

B. LF (30 to 300 kHz) and MF (300 kHz to 3 MHz)

At the lower end of the *low-frequency* (LF) region, transmission characteristics resemble VLF. As the frequency rises, the surface wave attenuation increases, and even though the noise decreases, the useful range of the surface wave is reduced.

During the daytime, ionospheric modes are attenuated in the D layer of the ionosphere. The waveguide mode representation of the waves can be replaced by a reflection representation. As the *medium-frequency* (MF) region is approached, the daytime sky wave reflections are too weak to use. The surface wave attenuation limits the daytime range to a few hundred kilometers at the low end of the MF band to about 100 km at the high end. Throughout this region, the range is limited by atmospheric noise. As the frequency increases, the noise decreases and is minimum during daylight hours. The receiver *noise figure* (NF) makes little contribution to overall noise unless the antenna and antenna coupling system are very inefficient. At night the attenuation of the sky wave decreases, and reception can be achieved up to thousands of kilometers. For ranges of one hundred to several hundred kilometers, where the single-hop sky wave has comparable strength to the surface wave, fading occurs. This phenomenon can become quite deep during those periods when the two waves are nearly equal.

At MF, the sky wave fades as a result of Faraday rotation and the linear polarization of antennas. At some ranges, additional fading occurs because of interference between the surface wave and sky wave or between sky waves with different numbers of reflections. When fading is caused by two (or more) waves that interfere as a result of having traveled over paths of different lengths, various frequencies within the transmitted spectrum of a signal can be attenuated differently. This phenomenon is known as *selective fading* and results in severe distortion of the signal. Because much of the MF band is used for AM broadcast, there has not been much concern about receiver designs that will offset the effects of selective fading. However, as the frequency nears the *high-frequency* (HF) band, the applications become primarily long-distance communications, and this receiver design requirement is encountered. Some broadcasting occurs in the LF band, and in the LF and lower MF bands medium-range narrow-band communications and radio navigation applications are prevalent.

C. HF (3 to 30 MHz)

Until the advent of satellite-borne radio relays, the HF band provided the only radio signals capable of carrying voiceband or wider signals over very long ranges (up to 10,000 km). VLF transmissions, because of their low frequencies, have been confined to narrow-band data transmission. The high attenuation of the surface wave, the distortion from sky-wave-reflected *near-vertical incidence* (NVI), and the prevalence of long-range interfering signals make HF transmissions generally unsuitable for short-range communications. From the 1930s into the early 1970s, HF radio was a major medium for long-range voice, data, and photo communications, as well as for overseas broadcast services, aeronautical, maritime and some ground mobile communications, and radio navigation. Even today, the band remains active, and long-distance interference is one of the major problems. Because of the dependence on sky waves, HF signals are subject to both broad-band and selective fading. The frequencies capable of carrying the desired transmission are subject to all of the diurnal, seasonal, and sunspot cycles, and the random variations of ionization in the upper ionosphere. Sunspot cycles change every 11 years, and so propagation tends to change as well. Significant differences are typically experienced between day and night coverage patterns, and between summer to winter coverage. Out to about 4000 km, E-layer transmission is not unusual, but most of the very long transmission—and some down to a few thousand kilometers—is carried by F-layer reflections. It is not uncommon to receive several signals of comparable strength carried over different paths. Thus, fading is the rule, and selective fading is very common. Atmospheric noise is still high at times at the low end of the band, although it becomes negligible above about 20 MHz.

Receivers must be designed for high sensitivity, and impulse noise reducing techniques must often be included. Because the frequency must be changed frequently to obtain even moderate transmission availability, most HF receivers require coverage of the entire band and usually of the upper part of the MF band. For many applications, designs must be made to combat fading. The simplest of these is *automatic gain control* (AGC), which also is generally used in lower-frequency designs. Diversity reception is often required, where signals are received over several routes that fade independently—using separated antennas, frequencies, and times, or antennas with different polarizations—and must be combined to provide the best composite output. If data transmissions are separated into many parallel low-rate channels, fading of the individual narrow-band channels is essentially flat, and good reliability can be achieved by using diversity techniques. Most of the data sent over HF use such multitone signals.

In newer receiver designs, adaptive equalizer techniques are used to combat multipath that causes selective fading on broadband transmissions. The bandwidth available on HF makes possible the use of spread-spectrum techniques intended to combat interference and, especially, jamming. This is primarily a military requirement. However, much of the future HF design is likely to be for military use, because commercial communicators have been deserting HF for more reliable transmissions, such as satellite relay and fiber optic cable.

D. VHF (30 to 300 MHz)

Most *very-high frequency* (VHF) transmissions are intended to be relatively short-range, using line-of-sight transmission with elevated antennas, at least at one end of the path. In addition to FM and television broadcast services, this band handles much of the land mobile and some fixed services, and some aeronautical and aeronavigation services. So long as a good clear line of sight with adequate ground (and other obstruction) clearance exists between the antennas, the signal will tend to be strong and steady. The wavelength is, however, becoming sufficiently small at these frequencies so that reflection is possible from ground features, buildings, and some vehicles. Usually reflection losses result in transmission over such paths that is much weaker than transmission over line-of-sight paths. In land mobile service, one or both of the terminals may be relatively low, so that the earth's curvature or rolling hills and gullies may interfere with a line-of-sight path. While the range can be extended slightly by diffraction, in many cases the signal reaches the mobile station via multipath reflections that are of comparable strength or stronger than the direct path. The resulting interference patterns cause the signal strength to vary from place to place in a relatively random matter.

There have been a number of experimental determinations of the variability, and models have been proposed that attempt to predict it. Most of these models apply also in the *ultra-high frequency* (UHF) region. For clear line-of-sight paths, or those with a few well-defined intervening terrain features, accurate methods exist for predicting field strength. In this band, noise is often simply thermal, although man-made noise can produce impulsive interference. For vehicular mobile use, the vehicle itself is a potential source of noise. In the United States, mobile communications have used FM, originally of a wider band than necessary for the information, so as to reduce impulsive noise effects. However, recent trends have reduced the bandwidth of commercial radios of this type so that this advantage has essentially disappeared. The other advantage of FM is that hard limiting may be used in the receiver to compensate for level changes with the movement of the vehicle. Such circuits are easier to design than AGC systems, whose rates of attack and decay would ideally be adapted to the vehicle's speed.

Elsewhere in the world AM has been used satisfactorily in the mobile service, and *single-sideband* (SSB) modulation, despite its more complex receiver implementation, has been applied to reduce spectrum occupancy. Communications receivers in this band are generally designed for high sensitivity, a high range of signals, and strong interfering signals. With the trend toward increasing data transmission rates, adaptive equalization is required in some applications.

Ground mobile military communications use parts of this band, so spread-spectrum designs are also found. At the lower end of the band, the ionospheric scatter and meteoric reflection modes are available for special-purpose use. Receivers for the former must operate with selective fading from scattered multipaths with substantial delays; the latter require receivers which can detect acceptable signals rapidly and provide the necessary storage before the path deteriorates.

E. UHF (300 MHz to 3 GHz)

The transmission characteristics of UHF are essentially the same as those of VHF, except for the ionospheric effects at low VHF. It is at UHF and above that tropospheric scatter links have been used. Nondirectional antennas are quite small, and large reflectors and arrays are available to provide directionality. At the higher portions of the band, transmission closely resembles the transmission of light, with deep shadowing by obstacles and relatively easy reflection from terrain features, structures, and vehicles with sufficient reflectivity. Usage up to 1 GHz is quite similar to that at VHF. Mobile radio usage includes both analog and digital cellular radiotelephones. Transmission between earth and space vehicles occurs in this band, as well as some satellite radio relay (mainly for marine mobile use, including navy communications). Because of the much wider bandwidths available in the UHF band, spread-spectrum usage is high for military communications, navigation, and radar. Some line-of-sight radio relay systems use this band, especially those where the paths are less than ideal; UHF links can be increased in range by diffraction over obstacles. The smaller wavelengths in this band make it possible to achieve antenna diversity even on a relatively small vehicle. It is also possible to use multiple antennas and design receivers to combine these inputs adaptively to discriminate against interference or jamming. With the availability of wider bands and adaptive equalization, much higher data transmission rates are possible at UHF, using a wide variety of data modulations schemes.

F. SHF (3 GHz to 30 GHz)

Communication in the *super-high frequency* (SHF) band is strictly line-of-sight. Very short wavelengths permit the use of parabolic transmit and receive antennas of exceptional gain. Applications include satellite communications, point-to-point wideband relay, radar, and specialized wideband communications systems. Other related applications include developmental research, space research, military support systems, radio location, and radio navigation. Given line-of-sight conditions and sufficient fade margin, this band provides high reliability. Environmental conditions that can compromise SHF signal strength include heavy rain and solar outages (in the case of space-to-earth transmissions).

The majority of satellite links operate in either the *C*-band (4 to 6 GHz) or the *Ku*-band (11 to 14 GHz). Attenuation of signals resulting from meteorological conditions, such as rain and fog, is particularly serious for Ku-band operation, but less troublesome for C-band systems. The effects of galactic and thermal noise sources on low-level signals require electronics for satellite service with exceptionally low noise characteristics.

1.2 Modulation

Communications are transmitted by sending time-varying waveforms generated by the source or by sending waveforms (either analog or digital) derived from those of the source. In radio communications the varying waveforms derived from the source are transmitted by changing the parameters of a sinusoidal wave at the desired transmission frequency. This process is referred to as *modulation,* and the sinusoid is referred to as the *carrier.* The radio receiver must be designed to extract (demodulate) the information from the received signal. There are many varieties of carrier modulation, generally intended to optimize the characteristics of the particular system in some sense—distortion, error rate, bandwidth occupancy, cost, and/or other parameters. The receiver must be designed to process and demodulate all types of signal modulation planned for the particular communications system. Important characteristics of a particular modulation technique selected include the

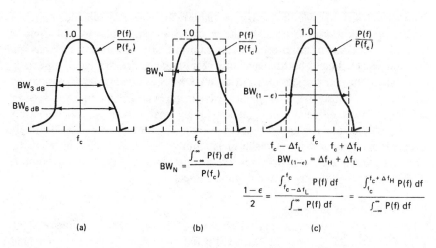

Figure 1.6 The relationship of various bandwidth definitions to power density spectrum: (*a*) attenuation bandwidth, (*b*) noise bandwidth, and (*c*) occupied bandwidth.

occupied bandwidth of the signal, the receiver bandwidth required to meet specified criteria for output signal quality, and the received signal power required to meet a specified minimum output performance criterion.

The frequency spectrum is shared by many users, with those nearby generally transmitting on different channels so as to avoid interference. Therefore, frequency channels must have limited bandwidth so that their significant frequency components are spread over a range of frequencies that is small compared to the carrier frequencies. There are several definitions of *bandwidth* that are often encountered. A common definition arises from, for example, the design of filters or the measurement of selectivity in a receiver. In this case, the bandwidth is described as the difference between the two frequencies at which the power spectrum density is a certain fraction below the center frequency when the filter has been excited by a uniform-density waveform such as white gaussian noise (Figure 1.6*a*). Thus, if the density is reduced to one-half, we speak of the 3 dB bandwidth; to 1/100, the 20 dB bandwidth; and so on.

Another bandwidth reference that is often encountered, especially in receiver design, is the *noise bandwidth*. This is defined as the bandwidth which, when multiplied by the center frequency density, would produce the same total power as the output of the filter or receiver. Thus, the noise bandwidth is the equivalent band of a filter with uniform output equal to the center frequency output and with infinitely sharp cutoff at the band edges (Figure 1.6*b*). This bandwidth terminology is also applied to the transmitted signal spectra. In controlling interference between channels, the bandwidth of importance is called the *occupied bandwidth* (Figure 1.6*c*). This bandwidth is defined as the band occupied by all of the radiated power except for a small fraction ε. Generally the band edges are set so that $\frac{1}{2}\varepsilon$ falls above the channel and $\frac{1}{2}\varepsilon$ below. If the spectrum is symmetrical, the band-edge frequencies are equally separated from the nominal carrier.

Every narrow-band signal can be represented as a *mean* or carrier frequency that is modulated at much lower frequencies in amplitude or angle, or both. This is true no matter what processes are used to perform the modulation. Modulation can be divided into two classes:

- *Analog modulation*: A system intended to reproduce at the output of the receiver, with as little change as possible, a waveform provided to the input of the transmitter.

- *Digital modulation*: A system intended to reproduce correctly one of a number of discrete levels at discrete times.

1.2.1 Analog Modulation

Analog modulation is used for transmitting speech, music, telephoto, television, and some telemetering. In some cases, the transmitter may perform operations on the input signal to improve transmission or to confine the spectrum to an assigned band. These may need to be reversed in the receiver to provide good output waveforms or, in some cases, it may be tolerated as distortion in transmission. There are essentially two pure modulations, amplitude and angle, although the latter is often divided into frequency and phase modulation. *Double-sideband with suppressed carrier* (DSB-SC), SSB, and *vestigial-sideband* (VSB) modulations are hybrid forms that result in simultaneous amplitude and angle modulation.

In amplitude modulation, the carrier angle is not modulated; only the envelope is modulated. Because the envelope by definition is always positive, it is necessary to prevent the modulated amplitude from going negative. Commonly this is accomplished by adding a constant component to the signal, giving rise to a transmitted waveform:

$$s(t) = A \left[1 + ms_{in}(t)\right] \cos\left(2\pi f t + \theta\right) \tag{1.5}$$

where A is the amplitude of the unmodulated carrier and $ms_{in}(t) > -1$. A sample waveform and a power density spectrum are shown in Figure 1.7. The spectrum comprises a line component, representing the unmodulated carrier power, and a power density spectrum that is centered on the carrier. Because of the limitation on the amplitude of the modulating signal, the total power in the two density spectra is generally considerably lower than the carrier power. The presence of the carrier,

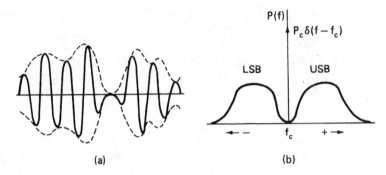

Figure 1.7 The process of amplitude modulation: (*a*) AM waveform and (*b*) power density spectrum. (*LSB* = lower sideband and *USB* = upper sideband.)

however, provides a strong reference frequency for demodulating the signal. The required occupied bandwidth is twice the bandwidth of the modulating signal.

The power required by the carrier in many cases turns out to be a large fraction of the transmitter power. Because this power is limited by economics and allocation rules, techniques are sometimes used to reduce the carrier power without causing negative modulation. One such technique is *enhanced carrier modulation,* which can be useful for communications using AM if the average power capability of the transmitter is of concern, rather than the peak power. In this technique, a signal is derived from the incoming wave to measure its strength. Speech has many periods of low or no transmission. The derived signal is low-pass filtered and controls the carrier level. When the modulation level increases, the carrier level is simultaneously increased so that overmodulation cannot occur. To assure proper operation, it is necessary to delay application of the incoming wave to the modulator by an amount at least equal to the delay introduced in the carrier control circuit filter. The occupied spectrum is essentially the same as for regular AM, and the wave can be demodulated by an AM demodulator.

Analog angle modulation is used in FM broadcasting, television audio broadcasting, and mobile vehicular communications. In FM, the instantaneous frequency of the waveform is varied proportionately to the signal so that the instantaneous frequency $f_i(t)$ and the instantaneous phase $\beta(t)$ are given by:

$$f_i(t) = f_o + ks_i(t) \tag{1.6}$$

$$\beta(t) = \beta_o + 2\pi k \int_{-\infty}^{t} s_i(x)\, dx \tag{1.6a}$$

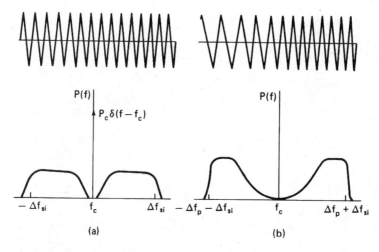

Figure 1.8 FM waveforms and spectra: (*a*) low-peak deviation and (*b*) high-peak deviation.

The bandwidth of the FM signal is a function of the multiplier k, and $ks_i(t)$ is the frequency deviation from the carrier. When the peak deviation Δf_p is small compared to unity, the bandwidth is approximately two times the input signal bandwidth $2\Delta f_{si}$. When the peak deviation is large compared to unity, the bandwidth is approximately $2(\Delta f_p + \Delta f_{si})$. This is known as the *Carson bandwidth*. Accurate predictions of the bandwidth are dependent on the details of the signal spectrum. Figure 1.8 illustrates FM waveforms having low and high deviations, and their associated spectra.

In *phase modulation* (PM), the instantaneous phase is made proportional to the modulating signal:

$$\beta(t) = ks_i(t) \tag{1.7}$$

The peak phase deviation β_p is the product of k and the maximum amplitude of $s_i(t)$. PM may be used in some narrow-band angle modulation applications. It has also been used as a method for generating FM with high stability. If the input wave is integrated before being applied to the phase modulator, the resulting wave is the equivalent of FM by the original input wave.

There are a variety of hybrid analog modulation schemes that are in use or have been proposed for particular applications. One approach to reducing the power required by the carrier in AM is to reduce or suppress the carrier. This is the DSB-SC modulation mentioned previously. It results in the same bandwidth requirement as for AM and produces a waveform and spectrum as illustrated in Figure 1.9.

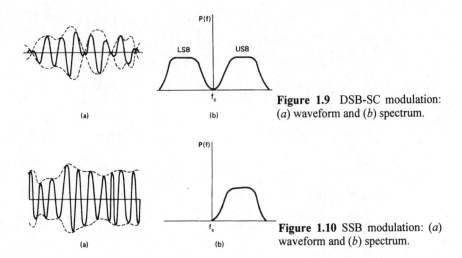

Figure 1.9 DSB-SC modulation: (*a*) waveform and (*b*) spectrum.

Figure 1.10 SSB modulation: (*a*) waveform and (*b*) spectrum.

Whenever the modulating wave goes through zero, the envelope of the carrier wave goes through zero with discontinuous slope, and simultaneously the carrier phase changes 180°. These sudden discontinuities in amplitude and phase of the signal do not result in a spreading of the spectrum because they occur simultaneously so as to maintain the continuity of the wave and its slope for the overall signal. An envelope demodulator cannot demodulate this wave without substantial distortion. For distortion-free demodulation, it is necessary for the receiver to provide a reference signal at the same frequency and phase as the carrier. To help in this, a small residual carrier may be sent, although this is not necessary.

The *upper sideband* (USB) and *lower sideband* (LSB) of the AM or DSB signal are mirror images. All of the modulating information is contained in either. The spectrum may be conserved by using SSB modulation to produce only one of these, either the USB or LSB. The amplitude and the phase of the resulting narrow-band signal both vary. SSB signals with modulation components near zero are imprac-tical to produce. Again, distortion-free recovery of the modulation requires the receiver to generate a reference carrier at the proper carrier frequency and phase. A reduced carrier may be sent in some cases to aid recovery. For audio transmission, accurate phase recovery is usually not necessary for the result to sound satisfactory. Indeed, small frequency errors can also be tolerated. Errors up to 50 Hz can be tolerated without unsatisfactory speech reception and 100 Hz or more without loss of intelligibility. Figure 1.10 illustrates the SSB waveform and spectrum. SSB is of value in HF transmissions because it is less affected by selective fading than AM and also occupies less bandwidth. A transmission that sends one SSB signal above the carrier frequency and a different one below it is referred to as having *inde-*

pendent sideband (ISB) modulation. SSB has found widespread use in voice multiplexing equipment for both radio and cable transmission.

For multiplexing channels in the UHF and SHF bands, various techniques of pulse modulation are used. These techniques depend upon the sampling theorem that any band-limited wave can be reproduced from a number of samples of the wave taken at a rate above the *Nyquist rate* (two times the highest frequency in the limited band). In PM schemes, the baseband is sampled and used to modulate a train of pulses at the sampling rate. The pulses have a duration much shorter than the sampling interval, so that many pulse trains can be interleaved. The overall pulse train then modulates a carrier using one of the standard amplitude or angle modulation techniques. Among the pulse modulation schemes are:

- PAM (*pulse-amplitude modulation*)
- PPM (*pulse-position* or *pulse-phase modulation*), in which the time position about an unmodulated reference position is changed
- PWM (*pulse-width modulation*), PLM (*pulse-length modulation*), and PDM (*pulse-duration modulation*), in which the width of the pulse is changed in response to the input signal

A modulated pulse train of this sort obviously occupies a much wider bandwidth than the modulation baseband. However, when many pulse trains are multiplexed, the ratio of pulse bandwidth to channel bandwidth is reduced. There are certain performance advantages to some of these techniques, and the multiplexing and demultiplexing equipment is much simpler than that required for frequency stacking of SSB channels.

It should be noted that PWM can be used to send a single analog channel over a constant-envelope channel such as FM. The usual approach to PWM is to maintain one of the edges of the pulse at a fixed time phase and vary the position of the other edge in accordance with the modulation. For sending a single channel, the fixed edge may be suppressed and the location of the leading and trailing edges are modulated relative to a regular central reference with successive samples. This process halves the pulse rate and, consequently, the bandwidth. It is an alternative approach to direct modulation for sending a voice signal over an FM, PM, or DSB-SC channel.

Pulse-code modulation (PCM) is another technique for transmitting sampled analog waveforms. Sampling takes place above the Nyquist rate. Commonly, a rate of 8 kHz is used for speech transmission. Each sample is converted to a binary number in an *analog-to-digital* (A/D) converter; the numbers are converted to a binary pulse sequence. They must be accompanied by a framing signal so that the proper interpretation of the code can be made at the receiver. Often PCM signals are multiplexed into groups of six or more, with one synchronizing signal to provide

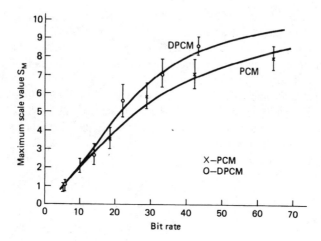

Figure 1.11 Performance comparison between PCM and DPCM systems. The length of the vertical bar through each point equals the variance in the scale value.

both channel and word synchronization. PCM is used extensively in telephone transmission systems, because the binary signals being encoded can be made to have relatively low error rates on any one hop in a long-distance relayed system. This permits accurate regeneration of the bit train at each receiver so that the cumulative noise over a long channel can be maintained lower than in analog transmission. Time division multiplexing permits the use of relatively small and inexpensive digital multiplexing and demultiplexing equipment.

Speech spectrum density tends to drop off at high frequencies. This has made the use of *differential PCM* (DPCM) attractive in some applications. It has been determined that when the difference between successive samples is sent, rather than the samples themselves, comparable speech performance can be achieved with the transmission of about two fewer bits per sample. This permits a saving in transmitted bandwidth with a slight increase in the noise sensitivity of the system. Figure 1.11 shows a performance comparison for various PCM and DPCM systems.

The ultimate in DPCM systems would offer a difference of only a single bit. This has been found unsatisfactory for speech at usual sampling rates. However, single-bit systems have been devised in the process known as *delta modulation* (DM). A block diagram of a simple delta modulator is shown in Figure 1.12. In this diagram the analog input level is compared to the level in a summer or integrator. If the summer output is below the signal, a 1 is generated; if it is above, a 0 is generated. This binary stream is transmitted as output from the DM and at the same time provides the input to the summer. At the summer, a unit input is interpreted as a positive unit increment, whereas a zero input is interpreted as a negative unit

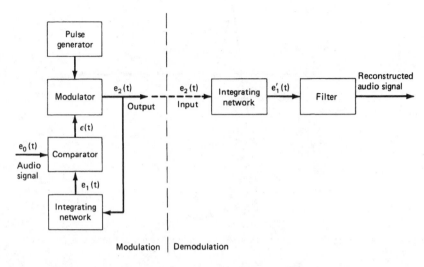

Figure 1.12 Block diagram of a DM modulator and demodulator.

input. The sampling rate must be sufficiently high for the summer to keep up with the input wave when its slope is high, so that slope distortion does not occur.

To combat slope distortion, a variety of adaptive systems have been developed to use saturation in slope to generate larger input pulses to the summer. Figure 1.13 shows the block diagram of *high-information* DM (HIDM), an early adaptive DM system. The result of a succession of 1s or 0s of length more than 2 is to double the size of the increment (or decrement) to the summer, up to a maximum size. This enables the system to follow a large slope much more rapidly than with simple DM.

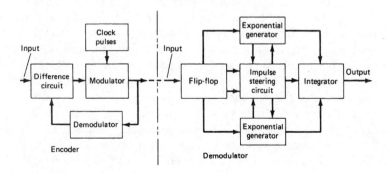

Figure 1.13 Block diagram of a HIDM system.

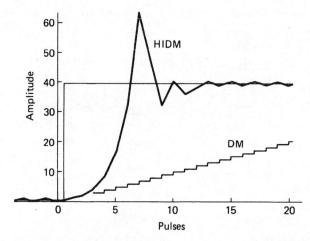

Figure 1.14 Comparison of responses of HIDM and DM to a step function.

Figure 1.14 illustrates this for the infinite slope of a step function. HIDM and other adaptive DM systems have been found to be of value for both speech and video communications.

1.2.2 Modulation for Digital Signals

With the explosive growth of digital data exchange, digital transmission has assumed ever greater importance in the design of communications equipment. Although the transmission of binary information is required, the method of transmission is still the analog radio transmission medium. Hence, the modulation process comprises the selection of one of a number of potential waveforms to modulate the transmitted carrier. The receiver must determine, after transmission distortions and the addition of noise, which of the potential waveforms was chosen. The process is repeated at a regular interval T, so that $1/T$ digits are sent per second. The simplest digital decision is binary, i. e., one of two waveforms is selected, so digital data rates are usually expressed in *bits per second* (b/s). This is true even when a higher-order decision is made (*m*-ary) among m different waveforms. The rate of decision is called the *symbol rate*; this is converted to bits per second by multiplying by the logarithm of m to the base 2. In most applications m is made a power of 2, so this conversion is simple.

AM and angle modulation techniques described previously can be used for sending digits, and a number of hybrid modulations are also employed. The performance of digital modulation systems is often measured by the ratio of energy required per bit to the white gaussian noise power density E_b/n_o required to produce specified bit error rates. In practical transmission schemes, it is also necessary to consider the occupied bandwidth of the radio transmission for the required digital rate. The measure bits per second per hertz may be used for modulation comparisons. Alternatively, the occupied bandwidth required to send a certain standard digital rate is often used.

Coding may be employed in communications systems to improve the form of the input waveform for transmission. Coding may be used in conjunction with the modulation technique to improve the transmission of digital signals, or it may be inserted into an incoming stream of digits to permit detection and correction of errors in the output stream. This latter use, *error detection and correction* (EDAC) coding, is a specialized field that is not normally considered a part of the receiver. Some techniques that improve the signal transmission, correlative coding, are considered modulation techniques. PCM and DM, discussed previously, may be considered source coding techniques.

A. Coding system basics

By using a binary input to turn a carrier on or off, an AM system for digital modulation known as *on-off keying* (OOK) is produced. This may be generalized to switching between two amplitude levels, which is then known as *amplitude-shift keying* (ASK). ASK, in turn, can be generalized to m levels to produce an m-ary ASK signal. Essentially this process represents modulating an AM carrier with a square wave or a step wave. The spectrum produced has carrier and upper and lower sidebands, which are the translation of the baseband modulating spectrum. As a result, zero frequency in the modulating spectrum becomes the carrier frequency in the transmitted spectrum. Because a discontinuous (step) amplitude produces a spectrum with substantial energy in adjacent channels, it is necessary to filter or otherwise shape the modulating waveform to reduce the side lobe energy. Because the modulation causes the transmitter amplitude to vary, binary ASK can use only one-half of the transmitter's peak power capability. This can be an economic disadvantage. An envelope demodulator can be used at the receiver, but best performance is achieved with a coherent demodulator. Figure 1.15 gives examples of ASK waveforms, power density spectra, and the locus in the Argand diagram. The emphasized points in the latter are the amplitude levels corresponding to the different digits. The diagram is simply a line connecting the points because the phase remains constant. The group of points is called a *signal constellation.* For ASK, this diagram is of limited value, but for more complex modulations it provides a useful insight into the process. Figure 1.16 shows the spectrum density of OOK for various transition shapes and tabulates noise and occupied bandwidths.

The digital equivalents of FM and PM are *frequency-shift keying* (FSK) and *phase-shift keying* (PSK), respectively. These modulations can be generated by using appropriately designed baseband signals as the inputs to a frequency or phase modulator. Often, however, special modulators are used to assure greater accuracy and stability. Either binary or higher-order m-ary alphabets may be used in FSK or PSK to increase the digital rate or reduce the occupied bandwidth. Early FSK

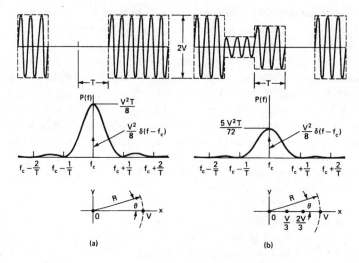

Figure 1.15 Example of waveforms, spectra and Argand plots: (*a*) binary modulation and (*b*) quaternary ASK modulation.

Shape	Discrete component power/V^2		Continuous spectrum BT		
	Direct current	All others	Noise	3 dB	0.99 Occupancy
Rectangular	0.250	0	1.000	0.8859	20.6
Triangular	0.0625	0.0208	1.333	1.2757	2.60
Sine	0.101	0.0237	1.238	1.1890	2.36
Raised cosine	0.0625	0.0625	1.500	1.4406	2.82

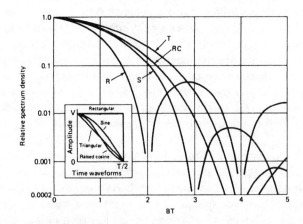

Figure 1.16 OOK power density specta. (*R* = rectangular, *S* = sine, *T* = triangular, and *RC* = raised cosine.)

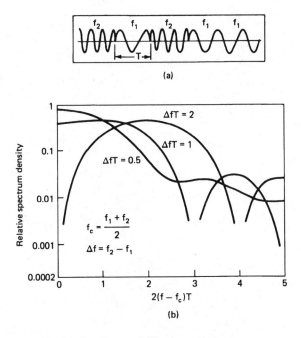

Figure 1.17 Binary FEK: (*a*) waveform and (*b*) spectrum.

modulators switched between two stable independent oscillator outputs. This resulted, however, in a phase discontinuity at the time of switching. Similarly, many PSK modulators are based on rapid switching of phase. In both cases, the phase discontinuity causes poor band occupancy because of the slow rate of out-of-band drop-off. Such signals have been referred to as *frequency-exchange keying* (FEK) and *phase-exchange keying* (PEK) to call attention to the discontinuities. Figure 1.17 illustrates a binary FEK waveform and its power spectrum density. The spectrum is the same as two overlapped ASK spectra, separated by the peak-to-peak frequency deviation. The Argand diagram for an FEK wave is simply a circle made up of superimposed arcs of opposite rotation. It is not easily illustrated. Figure 1.18 provides a similar picture of the PEK wave, including its Argand diagram. In this case, the Argand diagram is a straight line between the two points in the signal constellation. The spectrum is identical to the OOK spectrum with the carrier suppressed and has the same poor bandwidth occupancy.

The Argand diagram is more useful in visualizing the modulation when there are more than two points in the signal constellation. Quaternary modulation possesses four points at the corners of a square. Another four-point constellation occurs for binary modulation with 90° phase offset between even- and odd-bit transitions. This sort of offset, but with appropriately reduced offset angle, may also be used with *m*-ary signals. It can assist in recovery of the timing and phase

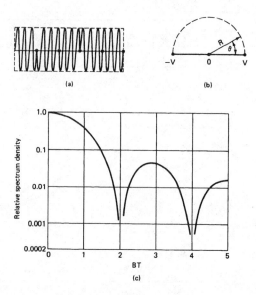

Figure 1.18 Binary PEK signal: (*a*) waveform, (*b*) Argand diagram, and (*c*) spectrum.

reference in the demodulator. In PEK, the transition is presumably instantaneous so that there is no path defined in the diagram for the transition. The path followed in a real situation depends on the modulator design. In Figure 1.19, where these two modulations are illustrated, the path is shown as a straight line connecting the signal points.

Continuous-phase constant-envelope PSK and FSK differ only slightly because of the basic relationship between frequency and phase. In principle, the goal of the PSK signals is to attain a particular one of *m* phases by the end of the signaling interval, whereas the goal of FSK signals is to attain a particular one of *m* frequencies. In the Argand diagram both of these modulation types travel around a circle—PSK from point to point and FSK from rotation rate to rotation rate (Figure 1.20). With constant-envelope modulation, a phase plane plot (tree) often proves useful. The spectrum depends on the specific transition function between states of

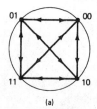

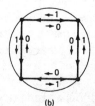

Figure 1.19 Argand diagrams of signal states and transitions: (*a*) quaternary and (*b*) phase-offset binary PEK.

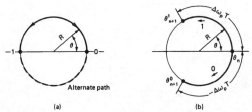

Figure 1.20 Argand diagrams: (*a*) binary PSK and (*b*) binary FSK.

frequency or phase. Therefore, spectra are not portrayed in Figures 1.21 and 1.22, which illustrate waveforms and phase trees for binary PSK and FSK, respectively.

The *m*-ary PSK with continuous transitions may have line components, and the spectra differ as the value of *m* changes. However, the spectra are similar for different *m* values, especially near zero frequency. Figure 1.23 shows spectra when the transition shaping is a raised cosine of one-half the symbol period duration for various values of *m*. Figure 1.24 gives spectral occupancy for binary PSK with several modulation pulse shapes. Figure 1.25 does the same for quaternary PSK. The spectrum of binary FSK for discontinuous frequency transitions and various peak-to-peak deviations less than the bit period is shown in Figure 1.26. Band occupancy for discontinuous-frequency binary FSK is shown in Figure 1.27. Figure 1.28 shows the spectrum occupancy for a binary FSK signal for various transition shapes but the same total area of $\pi/2$ phase change. The rectangular case corresponds to a discontinuous frequency transition with peak-to-peak deviation equal to 0.5 bit rate. This particular signal has been called *minimum-shift keying* (MSK) because it is the FSK signal of smallest deviation that may be demodulated readily using coherent quadrature PM.

The wide bandwidth and the substantial out-of-channel interference of PEK signals with sharp transitions can be reduced by placing a narrow-band filter after

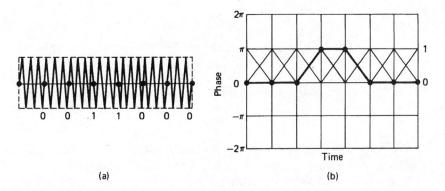

Figure 1.21 Binary PSK: (*a*) waveform and (*b*) phase tree.

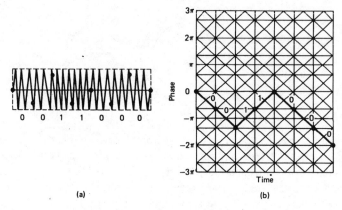

Figure 1.22 Binary FSK: (*a*) waveform and (*b*) phase tree.

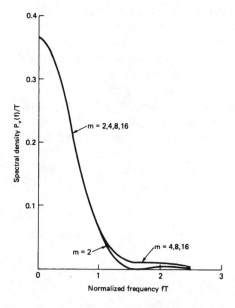

Figure 1.23 Spectra for *m*-ary PSK and half-symbol period raised cosine transition shaping.

the modulator. The filter tends to change the rate of transition and to introduce an envelope variation which becomes minimum at the time of the phase discontinuity. When the phase change is 180°, the envelope drops to zero at the point of discontinuity and the phase change remains discontinuous. For smaller phase changes, the envelope drops to a finite minimum and the phase discontinuity is eliminated. Thus, discontinuous PEK signals with 180° phase change, when passed

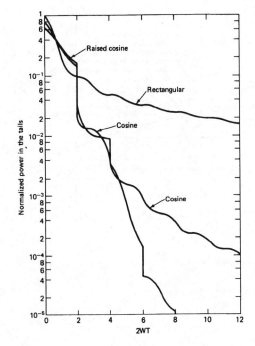

Figure 1.24 Spectrum occupancy of binary PSK with various transition shapings.

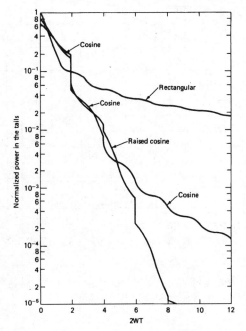

Figure 1.25 Spectrum occupancy for quaternary PSK with various transition shapings.

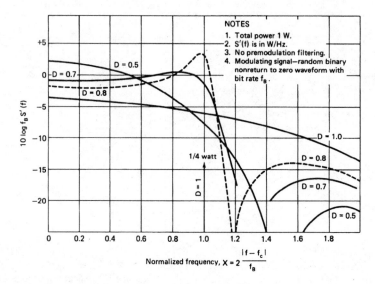

Figure 1.26 Spectra of binary FSK with sharp transitions.

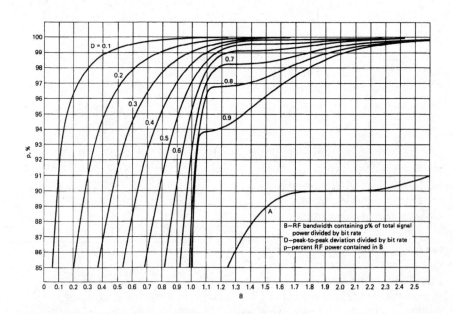

Figure 1.27 Band occupancy of binary FSK with sharp transitions at bit rate $1/T$. (Curve A = band occupancy of phase modulations with 180° peak-to-peak deviation.)

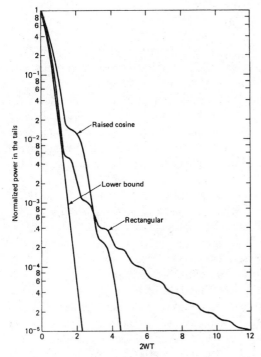

Figure 1.28 Band occupancy for *minimum-shift keying* (MSK) with transition shaping.

through limiting amplifiers, still have a sharp envelope notch at the phase discontinuity point, even after filtering. This tends to restore the original undesirable spectrum characteristics. To ameliorate this difficulty, offsetting the reference between symbols may be employed. This procedure provides a new reference for the measurement of phase in each symbol period—90° offset for binary, 45° for quaternary, and so on. In this way there is never a 180° transition between symbols, so that filtering and limiting can produce a constant-envelope signal with improved spectrum characteristics. In offset-keyed quaternary PSK, the change between successive symbols is constrained to ±90°. After filtering and limiting to provide a continuous-phase constant-envelope signal, the offset-keyed quaternary PSK signal is almost indistinguishable from MSK.

Another type of modulation with a constraint in generation is *unidirectional PSK* (UPSK), which also uses a quaternary PSK modulator. In this form of modulation, if two successive input bits are the same, there is no change in phase. However, if they differ, then the phase changes in two steps of 90°, each requiring one-half symbol interval. The direction of phase rotation is determined by the modulator connections and may be either clockwise or counterclockwise. The result is a wave that half the time is at the reference frequency and half the time at a lower or higher average frequency by one-half the input bit rate. The spectrum has a center 0.25 bit rate above or below the reference frequency. When it is narrow-band filtered and limited, the signal is almost indistinguishable from MSK offset from the reference frequency by 0.25 bit rate.

As with analog modulation, digital modulation may occur simultaneously in amplitude and angle. For example, an FSK or PSK wave can be modulated in amplitude to one of several levels. If the ASK does not occur at a faster rate and uses shaped transitions, there is little overall change in bandwidth and a bit or two may be added to the data rate. The performance of three types of signal constellations are illustrated in Figure 1.29. The *type II* system achieves better error performance than the *type I,* and both use synchronized amplitude and phase modulators. The *type III* system provides slightly better error performance than the *type II* and can be implemented easily using quadrature-balanced mixers (an approach often used to produce quaternary PSK signals). Because this is identical to DSB-SC AM using quadrature carriers, the name *quadrature AM* (QAM) has been applied to the *type III* signal constellation as well as quaternary ASK. Larger signal constellations are being used in digital microwave systems. At frequencies below 1 GHz, transmission impairments have generally kept transmissions to 8-ary or lower, where the advantages over FSK or PSK are not so significant.

Requiring continuous phase from angle modulation places a constraint on the process. Transition shaping to improve the spectrum is another type of constraint. Differential encoding of the incoming binary data so that a 1 is coded as no change in the outgoing stream and a 0 as a change is a different kind of constraint. This constraint does not affect bandwidth but assures that a 180° phase shift can be demodulated at the receiver despite ambiguity in the reference phase at the receiver. To eliminate receiver phase ambiguity, *m*-ary transmissions may also be encoded differentially. There has been a proliferation of angle modulation types with different constraints, with the primary objectives of reducing occupied bandwidth for a given transmission rate or improving error performance within a particular bandwidth, or both. A few of these systems are summarized here.

Partial response coding was devised to permit increased transmission rate through existing narrow-band channels. It can be used in baseband transmission or with continuous AM, PM, or FM. The initial types used ternary transmission to double the transmission rate through existing channels where binary transmission was used. These schemes, known as *biternary* and *duobinary* transmission, form constrained ternary signals that can be sent over the channel at twice the binary rate, with degraded error performance. The duobinary approach is generalized to *polybinary,* wherein the *m*-ary transmission has a number of states, every other one of which represents a 1 or a 0 binary state. For $m > 3$, this permits still higher transmission rates than ternary at further error rate degradation. Two similar modulation processes are referred to as *tamed frequency modulation* (TFM) and *gaussian filtered MSK* (GMSK).

When the response to a single digital input is spread over multiple keying intervals, it is sometimes possible to improve demodulation by using correlation

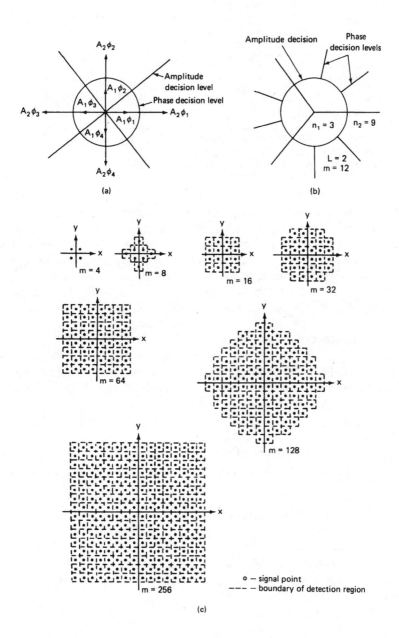

Figure 1.29 Examples of AM PSK constellations: (*a*) Type I, independent amplitude and phase decisions, (*b*) Type II, phase decision levels depend on amplitude, and (*c*) Type III, uniform square decision areas.

TABLE 1.1 *m*-ary CPFSK Bandwidth-Performance Tradeoffs

CPFSK scheme	Bandwidth $2BT_b$			$D^2_{min}/2E_b$	Gain over MSK, dB	N_B symbols
	90%	99%	99.9%			
$M = 2, h = 0.5$	0.78	1.20	2.78	2.0	0	2
$M = 4, h = 0.25$	0.42	0.80	1.42	1.45	-1.38	2
$M = 8, h = 0.125$	0.30	0.54	0.96	0.60	-5.23	2
$M = 4, h = 0.40$	0.68	1.08	2.08	3.04	1.82	4
$M = 4, h = 0.45$	0.76	1.18	2.20	3.60	2.56	5
$M = 8, h = 0.30$	0.70	1.00	1.76	3.0	1.76	2
$M = 8, h = 0.45$	1.04	1.40	2.36	5.40	4.31	5

over these intervals to distinguish among the possible waveforms. For this reason, the term *correlative coding* has been applied to such techniques. Table 1.1 shows some performance and bandwidth tradeoffs for *m*-ary *continuous-phase FSK* (CPFSK), without shaping filters. Generally speaking, by selecting a good set of phase trajectories longer than the keying period and by using correlation or Viterbi decoding in the demodulation process, both narrower bandwidths and better performance can be achieved than for conventional MSK.

B. Digital compression systems

Enhancements to the basic digital coding schemes described in the previous section have led to a family of high-quality coding and compression systems for audio, video, and data sources. Some of the more common high-performance coding systems for audio communications include:

- The audio portion of the MPEG (*Moving Pictures Experts Group*) standard
- *apt-X*, a digital audio compression system (Audio Processing Technology)
- *MUSICAM* audio compression
- *AC-2* audio compression (Dolby Labs)

Decades of research in psychoacoustics, the science of sound perception, have provided the following two fundamental principles upon which advanced compression schemes rely:

- **Threshold**: the minimum level of sound that can be heard. The *absolute threshold* is the sound level that is just detectable in the absence of all other sounds. As a result, if a sound is below the absolute threshold, a listener cannot hear it even under the best possible conditions. The sound, therefore, does not need to be part of the stored or transmitted signal. The *threshold of hearing* curve forms the lowest limit of digital encoding. Sounds below the limit simply are not encoded.

- **Masking**: the hiding of a low-level sound by a louder sound. The mechanisms of masking are sufficiently well understood that a model can be embedded into an encoder with good success. The model calculates the masking produced by a signal to determine what can be heard and what cannot. Masking, as discussed here, is more applicable to music signals than speech.

An additional principle of importance in audio compression is the redundancy of many audio waveforms. If a coding or compression system eliminates duplicate information, the bit error rate of the signal may be reduced with no measurable loss of signal quality.

Implementation of the foregoing principles vary from one compression scheme to the next. There are some techniques, however, that are common to many systems. It is common practice to divide the audible frequency range into subbands that approximate auditory critical bands. Bit allocation and quantization schemes are critical design elements. The choices to be made include assignment of the available bit rate to representations of the various bands. Differences in masking characteristics as a function of frequency are also significant. Proper filtering is of great importance. A filter bank confines coding errors temporally and spectrally in such a way as to allow the greatest compression at an acceptable performance limit. Many compression schemes represent a mathematical compromise between resolution and complexity. As the complexity increases, processing delays also increase.

Successful implementation of the foregoing principles require high-speed digital processing to examine the input waveform and adjust the sampling or coding parameters to maximize data throughput. Advanced signal processing chips, designed specifically for audio compression applications, have made implementation of a variety of coding schemes practical.

1.3 Radio Receiver Configurations

The first receivers were built with a tuned antenna and some iron dust in a glass tube to observe a tiny spark generated by activating the transmitter. Figure 1.30 shows such a test setup. The range of this transmission was very short, because substantial energy is required to generate enough voltage for the spark. An early

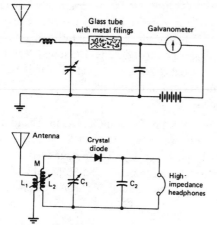

Figure 1.30 Test setup for spark detection.

Figure 1.31 Schematic diagram of a simple crystal receiver.

radiotelegraph detector based on the nonlinear properties of iron dust was known as a *coherer*. This somewhat insensitive and unreliable detection technique was gradually replaced by a crystal detector (an early use of semiconductors). A simple receiver with crystal detector is shown in Figure 1.31. It consists of an antenna, a tuned input circuit, a crystal detector, and a headset with an RF bypass condenser. At the time, the crystal detector, which is essentially a rectifying junction, was a piece of mineral with a tiny metal contact ("catwhisker") pressed against it. The operator adjusted the catwhisker on the mineral surface to get maximum sensitivity. This crystal detection receiver has a reception range of up to about 100 mi for some signals in the broadcast band or lower frequencies. Its basic disadvantages are lack of selectivity and sensitivity. However, the advantages are simplicity and absence of external power supply. The headphone, a high-impedance type of about 2 to 4 kΩ, delivers enough audio from the very minute radio energy to enable reception.

With the advent of the vacuum tube, receivers were improved by the addition of *radio frequency* (RF) preamplifiers, which isolated the antenna from the crystal detector's loading to provide higher selectivity. They also amplified the level of the signal and allowed additional selectivity circuits to be added. Amplifiers following the detector also increased the audio power level at the headset and eventually permitted the use of crude loudspeakers so that more than one person could hear the transmission without using headphones. The crystal detector was also retired shortly after the introduction of the vacuum tube when it was found that vacuum tubes could do a good job without such sensitive adjustments. Figure 1.32 is representative of these early vacuum-tube receivers. These receivers were relatively complicated to use because the variable capacitors had to be tuned individually to the same frequency. Gear systems were devised to provide simultaneous capacitor tuning and were soon replaced by common-shaft multiple variable capacitors

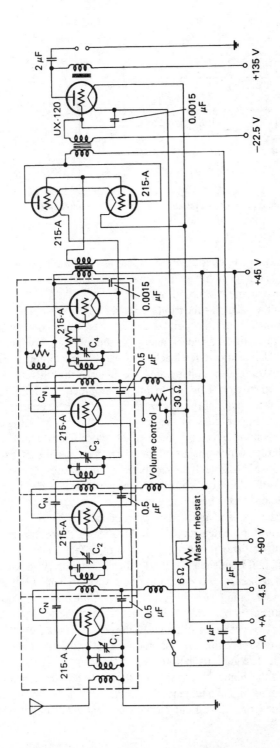

Figure 1.32 Schematic diagram of an early vacuum-tube receiver.

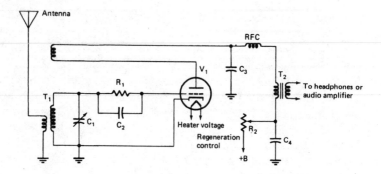

Figure 1.33 Schematic diagram of a classic, simple regenerative receiver.

(known as a ganged or gang condenser). Single-dial radio tuning made receivers much easier to use, and large numbers of vacuum-tube triode *tuned RF* (TRF) sets were produced for use in the AM broadcast band.

The instability of the vacuum-tube triode led to the discovery that when the tube was close to oscillation, the sensitivity and selectivity were tremendously increased, i. e., the addition of amplified feedback energy to the input energy increased the gain substantially. The resultant feedback also caused the circuit to present a negative resistance to the input circuit, tending to counteract the inherent resistances that broadened its selectivity. This led to the development of the regenerative detector where feedback was purposely introduced, under operator control to exploit these effects. Figure 1.33 is the schematic diagram of a regenerative detector. The regenerative detector and an audio amplifier provided a simple high-sensitivity radio set, which was useful for experimenters in the HF band where long-range reception was possible. If the feedback was increased sufficiently to just cause oscillation, on-off Morse code transmissions could be detected by offsetting the tuning slightly to produce a beat note output.

Armstrong subsequently invented the *superregenerative* receiver, which he hoped would be useful for the detection of FM broadcasts. Proper selection of the time constant of an oscillator produced a quenching effect whereby the circuit oscillates at two different frequencies. Initially, the circuit with large RF feedback increases gain until it breaks into oscillation at a strong signal level at an RF determined by the input tuning. However, after a short time the rectification in the grid circuit causes sufficient bias to be built up across the RC circuit to quench the oscillation. When the bias decreases sufficiently, the process repeats. By adjustment of the time constant, this relaxation oscillation can be set at 20 kHz or above, where the ear cannot hear it. The gain just before oscillation is extremely high. The rate and extent to which the higher-frequency oscillation builds up depends on the strength of the incoming signal and hence the frequency relative to the incoming tuned circuit. By appropriate tuning, the circuit can demodulate either AM or FM signals. The bandwidth is extremely wide, so that superregeneration is not suitable for many applications. Because the circuit oscillates and is connected through its input circuit to the antenna, it produces an unacceptable amount of radiation unless preceded by an isolating amplifier.

The circuit may be quenched by relaxation oscillations in the grid circuit, as described above, or by a separate supersonic oscillator. The latter provides much better control of the process. The first FM receiver consisted of an RF preamplifier, a superregenerative section, and an audio amplifier. The superregenerative principle was also used by radio amateurs exploring the capabilities of the VHF and UHF bands. Some early handie-talkies were built around the superregenerative receiving principle. They used the same tube in a revised configuration as a transmitter, with the audio amplifier doing double duty as a transmitter modulator. A change in the *resistor-capacitor* (RC) time constant in the grid circuit permitted generation of a narrow-band transmission of adequate stability. Amateurs long used such two-tube hand-held transceivers for operation up to 250 MHz. Even today, some remote control systems for model airplanes, model ships, and garage door openers, where little power consumption is permitted, will take advantage of the superregeneration principle. Such receivers have been built using either bipolar or field effect transistors, and show surprisingly good results. However, the development of superheterodyne-based receivers relegated TRF, regenerative, and superregenerative receivers to occasional special-purpose roles.

The *homodyne* or coherent detection receiver is related to the regenerative receiver. When two signals of different frequencies are input to a nonlinear element such as a detector, they produce new outputs at the sum and difference of the original signals (as well as higher-order sums and differences). The sum frequency is at RF, at about twice the signal frequency, and is filtered from the receiver output. The difference frequency, however, can be set in the audio band to produce a tone that, as mentioned previously, can be used to detect on-off coded signals. The *beat frequency oscillator* (BFO) of modern communications receivers produces a signal that is mixed with the incoming signal to obtain an audible beat note. A receiver using this principle is called a *heterodyne* receiver. When a heterodyne receiver is tuned to the frequency of an incoming AM signal, the tone can be reduced to zero, and only the modulation remains. The homodyne receiver makes use of this principle.

Figure 1.34 shows the schematic diagram of a homodyne receiver. The antenna signal, after filtering (and often, RF amplification) is fed to a detector where it is mixed with a *local oscillator* (LO) signal and beat directly to the audio band. The audio circuits can provide a narrow bandwidth to reject adjacent channel beats, and substantial gain can be provided at audio frequencies. Such receivers provide a potential improvement and cost reduction over receivers employing many tuned RF stages. The disadvantage is that the LO signal must be precisely on frequency and in phase with the desired received signal. Otherwise a low-frequency beat note corrupts the reception. Such receivers have limited current use. However, for some

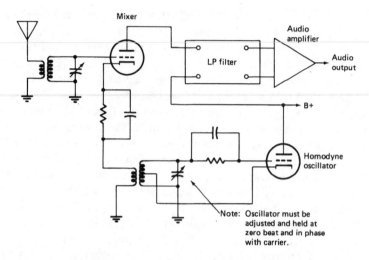

Figure 1.34 Schematic diagram of a simple vacuum tube-based homodyne receiver.

radar applications and special purposes they can provide good results. These receivers may be considered precursors of the coherent demodulators used in modern PSK and QAM digital data demodulation.

1.3.1 Superheterodyne Receivers

The *superheterodyne* receiver makes use of the heterodyne principle of mixing an incoming signal with a signal generated by a LO in a nonlinear element (Figure 1.35). However, rather than synchronizing the frequencies, the superheterodyne receiver uses a LO frequency offset by a fixed *intermediate frequency* (IF) from the desired signal. Because a nonlinear device generates identical difference frequencies if the signal frequency is either above or below the LO frequency (and also a number of other spurious responses), it is necessary to provide sufficient filtering prior to the mixing circuit so that this undesired signal response (and others) is substantially suppressed. The frequency of the undesired signal is referred to as an *image frequency,* and a signal at this frequency is referred to as an *image*. The image frequency is separated from the desired signal frequency by a difference equal to twice the IF. The preselection filtering required at the signal frequency is much broader than if the filtering of adjacent channel signals were required. The channel filtering is accomplished at IF. This is a decided advantage when the receiver must cover a wide frequency band, because it is much more difficult to maintain constant bandwidth in a tunable filter than in a fixed one. Also, for receiving different signal types, the bandwidth may be changed relatively easily at a fixed frequency by switching filters of different bandwidths. Because the IF at which channel selectivity is provided is often lower than the signal band

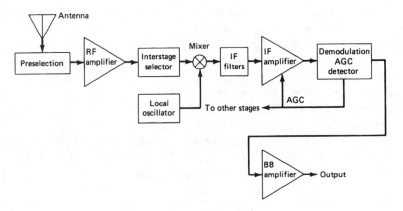

Figure 1.35 Block diagram of a superheterodyne receiver.

frequencies, it may be easier to provide selectivity at IF, even if wide-band RF tuning is not required.

Because of the nature of active electronic devices, it is generally easier to provide high stable gain in a fixed-frequency amplifier than in a tunable one, and gain is generally more economical at lower frequencies. Thus, although the superheterodyne receiver does introduce a problem of spurious responses not present in the other receiver types, its advantages are such that it has replaced other types except for special applications. For this reason, we shall discuss the superheterodyne block diagram at somewhat more length than the other types, and the detailed design chapters later in the book are based on this type of receiver (although, except for a few specifics, they are equally applicable to the other receiver types).

Referring to Figure 1.35, the signal is fed from the antenna to a preselector filter and amplifier. The input circuit is aimed at matching the antenna to the first amplifying device so as to achieve the best sensitivity while providing sufficient selectivity to reduce the probability of overload from strong undesired signals in the first amplifier. Losses from the antenna coupling circuit and preselection filters decrease the sensitivity. Because sufficient selectivity must be provided against the image and other principal spurious responses prior to the mixing circuit, the preselection filtering is often broken into two or more parts with intervening amplifiers to minimize the effects of the filter loss on the NF. The LO provides a strong stable signal at the proper frequency to convert the signal frequency to IF. This conversion occurs in the block called *mixer* in Figure 1.35. However, this type of element has also been called first detector, converter, or frequency changer.

The output from the mixer is applied to the IF amplifier, which amplifies the signal to a suitable power level for the demodulator. This circuit derives from the IF signal the modulation signal, which may be amplified by the baseband amplifier to the level required for output. Normally the output of an audio amplifier may be fed to a headphone or loudspeaker at the radio, or coupled to a transmission line for remote users. A video signal requires development of sweep, intensity, and usually color signals from the amplified video demodulation prior to display. In other cases, the output may be supplied to a data demodulator to produce digital data signals from the baseband signal. The data demodulator may be part of the receiver or may be provided separately as part of a data modem. The data modem may also be fed directly from the receiver at IF. Often data demodulation will be accomplished using digital processing circuits rather than analog demodulators and amplifiers. In this case the IF amplifier must be designed to provide the appropriate level to an A/D converter so that digital processing may be carried out. Additional IF filtering, data demodulation, and error control coding may all be done by digital circuits or a microprocessor, either in the radio or as part of an external modem.

An alternative to IF sampling and A/D conversion is the conversion of the signal to baseband in two separate coherent demodulators driven by quadrature LO signals at the IF. The two outputs are then sampled at the appropriate rate for the baseband by two A/D converters or a single multiplexed A/D converter, providing the in-phase and quadrature samples of the baseband signal. Once digitized, these components can be processed digitally to provide filtering, frequency changing, phase and timing recovery, data demodulation, and error control. Digital processing of audio and/or video signals is commonly found in such receivers using specialized *digital signal processing* (DSP) chips and chip sets.

There are a number of other functions required in an operating receiver, beyond those shown in the block diagram. These include gain control, manual and automatic; nonlinear impulse noise processing; BFO and heterodyne detector for OOK; adaptive signal processing; and diversity combining.

1.3.2 Tube Versus Semiconductor-Based Receivers

Rapid advances in the development of semiconductor devices have allowed design engineers to build transistor- and integrated circuit-based receivers that work well into the millimeter wave region. Looking back to 1960, when tubes such as the 6CW4 and 417A were the dominant devices for building VHF and UHF receivers, the improvement in NF and sensitivity has been dramatic. It is sometimes debated, however, whether the actual dynamic range, which is the ratio of the maximum input level (close to the 1-dB compression point) to the noise floor, really has been improved.

The major differences between tube designs and bipolar transistor-gallium arsenide designs lie in the matching techniques between the stages and compromises between selectivity, NF, and losses. Experience shows that the weakest point in past designs was the difficulty of building good mixers. One difficulty suffered by tube designs was the inability to test a circuit before it was built.

In the "old days," it was necessary to build all the circuits prior to evaluating them. Today's technologies allow designers to conduct feasibility studies. Because preamplifiers and mixers are part of the chain of the system that largely effects the overall performance, the following modeling capabilities are mandatory:

- Modeling the NF of low-noise amplifiers
- Modeling the 3 rd-order intercept point of amplifiers
- Modeling the insertion gain/loss of mixers, including the NF
- Predicting the phase noise of the oscillator

Engineers can use modern *computer-aided design* (CAD) programs to perform these and other characterization steps as exercises in software, usually long before the circuit is committed to hardware. Generally speaking, measurements and simulations agree quite well. The ability to model circuits in software, perhaps more than any other single development, has permitted the design of better receiver systems.

1.4 Typical Radio Receivers

We have described the systems and environments in which radio communications receivers operate as well as the general configurations of such receivers. In later chapters, we review in detail the processes and circuits involved in radio receiver design. Here we feel it desirable to provide a description of a typical modern radio communications receiver, so that the "forest" may be appreciated lest we become lost among the "trees." There are a large number of excellent receivers available from many manufacturers both in the United States and overseas. For the purposes of this book, however, we have confined the discussion to those most familiar to the authors. The following discussion of a representative design will serve as a backdrop for the subsequent detailed treatment of the elements of receiver technology. Modern communications receivers consist of a variety of analog and digital circuits. As an example of a typical communications receiver we have selected the Rohde and Schwarz EK895 digital VLF-HF receiver, which covers frequencies from 10 kHz to 30 MHz.

1.4.1 Design Example: The EK895

The capabilities of a receiver are best appreciated from its electrical specifications. Table 1.2 lists the electrical characteristics of the EK895. From a user point of view, desirable characteristics include good sensitivity and operational convenience in front panel or computer controls. Also very important is the ability of the receiver to provide interference-free signal detection in hostile environments, especially in the presence of a large number of simultaneous strong interfering signals. Reliability of the receiver cannot be overemphasized. A receiver that fails

Table 1.2 Specifications for the EK895 Digital VLF-HF Receiver
Source: *Rohde and Schwarz.*

Frequency range	10 kHz to 30 MHz	
Resolution	1 Hz	
Frequency drift	-10 to $+45\,°C$	aging/year
Frequency standard	5×10^{-7}	1×10^{-6}
Oven-controlled frequency standard	1×10^{-7}	1×10^{-7}
External frequency standard	$1/5/10$ MHz, 0.2 to $1\ V_{rms}$	
Antenna input	BNC connector, 50 Ω	
VSWR	<3	
Max. input voltage (≤30 MHz)	100 V EMF	
Oscillator reradiation into		
50-Ω termination	≤10 µV	
Demodulation modes	CW/MCW (A1A, A1B, A2A, A2B)	
	FAX1 (F1C)	
	AM/AME (A3E, H2A, H2B, H2E)	
	USB/LSB (R2A, R3E, J2A, J3E)	
	ISB (B8E)	
	FSK/AFSK (F1A, F1B), F6 (F7B)	
	FAX2 (F3C), FM (F3E)	
	DATA LINK acc. to MIL STD 188-203-1A (on request)	

IF bandwidth	**3 dB**	**60 dB**
(standard values)	± 75 Hz	± 150 Hz
	± 150 Hz	± 225 Hz
	± 300 Hz	± 430 Hz)
	± 500 Hz	± 770 Hz)
	± 750 Hz	± 990 Hz
	±1050 Hz	±1600 Hz
	±1200 Hz	±1760 Hz
	±1350 Hz	±1900 Hz
	±1550 Hz	±2100 Hz
	±3000 Hz	±4200 Hz)
	±4000 Hz	±5200 Hz
	(quasicontinuous bandwidth on request)	

Sensitivity (for S/N=10 dB, f=0.1 to 30 MHz)	
A1A (CW)	0.4 µV EMF (-121 dBm), BW=300 Hz
J3E (SSB), J7B	1.0 µV EMF (-113 dBm), BW=2.7 kHz
H3E (AME), 1 kHz, m=60%	2.7 µV EMF (-104 dBm), BW=6 kHz
with preamplifier, f=0.2 to 30 MHz	
A1A (CW)	0.2 µV EMF (-127 dBm), BW=300 Hz
J3E (SSB), J7B	0.4 µV EMF (-121 dBm), BW=2.7 kHz
H3E (AME), 1 kHz, m=60%	1.0 µV EMF (-113 dBm), BW=6 kHz
BFO	-5.00 to $+5.00$ kHz
Resolution	10 Hz
AFSK/FSK demodulator	transfer rate (50 to 600 baud) and deviation range (±42.5 to ±425 Hz) adjustable; V.28 interface and audible tone circuit
Diplex telegraphy demodulator (F7B)	2 x V.28 interface
Channel memory	for 1000 channels, nonvolatile, storage of complete receiver setup for each channel
Data interface	RS-232-C, RS-485 (bus-compatible)
Transfer rate	50 to 38,400 baud

TABLE 1.2 (continued)

Immunity to interference, non-linearities
Intermodulation (1.5 to 30 MHz;
$\Delta f \geq 30$ kHz; interfering signal 0 dBm)
$\quad IP_2$ >60 dBm (typ. 70 dBm)
$\quad IP_3$ >30 dBm (typ. 35 dBm)

Crossmodulation (0.1 to 30 MHz,
interfering signal 5 V EMF (+21 dBm);
$\Delta f \geq 30$ kHz; m = 0.3; f = 1 kHz;
signal level 10 mV EMF (−33 dBm))   ≤10% modulation transfer

Blocking (0.1 to 30 MHz;
interfering signal 6.3 V EMF
(+23 dBm); $\Delta f \geq 30$ kHz;
signal level 1 mV EMF (−53 dBm);
m = 0.3; f = 1 kHz)   ≥1 dB signal attenuation

Desensitization
(interfering signal 300 mV EMF;
$\Delta f \geq 30$ kHz; signal level
30 µV EMF; bandwidth 3.1 kHz)   ≥20 dB SINAD

Inherent spurious signals
(f >100 kHz)   <−113 dBm (nominal −124 dBm)
Image frequency rejection   >90 dB
IF rejection   >90 dB
Weighted S/N ratio
for 1 mV EMF   >46 dB SINAD

Gain control   automatic (AGC), manual (MGC) or
                                 remote (DGC)
AGC error   ≤3 dB (1 µV to 1 V EMF)
Time response constants
    Attack time   <10 ms
    Decay time   25/150/500 ms, 1 s, 3 s
    DGC range   0 to 120 dBµV EMF in 1-dB steps

Outputs
AF output 1   0.3 to 3.4 kHz; floating; $Z_{out} = 600\ \Omega$
    Settling range   −10 to +10 dBm
    FAX1 (F1C)   1.9 kHz ± 150 Hz in VLF range
                                (f <500 kHz)
                                1.9 kHz ±400 Hz in HF range
                                (f ≥500 kHz)
    FAX2 (F3C)   1.9 kHz modulated
    FM (F3E)   NBFM (3-dB bandwidth: ±4 kHz)
AF output 2 (LSB in ISB mode)   0.3 to 3.4 kHz, floating, $Z_{out} = 600\ \Omega$
    Settling range   −10 to +10 dBm
Monitoring output   500 mV, $Z_{out} = 332\ \Omega$
FM video output   1 V/kHz, $Z_{out} = 1\ k\Omega$
IF output (analog)   0 to 40 kHz in 100-Hz steps, 0 dBm
                                into 600 Ω or 455 kHz, 0 dBm into
                                50 Ω (optional)
IF output (digital)   serial data (clock, data, frame),
                                100 kSPS
I/Q output (digital)   baseband signal, multiplexed,
                                5 V CMOS

during crucial communication traffic periods is useless. A modern receiver, therefore, must incorporate built-in test equipment and diagnostic routines to allow the user to check its characteristics and thus anticipate potential failures. The EK895 has been designed to satisfy these requirements.

The receiver features high resolution of tuning frequency, down to 1 Hz, with 1000 programmable channel memory locations. In addition, a scan mode is provided for programmable frequency ranges and desired channel sequences. Demodulators for AM, *continuous wave* (CW), LSB, USB, ISB, FM, FSK, and facsimile (FAX) are included in the basic configuration. Eleven bandwidths from 150 Hz to 8 kHz are provided, with quasicontinuous bandwidths on request. An internal microprocessor is used for bus interfacing, menus, and user programs. The control interface complies with international standards. All settings of the receiver can be directed remotely via an ASCII (*American Standard Code for Information Interchange*) terminal, a personal computer, or over long distances by modem. Using line drivers, a master receiver can control up to 99 detached receivers in master-slave operation.

A basic block diagram of the digital VLF-HF receiver is shown in Figure 1.36. As shown, the system is divided into five major elements:

- RF unit
- Synthesizer
- IF/AF processor
- System control processor
- Power supply

In addition, provisions have been made for two option modules. Figure 1.37 shows the front panel of the EK895.

A. RF unit

The antenna signal is routed via a low-pass filter, which is provided for rejecting image frequencies and suppressing oscillator reradiation, and applied to the input mixer where it is converted to the first IF of 41.44 MHz by means of an oscillator variable in 1 Hz steps. The crystal filter that follows determines the maximum receiving bandwidth of 10 kHz and provides for rejection of the second image frequency. A fixed frequency of 40 MHz is used for conversion into the second IF of 1.44 MHz.

B. Synthesizer

The synthesizer supplies all of the conversion frequencies required for the RF and IF demodulator units. The first conversion oscillator, derived from direct digital frequency synthesis, can be varied in 1 Hz steps. The settling time of the oscillator is 5 ms for any frequency variation. Two *phase-locked loops* (PLLs) produce 40 MHz and 5.66 MHz fixed frequencies. Operation of the total of four PLLs in the synthesizer is continuously monitored for adherence to specifications.

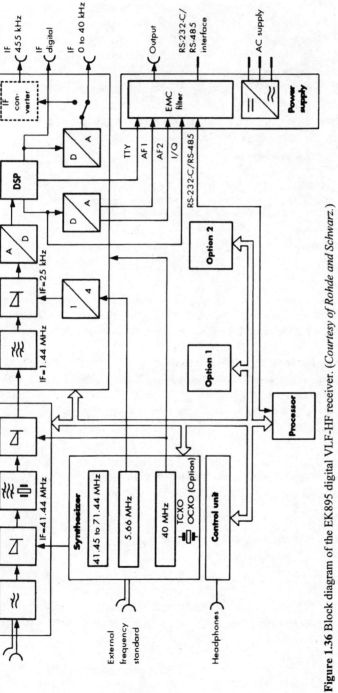

Figure 1.36 Block diagram of the EK895 digital VLF-HF receiver. (*Courtesy of Rohde and Schwarz.*)

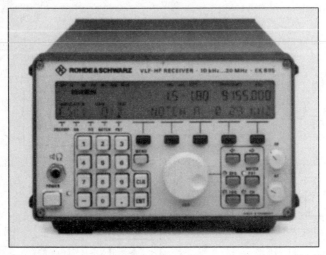

Figure 1.37 Front panel display of the EK895. (*Courtesy of Rohde and Schwarz.*)

In the basic version of the receiver, all frequencies are derived from a temperature-compensated crystal oscillator. Higher accuracy requirements can be obtained by including a heated crystal oscillator or an external frequency standard.

C. IF and *audio frequency* (AF) processor
The second IF is converted to the third IF of 25 kHz using the 5.66 MHz fixed frequency source. After digitization of the third IF frequency in a 16 bit A/D converter, the digital signal processor assumes all signal generation and processing tasks, including:

- Automatic, remote, or manual control of operating parameters
- Measurement of received levels
- Filtering with 11 fixed or quasicontinuously adjustable bandwidths
- Demodulation, passband tuning, and filtering
- Noise blanking and squelch
- Generation of the BFO frequency analog IF from 0 to 40 kHz
- Generation of the digital IF as serial data
- Generation of the I/Q data stream

D. System control processor

The 16-bit *complementary metal oxide semiconductor* (CMOS) microprocessor is responsible for setting and managing the other module functions, and communicates with the outside world via the front panel controls and data interface. The processor executes the internal programs and ensures operational reliability through the following routines:

- Nonvolatile storage of all settings
- Continuous testing of *central processing unit* (CPU), *random-access memory* (RAM), and *programmable read-only memory* (PROM) functions
- Continuous monitoring of the synthesizer circuits
- Built-in testing functions for each module

ASCII command syntax is used for controlling the receiver via a serial interface. Functions available remotely include basic settings, search operations, test operations, system functionality, and channel control operations.

1.4.2 VHF and UHF Receivers

Communications receivers for the VHF and UHF bands typically cover portions of the bands from 30 to 1000 MHz (sometimes overlapping HF by operating as low as 20 MHz), and typically have designs similar to that described in the previous section. Services in the VHF and UHF bands include the following:

- 30 to 54 MHz: police radios and pagers
- 54 to 88 MHz: low-band broadcast television
- 88 to 108 MHz: the FM broadcast band
- 118 to 136 MHz: aircraft communications
- 148 to 174 MHz: mobile ground radio applications
- 174 to 225 MHz: high-band broadcast television assignments
- 225 to 400 MHz: military aircraft communications
- 450 to 470 MHz: ground mobile operations
- 470 to 806 MHz: UHF broadcast television
- 806 to 902 MHz: ground mobile applications
- 902 to 960 MHz: other fixed and land mobile services

- Above 1000 MHz: specialized military communications, radar, radio-navigation facilities, point-to-point microwave, and satellite

Many of the frequencies described in this very broad discussion have shared usage; some bands are reserved for radio amateur use and scientific studies. In other parts of the world, the allocations differ.

Some of the communications in these frequency regions, especially toward 1000 MHz and above, have very complex modulation capabilities, including time and frequency multiplexing of many channels, burst transmission modes, frequency hopping, or other *spread-spectrum* waveforms. The dynamic range of a receiver may need to be greater than that of an HF receiver because of an even wider range of received field strengths. This is because high-gain antennas with good directivity can be constructed easily, especially at the higher frequencies. Gains of 10 to 13 dB over a dipole are not uncommon. In remote low-noise listening areas, low-noise preamplifiers may be necessary because of the reduced atmospheric and man-made noise in these frequency bands. NF of 4 dB and lower are often needed, although in many applications a NF of 8 dB will suffice. In the vicinity of cities, there may be a large number of high-power transmitters that can result in input signals of up to 100 mV or more. Such low-noise conditions and high-signal levels seldom occur at HF. Hence, the dynamic range of VHF/UHF receivers must generally be significantly higher.

Receiver systems can at times be very complex. A particular example of this is a signal surveillance system, such as illustrated in the block diagram of Figure 1.38. The circularity disposed antenna array, with its beam formers, allows the direction of incoming signals to be identified as well as the frequency and character of transmissions determined by the channelized and sweeping receivers. A large installation of such receivers feeds information to switching systems and an information processing system that can configure the equipment and perform the analysis of all significant received information.

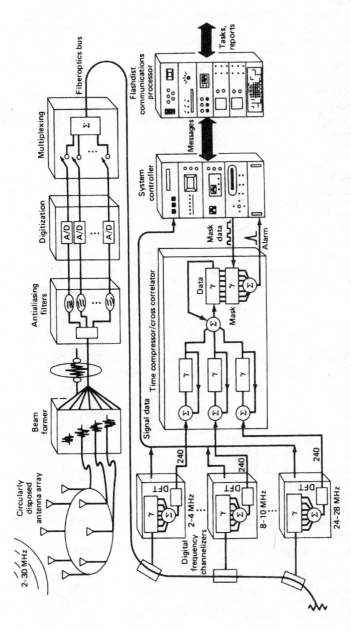

Figure 1.38 Block diagram of a complete surveillance system.

1.5 Bibliography

"Ground-Wave Propagation Curves for Frequencies between 10 kHz and 30 MHz," Recommendation 368-4, CCIR, vol. V, sec. 5B, p. 23, ITU, Geneva, 1982.

"World Distribution and Characteristics of Atmospheric Radio Noise," CCIR Rep. 322, ITU, Geneva, 1964.

Abate, J. E., "Linear and Adaptive Delta Modulation," *Proc. IEEE,* vol. 55, p. 298, Mar. 1967.

Abramson, N., "The Aloha System—Another Alternative for Computer Communications," *AFIPS Proc.,* p. 543, SJCC, 1970.

Akass, J. L., and N. C. Porter, "The PTARMIGAN System," IEE Conf. Pub. 139, Communications, 1976.

Anderson, R. R., and J. Salz, "Spectra of Digital FM," *Bell Sys. Tech. J.,* vol. 44, p. 1165, July-Aug. 1965.

Aulin, T., and C. E. W. Sundberg, "Continuous Phase Modulation—Part II: Full Response Signaling," *IEEE Trans.,* vol. COM-29, p. 196, Mar. 1981.

Aulin, T., N. Rydbeck, and C. E. W. Sundberg, "Continuous Phase Modulation—Part II: Partial Response Signalling," *IEEE Trans.,* vol. COM-29, p. 210, Mar. 1981.

Blecher, H., "Advanced Mobile Phone Service," *IEEE Trans.,* vol. VT-29, p. 238, May 1980.

Brogle, A. P., "A New Transmission Method for Pulse-Code Modulation Communication Systems," *IRE Trans.,* vol. CS-8, p. 155, Sept. 1960.

Bullington, K., "Radio Propagation at Frequencies above 30 Megacycles," *Proc. IRE,* vol. 35, p. 1122, Oct. 1947.

Campopiano, C. N., and B. G. Glazer, "A Coherent Digital Amplitude and Phase Modulation Scheme," *IRE Trans.,* vol. CS-10, p. 90, Mar. 1962.

deJager, F., "Deltamodulation, A Method of P.C.M. Transmission Using the 1-Unit Code," *Philips Res. Rep.,* vol. 7, p. 442, 1952.

deJager, F., and C. B. Decker, "Tamed Frequency Modulation, a Novel Technique to Achieve Spectrum Economy in Digital Transmission," *IEEE Trans.,* vol. COM-26, p. 534, May 1978.

Donaldson, R. W., and D. Chan, "Analysis and Subjective Evaluation of DCPM Voice Communication System," *IEEE Trans.,* COM-17, Feb. 1969.

Egli, J. J., "Radio Propagation above 40 MHz, over Irregular Terrain," *Proc IRE,* vol. 45, p. 1383, Oct. 1957.

Hancock, J. C., and R. W. Lucky, "Performance of Combined Amplitude and Phase-Modulated Communication Systems," *IRE Trans.,* vol. CS-8, Dec. 1960.

Jenks, F. G., P. D. Morgan, and C. S. Warren, "Use of Four-Level Phase Modulation for Digital Radio," *IEEE Trans.,* vol. EMC-14, p. 113, Nov. 1972.

Kahn, L. R., "Compatible Single Sideband," *Proc. IRE,* vol. 49, p. 1503, Oct. 1961.

Kahn, R. E., S. A. Gronemeyer, J. Burchfiel, and R. C. Kunzelman, "Advances in Packet Radio Technology," *Proc. IEEE,* vol. 66, p. 1468, Nov. 1978.

Keen, R., *Wireless Direction Finding* (Iliffe and Sons, London, 1938).

Lender, A., "Correlative Digital Communication Techniques," *IEEE Trans.,* vol. COM-12, p. 128, Dec. 1964.

Lender, A., "The Duobinary Technique for High-Speed Data Transmission," *AIEE Trans.,* vol. 82, pt. 1, p. 214, Mar. 1963.

LongLey, A. G., and P. L. Rice, "Prediction of Tropospheric Radio Transmission Loss over Irregular Terrain—A Computer Method—1968," ESSA Tech. Rep. ERL-79-ITS-67, July 1968.

Murota, K., and K. Hirade, "GMSK Modulation for Digital Mobile Radio Telephony," *IEEE Trans.,* vol. COM-29, p. 1044, July 1981.

Nossen, E. J., "The RCA VHF Ranging System for Apollo," *RCA Engineer,* vol. 19, p. 75, Dec. 1973/Jan. 1974.

Nossen, E. J., and V. F. Volertas, "Unidirectional Phase Shift Keyed Communication Systems," U.S. Patent 4130802, Dec. 19, 1978.

Nyquist, H., "Certain Topics in Telegraph Transmission Theory," *Trans. of the AIEE,* vol. 7, p. 617, Apr. 1928.

Okamura, Y. E., E. Ohmori, T. Kawano, and K. Fukudu, "Field Strength and Its Variability in VHF and UHF Land-Mobile Service," *Rev. Tokyo Elec. Commun. Lab.,* vol. 16, p. 825, Sept./Oct. 1968.

Oliver, B. M., J. R. Pierce, and C. E. Shannon, "The Philosophy of PCM," *Proc. IRE,* vol. 36, p. 1324, Nov. 1948.

Panter, P. F., *Modulation, Noise and Spectral Analysis* (McGraw-Hill, New York, 1965).

Pelchat, M. G., "The Autocorrelation Function and Power Spectra of PCM/FM with Random Binary Modulating Waveforms," *IEEE Trans.,* vol. SET-10, p. 39, Mar. 1964.

Prabhu, V. K., "Spectral Occupancy of Digital Angle-Modulation Signals," *Bell Sys. Tech. J.,* vol. 55, p. 429, Apr. 1976.

Prabhu, V. K., and H. E. Rowe, "Spectra of Digital Phase Modulation by Matrix Methods," *Bell Sys. Tech. J.,* vol. 53, p. 899, May-June 1974.

Pratt, H., and H. Diamond, "Receiving Sets for Aircraft Beacon and Telephony," *Proc. IRE,* vol. 17, Feb. 1929.

Rick, P. L., A. G. Longley, K. A. Morton, and A. P. Barsis, "Transmission Loss Predictions for Tropospheric Communications Circuits," NBS Tech. Note 101, vols. I and II (revised), U.S. Dept. of Commerce, 1967.

Schiff, L., "Traffic Capacity of Three Types of Common-User Mobile Radio Communication Systems," *IEEE Trans.,* vol. COM-18, Feb. 1970.

Scholtz, R. A., "The Origins of Spread Spectrum Communications," *IEEE Trans.* vol. COM-30, p. 822, May 1982.

Schulte, H. J., Jr., and W. A. Cornell, "Multi-Area Mobile Telephone System," *IRE Trans.,* vol. VC-9, p. 49, May 1960.

Spaulding, A. D., and R. T. Disney, "Man-Made Radio Noise—Part 1: Estimates for Business, Residential, and Rural Areas," OT Rep. 74-38, U.S. Dept. of Commerce, June 1974.

Tjhung, T. T., "The Band Occupancy of Digital FM Signals," *IEEE Trans.,* vol. COM-12, p. 211, Dec. 1964.

van de Weg, H., "Quantizing Noise of a Single Integration Delta Modulation System with an *N*-Digit Code," *Philips Res. Rep.,* vol. 8, p. 367, 1953.

Winkler, M. R., "High Information Delta Modulation," *IEEE Int. Conv. Rec.,* pt. 8, p. 260, 1963.

Winkler, M. R., "Pictorial Transmission with HIDM," *Globecom VI Rec.* (June 2-4, 1964).

Additional Suggested Reading

"Effects of Noise in Receiving Systems," *Ham Radio* magazine, Nov. 1997.

"Eight Ways to Better Radio Receiver Design," *Electronics,* Feb. 20, 1975.

Chapter 2

Radio Receiver Characteristics

2.1 Introduction

Communications receivers, with which we are primarily concerned in this book, have much in common with receivers for other applications, such as direction finders, radio altimeters, radar, and related systems. As mentioned in Chapter 1, the superheterodyne configuration is used for almost all applications and, therefore, the remainder of the book assumes such a design. Nevertheless, many of the receiver characteristics described in this chapter apply to other configurations as well.

User groups often specify required characteristics and test procedures as appropriate for their particular applications. Therefore, we have limited this chapter to a general review of important characteristics of communications receivers. Each new design has its own detailed requirements, and compromises among these may be required for practical implementation. Thus, the engineer must analyze the required characteristics carefully before undertaking a new design. In setting specifications, we need to note that overall performance also can be affected by other components of the receiving system and by the electromagnetic environment in which the system operates.

2.1.1 Input Characteristics

The first essential function of any radio receiver is to effect the transfer of energy picked up by the antenna to the receiver itself through the input circuits. Maximum energy is transferred if the impedance of the input circuit matches that of the antenna (inverse reactance and same resistance) throughout the frequency band of the desired signal. This is not always feasible, and the best energy transfer is not essential in all cases. A receiver may also be connected with others through a *hybrid* or *active multicoupler* to a single antenna. Such arrangements are sometimes very sensitive to mismatches.

There are at least three antenna matching problems in a receiver. The first and, in many cases, most crucial problem is that the receiver may be used from time to time with different antennas whose impedances the potential users cannot specify fully. Second, antennas may be used in mobile applications or in locations subject to changing foliage, buildings, or waves at sea, so that the impedance, even if measured accurately at one time, is subject to change from time to time. Third, at some frequencies, the problems of matching antennas are severely limited by available components, and the losses in a matching network may prove greater than for a simpler lower-loss network with poorer match.

When antenna matching is important over a wide band, it may be necessary to design a network that can be tuned mechanically or electrically under either manual or automatic control in response to a performance measure in the system. In older receivers with a wide tuning range, it was common to have a mechanically tuned preselector that could be adjusted by hand and was generally connected directly to the *variable-frequency oscillator* (VFO) used for the first conversion. At times a special trimmer was used in the first circuit to compensate for small antenna mismatches. Thus, tuning of the circuit could be modified to match the effects of the expected antenna impedance range. Modern wide tuning receivers often use one-half-octave switchable filters in the preselector, which may be harder to match, but are much easier to control by computer. Similarly, the first oscillator is generally a microprocessor-controlled synthesizer.

Often the problem of antenna matching design is solved by the user specification that defines one or more "dummy antenna" impedances to be used with a signal generator to test the performance of the receiver for different receiver input circuits. In this case, the user's system is designed to allow for the mismatch losses in performance that result from the use of actual antennas. When it is necessary to measure receiver input impedance accurately, it is best accomplished through a network analyzer.

A number of other receiver input considerations may occur in some cases. The input circuits may be balanced or unbalanced, or may need to be connectable either way. The input circuits may require grounding, isolation from ground, or either connection. The circuits may need protection from high-voltage discharges or from impulses. They may need to handle, without destruction, high-power nearby signals, both tuned to the receiver frequency and off tune. Thus, the input circuit can—at times—present significant design challenges.

2.1.2 Gain, Sensitivity, and NF

Communications receivers are required to receive and process a wide range of signal powers, but in most cases it is important that they be capable of receiving distant signals whose power has been attenuated billions of times during transmission. The extent to which such signals can be received usefully is determined by the noise levels received by the antenna (natural and man-made) and those generated within the receiver itself. It is also necessary that the receiver produce a level of output power suitable for the application. Generally the ratio of the output power of a device to the input power is known as the gain. The design of a receiver includes gain distribution (see Chapter 3) among the various stages so as to provide adequate receiver gain and an optimum compromise among the other characteristics.

While there are limits to the amount of gain that can be achieved practically at one frequency because of feedback, modern receivers need not be gain-limited.

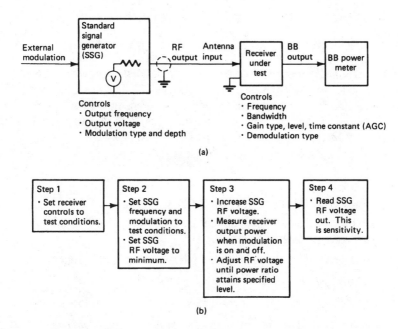

Figure 2.1 Receiver sensitivity measurement: (*a*) test setup and (*b*) procedure.

When the gain is sufficiently high, the weakest signal power that may be processed satisfactorily is noise-limited. This signal level is referred to as the *sensitivity* of the system at a particular time and varies depending on the external noise level. It is possible in some systems for the external noise to fall sufficiently so that the system sensitivity is established by the internal noise of the receiver. A receiver's sensitivity is one of its most important characteristics. There are no universal standards for its measurement, although standards have been adopted for specific applications and by specific user groups. Figure 2.1 shows a block diagram of the test setup and the typical steps involved in determining receiver sensitivity.

A. AM sensitivity

The typical AM sensitivity definition requires that when the input signal is sinusoidally modulated w percent at x hertz, the receiver bandwidth (if adjustable), having been set to y kilohertz, shall be adjusted to produce an output S/N of z decibels. The resulting signal generator open-circuit voltage level shall be the sensitivity of the receiver. The values of w, x, y, and z vary; common values are:

- $w = 30$ percent
- $x = 1000$ Hz

- $y = 6$ kHz
- $z = 10$ dB

Also, it is assumed that the noise in question is random thermal noise without any associated squeals or whistles, and the signal received, like that generated, is an undistorted sinusoid. Let us consider how the measurement is made using the test setup of Figure 2.1. Assume that we wish to measure the AM sensitivity of an HF receiver at 29 MHz, using the numerical values listed previously. Having set the carrier frequency, the following steps are taken:

- Select the 6 kHz bandwidth setting of the receiver.
- Set the gain control to *manual gain control* (MGC) or AGC, depending on which sensitivity is to be measured.
- Set the AGC time constant appropriate to normal AM reception.
- Turn off the BFO.
- Set the gain control (if it is manual) so that the receiver does not overload.
- Set the audio level control to an appropriate level for the power meter or *root-mean-square* (rms) voltmeter measuring the output.
- With the signal generator modulation turned on and set to 30 percent, increase the generator voltage from its minimum output of less than 0.1 µV to a level of 1 µV or more.
- While observing the audio output meter, switch off the signal generator modulation; observe the reduction in output level.
- Repeat this process several times while adjusting the signal generator output level until the difference between the two readings is precisely 9.5 dB.

A good receiver has a sensitivity of about 1.5 µV in this test. For a 20-dB S/N, a value of about 5 µV should be obtained.

It should be noted that to achieve a 10-dB S/N, the output readings were carefully adjusted to a difference of 9.5 dB. This is because the reading with modulation on includes not only the signal, but also the noise components. Thus, what is being measured is the *signal-plus-noise-to-noise ratio* $(S + N)/N$, and an adjustment must be made to obtain S/N. If the ratio is sufficiently high, the difference becomes negligible. Even at 10 dB the difference is only 0.5 dB, as indicated in this example. In some cases, the user specifies $(S + N)/N$ rather than S/N. A similar consideration applies in the case of *signal-plus-noise-plus-distortion* (SINAD) measurements.

While making sensitivity measurements, the accuracy of the instrument used at the receiver to indicate antenna input voltage can also be checked. This sort of instrument is not intended as an accurate device, and an error of ±3 dB would not be considered unusual.

Another method of sensitivity measurement that is often used takes into account distortion and all internal noises that, as a practical matter, can interfere with reception. In this measurement, the SINAD ratio is adjusted rather than the S/N alone. A selective band-reject filter at the modulation frequency is switched in to remove the fundamental frequency. Harmonics and other nonlinear distortion are thus added to the noise. Depending upon the particular test conditions, the S/N can be achieved at a lower input voltage than the equivalent SINAD value. Modern test equipment includes tools for such measurements.

B. SSB and CW sensitivity

The SSB mode of reception translates the sideband to the audio frequency, rather than using an envelope demodulator as in conventional AM. This eliminates the nonlinear transformation of the noise by the carrier that occurs in the measurement of AM sensitivity. Also, there is no carrier power in SSB, and the required bandwidth is about half that needed for AM, so that substantially improved sensitivity can be expected. The sensitivity test is performed with the signal generator frequency offset from the nominal (suppressed) carrier frequency by a selected frequency, again often selected as 1000 Hz. In this case, it is necessary to make the measurement with AGC off. Otherwise, turning the signal off would increase gain, changing the resulting noise level. In other aspects, the characteristic is measured much the same as AM sensitivity. Specific differences for SSB include:

- Set the signal generator frequency to 29.1 MHz if it is an USB signal; modulation is not required.

- Increase the signal generator level from the minimum until signal output is apparent; check power outputs with the generator on and off.

- Adjust the generator until the output ratio in the two cases becomes 9.5 dB.

The SSB sensitivity for a good receiver at the 50 Ω input is 0.1 to 0.3 μV. A similar measurement can be made using SINAD rather than S/N.

For coded CW signals, the appropriate bandwidth must be selected, typically 150 to 500 Hz. The signal generator must be set on-frequency, and the BFO must be tuned to a 1000 Hz beat note. Otherwise the sensitivity measurement is the same as for SSB. The CW sensitivity should be about 0.03 to 0.1 μV for a 9.5-dB S/N.

C. FM sensitivity

For FM sensitivity measurements an FM signal generator must be available (Figure 2.1). Sensitivity is measured at a specified deviation. Often there is no manual or automatic gain control. Deviation settings vary substantially with the radio usage and have been gradually becoming smaller as closer channel assignments have been made over the years to alleviate spectrum crowding. For commercial communications receivers, a deviation of 2.1 kHz rms (3.0 kHz peak) sinusoidal modulation at a 1000 Hz rate is customary. The sensitivity measurement is generally performed in the same manner as before and a 12-dB SINAD measurement is often used. A good receiver will provide about 0.1 to 0.2 μV sensitivity. For a 20-dB S/N, a value of about 0.5 μV results.

Another measure that is sometimes used is the *quieting sensitivity* in which the noise output level is first measured in the absence of a signal. The signal generator is unmodulated and gradually increased in level until the noise is reduced by a predetermined amount, usually 20 dB. This value represents the quieting sensitivity. A typical value would be 0.15 to 0.25 μV. As the signal level is further increased, with the modulation being switched on and off, the ultimate S/N level occurs. This can provide information on residual system noise, particularly frequency synthesizer close-in noise. For example, if the measurement is performed with a 3-kHz base bandwidth and the synthesizer has a residual FM of 3 Hz, the ultimate S/N is limited to 60 dB.

D. NF

Sensitivity measures depend upon specific signal characteristics. The NF measures the effects of inherent receiver noise in a different manner. Essentially it compares the total receiver noise with the noise that would be present if the receiver generated no noise. This ratio is sometimes called the noise factor F, and when expressed in dB, the NF. F is also defined equivalently as the ratio of the S/N of the receiver output to the S/N of the source. The source generally used to test receivers is a signal generator at local room temperature. An antenna, which receives not only signals but noises from the atmosphere, the galaxy, and man-made sources, is unsuitable to provide a measure of receiver NF. However, the NF required of the receiver from a system viewpoint depends on the expected S/N from the antenna. The effects of external noise are sometimes expressed as an equivalent antenna NF.

For the receiver, we are concerned with internal noise sources. Passive devices such as conductors generate noise as a result of the continuous thermal motion of the free electrons. This type of noise is referred to generally as thermal noise, and is sometimes called *Johnson noise* after the person who first demonstrated it. Using the statistical theory of thermodynamics, Nyquist showed that the mean-square thermal noise voltage generated by any impedance between two frequencies f_1 and f_2 can be expressed as:

$$\overline{V_n^2} = 4\,k\,t \int_{f_1}^{f_2} R(f)\,df \qquad (2.1)$$

where $R(f)$ is the resistive component of the impedance.

Magnetic substances also produce noise, depending upon the residual magnetization and the applied dc and RF voltages. This is referred to as the *Barkhausen effect*, or *Barkhausen noise*. The greatest source of receiver noise, however, is generally that generated in semiconductors. Like the older thermionic tubes, transistors and diodes also produce characteristic noise. *Shot noise* resulting from the fluctuations in the carrier flow in semiconductor devices produces wide-band noise, similar to thermal noise. Low-frequency noise or $1/f$ noise, also called *flicker effect*, is roughly inversely proportional to frequency and is similar to the contact noise in contact resistors. All of these noise sources contribute to the "excess noise" of the receiver, which causes the NF to exceed 0 dB.

The NF is often measured in a setup like that of Figure 2.1, using a specially designed and calibrated white-noise generator as the input. The receiver is tuned to the correct frequency and bandwidth, and the output power meter is driven from a linear demodulator or the final IF amplifier. The signal generator is set to produce no output, and the output power is observed. The generator output is then increased until the output has risen 3 dB. The setting on the generator is the NF in decibels.

The NF of an amplifier can also be calculated as the ratio of input to output S/N, per the equation:

$$NF = 10 \log \left[\frac{(S/N)_1}{(S/N)_0} \right] \qquad (2.2)$$

where NF is the noise figure in dB and $(S/N)_1$ and $(S/N)_2$ are the amplifier input and output SNR, respectively.

The NF for a noiseless amplifier or lossless passive device is 0 dB; it is always positive for nonideal devices. The NF of a lossy passive device is numerically equal to the device insertion loss. If the input of a nonideal amplifier of gain G (dB) and noise figure NF (dB) were connected to a matched resistor, the amplifier output noise power P_{No} (dB) would be:

$$P_{No} = 10 \log(kT) + 10 \log(B) + G + NF \qquad (2.3)$$

where k is Boltzmann's constant (1.38×10^{-20} mW/°K), T is the resistor temperature in °K, and B is the noise bandwidth in Hz.

When amplifiers are cascaded, the noise power rises toward the output as noise from succeeding stages is added to the system. Under the assumption that noise powers add noncoherently, the noise figure NF_T of a cascade consisting of two stages of numerical gain A_1 and A_2 and noise factor N_1 and N_2, is given by Friis' equation:

$$NF_T = 10 \log \left[\frac{N_1 + (N_2 - 1)}{A_1} \right] \qquad (2.4)$$

where the noise factor is $N = 10^{(NF/10)}$ and the numerical gain is $A = 10^{(G/10)}$. The system NF, therefore, is largely determined by the first stage NF when A_1 is large enough to make $(N_2 - 1)/A_1$ much smaller than N_1.

E. Minimum detectable signal

Another measure of sensitivity is the *minimum detectable signal* (MDS). The setup for this type of measurement is generally the same as for CW or SSB sensitivity. However, if available, the IF output of the receiver is used to feed an rms voltmeter. This avoids potential nonlinearity in the CW or SSB mixer. The measurement is similar to the NF measurement except that a sinusoidal signal generator replaces the noise generator to produce the doubling of output power over noise power alone. This signal power, just equal to the noise power, is defined as the MDS. Because receiver noise varies with bandwidth, so does the MDS, which may be expressed as:

$$MDS = k\,T\,B_n\,F \qquad (2.5)$$

In dBm, $MDS = -174 + 10\log B_n + NF$, where B_n is the noise bandwidth of the receiver. (0 dbm = decibels referenced to 1mW.)

The available thermal noise power per hertz is -174 dBm at 290°K (63°F), an arbitrary reference temperature near standard room temperatures. When any two of the quantities in the expression are known, the third may be calculated. As in the case of NF measurements, care is required in measuring MDS, because a large portion of the power being measured is noise, which produces MDS' typical fluctuations.

2.2 Selectivity

Selectivity is the property of a receiver that allows it to separate a signal or signals at one frequency from those at all other frequencies. At least two characteristics must be considered simultaneously in establishing the required selectivity of a receiver. The selective circuits must be sufficiently sharp to suppress the interference from adjacent channels and spurious responses. On the other hand, they must be broad enough to pass the highest sideband frequencies with acceptable distortion in amplitude and phase. Each class of signals to be received may require different selectivity to handle the signal sidebands adequately while rejecting interfering transmissions having different channel assignment spacings. However, each class of signal requires about the same selectivity throughout all the frequency bands allocated to that class of service. Older receivers sometimes required broader selectivity at their higher frequencies to compensate for greater oscillator drift. This requirement has been greatly reduced by the introduction of synthesizers for control of LOs and economical high-accuracy and high-stability crystal standards for the reference frequency oscillator. Consequently, except at frequencies above VHF, or in applications where adequate power is not available for temperature-controlled ovens, only the accuracy and stability of the selective circuits themselves may require selectivity allowances today.

Quantitatively the definition of selectivity is the bandwidth for which a test signal x decibels stronger than the minimum acceptable signal at a nominal frequency is reduced to the level of that signal. This measurement is relatively simple for a single selective filter or single-frequency amplifier, and a selectivity curve may be drawn showing the band offset both above and below the nominal frequency as the selected attenuation level is varied. Ranges of 80 to 100 dB of attenuation can be measured readily, and higher ranges—if required—may be achieved with special care. A test setup similar to Figure 2.1 may be employed with the receiver replaced by the selective element under test. Proper care must be taken to achieve proper input and output impedance termination for the particular unit under test. The power output meter need only be sufficiently sensitive, have uniform response over the test bandwidth, and have monotonic response so that the same output level is achieved at each point on the curve. A typical IF selectivity curve is shown in Figure 2.2.

The measurement of overall receiver selectivity, using the test setup of Figure 2.1, presents some difficulties. The total selectivity of the receiving system is divided among RF, IF, and baseband selective elements. There are numerous amplifiers and frequency converters, and at least one demodulator intervening between input and output. Hence, there is a high probability of nonlinearities in the nonpassive components affecting the resulting selectivity curves. Some of the effects which occur are overload, modulation distortion, spurious signals, and spurious responses. (Some of these effects are discussed in later sections on dynamic range and spurious outputs.) If there is an AGC, it must be disabled so that it cannot change the amplifier gain in response to the changing signal levels in various stages of the receiver. If there is only an AM or FM demodulator for use in the measurement, distortions occur because of the varying attenuation and phase shift of the circuits across the sidebands. Many modern receivers have frequency converters for SSB or CW reception, so that measurements can be made without modulation. Also, final IF outputs are often available, so that selectivity measure-

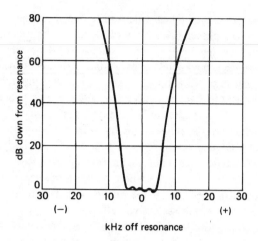

Figure 2.2 Example of an IF selectivity curve.

ments can be made of the combined RF and IF selectivity without worrying about the demodulator or baseband circuits.

When measuring complete receiver selectivity, with either modulated or non-modulated signal, it is wise to use an output power meter calibrated in decibels. The measurement proceeds as described previously. However, if any unusual changes in attenuation or slope are observed, the generator level may be increased in calibrated steps; it should be noted whether the output changes decibel for decibel. If not, what is being observed at this point is not the selectivity curve, but one of the many nonlinearities or responses of the system.

2.3 Dynamic Range

The term *dynamic range*, especially in advertising literature, has been used to mean a variety of things. We must be especially careful in using a common definition when comparing this characteristic of receivers. In some cases, the term has been used to indicate the ratio in decibels between the strongest and weakest signals that a receiver could handle with acceptable noise or distortion. This is the ratio between the signal that is so strong that it causes maximum tolerable distortion and the one that is so weak that it has the minimum acceptable S/N. This measure is of limited value in assessing performance in the normal signal environment where the desired signal may have a range of values, but is surrounded by a dense group of other signals ranging from very weak to very strong. The selective circuits of a receiver can provide protection from many of these signals. However, the stronger ones, because of the nonlinearity of the active devices necessary to provide amplification and frequency conversion, can degrade performance substantially. In modern parlance, dynamic range refers to the ratio of the level of a strong out-of-band signal that in some manner degrades signal performance of the receiver to a very weak signal. The most common weak signal considered is the MDS, and differing strong degrading signal levels may be used. It is, therefore, important to know which definition is meant when evaluating the meaning of the term *dynamic range*. This will be discussed further as the various degradations are considered.

If the foregoing discussion of dynamic range seems vague, it is because there is not one characteristic but several that is encompassed by the term. Each may have a different numeric value. A receiver is a complex device with many active stages separated by different degrees of selectivity. The response of a receiver to multiple signals of different levels is extremely complex, and the results do not always agree with simple theory. However, such theory provides useful comparative measures. If we think of an amplifier or mixer as a device whose output voltage is a function of the input voltage, we may expand the output voltage in a power series of the input voltage:

$$V_o = \Sigma \, a_n \, V_i^n \qquad\qquad\qquad (2.6)$$

where a_1 is the voltage amplification of the device and the higher-order a_n cause distortion.

Because the desired signal and the undesired interference are generally narrow-band signals, we may represent V_i as a sum of sinusoids of different amplitudes and frequencies. Generally $(A_1 \sin 2\pi f_1 t + A_2 \sin 2\pi f_2 t)^n$, as a result of trigonometric identities, produces a number of components with different frequencies, $m f_1 \pm (n - m) f_2$, with m taking on all values from 0 to n. These *intermodulation* (IM) products may have the same frequency as the desired signal for appropriate choices of f_1 and f_2. When n is even, the minimum difference between the two frequencies for this to happen is the desired frequency itself. This type of even IM interference can be reduced substantially by using selective filters.

When n is odd, however, the minimum difference can be very small. Because m and $n - m$ can differ by unity, and each can be close to the signal frequency, if the adjacent interferer is δf from the desired signal, the second need be only $2\delta f/(n - 1)$ further away for the product to fall at the desired frequency. Thus, odd-order IM products can be caused by strong signals only a few channels removed from the desired signal. Selective filtering capable of reducing such signals substantially is not available in most superheterodyne receivers prior to the final IF. Consequently, odd-order IM products generally limit the dynamic range significantly.

Other effects of odd-order distortion are desensitization and cross modulation. For the case where n is odd, the presence of the desired signal and a strong

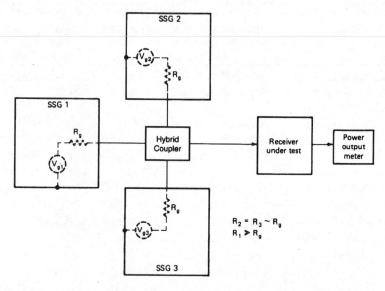

Figure 2.3 Test setup for measuring the dynamic range properties of a receiver.

interfering signal results in a product of the desired signal with an even order of the interfering signal. One of the resulting components of an even power of a sinusoid is a constant, so the desired signal is multiplied by that constant and an even power of the interferer's signal strength. If the interferer is sufficiently strong, the resulting product will subtract from the desired signal product from the first power term, reducing the effective gain of the device. This is referred to as *desensitization*. If the interferer is amplitude-modulated, the desired signal component will also be amplitude-modulated by the distorted modulation of the interferer. This is known as *cross modulation* of the desired signal by the interfering signal.

This discussion provides a simple theory that can be applied in considering strong signal effects. However, the receiver is far more complicated than the single device, and strong signal performance of single devices by these techniques can become rapidly intractable as higher-order terms must be considered. Another mechanism also limits the dynamic range. LO noise sidebands at low levels can extend substantially from the oscillator frequency. A sufficiently strong off-tune signal can beat with these noise sidebands in a mixer, producing additional noise in the desired signal band. Other characteristics that affect the dynamic range are spurious signals and responses and blocking. These are discussed in later sections.

The effects described here occur in receivers, and tests to measure them are essential to determining the dynamic range. Most of these measurements involving the dynamic range require more than one signal input. They are conducted using two or three signal generators in a test setup such as that indicated in Figure 2.3.

2.3.1 Desensitization

Desensitization measurements are related to the 1-dB compression point and general linearity of the receiver. Two signal generators are used in the setup of Figure 2.3. The controls of the receiver under test are set as specified, usually to one of the narrower bandwidths and with MGC set as in sensitivity measurements so as to avoid effects of the AGC system. The signal in the operating channel is modulated and set to a specified level, usually to produce an output S/N or SINAD measurement of a particular level, for example, 13 dB. The interfering signal is moved off the operating frequency by a predetermined amount so that it does not affect the S/N measurement because of beat notes and is then increased in level until the S/N measurement is reduced by a specified amount, such as 3 dB. More complete information may be obtained by varying the frequency offset and plotting a desensitization selectivity curve. In some cases, limits for this curve may be specified. The curve may be carried to a level of input where spurious responses, reciprocal mixing, or other effects may prevent an unambiguous measurement. Measurements to 120 dB above sensitivity level may often be achieved.

If the degradation level at which desensitization is measured is set to 1-dB, and the desensitizing signal is well within the passband of the preselector filters, the desensitization level corresponds to the 1 dB *gain compression* (GC), which is experienced by the system up to the first mixer. (See the subsequent discussion of intermodulation and intercept points.) A gain compression (or *blocking*) dynamic range can be defined by comparing the input signal level at 1-dB GC to the MDS, i. e., dynamic range (dB) equals the GC (input dBm) minus the MDS (input dBm). This is sometimes referred to as the *single-tone dynamic range*, because only a single interfering signal is needed to produce GC.

2.3.2 AM Cross Modulation

Although many saturation effects in receivers have been called cross modulation, SSB and FM are not cross-modulated in the same sense as described previously. Cross modulation occurs in AM and VSB signals by a strong modulated signal amplitude-modulating a weak signal through the inherent nonlinearities of the receiver. Cross modulation typically occurs in a band allocated for AM use and requires a much higher interfering signal level than for the generation of IM products. The typical measurement setup is similar to that for overload measurements, except that the interfering signal is amplitude-modulated, usually at a high level, such as 90 percent. The modulation is at a different frequency than that for the operating channel (if it is modulated), and a band-pass filter is used in the output to assure that the transferred modulation is being measured. The out-of-channel interfering signal is increased in level until the desired signal has a specified level of output at the cross modulation frequency, for example, the equivalent of 10 percent modulation of the desired carrier. One or more specific offsets may be specified for the measurement, or a cross-modulation selectivity curve may be taken by measuring carrier level versus frequency offset to cause the specified degree of cross modulation.

In television systems, cross modulation can result in a ghost of an out-of-channel modulation being visible on the operating channel. The so-called three-tone test for television signals is a form of cross-modulation test. Since most cross-modulation problems occur in the AM broadcast and television bands, cross modulation is not of much interest in most communications receivers whereas other nonlinear distortions are of interest. It would be possible to define a cross-modulation dynamic range, but we are not aware of any use of such a measure.

2.3.3 IM

As described in previous sections, IM produces sum and difference frequency products of many orders that manifest themselves as interference. The measurement of the IM distortion performance is one of the most important tests for a communications receiver. No matter how sensitive a receiver may be, if it has poor immunity to strong signals, it will be of little use. Tests for even-order products determine the effectiveness of filtering prior to the channel filter, while odd-order products are negligibly affected by those filters. For this reason, odd-order products are generally much more troublesome than even-order products and are tested for more frequently. The second- and third-order products are generally the strongest and are the ones most frequently tested. A two-signal generator test set is required for testing, depending on the details of the specified test.

For IM measurements, the controls of the receiver under test are set to the specified bandwidths, operating frequency, and other settings as appropriate, and the gain control is set on manual (or AGC disabled). One signal generator is set on the operating frequency, modulated and adjusted to provide a specified S/N (that for sensitivity, for example). The modulation is disabled, and the output level of this signal is measured. This must be done using the IF output, the SSB output with the signal generator offset by a convenient audio frequency, or with the BFO on and offset. Alternatively, the dc level at the AM demodulator can be measured, if accessible. The signal generator is then turned off. It may be left off during the remainder of the test or retuned and used to produce one of the interfering signals.

For second-order IM testing, two signal generators are now set to two frequencies differing from each other by the operating frequency. These frequencies can be equally above and below the carrier frequency at the start, and shifted on successive tests to assure that the preselection filters do not have any weak regions. The signal with the frequency nearest to the operating frequency must be separated far enough to assure adequate channel filter attenuation of the signal (several channels). For third-order IM testing, the frequencies are selected in accordance with the formula given previously so that the one further from the operating frequency is twice as far as the one nearer to the operating frequency. For example, the nearer interferer might be three channels from the desired frequency and the further one, six channels in the same direction.

In either case, the voltage levels of the two interfering signal generators are set equal and are gradually increased until an output equal to the original channel output is measured in the channel. One of several performance requirements may be specified. If the original level is the sensitivity level, the ratio of the interfering generator level to the sensitivity level may have a specified minimum. Alternatively, for any original level, an interfering generator level may be specified that must not produce an output greater than the original level. Finally, an *intercept point* (IP) may be specified.

The IP for the nth order of intermodulation occurs because the product is a result of the interfering signal voltages being raised to the nth power. With equal voltages, as in the test, the resultant output level of the product increases as:

$$V_{dn} = c_n V^n \tag{2.7}$$

where c_n is a proportionality constant and V is the common input level of the two signals. Because the output from a single signal input V at the operating frequency is $G_v V$, there is a theoretical level at which the two outputs would be equal. This value V_{IPn} is the nth IP, measured at the input. It is usually specified in dBm. In practice the IPs are not reached because as the amplifiers approach saturation, the voltage at each measured frequency becomes a combination of components from various orders of n. Figure 2.4 indicates the input-output power relationships in second- and third-order IPs.

In Equation 2.7 we note that at the IP,

$$V_{dn} = c_n \left(V_{IPn}\right)^n \text{ and } \left(V_{IPn}\right)_{out} = G_v \left(V_{IPn}\right)_{in} \tag{2.8}$$

This leads to:

$$c_n = G_v \left(V_{IPn}\right)^{1-n} \text{ and } V_{dn} = G_v V \left[\frac{V}{V_{IPn}}\right]^{1-n} \tag{2.9}$$

The ratio of signal to distortion becomes $(V_{IPn}/V)^{n-1}$. In decibels it becomes:

$$R_{dn} = 20 \log \left[\frac{V}{V_{dn}}\right] = (n-1)\left[20 \log V_{IPn} - 20 \log V\right] \tag{2.10}$$

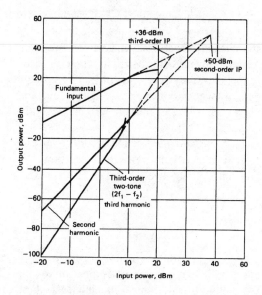

Figure 2.4 Input-output power relationships for second- and third-order intercept points.

If the intercept level is expressed in dBm rather than voltage, then the output power represented by V must be similarly expressed.

The IM products we have been discussing originate in the active devices of the receiver, so that the various voltages or power levels are naturally measured at the device output. The IP is thus naturally referred to the device output and is so specified in most data sheets. In the foregoing discussion, we have referred the IP to the voltage level at the device input. If the input power is required, we subtract from the output intercept level in decibels, the amplifier power gain or loss. The relationship between input and output voltage at the IP is given in Equation 2.8. Reference of the IP to the device input is somewhat unnatural but is technically useful because the receiver system designer must deal with the IP generation in all stages and needs to know at what antenna signal level the receiver will produce the maximum tolerable IM products.

Consider the input power (in each signal) that produces an output IM product equal to the MDS. The ratio of this power to the MDS may be called the *third-order IM dynamic range*. It also is sometimes referred to as the *two-tone dynamic range*. Expressing Equation 2.10 in terms of input power and input IP measured in dBm, we have:

$$R_{dn} = (n-1)\,[IP_{n\,(in)} - P_{(in)}] \qquad\qquad (2.10a)$$

When we substitute MDS for the distortion and $MDS + DR$ for $P_{(in)}$ we obtain:

$$DR = (n-1)\,[IP_{n\,(in)} - MDS - DR],$$
$$nDR = (n-1)\,[IP_{n\,(in)} - MDS]$$

When n is 3, we find the relationship:

$$DR = \frac{2\,[IP_{3\,(in)} - MDS]}{3}$$

A dynamic range could presumably be defined for other orders of IM, but it is not common to do so. From the three different definitions of dynamic range described in this section, it should be clear why it is important to be careful when comparing receiver specifications for this characteristic.

2.3.4 Reciprocal Mixing

Dynamic range can also be limited by LO noise sidebands translated to the IF by a strong signal that is outside of the IF bandwidth. This process is called *reciprocal mixing* since it is caused by the mixing of the weak LO sidebands with a strong interfering signal, which is subsequently eliminated from the passband by the IF filter.

The effect may be better understood by referring to the spectrum sketches in Figure 2.5. The mixer input spectrum includes the desired signal at frequency f_s, an interferer at f_i, both portrayed as spikes, and the noise floor of the receiver. The LO spectrum is the desired spike at f_o with its accompanying noise sidebands, resulting primarily from random phase fluctuations in the oscillator. At the output of the mixer, the difference spectrum includes the translated input signal, interferer, and noise floor, but with the noise sidebands transferred to each of the other signals proportionately to its level. Only the noise sidebands of the strong interfering signal are of significance. The IF filter removes the interfering signal, but cannot remove its sidebands that fall in the IF pass band. Therefore, a substantial increase in noise over the receiver noise floor may occur. This effect can show up in any of the strong signal tests, desensitization, cross-modulation, or IM.

In the desensitization test, instead of the interfering signal reducing the audio output from the receiver, the output may be increased because of the noise of reciprocal mixing. It is possible to use a narrow band audio filter to separate the interfering audio output and determine the actual desensitization level. However, the measurement has no real significance, because the output S/N has been reduced

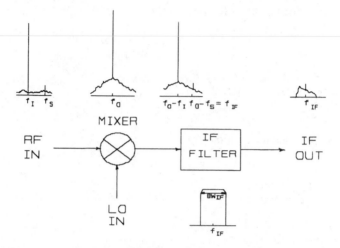

Figure 2.5 Schematic diagram illustrating reciprocal mixing.

below usable, long before desensitization occurs. Analogously, in a cross-modulation measurement, the noise level from reciprocal mixing may produce an unusable S/N long before the interfering signal is strong enough to cause interference from the undesired modulation. IM testing can also be disrupted by the noise from reciprocal mixing; however, this is less likely in a receiver with a well designed oscillator. In this case, the signal level required for third-order IM interference is likely to be below that required to produce excessive reciprocal mixing noise.

The only effective way to reduce reciprocal mixing noise is by improving the spectral purity of the LO, so that the noise sidebands are minimized. While RF selectivity can reduce the level of interferers at frequencies sufficiently distant from the desired signal, it has negligible effect at more moderate separations. Also, there is little value in seeking an IF filter with out-of-band rejection more than approximately 10 dB greater than that which results in an acceptable reciprocal mixing level. Design techniques for minimizing LO noise sidebands are discussed in Chapter 7.

2.4 Spurious Outputs

Because a modern superheterodyne receiver has a synthesizer and may have several LOs, it is possible that at some frequencies the unit may produce outputs without any inputs being present. These are referred to as *spurious signals*. Other sources of spurious signals include the following:

- Power supply harmonics
- Parasitic oscillations in amplifier circuits
- IF subharmonics (for receivers with an IF above the signal band)

Tests must be performed to determine whether the receiver has such inherent spurious signals. The test is best done under computer control. The receiver must be tuned over the entire frequency range in each receive mode, with the baseband output monitored. Any sudden change in noise other than switching transients of the synthesizer and filters could be the result of a spurious signal. Some receiver data sheets indicate that they are 99.99-percent spurious-free. This is a somewhat imprecise technical description. A sounder specification will require that no spurious signal be higher than a particular level, for example, the equivalent of the specified sensitivity. Alternatively, a specification may require all but a specified number of spurious signals to be below the specified level. This is often a sound economic compromise. It is possible to build receivers with fewer than five such spurious signals.

Spurious responses occur when a signal at a frequency to which the receiver is not tuned produces an output. The superheterodyne receiver has two or more inherent spurious responses, IF and image, and a large number of other responses because of device nonlinearities. Each IF that is in use in a superheterodyne configuration has the potential of causing a response if a sufficiently high input signal is applied to exceed the rejection of the selective circuits (or to leak around these circuits via unsuspected paths). The first IF response usually has less rejection than the subsequent IF responses, because the input preselector filters tend to be the broadest. If necessary, special rejection circuits (*traps*) can be built to provide extra rejection. A good communications receiver design should have more than 80-dB IF rejection, and in most cases, over 120 dB is not unreasonable.

It was pointed out that spurious signals can be generated at a subharmonic of an IF if there is sufficient feedback between output and input. Spurious outputs may also occur at subharmonics of the IF because of nonlinearities. If the receiver tunes through a subharmonic of the IF, even if it does not oscillate to produce a spurious signal, the harmonic generation and feedback can cause spurious responses. When the receiver is tuned to a harmonic of the IF, nonlinearities from the later amplifier stages combined with feedback to the input circuits can cause spurious responses. If all the signals in these cases are precisely accurate, the resultant may simply show up as a little extra distortion in the output. But if there are slight differences between the signal and IF, an in-band beat note can occur. Most of these problems can be cured by good design, but it is essential to make a careful survey of spurious responses at these special frequencies.

As explained in Chapter 1, the superheterodyne receiver converts the incoming RF to the IF by mixing it with a locally generated signal in a frequency converter circuit. The IF is either the sum or the difference of the RF and the LO frequency.

For a selected LO frequency, there are two frequencies that will produce the same IF. One is the selected RF and the other, which is generally discriminated against by selective circuits favoring the first, is the image frequency. If the IF is below the oscillator frequency, the image frequency and the RF are separated by twice the IF; if the IF is above the oscillator frequency, they are separated by twice the LO frequency. In most cases, the former condition applies. When there is more than one IF in a receiver, there are images associated with each. Good receivers have image rejection from 80 to 100 dB.

Because frequency converters are not simply square-law or product-law devices, higher-order nonlinearities produce outputs at frequencies $mf_i \pm nf_o$. If any one of these frequency values happens to fall at the IF, a spurious response results. As the orders m and n increase, the signal levels tend to decrease, so that the higher-order images generated tend to become much lower than the direct image. However, some of the combinations tend to result from input frequencies near the selected operating frequency. In this case, they are afforded only small protection by the selective circuits. Because sometimes high-order images are generated, it is necessary to make thorough measurements for all spurious responses. The test setup required is the same as that shown in Figure 2.1.

Because of the changing pattern of spurious responses as the LO frequency is changed, it is customary to test their levels at many frequency settings of the receiver. If possible, this test is best done automatically under computer control. In any case, no fewer than three measurements are desirable in any frequency band of the RF preselector—one in the center and one near either end. With automatic testing the total coverage may be scanned more thoroughly. Before commencing the test, the receiver controls should be set to the appropriate selectivity, to MGC (or AGC disabled if there is no gain control), and to an appropriate signal mode. The test may be made with either modulated or unmodulated signals, with a small change in the results. If only one mode is to be used, the receiver should probably be set to the most narrow bandwidth and a mode that allows measurements with unmodulated signals.

At each frequency setting of the receiver, a sensitivity measurement is first made to establish a reference signal level. Then the signal generator is tuned out of channel, and the level is increased to a large value. This would normally be specified and might be a level relative to the sensitivity measurement, such as 120 dB greater, or simply a high voltage, such as 1 or 2 V. The signal generator is then swept (first on one side of the channel and then on the other side) until a response is detected. The response is tuned to maximum and the level is backed off until the output conditions are the same as for the sensitivity measurement. The ratio of the resultant signal level to the sensitivity is the *spurious response rejection*. The scanning proceeds until all of the spurious responses between specified limits, such as 10 kHz and 400 MHz, have been cataloged and measured. The receiver is then retuned and the process repeated until measurements have been made with the receiver tuned to all of the test frequency settings. A good communications receiver will have all but a few of its spurious response rejections more than 80 to 90 dB down.

2.5 Gain Control

Communications receivers must often be capable of handling a signal range of 100 dB or more. Most amplifiers remain linear over only a much smaller range. The later amplifiers in a receiver, which must provide the demodulator with about 1 V on weak signals, would need the capability to handle thousands of volts for strong signals without some form of gain control. Consequently, communications receivers customarily provide means for changing the gain of the RF or IF amplifiers, or both.

For applications where the received signal is expected to remain always within narrow limits, some form of manually selectable control can be used, which may be set on installation and seldom adjusted. There are few such applications. Most receivers, however, even when an operator is available, must receive signals that vary by tens of decibels over periods of fractions of seconds to minutes. The level also changes when the frequency is reset to receive other signals that may vary over similar ranges but with substantially different average levels. Consequently an AGC is very desirable. In some cases, where fading and modulation rates may be comparable, better performance can be achieved by an operator, using MGC, so both types of control circuits are common.

Some angle modulation receivers provide gain control by using amplifiers that limit on strong signals. Since the information is in the angle of the carrier, the resulting amplitude distortion is of little consequence. Receivers that must preserve AM or maintain very low angle modulation distortion use amplifiers that can be varied in gain by an external control voltage. In some cases, this has been accomplished by varying the operating points of the amplifying devices, but most modern communications sets use separate solid-state circuits or switched passive elements to obtain variable attenuation between amplifier stages with minimum distortion. For manual control, provision can be made to let an operator set the control voltage for these variable attenuators. For automatic control, the output level from the IF amplifiers or the demodulator is monitored by the AGC circuit and a low-pass negative-feedback voltage is derived to maintain a relatively constant signal level.

A block diagram of a dual AGC loop system for a communications receiver is illustrated in Figure 2.6. One loop is driven by first IF energy that is band-limited, and the other loop is driven by second IF energy that is band-limited by optional second IF filters. The first loop controls a PIN diode pi attenuator ahead of the first mixer. The second loop controls the second IF amplifier stages. In this design, a microprocessor adjusts the time constants of both loops so that time delays introduced by the filters do not cause AGC oscillation.

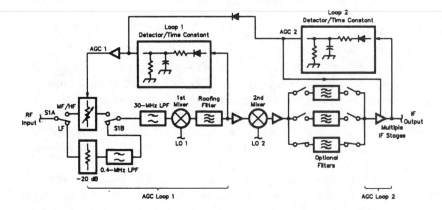

Figure 2.6 Block diagram of a dual loop AGC system for a communications receiver.

A number of tests of gain control characteristics are customarily required. MGC may be designed to control gain continuously or in steps. It is important that the steps be small enough that operators do not detect large jumps as they adjust the gain. Since the gain must be controlled over a very wide range, the MGC is easiest to use if it tends to cause a logarithmic variation. Usually the testing of the MGC is confined to establishing that a specified range of gain control exists and measuring the degree of decibel linearity versus control actuation.

The principal AGC characteristics of importance are the steady-state control range and output-input curve, and the attack and decay times. In a good communications set, a variety of time constants are provided for the AGC to allow for different modulation types. For AM voice modulation, the radiated carrier is constant and the lowest sidebands are usually several hundred hertz removed from the carrier. At the receiver, the carrier component can be separated from the demodulated wave by a low-pass filter and can serve as the AGC control voltage. The response time of the filter, which is often just an RC network, need only be fast enough to respond to the fading rate of the medium, which is a maximum of 5 or 10 dB/s in most AM applications. A response time of 0.1 to 0.2 s is required for such a fading rate. For the more common slower rates, responses up to a second or more can be used.

For SSB applications, there is no carrier to provide control. It is therefore necessary for the receiver to respond rapidly to the onset of modulation. To avoid a transient peak after every syllable of speech, the gain should increase slowly after the modulation drops. If the AGC decay time is too slow, the AGC may not help

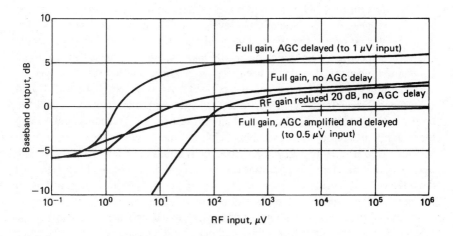

Figure 2.7 Representative input-output AGC curves.

at the higher fading rates; if it is too fast, each new syllable will start with a roar. The need is for a rapid AGC attack time and longer release time, such as a 0.01 s attack and 0.2 s release. Thus, each modulation type may have its different requirements, and it is common to adapt the AGC response times to the different modulation types that must be handled.

To test for the AGC range and input-output curve, a single signal generator is used (as in Figure 2.1) in the AM mode with the receiver's AGC actuated. The signal generator is set to several hundred microvolts, and the baseband output level is adjusted to a convenient level for output power measurement. The signal generator is then tuned to its minimum level and the output level is noted. The signal is gradually increased in amplitude, and the output level is measured for each input level, up to a maximum specified level, such as 2 V. Figure 2.7 shows some typical AGC curves. In most cases, there will be a low-input region where the signal output, rising out of the noise, varies linearly with the input. At some point, the output curve bends over and begins to rise very slowly. At some high level, the output may drop off because of saturation effects in some of the amplifiers. The point at which the linear relationship ends is the *threshold* of the AGC action. The point at which the output starts to decrease, if within a specified range, is considered the upper end of the AGC control range. The difference between these two input levels is the AGC control range. If the curve remains monotonic to the maximum input test level, that level is considered the upper limit of the range. A measure of AGC effectiveness is the increase in output from a specified lower input voltage level to an upper input voltage level. For example, a good design might have an AGC with a threshold below 1 μV that is monotonic to a level of 1 V and has the 3 dB increase in output between 1 μV and 0.1 V.

Measurement of AGC attack and release times requires a storage oscilloscope and a signal generator capable of rapid level change. The simplest test is to switch the generator on or off, but testing by switching a finite attenuation in and out is sometimes specified. The switching should take place in less than 1 ms (without bounce if a mechanical switch is used). The oscilloscope sweep may be keyed by the switching signal or by an advanced version of it. A sweep rate of about 10 ms/cm is a reasonable one. If it is available, the test voltage may be the AGC control voltage or alternatively the baseband output of an AM test signal. The attack and release times should be measured for switching between a number of different levels, such as 10 to 1000 μV, 10 to 100,000 μV, 100 to 10,000 μV, or 0 to 10,000 μV. The output wave is measured to determine the time required from the input change until the observed output reaches a certain fraction of its steady-state value (such as 90 percent). At the same time, the waveform is observed to ensure that the transition is relatively smooth and without ringing. The attack and decay times may be measured in seconds or milliseconds. Another useful measure of decay time is the rate of gain increase in decibels per second.

Another related stability test is often performed in which the signal generator is set to the 1 mV level and tuned to one of the 12-dB attenuation frequencies of the IF selectivity curve. The AGC voltage or output baseband voltage should stabilize smoothly, without signs of instability.

2.6 BFO

The BFO is used to produce output tones from an on-off or frequency-keyed signal. It must provide a certain tuning range, which is generally specified. This range must be tested. The product demodulator generally used for introducing the BFO signal has a finite carrier suppression and may have leakage into other stages. For some purposes, such as recording the IF output, BFO leakage could cause unwanted IM distortion. IF and baseband outputs should be tested with a selective microvoltmeter to assure that at these outputs the BFO level is adequately below the signal levels. A level at least 50 dB below the IF signal level is reasonable in practice, and the baseband attenuation should be at least that much.

2.7 Output Characteristics

Receiver outputs are taken at either baseband or IF. Usually there are baseband outputs, providing amplified versions of the demodulated signal. Often IF outputs are also provided for the connection of external demodulators and processors. In some cases, the receiver output is a digital signal that has already been processed for use by an external system, or simply the output from the A/D converter, which represents the sampled values of the receiver IF or baseband.

While there are a number of receiver characteristics that relate to all of these outputs, this discussion is primarily concerned with the baseband because it appears in most receivers. The output impedance is generally specified at some frequency, and its variation over the band of output frequencies also may be specified. A value of 600 Ω resistive at 1000 Hz is common, and there may be several impedances required to permit interfacing with different types of external devices. Maximum undistorted power output is a characteristic frequently specified and usually refers to the maximum power output level that can be attained at the reference frequency without generating harmonic distortion in excess of some low fraction of the output power (usually specified in percent). Different characteristics are of importance, depending upon the anticipated receiver usage. When the output is to be used by the human ear, the amplitude variation with frequency and the nonlinear distortion are most important. When the output is to be used for generating pictures for the eye, the phase variation with frequency and the transient response of the system are also important. If the output is a processed digital data stream, the specific waveforms of the symbols, the number of symbols per second, and the fraction of errors become significant. In this discussion, only some of the more common baseband characteristics will be reviewed.

2.7.1 Baseband response and noise

The sensitivity of a receiver can be influenced by both baseband and IF band-widths. It is desirable that the overall baseband response should be adapted to the particular transmission mode. To measure the baseband frequency response, a single signal generator test setup is used. For AM- or FM-receiving modes, the generator is tuned to the selected RF, adjusted to a sufficiently high level (for example, 1 mV) that the S/N is limited by residual receiver noise and modulated to a specified level at the reference baseband frequency. The output level is adjusted to a value that permits the output meter to provide an adequate range above the residual noise (at least 40 dB and preferably more). The modulating frequency is then varied over the specified range appropriate to the particular mode (100 to 4000 Hz for speech, 10 to 20,000 Hz for high-fidelity music, and 10 Hz to 4.5 MHz for video). The change in output level in decibels is plotted against the frequency, and the resulting curve is compared against the specified requirements. Figure 2.8 is a typical audio response curve of a communications receiver for voice reception.

In the SSB mode, audio response measurements must be made somewhat differently. The signal generator must be offset at RF to produce the proper output frequency. The AGC must be disabled, the audio gain control set near maximum, and the output reference level adjustment made with the MGC. In an AM or FM mode, either MGC or AGC may be used. In the AM mode, there could be differences in the response measurements, since the AGC time constants may permit some feedback at the low frequencies. Measurements of the baseband amplifier alone require in all cases feeding that amplifier from a separate baseband signal generator rather than the demodulator, but are otherwise the same.

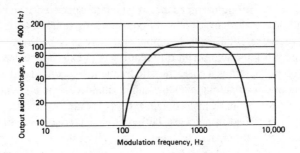

Figure 2.8 Typical audio response curve for a communications receiver.

The ultimate S/N available is limited by the noise and hum in the baseband amplifier and the low-frequency amplitude and phase noise of the various LOs used in frequency conversion. When AGC is in use, low-frequency hum and noise on the AGC control line can also contribute to the noise level. To measure the ultimate S/N with AGC on, using the test setup of Figure 2.1, the signal generator is tuned to the selected RF and modulated (as in the sensitivity test). The output level is adjusted to a setting on the output meter that provides substantial downward range (60 dB or more). The signal is then increased to a high value, such as 1 mV, and the output levels are measured with modulation switched on and off. The signal level is then increased by factors of 10 and 100, and the measurement is repeated. The S/N measured should be the same for the three measurements and represents the ultimate value attainable.

For measurements without AGC, the MGC attenuation is set to provide the desired output level with the signal generator set to its initial high signal level. When the signal level is increased, the RF and IF gain is reduced manually to maintain the initial output level. Otherwise, the test is the same. With MGC the test can also be run in the SSB mode, using signal generator offset to produce the reference output frequency and turning the generator off to determine the residual noise level. In this case, the LO noise does not contribute to the measured values of residual noise. The residual hum and noise can be measured at the baseband output with the demodulator output replaced by an appropriate terminating resistance. This should be done for several settings of the baseband level control to establish its effect on the noise level. A well-designed receiver can provide a 55- to 60-dB ultimate S/N in the AM, SSB, CW, and FM modes. If measurements this high are required, care should be taken to ensure that the signal generator used has a substantially lower residual hum and noise characteristic with the modulation off.

2.7.2 Harmonic Distortion

To measure harmonic distortion, the same test setup as for measuring the frequency response may be used, except that the output meter must be augmented by a spectrum analyzer or a distortion meter. At various levels of signal output, a spectrum analyzer can measure the amplitude of the fundamental and all the harmonics that are generated. The harmonics can be *root-sum-squared* (rss) to yield the total harmonic distortion, which is generally specified. A simple distortion meter may be made by using a narrow-band reject filter that may be inserted or removed from the circuit without change in loss at frequencies as high as the harmonics. When the filter is out of the circuit, the output meter measures the fundamental plus harmonic power; when the filter is in, harmonic power alone is measured. From these measurements the percent harmonic distortion can be calculated.

Because of the difficulty of making such a filter that is tunable over a wide range, total harmonic distortion measurement sets are available using a fundamental cancellation principle. The set includes a baseband signal generator to provide the modulating signal. The same generator is used to drive a phase shifting and attenuation network that supplies the cancellation signal. The receiver output signal and the cancellation signal are added in a fixed attenuator network whose output is fed to a true-rms voltmeter. When the cancellation signal is fully attenuated, the output meter measures the signal plus distortion power output. The attenuator is then decreased and the phase adjusted until the fundamental signal from the receiver output is exactly canceled. The remaining harmonic distortion is then read on the output meter. Most modern test sets provide automatic nulling. Computer-based instruments offer completely automated measurements and detailed plots of the results.

2.7.3 IM Distortion

IM distortion measurement requires a signal generator capable of being modulated by a composite baseband signal comprising two tones of different frequencies and equal level. The output is measured using a spectrum analyzer. The level of the two tones is adjusted to provide a specified peak modulation, usually a substantial fraction of maximum allowable modulation for AM or FM. For SSB, where the test is frequently used, the peak level is calculated and is 3 dB above the composite power. As in the harmonic distortion test, the signal generator power is set to a level high enough to reach ultimate S/N. In this case, the spectrum analyzer provides the amplitudes of the two modulation frequencies, their harmonics within the baseband, and the various IM frequencies generated within the baseband. Usually only the third-order products $2f_1 - f_2$ and $2f_2 - f_1$ are of significance and are specified. In most cases the transmitter IM is much greater than that in the receiver. IM tests are also important in high-fidelity receivers intended for music reception.

Another form of IM test is performed in receivers intended for multichannel FDM applications. This is known as the *noise power ratio* (NPR) test. In this test, the signal generator is modulated with noise having a uniform spectrum over the portions of the baseband planned for use. However, a band-reject filter is used to eliminate the modulation components in one channel. At the receiver output, a bandpass filter is employed to allow the output power in this channel, resulting from IM, to be separately measured. The ratio of the power in the whole band to the power in the selected channel is the NPR. This test is used primarily for multichannel voice circuits and is not often used in the frequency range we are discussing in this book.

2.7.4 Transient Response

When a receiver must handle picture or data waveforms, the transient response to step changes in modulation is more important than the frequency response or the harmonic or IM production. The transient response can be measured directly by imposing a step in modulation on the signal generator and observing the receiver output with a storage oscilloscope. Generally, the transient response should be sufficiently limited in time that square-wave modulation at a sufficiently low frequency can be observed with a standard oscilloscope. Characteristics of the waveform that are of interest include:

- Rise time (usually measured between 10 and 90 percent of steady-state change)
- Amount of overshoot (if any)
- Duration of ringing (if any)
- Time required to settle (within a small percentage) to the steady-state value

Care must be taken that the signal generator transient modulation characteristics are such that they do not affect the waveform significantly.

Because the transient response and the total frequency response are related through the Fourier transform, as long as the receiver is in a linear operating region, it is often more convenient to measure the phase change and gain change accurately over the frequency band. Usually when this is done, specifications of differential phase change and amplitude change per unit frequency change are given. Test sets are available in some frequency ranges to sweep the baseband at a low rate, using small frequency variations at a higher rate to generate differential phase and amplitude changes, which may be shown on a calibrated output meter or oscilloscope. Differential phase per unit frequency is known as *envelope* or *group delay* because it corresponds to the delay of the envelope of a signal comprising a small group of waves at frequencies close to the measuring point. This is in contrast to the *phase delay*, which is the ratio of phase change divided by frequency change, both referenced to the carrier frequency. Phase delay identifies the delay of a single sinusoidal component at the frequency of measurement. Limits are often set on differential amplitude and differential phase variations over the baseband, or a substantial portion of it, in order to assure that the receiver will handle transients properly.

2.8 Frequency Accuracy and Stability

Modern communications receivers generally have LOs controlled by a frequency synthesizer, rather than the free-running oscillators that were prevalent for many years. In either event, it must be possible to set the oscillator on a selected frequency with sufficient accuracy that a transmission on the corresponding receiver frequency can be received properly. It is also necessary that, once set, the frequency remains unchanged for a sufficient period to allow the communication to take place, despite temperature changes, mechanical changes (tilt, vibration, and sudden shock), and general aging of circuit components. The substantially improved frequency accuracy and stability, combined with the availability of economical digital circuits, is the reason for the ascendancy of the synthesizer. Similar tests are run on both synthesizer-controlled receivers and receivers having free-running oscillators, although the specified performance is poorer for the latter.

The accuracy of receiver frequency settings may be measured by using a signal generator whose frequency is compared to an accurately calibrated frequency standard by a frequency counter. The receiver is set successively to a number of test frequencies, and the generator is tuned so that the signal is in the center of the pass band. The signal generator frequency is then measured and the error in the receiver setting is recorded. An alternative is to measure the frequency of all of the receiver's LOs since the received frequency is determined by them and the IF. In this case, the center frequency of the IF filters needs to be checked initially so that it may be combined properly with the oscillator frequencies. The specifications for frequency accuracy may vary depending on the particular application of the receiver and compromises required in the design. When very good crystal standards in ovens can be used to provide the synthesizer reference, frequency accuracies of 1 part in 10^7 or greater can be achieved.

All oscillators have some temperature drift; although when adequate power is available, most modern sets use crystal standards in thermostatically controlled ovens. Their only drawback is the time required for initial stabilization when power is first applied. In testing for temperature stability, we first determine the required temperature range of operation from the receiver specification. For a good-quality commercial receiver, a typical temperature range is –25 to +55°C. Where the intended use is always in a heated shelter, the lower limit is often relaxed. For military field use, however, the temperature range may be extended at both ends.

The first test required when power is applied to the receiver is the warm-up time to frequency stabilization. Receivers that use temperature-compensated crystal oscillators may require only seconds of warm up, but have an accuracy of only 1 *part per million* (ppm). Oven-stabilized crystal oscillators require more time to warm up, but provide higher ultimate accuracy by at least an order of magnitude.

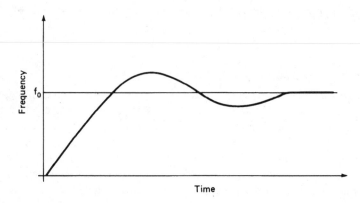

Figure 2.9 Plot of receiver local oscillator warm-up drift.

Many receivers have a LO port available at the rear of the receiver, which facilitates the measurements. At this point, the internal reference is multiplied to a high frequency, which provides higher resolution than a direct measurement of the frequency of the standard. For example, in an HF set with an internal frequency standard of 10 MHz, assume that the first IF is 81.4 MHz. If we set the receiver frequency to 18.6 MHz, the LO will be at 100 MHz. This provides about 10 times greater resolution on the frequency counter than direct measurement of the standard.

A plot of the warm-up drift (Figure 2.9) can be made using a variety of instruments. Such measurements should be made at both short- and long-term. A continuous measurement should be made for the first 10 to 15 min, followed by samples every half-hour for several hours or more if substantial variation is detected. The internal frequency standards in most common counters have an accuracy comparable to or poorer than the standard in some high-quality communications receivers. If higher accuracy of measurement is required, it is advisable to use a high-stability external frequency standard.

If the receiver being tested does not have a frequency synthesizer, a slightly different test can be performed using the test setup of Figure 2.10. A signal generator with synthesizer control is set to the test frequency, such as 29 or 30 MHz, and the receiver, with BFO on, is tuned to produce an audio beat note, such as 2 kHz. The beat note is fed from the audio output to a frequency counter so that the same type of plot can be made as a function of time. If the change is sufficiently great so as to exceed the audio range of the receiver, the signal generator frequency may be adjusted to bring the beat note back within range. In this case, the same standard should be used for the synthesized signal generator and frequency counter. Because of the poorer stability of the nonsynthesized receiver, the internal standard of either the signal generator or counter will suffice. In this case, it is important to

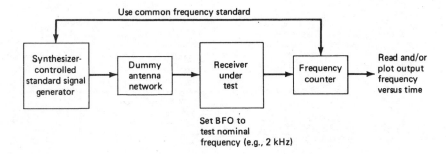

Figure 2.10 Test setup for stability measurements on a nonsynthesized receiver.

make measurements at a number of points throughout the range of the receiver, because the drift may change substantially.

Similar test setups are used to measure temperature stability. In this case, the receiver is placed in a temperature-controlled chamber, and the test instrumentation is located outside the chamber. The receiver is first allowed to warm up at room temperature until the frequency is stable. The temperature of the chamber is then raised to the maximum specified operating temperature and allowed to stabilize. Subsequently it is returned to room temperature to stabilize, then lowered to the minimum specified temperature, and finally returned to room temperature. Throughout the temperature cycle, frequency and temperature are recorded to assure that the chamber temperature has stabilized and to determine that transient and steady-state temperature changes are within the required limits.

Similar test setups are also used to measure frequency stability under various mechanical stresses. The receiver under test is mounted to a test table where the mechanical stress is applied. The test equipment is isolated from the test environment. A test may subject the receiver to slow or steady-state pitch, roll, or yaw. Another test may vibrate it in different directions using different vibration frequencies and waveforms. Still another may subject the receiver to heavy shocks. In each case, when it is required, the receiver must operate through the test and maintain frequency to specified limits. Limited tests of this sort are applied to most high-grade commercial equipment designs. Severe tests must be applied to military field equipment or other equipment intended for use in severe environments.

2.9 Frequency Settling Time

The tuning control of a synthesized receiver may be made quasicontinuous. Modern receivers that cover large frequency ranges in small steps use several phase-locked loops, and at least one of the loops usually has several bands. Whenever a PLL goes through its frequency range and must jump from one end to

86   Chapter 2

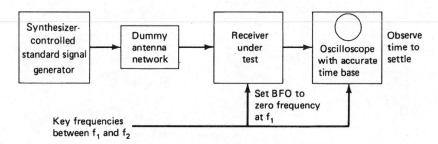

Figure 2.11 Test setup for measuring receiver settling time after a frequency change.

the other, the loop is out of lock for a short period. Whenever a loop must change bands, the same occurs. When the receiver has a digitally set tuning control, any change except for those that change tuning by a few channels can result in momentary loss of lock in one or more of the PLLs. Often, the frequency setting is controlled over a bus driven by a microprocessor, and the changes that are made from time to time may be substantial.

In frequency-hopping spread-spectrum and related applications, channels may be changed pseudorandomly over a 5 to 10 percent frequency range, with a possibility of one or more of the loops losing lock. In this case, the time required for the oscillator to settle is most important. The changes from one end of the loop band to the other, or from one band in a loop to another, result in the slowest periods of settling; in some designs, the PLL can require several hundred milliseconds settling time for the worst cases. This can result in disturbing clicks when tuning the receiver, and could make frequency hopping undesirably slow when those points have to be encompassed in the hop band. The time required for the oscillator to lock up and settle to the new frequency, thus, can be an extremely important receiver characteristic.

As an example of the measurement of this characteristic, let us assume that a receiver is tuned to 10 MHz, and a 10 MHz signal is applied to the input terminal from a synthesized signal generator. When the receiver control is set 1 kHz lower, let us assume that at least one receiver synthesizer loop must make a range or band change and lose lock during the transition before settling to the new frequency. Once such critical points in the synthesizer range are determined, we can use an oscilloscope to measure the time required to achieve a 1 kHz beat note (in SSB or CW mode), after the command to change has been received.

Figure 2.11 shows the test setup that is probably easiest for determining the settling time. Initially the receiver is tuned to carrier frequency so that the beat note is zero. If the oscilloscope is keyed by the command signal, we can observe on the oscilloscope how long it requires for a beat note to be observed and lock to 1 kHz.

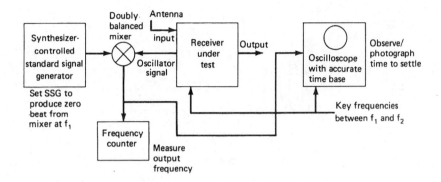

Figure 2.12 Test setup for measuring synthesizer settling time, modified for greater measurement precision.

If this time is long, a long-persistence screen or storage oscilloscope may be needed. A shift back should also be employed to determine whether a difference exists in the two acquisition times. In this case, after lockup the signal will rapidly approach zero frequency and, because of the audio amplifier low frequency characteristic, similarly approach zero amplitude.

This rather crude method of measurement is probably adequate unless we are concerned with frequency hopping of digital signal modulation. For this case or others where higher time precision is required, the test setup is modified as indicated in Figure 2.12. Here we measure the synthesizer frequency directly as obtained from a special receiver LO output. The signal is mixed in a doubly balanced mixer with the output of the synthesized signal generator, and the resultant beat note is applied to both the oscilloscope and the frequency counter. Initially the receiver is set to one of the two frequencies being used for test, and the signal generator is adjusted until the beat note is zero. The receiver is then keyed to the second frequency and allowed to settle into a steady beat note, which may be checked for frequency accuracy by the counter. The change back to the original frequency triggers the oscilloscope simultaneously with the synthesizer reset control signal. The beat note may vary wildly for a short period, but then it gradually returns to zero, as shown in Figure 2.13. The time required to reach a steady-state direct current may be measured with reasonable accuracy. If desired, timing pips can be superimposed on the trace to avoid dependence on the oscilloscope calibration and linearity.

Dependent on the type of demodulation used for the digital signal, settling time can be considered to be reached when the beat note has come within a few hertz of zero frequency or when it reaches steady direct current. It should be remembered that not every frequency jump requires so long to settle. When the change is a

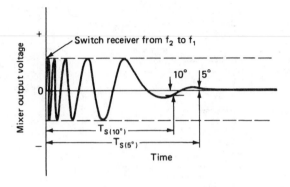

Figure 2.13 Typical oscilloscope pattern for measuring settling time.

relatively small increment in the internal frequency loop, frequency lock may not be lost, and the acquisition of the new frequency and phase lock upon it is much faster. Only if an internal loop has to jump—for example, from 80 to 70.01 MHz rather than from 79.99 to 80 MHz—will the loop go out of lock for a short period and need to reacquire lock.

2.10 Electromagnetic Interference

As a piece of electric equipment, a radio receiver can be a source of interference to other equipment. Similarly, RF voltages picked up on the input and output lines can be coupled to the signal circuits to interfere with radio reception. It is also possible for the all-pervasive electromagnetic fields of our environment to penetrate the receiver cabinet and couple to the signal circuits. All of these phenomena are lumped under the term *electromagnetic interference* (EMI), and the ability of the receiver to perform satisfactorily in the environment is referred to as *electromagnetic compatibility* (EMC). Before a design is complete, it must be evaluated for its EMC.

The most significant EMI produced by most receivers are the signals on the antenna line at the first LO and other LO frequencies. In addition, in a synthesized receiver it is possible for various other frequencies to be produced. If a number of receivers are connected to a common antenna, such oscillator interference can generate spurious interfering signals. Similarly, although the power is small, the radiation from the antenna may produce interference in other nearby receivers. In some military situations, an enemy might be able to use oscillator radiation to pinpoint the position of a receiving station and tell when it is operational. Tests for LO voltages on the antenna line as well as any other spurious signals, such as power supply harmonics and noise or microprocessor clock harmonics and noise, are generally measured in accordance with test specifications from the governing communications agency; in the United States this is the *Federal Communications Commission* (FCC).

These signals as well as those on other input and output lines of the receiver are measured by using a spectrum analyzer or a scanning receiver at maximum sensitivity. The scan is often made from very low to very high frequencies (10 kHz to 10 or more times the highest LO frequency of the receiver, for example). A good receiver should produce no more than a few picowatts in the nominal antenna impedance at any frequency.

In addition to the measurement of spurious signals on the input and output lines, measurement of direct radiation from the receiver is required. Measurements are made in a standard configuration with field measurement equipment located a specified distance from the receiver. Again, the governing agency generally specifies the maximum acceptable field at that distance and often indicates what field measurement equipment is acceptable for the test.

The inverse tests are also of importance. Susceptibility of the set to radio waves on the power line, output lines, and other input lines can be of significance, especially in an environment with many transmitters or other radiators nearby. Response to both CW signals and broad-band noise on the lines is appropriate. Susceptibility at the tuned frequency of the receiver and the various IFs is most important and should be 80 dB or more above the sensitivity of the receiver at the antenna.

The receiver, with power line and baseband output carefully filtered and all other input and output ports shielded, should also be tested for susceptibility to electromagnetic fields. The governing agency specifies the test setup for the generation of the field, which may be a terminated transmission line a specified distance from the receiver in a shielded cage of specific dimensions. This test is especially important if the receiver is to be used in a station with many powerful transmitters or at a confined site where transmission lines from transmitters pass nearby or transmitting antennas are relatively close. The field is modulated appropriately to the receiver mode setting and set to a high level, for example, 10 V/m. The carrier frequency of the field is then swept over a range encompassing any likely receiver responses. If there is an output, the field is reduced until the S/N is that specified for sensitivity measurements, and the field strength is measured or calculated. In a well-shielded design, the tuned frequency of the set at the maximum gain control setting should be the only significant output. The field level required at this frequency should be on the order of volts per meter when the set has been especially designed for service in high fields.

2.11 Digital Receiver Characteristics

The foregoing has not exhausted the analog receiver characteristics that may be of interest but has reviewed some of the more significant ones. For example, in FM sets, the *capture ratio* is important. Tests of special features such as squelch

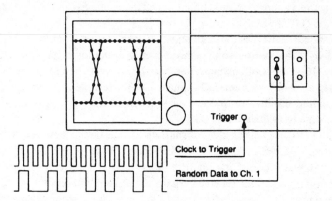

Figure 2.14 Test setup for eye pattern measurement.

sensitivity and threshold also are important in many applications. Clearly, an area of increasing interest is the characterization of systems utilizing digital modulation techniques. Because a digital radio system is a hybrid A/D device, many of the test procedures outlined previously for analog receivers are useful and important in characterizing a digital radio. Additional tests, primarily involving the analysis of *bit error rates* (BER), must also be run to properly identify any weak points in a receiver design.

2.11.1 BER Testing

The primary method for testing the quality of transmission over a high speed digital communications link is the BER, defined as the number of bit errors divided by the number of bits transmitted. The BER is also used to qualify the sensitivity and noise characteristics of a receiver. The major contributor to BER is *jitter*, which results from noise in the system. This noise causes the output comparator to vary the time of its transition relative to the data clock. If the transition time changes too much, an error will result.

Using a communications signal analyzer specifically designed for BER testing, jitter can be displayed directly, or the BER is simply tabulated by the analyzer. The format of the testing signal is determined by the application to which the digital radio system is applied. A variety of formats and protocols are used. For end-to-end testing, the analyzer feeds a reference RF signal generator whose characteristics are known, and the output signal is applied to the receiver antenna input.

The noise and jitter on a data waveform provides vital information about the quality of the signal. A typical setup for capturing an *eye pattern* is shown in Figure 2.14. Eye patterns are the traditional method of displaying high-speed digital data (Figure 2.15). Some communications signal analyzers augment this information with a built-in statistical database, which allows additional analysis, including

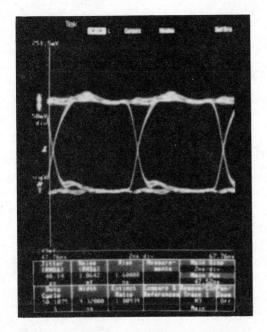

Figure 2.15 Eye pattern display of BER measurement. (*Courtesy of Tektronix.*)

automated noise and jitter measurements on random data. Sophisticated software can also analyze the form of the distribution, providing mean, rms, and standard deviation results.

Proper receiver design involves identifying the areas of the system that are likely to cause problems. LO phase noise is one such area. Phase noise can seriously impair the performance of a digital receiver. Excessive phase noise can increase the BER, especially in systems using phase-based modulation schemes, such as binary PSK and quadrature PSK. For a given statistical phase-error characteristic, BER is degraded according to the percentage of time that the phase error causes the signal position in signal space to cross a decision boundary.

As mentioned in Chapter 1, computer-aided simulation of receiver circuits has added new features and flexibility to radio system design. Suppliers of high-frequency analog *computer-aided engineering* (CAE) software are beginning to add features for RF digital communications systems design. Specific types of digital modulation are now built into simulators, in addition to special-purpose routines to determine the expected BER of the proposed system.

2.12 Bibliography

"Standards Testing: Bit Error Rate," application note 3SW-8136-2, Tektronix, Beaverton, OR, July 1993.
"Waveform Analysis: Noise and Jitter," application note 3SW8142-2, Tektronix, Beaverton, OR, March 1993.

Engelson, M., and J. Herbert, "Effective Characterization of CDMA Signals," *Microwave Journal*, p. 90, January 1995.

Johnson, J. B., "Thermal Agitation of Electricity in Conduction," *Phys. Rev.*, vol. 32, p. 97, July 1928.

Howald, R., "Understand the Mathematics of Phase Noise," *Microwaves & RF*, p. 97, December 1993.

Nyquist, H., "Thermal Agitation of Electrical Charge in Conductors," *Phys. Rev.*, vol. 32, p. 110, July 1928.

Pleasant, D., "Practical Simulation of Bit Error Rates," *Applied Microwave and Wireless*, p. 65, Spring 1995.

Rohde, U., "Key Components of Modern Receiver Design—Part 1," *QST*, p. 29, May 1994.

Watson, R., "Receiver Dynamic Range; Pt. 1, Guidelines for Receiver Analysis," *Microwaves & RF*, vol. 25, p. 113, Dec. 1986.

Wilson, E., "Evaluate the Distortion of Modular Cascades," *Microwaves*, vol. 20, Mar. 1981.

Additional Suggested Reading

"Eight Ways to Better Radio Receiver Design," *Electronics*, Feb. 20, 1975.

Chapter 3

Receiver System Planning

3.1 The Receiver Level Plan

The most important performance characteristics of a receiver are its sensitivity and dynamic range. While these characteristics may be specified in a number of ways, the NF and the second- and third-order IPs are excellent measures that generally can be converted to any required specification for these characteristics. For a superheterodyne receiver, other important characteristics that must be carefully planned include the number, location, and strength of spurious responses; the selectivities to be available for different services; and the method of tracking RF preselector tuning to the LO frequency.

The ideal receiver would have 0-dB NF, very high IPs (30 to 50 dBm), and no spurious responses in excess of the thermal noise level in the most narrow available channel bandwidth of the receiver. Such ideals are not attainable in the physical world. The closest possible approach to their attainment, given the state of the art when the receiver is designed, would result in a cost that few if any customers would be willing to pay. Consequently, the design must effect a compromise between physics and economics. The most useful tool to help with these tradeoffs is the *gain* or *level diagram*.

A complete level diagram identifies each stage of the receiver from antenna input to baseband output. The impedance levels at various points are identified where significant; the power (or in some cases, voltage) gain of each stage is indicated; and the NF for each active stage is recorded at its prospective operating point, as are the second- and third-order IPs for that stage. For each frequency-changing circuit, the mixer type, NF, gain, and IPs are established for the operating conditions, including the LO input level. These conditions also determine the levels of various orders of spurious responses.

One of the first decisions that must be made in the design of a superheterodyne receiver is the number and position of IF conversions. Next the frequency range of each LO must be determined because this establishes the locations of the spurious responses of various orders. There are two choices for each LO frequency, defined by the equation $|f_s \pm f_{IF}| = f_o$. These selections are not subject to easy generalization. Depending on the number of RF bands chosen and their frequencies, and on the availability of stable, fixed-bandwidth filters at potential IFs, a number of alternatives may need to be evaluated before a final choice is made.

Another important decision is the gain distribution throughout the system, because this determines the NF and the signal levels at various points in the system. To obtain the best NF, adequate gain is required prior to the first mixer stage, since mixers tend to have poor NFs and the mixer design with lowest spurious responses may well exhibit a loss. However, minimum IM product levels occur when the level is as low as possible prior to the final channel bandwidth selection. Usually this implies a minimum gain prior to the final IF amplifier, where channel bandwidths are likely to be established. Minimization of the signal level at the mixer input also reduces the level of spurious responses. In some systems, it may be necessary to accept lowered sensitivity to avoid high spurious response and IM levels. In such cases, the preselection filter outputs may be fed directly to the mixer, without RF amplification, and filter and mixer losses prior to the first IF amplifier must be minimized.

Receiver planning, thus, is a cut-and-try process centered around the receiver level diagram. Initial selections are made; the NF, IPs, and levels of spurious responses closest to the RF are evaluated and compared to the specified goals. This leads to a second set of selections, and so on until the appropriate compromise has been achieved. In this process, as will be seen later, the latter stages of the receiver may generally be neglected until these initial selections have been made. Once the choices are made, the diagram may be expanded to encompass all the stages. As the diagram grows, other performance characteristics can be evaluated, until finally the complete level diagram serves as a road map for detailed receiver design.

As an example, Figure 3.1 is a partial level diagram of an HF receiver showing the input circuit, the first and second mixers, the first IF amplifier, and the second IF input amplifier and channel selection filters. The remainder of the second IF amplifier is indicated simply as a variable-gain amplifier with 4-dB NF at maximum gain. The IPs of this amplifier are not shown, because signals strong enough to cause IM at this point in the circuit would be in the channel bandwidth and would prevent useful signal reception even without IM. For each earlier stage, the NFs, gains, IPs, and representative signal levels are indicated.

The following sections discuss calculations of the overall receiver NF and IPs, using this diagram as an illustrative example. We also discuss the spurious response locations, using the band selection, IFs, and LO frequencies of Figure 3.1 to illustrate the points. The design of selective circuits is then briefly reviewed, and we end the chapter with a discussion of tracking of the preselector and LO.

3.2 Calculation of NF

Recall from Chapter 2 that the noise factor F is the ratio of the S/N available at the receiver input to that available at the output of the IF amplifier. The demodulator, unless a product demodulator such as that used for SSB, may be inherently nonlinear. Because different modulator types connected to the same amplifier can show different output S/N values for the same input S/N, it is best to measure NF

Power gain	dB	-0.5	-6	-1	10	-3.5	12	-6	8	-4	>90
Noise figure	dB	—	-6	—	2	—	3	-6	5	—	4
Cumulative noise figure	dB	10.81	10.31	4.31	3.31	8.18	4.68	12.02	6.02	8	4
IP_2	dBm	—	80	—	*	*	*	*	*	—	* (out)
Cumulative IP_2	dBm	86.5	86	—	—	—	—	—	—	—	* (in)
IP_3	dBm	—	32	—	30	—	30	27	30	—	* (out)
7-kHz in-band IP_3	dBm	13.9	13.4	7.4	6.4	16.6	13.1	26.8	22	—	* (in)
25-kHz in-band IP_3	dBm	27.2	26.7	21	20	—	*	*	*	—	* (in)
Out-of-band IP_3	dBm	38.5	38	—	*	—	*	*	*	—	* (in)

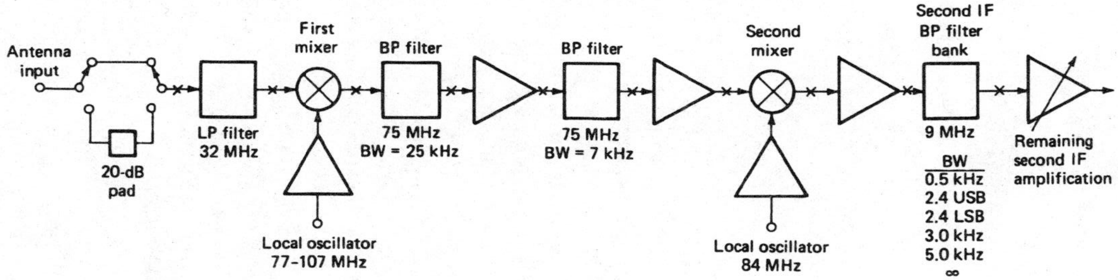

* Irrelevant because of first IF filter.
X Locations of transmit-receive and filter bank switches

Figure 3.1 Level diagram of a double superheterodyne HF receiver.

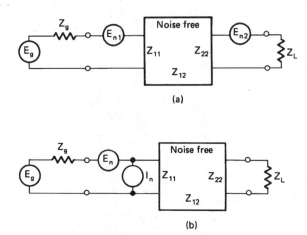

(a)

(b)

Figure 3.2 Equivalent circuits for representing noise in a two-port network.

prior to the demodulator. The NF is greater than unity because of signal losses and thermal noise in passive circuits, and because of the introduction of noises in addition to thermal noise in the active circuits (and some passive components). The definition of F applies equally to every two-terminal-pair network, whether active or passive, or a simple amplifier, filter, or cascade of several simple networks. For passive networks without excess noise sources, the value of F is simply the loss of the circuit driven by the particular generator because the available signal power at the output is reduced by the loss, while the same thermal noise power kTB is available at both input and output.

For active devices we have not only thermal, but also shot and flicker noise. At a particular frequency and set of operating conditions it is possible to define a circuit model that applies to any linear two-port device [3.1]. Figure 3.2 shows two forms that this model may take. The form in Figure 3.2b is most useful in noise calculations because it refers the noise effects to the input terminal pair, followed by a noiseless circuit with gain G and output impedance as determined by the two-port parameters and the impedance of the input circuit. It will be noted that the model represents the input noise by a serial noise voltage generator and a shunt noise current generator. Both are necessary for the network noise characterization.

For a known generator impedance, the current source can be converted to a voltage source by multiplying it by the generator impedance. It is therefore obvious that the value of F for the network depends on the impedance of the driving source. For a specific source, the equivalent current source can be converted to a voltage source and added rms (assuming no correlation) to the network serial noise voltage generator, thus resulting in a single noise source in the model. Conversely, the serial source could be converted to a shunt source by dividing its voltage by the generator

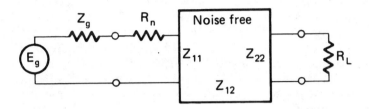

Figure 3.3 Simplified equivalent circuit for representing noise in a two-port network when the driving impedance is known.

impedance. Then the amplifier noise could be represented by a single shunt source. Similarly, the noise could be represented by a single equivalent serial resistance or shunt conductance by using the relationships $E_n^2 = 4kTBR$ or $I_n^2 = 4kTBG$. It must be realized, however, that the resistance or conductance is not a circuit component, but only a noise generator.

To illustrate, Figure 3.3 shows the block diagram of a signal generator with 1 μV EMF and a purely resistive impedance R_g, feeding an amplifier with equivalent noise resistor R_n, and having a noiseless amplifier with noiseless output impedance R_L assumed to be purely resistive. From the Nyquist formula, we calculate the equivalent mean-square noise voltage as:

$$E_n^2 = 4\,k\,T_0\,B(R_n + R_p) \tag{3.1}$$

where $R_p = R_g R_L/(R_g + R_L)$ and R_n represents the noise contribution from the amplifier. When, for example, $B = 2000$ Hz, $R_n = 200\ \Omega$, $R_g = 1000\ \Omega$, $R_L = 10,000\ \Omega$, $k = 1.38 \times 10^{-23}$, $T_0 = 300$ K, and $R_p = 909.1\ \Omega$, then:

$$E_n = [4 \times 1.38 \times 10^{-23} \times 300 \times 2000 \times (200 + 909.1)]^{1/2} \tag{3.1a}$$

$$= 0.1917\ \mu V$$

If the amplifier were noiseless, the equivalent rms noise voltage would be 0.1735 μV. The ratio between the two voltages is 1.1:1. The amplifier has increased the noise by 10 percent.

Because of the amplifier load, the EMF of the generator produces an input voltage $V_m = 1 \times (10,000/11,000) = 0.909\ \mu$V. The S/N from the amplifier under this input condition is 0.909/0.191 = 4.742 (voltage ratio), which is 13.52 dB. For the noiseless amplifier, the S/N would be 0.909/0.1735 = 5.240, or 14.39 dB. In this case, the noise factor for the amplifier can also be calculated, $F = (R_p + R_n)/R_p = 1 + R_n/R_p = 1 + 200/909.1 = 1.22$ (power ratio) or 0.864 dB NF, the same as the difference between 14.39 and 13.52, except for rounding errors. The load resistor is substantially higher (10 times) than that of the generator. For a perfect match (if the noise resistor were the same), the noise factor $F = 1 + 200/500 = 1.4$, or an NF of 1.461 dB. From this simple example, it is apparent that matching for optimum transfer of energy does not necessarily mean minimum NF.

The typical noise resistor for tubes ranges from $3/g_m$ to $5/g_m$, while for bipolar transistors, $1/g_m$ is a good approximation. This is only approximate, because the noise resistor can vary with changes in the generator impedance. At higher frequencies, the input capacitance, feedback from output to input, and the question of correlation of voltages come into play. The simple equivalent noise resistor is no longer usable. Therefore, we require a more complete equivalent model (discussed in Chapter 5).

3.2.1 Noise Factor for Cascaded Circuits

A receiver includes many circuits connected in cascade. To arrive at the overall NF, it is necessary to consider the contribution of all stages. We have seen that the noise factor of a passive circuit with parts generating only thermal noise is equal to the loss:

$$F = L_p = \frac{1}{G_p} \tag{3.2}$$

For an active circuit, there is invariably some excess noise, and the NF referred to the input may be expressed in terms of an equivalent resistor R_n in series with the input circuit:

$$F = \frac{R_p + R_n}{R_p} = 1 + \frac{R_n}{R_p} \tag{3.3}$$

Thus, the excess noise added by the nonthermal sources is $F - 1$.

As pointed out by Friis [3.2], since the noise factor is the ratio of the available output S/N to the available S/N of the source, it is unaffected by the value of the output impedance. However, as noted before, it is affected by the value of the input impedance. Consequently, the NF of each of the cascaded two-ports must be measured using as input impedance the output impedance of the preceding stage.

Again, following Friis, consider two cascaded circuits a and b. By definition, the available output noise from b is:

$$N_{ab} = F_{ab} G_{ab} k T B \qquad (3.4)$$

where B is the equivalent bandwidth in which the noise is measured. The total gain G_{ab} is the product of the individual gains, so:

$$N_{ab} = F_{ab} G_a G_b k T B \qquad (3.5)$$

The available noise from network a at the output of network b is:

$$N_{b|a} = N_a G_b = F_a G_a G_b k T B \qquad (3.6)$$

The available noise added by network b (its excess noise) is:

$$N_{b|b} = (F_b - 1)G_b k T B \qquad (3.7)$$

The total available noise N_{ab} is the sum of the available noises contributed by the two networks. Therefore, we obtain:

$$N_{ab} = N_{b|a} + N_{b|b} = F_a G_a G_b k T B + (F_b - 1) G_b k T B \qquad (3.8)$$

$$N_{ab} = \left[F_a + \frac{F_b - 1}{G_a} \right] G_a G_b k T B \qquad (3.8a)$$

and, comparing with Equation 3.5:

$$F_{ab} = F_a + \frac{F_b - 1}{G_a} \qquad (3.9)$$

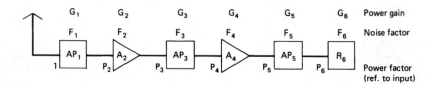

Figure 3.4 Block diagram of cascaded two-port circuits with attenuator pads AP, amplifiers A, and receiver R.

This may clearly be extended to any number of circuits:

$$F = F_1 + \frac{F_2 - 1}{G_1} + \frac{F_3 - 1}{G_1\,G_2} + \frac{F_4 - 1}{G_1\,G_2\,G_3} + \ldots \tag{3.10}$$

Figure 3.4 shows a cascaded circuit made up of several different components. The overall NF of the configuration may be calculated using Equation 3.10. By rearrangement, the number of terms may be reduced by the number of attenuator pads. Substituting the noise factor of a passive circuit $1/G_p$ for each of the pads and collecting terms:

$$F_{tot} = \frac{F_2}{P_2} + \frac{F_4 - G_3}{P_4} + \frac{F_6 - G_5}{P_6} \tag{3.11}$$

where $P_n = G_1 G_2 G_3 \ldots G_{n-1}$ is the power gain product at the input to the component n.

Every term contributes to the overall noise factor the equivalent of an active two-port with a preceding attenuation pad. The power gain (< 1) of the preceding attenuator pad is subtracted from the noise factor of each amplifier. The difference is divided by the power gain product P_n at the input of the amplifier.

3.3 Noise Correlation Matrix

The simplified noise factor calculations work well for many circuits, especially at frequencies below 100 MHz, and are adequate for planning purposes in most cases. Calculation of the noise factor of cascaded circuits can go awry if the individual noise factors are based on available S/R at the circuit input divided by that available at the output. In fact, the actual circuit impedances are not appropriate to achieving these available S/Ns. Often, information is available on the NF and gain of a circuit or device for some specific termination(s), which may not be convenient for use in a particular design. Where accurate calculations are required, noise representations such as those shown in Figure 3.2 must be used.

A single-noise voltage or current source at the input is adequate for many lower frequency applications, but we must use both sources for complete accuracy. This is especially important as we approach microwave frequencies. There may be correlation between the two noise sources (current or voltage), because many different internal sources may be represented by just the two terminal sources. Calculation of NF requires signal and noise powers; hence, second-order noise statistics suffice.

The noise sources shown in the representations of Figure 3.2 depend only on the characteristics of the network and are independent of the terminations. The noise-free network adds no noise output. Moreover, the definition of noise factor is independent of the output termination. Thus, in a representation such as Figure 3.2b, we can calculate NF by considering just the input noise sources. A similar approach may be taken by replacing the output voltage source in Figure 3.2a by an input voltage source that will produce the same output. The calculation results in an additional input voltage source, $-(E_{11} + Z_g) E_{n2}/Z_{21}$, which is dependent not only on the network, but also the source impedance. The (open circuit) voltage $-Z_g E_{n2}/Z_{21}$ is just that from an input current source $-E_{n2}/Z_{21}$ independent of the source impedance. This is equivalent to the current source I_n in Figure 3.2b.

By considering just the portions of the circuit preceding the noise-free network, the NF and its dependence on the input impedance can be determined. If we consider first the case where there is no correlation between E_n and I_n, the noise power input to the network is proportional to $[(E_g)^2 + (E_n)^2 + (I_n Z_g)^2]$, whereas the source noise power is E_g^2. Consequently, we obtain:

$$F = 1 + \frac{E_n^{\,2} + (I_n Z_g)^2}{E_g^{\,2}} \tag{3.12}$$

Substituting $Z_g = R_g + jX_g$ and $E_g^2 = 4kTBR_g$, we find:

$$F = 1 + \frac{E_n^{\,2}}{4\,k\,T\,B\,R_g} + I_n^{\,2}\,\frac{R_g^{\,2} + X_g^{\,2}}{4\,k\,T\,B\,R_g} \tag{3.12a}$$

Note that $X_g = 0$ minimizes F with regard to variation of X_g. Because R_g affects both the second and third terms of F, and in different ways, we set the partial derivative of F with regard to R_g equal to zero to determine the optimum (minimum) value of F, and find $R_{go}^2 = E_n^2 / I_n^2$. If we adopt the usual representation of the noise voltage (current) as impedance (admittance), $E_n^2 = 4kTBR_n$, and $I_n^2 = 4kTBG_n$, then $R_{go}^2 = R_n / G_n$ and:

$$F_o = 1 + 2(R_n \, G_n)^{1/2} \tag{3.13}$$

When correlation exists between E_n and I_n, Equation 3.12 becomes more complex:

$$F = 1 + \frac{(E_n + I_n Z_g)(E_n^* + I_n^* Z_g^*)}{E_g^2} \tag{3.12b}$$

Because E_n and I_n are correlated, their cross products no longer vanish (when averaged over time and the ensemble of noise functions). The product $E_n I_n^*$, when so averaged, produces a complex value, $C = C_r + jC_x$, and $E_n^* I_n$, the complex conjugate, $C^* = C_r - jC_x$. Then:

$$F = 1 + \frac{E_n^2 + I_n^2 (R_g^2 + X_g^2) + 2(C Z_g^* + C^* Z_g)}{4 \, k \, T \, B \, R_g} \tag{3.12c}$$

If we break E_n into two parts, one uncorrelated with I_n and the other correlated, $E_n = E_u + E_c \equiv E_u + Z_c I_n$, where Z_c, the complex constant that relates the correlated part of E_c to I_n, may be considered a correlation impedance. With this substitution, Equation 3.12b becomes:

$$F = 1 + \frac{[E_u + I_n(Z_c + Z_g)][E_u^* + I_n^*(Z_c^* + Z_g^*)]}{E_g^2} \tag{3.12d}$$

When the noise resistance and conductance equivalents are substituted, as before, we find:

$$F = 1 + \frac{R_u + G_n[(R_c + R_g)^2 + (X_c + X_g)^2]}{R_g} \qquad (3.12e)$$

In this case, the input impedance that produces an optimum NF is found to be $X_{go} = -X_c$ and $R_{go}^2 = R_u/G_n + R_c^2$. As a result we obtain:

$$F_o = 1 + 2 R_c G_n + 2(R_u G_n + R_c^2 G_n^2)^{1/2} \qquad (3.13a)$$

The analogous results obtained by conversion of all noise sources to parallel current sources are $B_{go} = -B_c$; $G_{go}^2 = G_u/R_n + G_c^2$ and:

$$F_o = 1 + 2 R_n G_c + 2(G_u R_n + G_c^2 R_n^2)^{1/2} \qquad (3.13b)$$

Using the foregoing results, the nonoptimum F may be expressed:

$$F = F_o + \frac{G_n[(R_g - R_o)^2 + (X_g - X_o)^2]}{R_g} \qquad (3.14)$$

$$F = F_o + \frac{R_n[(G_g - G_o)^2 + (B_g - B_o)^2]}{G_g} \qquad (3.14a)$$

Either of these expressions allows us to measure the noise parameters of a two pole-pair network if we have a noise measurement test arrangement, such as described in Section 2.3, in which the noise generator input impedance (admittance) may be varied in a controllable manner. If we make measurements of F for four appropriately chosen values of the input impedance (admittance), we obtain four equations for F that may be used to determine the four noise unknowns (F_o, G_n, R_o, and X_o, for example) in Equation 3.14. From these values, the four parameters by which the noise performance of the network is defined can be determined from the earlier relationships given.

For cascading two pole-pair networks, it is convenient to use ABCD notation:

$$V_1 = A V_2 - B I_2; \; I_1 = C V_2 - D I_2; \; \text{or, in matrix form:}$$

$$\left\|\begin{matrix} V_1 \\ I_1 \end{matrix}\right\| = \left\|\begin{matrix} A\ B \\ C\ D \end{matrix}\right\| \times \left\|\begin{matrix} V_2 \\ -I_2 \end{matrix}\right\| \tag{3.15}$$

If we represent the input and output vectors as $\|W_1\|$ and $\|W_2\|$, respectively, and the matrix as $\|E\|$, then a cascade of such circuits can be represented by $\|W_1\| = \|E_1\| \times \|E_2\| \times \ldots \|E_n\| \times \|W_{n+1}\|$. This is the equivalent of $\|W_1\| = \|E_{tot}\| \times \|W_{n+1}\|$, where $\|E_{tot}\| = \|E_1\| \times \|E_2\| \times \ldots \|E_n\|$ is the equivalent ABCD matrix for the entire cascade. This process is useful for a computer program to produce the overall performance of a cascade of individual circuits (as in a receiver). All that is required is a succession of matrix multiplications, as each circuit is added to the cascade.

When the noise vector from Figure 3.2b is considered:

$$\left\|\begin{matrix} V_1 \\ I_1 \end{matrix}\right\| = \left\|\begin{matrix} A\ B \\ C\ D \end{matrix}\right\| \times \left\|\begin{matrix} V_2 \\ -I_2 \end{matrix}\right\| + \left\|\begin{matrix} V_n \\ I_n \end{matrix}\right\| \tag{3.15a}$$

When considering a cascade of two circuits we find that $\|W_1\| = \|E_1\| \times \|E_2\| \times \|W_3\| + \|E_1\| \times \|N_2\| + \|N_1\|$. Here the N-vectors represent the noise vectors of the two circuits. The product of the matrices produces the appropriate total matrix for the input and output signal voltages, and the resulting noise output vector is converted to a single noise input vector. However, this formulation of the noise does not provide the information on the resulting correlation between the sources to complete noise factor calculations. Because there is no correlation between the noise sources in the first network and those in the second, we can compute the correlation of the two terms in vector $\|E_1\| \times \|N_2\|$ from the correlation of those in $\|N_2\|$. However, it would be convenient to have a conversion that does this for us automatically. For this purpose, the noise correlation matrix has been introduced. (See references [3.3, 3.4, and 3.5].)

In the foregoing, the vector

$$\|N_1\| = \left\|\begin{matrix} V_{n1} \\ I_{n1} \end{matrix}\right\|$$

Its conjugate transpose is $\|N_1{}^*\|^T = \|V_{n1}{}^*\ I_{n1}{}^*\|$. Their matrix product is:

$$\|C_1\| = \|N_1\| \times \|N_1{}^*\|^T \equiv \left\|\begin{matrix} C_{11}\ C_{12} \\ C_{21}\ C_{22} \end{matrix}\right\| = \left\|\begin{matrix} V_{n1} \times V_{n1}{}^*\ \ V_{n1} \times I_{n1}{}^* \\ I_{n1} \times V_{n1}{}^*\ \ I_{n1} \times I_{n1}{}^* \end{matrix}\right\| \tag{3.15a}$$

Recalling that the Vs and Is are the spectral densities of the actual random noise variables averaged over the ensembles of noise functions, and introducing the parameters from our prior discussion, we obtain:

$$\|C_1\| = \left\| \begin{matrix} 4\,k\,T\,R_{n1} & C_{r1} + jC_{x1} \\ C_{r1} - jC_{x1} & 4\,k\,T\,G_{n1} \end{matrix} \right\| \tag{3.16}$$

$$\|C_1\| = 4kT \left\| \begin{matrix} R_n 1 & F_{o1} - 1/2 - R_{n1}Y_{o1}^* \\ F_{o1} - 1/2 + R_{n1}Y_{o1} & R_{n1}\,|Y_{o1}|^2 \end{matrix} \right\| \tag{3.16a}$$

$$\|C_1\| = 4kT \left\| \begin{matrix} G_{n1}\,|Z_{o1}|^2 & F_{o1} - 1/2 - G_{n1}Z_{o1}^* \\ F_{o1} - 1/2 + G_{n1}Z_{o1} & G_{n1} \end{matrix} \right\| \tag{3.16b}$$

The multiplier $4kT$ assumes a single-sided spectrum density. If the two-sided density were used, the multiplier would drop to $2kT$.

The total noise vector is $\|N_1\| + \|E_1\| \times \|N_2\|$; its transpose conjugate is $\|N_1^*\| + \|N_2^*\| \times \|E_1^*\|^T$. The overall correlation matrix is found by multiplying the overall noise vector by its conjugate transpose:

$$\|C_{tot}\| = [\,\|N_1\| + \|E_1\| \times \|N_2\|\,] \times [\,\|N_1^*\|^T + \|N_2^*\|^T \times \|E_1^*\|^T\,] \tag{3.16c}$$

The noise of the first and second two-poles is uncorrelated, so this becomes:

$$\|C_{tot}\| = \|N_1\| \times \|N_1^*\|^T + \|E_1\| \times \|N_2\| \times \|N_2^*\|^T \times \|E_1^*\|^T \tag{3.17}$$

$$\|C_{tot}\| = \|C_1\| + \|E_1\| \times \|C_2\| \times \|E_1^*\|^T \tag{3.17a}$$

Thus, we have the rule for combining cascaded noise matrices to obtain an overall noise matrix.

As mentioned previously, the use of the ABCD-matrix provides a simple procedure for calculating the gain and noise performance of cascaded stages that is especially useful in computer programs. The first stage matrix is multiplied by the second to produce an equivalent overall single matrix. This matrix, in turn, is multiplied by the third stage matrix, and so on. After each multiplication the resultant matrix to be used in the next multiplication is stored. Similarly, we start with the first noise matrix $\|C_1\|$, the second, $\|C_2\|$, and the first stage ABCD-matrix, $\|E_1\|$, and produce a resultant noise matrix, using Equation 3.17. The resultant noise matrix is then combined with that of the third stage, using the resultant ABCD-matrix of the first two stages to produce the three-stage resultant noise matrix, and so on.

It is not necessary in all cases to use the ABCD-matrix to calculate the overall performance. In some cases, a Z-matrix, Y-matrix, S-matrix, or other matrix may prove most desirable in some circumstances, in which case, it is necessary to develop appropriate formulas to replace Equation 3.17 to calculate the equivalent noise matrix. Also, the noise matrix based on the representation of Figure 3.2b may not prove best in all cases. A matrix based on the two-voltage representation of Figure 3.2a, or the two-current representation of Figure 3.2b may be most useful in particular cases. These noise matrices are expressed as follows:

$$\|C_{1v}\| = \|N_{1v}\| \times \|N_{1v}^{*}\|^{T} \equiv \left\| \begin{matrix} C_{11} & C_{12} \\ C_{21} & C_{22} \end{matrix} \right\| = \left\| \begin{matrix} V_{n1} \times V_{n1}^{*} & V_{n1} \times V_{n2}^{*} \\ V_{n2} \times V_{n1}^{*} & V_{n2} \times V_{n2}^{*} \end{matrix} \right\| \tag{3.18}$$

$$\|C_{1i}\| = \|N_{1i}\| \times \|N_{1i}^{*}\|^{T} \equiv \left\| \begin{matrix} C_{11} & C_{12} \\ C_{21} & C_{22} \end{matrix} \right\| = \left\| \begin{matrix} I_{n1} \times I_{n1}^{*} & I_{n1} \times I_{n2}^{*} \\ I_{n2} \times I_{n1}^{*} & I_{n2} \times I_{n2}^{*} \end{matrix} \right\| \tag{3.18a}$$

It is necessary, of course, to develop the detailed expressions for these matrices in terms of measured noise characteristics, analogous to Equation 3.16, in order to use them. Also, the transformations to single overall matrices must be developed, based on the specific multiple circuit interconnection configurations and the form of the circuit matrices.

We shall not pursue these matters further here, because for most applications considered in this book we are interested in cascaded interconnections. At the lower frequencies, the assumption of no correlation between the noise sources of a circuit suffices, so that a single noise voltage or current may be used, as in Figure 3.3.

3.4 Calculation of IP

Prediction of IM distortion is an important consideration in planning the receiver design. As indicated earlier, a good measure of performance for any order of IM is the IP for that order. Usually only second- and third-order IPs are calculated; however, the technique may be extended to any order.

Figure 3.5 shows a configuration of two amplifiers with their voltage gains G_v and second- and third-order IPs. If we assume that a signal traversing the amplifiers encounters no phase shift, the composite IM performance may be calculated by

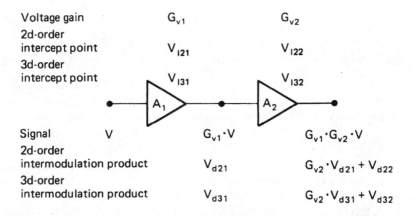

Figure 3.5 Block diagram of cascaded amplifiers with IM distortion.

assuming in-phase addition of the individual contributions. For example, the second-order product generated in amplifier A_1 is V_{d21} and that in A_2 is V_{d22}. Because V_{d21} is applied to the input of A_2, the overall IM product obtained at the output of A_2 is $(G_{v2} V_{d21} + V_{d22})$. The effect is the same as if an interfering signal of value

$$V_d = \frac{G_{v2}\, V_{d21} + V_{d22}}{G_{v1}\, G_{v2}} = \frac{V_{d21}}{G_{v1}} + \frac{V_{d22}}{G_{v1}\, G_{v2}} \tag{3.19}$$

were at the input. At the intercept point, this is equal to the input voltage V_{I2}. Generally (see Equation [2.5]), $V_{d2} = V^2/V_{I2}$, referred to the output of an amplifier. Thus, $V_{d2j} = V^2/V_{I2j}$ at the output of amplifier j. To place this discussion on a common footing, we can refer the signal level to the input, $V_{d2j} = (VG_{vj})^2/V_{I2j}$, and note that V_d can be expressed as V^2/V_{I2tot}. Collecting terms, we find:

$$\frac{1}{V_{I2tot}} = \frac{G_{v1}}{V_{I21}} + \frac{G_{v1}\, G_{v2}}{V_{I22}} \tag{3.20}$$

This may be extended to any number of amplifiers in cascade. It shows that the greater the gain to the indicated point, the more important it is to have a high IP. To reduce the problems associated with IM, selective filters should be provided as

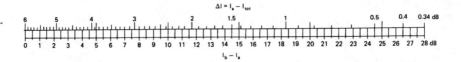

Figure 3.6 Nomogram for calculating the second-order IP for cascaded amplifiers.

near the front of the receiver as possible to reduce the gain to signals likely to cause IM.

While this formula is relatively easy to calculate, the IPs are generally available in dBm, which must first be converted to power before the formula may be used. The nomogram of Figure 3.6 allows the combination of values directly. For this, we rewrite Equation 3.20 as follows:

$$\frac{1}{V_I} = \frac{1}{V_a} + \frac{1}{V_b} \quad V_a \leq V_b \tag{3.21}$$

The various V quantities are those referred to the receiver input. It is irrelevant from which amplifier V_a and V_b are derived, but if we choose V_a to be the smaller as indicated and express the other as a ratio of V_a,

$$\frac{V_I}{V_a} = \frac{1}{1 + V_a/V_b} \tag{3.22}$$

the denominator on the right remains less than 2. The resultant is a relationship between two voltage ratios. The values I_J in Figure 3.6 correspond to the equivalent intercept levels V_I in Equation 3.22, measured in dBm. The use of this tool is quite simple:

- Using the gains, recompute the IPs to the system input. ($I_b = I_{bO}/G_{bI}$, where I_{bO} is measured at the amplifier output and G_{bI} is the gain between system input and amplifier output.)
- Form the difference value between the two recalculated IPs [I_b (dBm) − I_a (dBm)].

- In the nomogram, determine the value I and subtract it from I_a to get I_{tot}.

- If there are more than two amplifiers, select the resultant I from the first two and combine similarly with the third, and so on, until all amplifiers have been considered.

The procedure to determine the third-order IP is analogous to that for the second-order, noting, however, that $V_{d3} = V^3/V_{I3}^2$. In this case, after manipulating the variables, we find:

$$\frac{1}{V_{I3tot}} = \left[\left(\frac{G_{v1}}{V_{I31}}\right)^2 + \left(\frac{G_{v1}\,G_{v2}}{V_{I32}}\right)^2\right]^{1/2}$$

(3.23)

This can be simplified analogously to Equation 3.22 as follows:

$$\frac{1}{V_I^2} = \frac{1}{V_a^2} + \frac{1}{V_b^2} \quad V_a \leq V_b$$

(3.24)

or

$$\left(\frac{V_I}{V_a}\right)^2 = \frac{1}{1 + (V_a/V_b)^2}$$

(3.25)

Just as Figure 3.6 was used to evaluate Equation 3.22, so the nomogram in Figure 3.7 can be used to evaluate Equation 3.25.

All of these calculations need to be made with care. Some amplifiers invert the signal. In that case, the IM components can subtract rather than add. At RF, there are generally other phase shifts that occur either in the amplifiers or in their coupling circuits, so that without thorough analysis it is impossible to determine how the IM powers add vectorially.

Certainly the assumption of in-phase addition made in the preceding equations is the worst-case situation. In many practical cases, however, most of the IM distortion is confined to the stage prior to the selective filtering, so that the contributions of earlier stages may be neglected. Nonetheless, the matter needs careful attention.

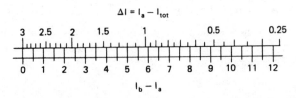

Figure 3.7 Nomogram for calculating the third-order IP for cascaded amplifiers.

3.4.1 Example of NF and IP Calculation

Referring now back to Figure 3.1, we shall use the level diagram to calculate the expected NF and IPs referred to the receiver input. First we consider the overall noise factor. The values shown in the diagram are substituted in Equation 3.10:

$$F_{tot} = 5.623 + \frac{0.585}{0.178} + \frac{1.239}{1.778} + \frac{0.995}{0.794} + \frac{2.981}{12.587} + \frac{2.162}{3.162} + \frac{1.512}{19.949} + \frac{1.512}{7.942}$$

$$= 12.05 \ or \ 10.81 dB$$

Second-order products can be generated only in the input mixer, because the following filter, with 25 kHz bandwidth, does not permit sufficient frequency separation. The 80-dBm output IP_2 is converted to an input IP_2 of 86.5 dBm by the mixer and filter losses, since the results are referred to the input.

We will next consider third-order products. All the equivalent input values of V_j^2 are developed across the same input resistor, so that the value of IP_{3j} in milliwatts may be substituted throughout Equation 3.24. There is but one source of IM in the RF chain, i. e., the mixer, with a 32-dBm IP_3. The gain to its output is −6.5 dB, or 0.2239. The contribution from this circuit alone to the IP is, therefore, at a level of 38.5 dBm, or 7079.5 mW. This is the out-of-band IP_3. The first IF input filter allows only signals in or close to its 25 kHz passband to produce IM products. For signals within this band, but not within the 7-kHz bandwidth of the following filter, the total (maximum) "25 kHz in-band" IP is given by 1/(0.2239/1584.9 + 1.778/1000) = 521.03 mW, or 27.17 dBm. For signals in or adjacent to the 7-kHz band, the second amplifier, the mixer, and the first amplifier at the second IF must be included. Thereafter, the final selectivity is provided. IM caused by near passband signals is not of importance because of their direct interference. We will calculate the 7-kHz in-band IP_3 by using Equation 3.24 and also by using the third-order nomogram of Figure 3.7. Using Equation 3.24, continuing as above, the overall IP may be obtained as:

$$\frac{1}{IP_3} = \frac{0.2239}{1584.9} + \frac{1.778}{1000} + \frac{12.589}{1000} + \frac{3.162}{501.2} + \frac{19.952}{1000} = 0.04077$$

$IP_3 = 24.528$ mW, or 13.90 dBm.

To use the nomogram, we must convert each of the five contributors to the total IP to their equivalent IP at the input. These become, respectively, 38.5, 27.5, 19, 22, and 17 dBm. We will proceed with the combination in the indicated order. For the first pair, 27.5 dBm corresponds to I_a and 38.5 dBm to I_b in Figure 3.7. The difference is 11 dB, resulting in an I of 0.33 that, when subtracted from 27.5, yields a net of 27.2 dBm. This, in combination with 19 dBm, produces a difference of 8.17 dB, an I of 0.62, and a net of 18.4 dBm. Proceeding in this manner, we get 16.8 dBm and, finally, 13.9 dBm.

3.5 Spurious Response Locations

Frequency changing occurs as a result of a second-power term in the mixer, giving rise to a product term when the input is the sum of two signals. Some mixers are designed to achieve the product of two inputs applied directly to separate terminals, rather than using the second-order nonlinearity at a single terminal. Either way, the resultant output term of the form $a_2 V_a V_b$ produces the desired frequency change. Here V_a and V_b represent the two input signals, and a_2 determines the mixing effectiveness of the device. If we take V_a as the signal whose frequency is to be changed and V_b as a sinusoid from the LO that is set to accomplish the desired change, simple multiplication, combined with trigonometric identities, shows that the output consists of two terms at different frequencies. (Because V_a is a narrow-band signal, we can represent it as a sinusoid, whose envelope and phase variations with time are not specifically indicated.)

$$a_2 V_a V_b = a_2 V_s \cos(2\pi f_s t + \phi) \times V_o \cos(2\pi f_o t + \theta) \tag{3.26}$$

$$= a_2 V_s V_o \frac{\cos[2\pi(f_o + f_s)t + \phi + \theta] + \cos[2\pi(f_o - f_s)t + (\theta - \phi)]}{2} \tag{3.26a}$$

The frequencies of these two terms are at the sum and the difference of the input frequencies, $|f_o \pm f_s|$. The absolute value is indicated because either f_o or f_s may be the higher frequency. Either frequency may be selected by a filter for further use as an IF of the receiver.

When the LO is set to f_o, an input at f_s produces an output at the IF. However, by the nature of Equation 3.26, other inputs may also produce an IF output. If the IF is the sum frequency, a signal of frequency $f_s = f_o + f_{IF}$ can also produce an output. If the IF is the difference frequency, the signal frequency may be either higher or lower than the signal frequency. In this case, a second frequency that is below or above the oscillator frequency by twice the IF will also produce an output at the IF. This unwanted response resulting from the product is called the *image* of the desired signal. Because the output cannot distinguish it from the desired signal, it is necessary to filter the image from the input prior to the mixer stage.

Another undesired response that must be guarded against is a signal at the IF. While the second-order mixing response does not produce an output at this frequency, many mixers have equal or higher first-order gain. Even when the circuit is balanced to cancel the first-order response, there is a practical limit to the cancellation. Usually the image and IF responses are the strongest undesired outputs of a mixer. The first step in the selection of an IF is to assure that its value permits the IF and image responses to be adequately filtered prior to the mixer. In some cases, when a very wide signal band is to be covered, it may be necessary to use more than one receiver configuration with different IF frequencies to achieve adequate IF and image frequency rejection.

While the IF and the image are the initial spurious responses that need to be considered in selecting the IF, they are, unfortunately, not the only such responses. No device has been found that produces only second-order output. Most devices, when driven with sufficiently strong inputs, have nonlinearities of very high order. An nth-order nonlinearity produces outputs at frequencies $|(n-m)f_o \pm mf_s|$, where m ranges from 0 to n. Thus, a mixer can produce outputs resulting from many orders of nonlinearity. Some of these products fall closer to the desired signal than the image and the IF, and some can, at certain desired frequencies, produce higher-order products falling at the desired frequency. Such spurious responses cannot be filtered from the input prior to the mixer because that would require filtering the desired signal. The steps necessary for optimum circuit performance include the following:

- Select a mixer with low response to high-order products
- Select the receiver band structure and IFs to minimize the number of products that fall within the IF filter passband and to have these of as high an order as possible

While the mixer output tends to reduce as the order increases, not all data sheets provide adequate information on the extent of high-order responses. The responses also are dependent on the operating conditions of the mixer, which may change throughout the tuning range. Consequently, a final check by measurement of the responses is essential.

While it is not always possible to predict the strength of spurious response signals, their frequencies may be predicted quite precisely from the general expression $|\,nf_o \pm mf_s\,| = f_{IF}$. Here m and n can take on all positive integer values and 0. When only positive frequencies are considered, this expression gives rise to three separate relationships:

$$nf_o + mf_s = f_{IF} \tag{3.27a}$$

$$nf_o - mf_s = f_{IF} \tag{3.27b}$$

$$nf_o - mf_s = -f_{IF} \tag{3.27c}$$

When m and n equal unity, three possible relationships between the desired signal and oscillator frequencies exist:

$$f_o + f_t = f_{IF} \tag{3.28a}$$

$$f_o - f_t = f_{IF} \tag{3.28b}$$

$$f_o - f_t = -f_{IF} \tag{3.28c}$$

Here the expression f_t has been used to designate the tuned frequency of the desired signal and to distinguish it from the spurious response signals of frequency f_s. Only one of the three cases in Equation 3.28 applies for a particular calculation.

The preceding relationships are all linear relationships, so it is comparatively easy to plot them and examine where spurious responses will occur. Figure 3.8 shows a chart based on typical AM broadcast band frequency selections. The line $F = S$ represents the desired signal. (Here $F = f_s$, $S = f_t = f_o + f_{IF}$.) The broadcast band runs from 540 to 1600 kHz. Note that the IF response (horizontal line at 456 kHz), if extended, would intersect the tuning line at the intersection of that line with the $F = (2S + I)/3$ and $F = (S + 2I)/3$ lines. The line $F = S + 2I$ is the image and remains substantially separated from $F = S$. However, the $F = S + I/2$ response is parallel to the tuning line and much closer (228 kHz) than the image (912 kHz). This is typical of difference mixers. A third-order response $F = S/2 + I$ coincides with the desired response at 912 kHz ($2f_{IF}$), and a fifth-order coincides with the

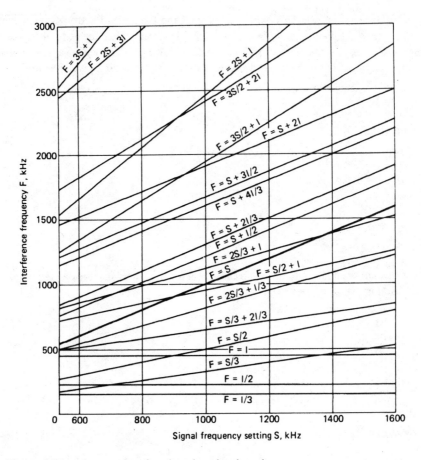

Figure 3.8 Interference chart for a broadcast band receiver.

desired response at 1368 kHz ($3f_{IF}$). The highest-order responses plotted are sixth-order, $F = S + 4I/3$ and $F = S + 2I/3$. Except for the beat notes, which are likely to occur if there is a reasonably strong station at 912 or 1368 kHz, it should be possible to provide preselection filtering to protect against the effects of the other "spurs," as long as the specifications on spurious response rejection are not too stringent. Usually they are not for broadcast receivers, where price is often more important than high-performance capability.

By using the IF as a normalizing factor, universal charts may be prepared. These can be helpful in selecting the IF because they allow visualization of the locations of the lower-order spurs. When the charts include too many orders, their use becomes more difficult. The normalized equations are:

$$nO + mS = 1 \qquad\qquad (3.29a)$$

$$nO - mS = 1 \qquad\qquad (3.29b)$$

$$nO - mS = -1 \qquad\qquad (3.29c)$$

and

$$O + T = 1 \qquad\qquad (3.30a)$$

$$O - T = 1 \qquad\qquad (3.30b)$$

$$O - T = -1 \qquad\qquad (3.30c)$$

Here O represents the oscillator frequency, T the tuning frequency, and S the spurious frequency, all measured in units of the IF ($f_s = S f_{IF}$, etc.). For each type of mixer selection, we may express O in terms of T, using the proper expression in Equations 3.30a to c, and substitute the expressions in Equation 3.27a. Charts may then be plotted to show the relative locations of the spurious frequencies to the order $m + n$ relative to the tuning curve ($S = T$). The tuning band, which has a width with fixed ratio to the lower frequency, may be moved along the T axis until the position is judged to be the best compromise possible, and the resulting IF is calculated. After some cut and try, it should be possible to select an IF, or a number of IFs, to use in further design evaluations.

Some typical charts of this sort are shown in Figures 3.9 through 3.12. Figure 3.9 is for a difference mixer with a low-side oscillator ($O - T = -1$). Most responses up to the sixth order have been plotted. Only the region greater than $T = 1$ is of interest. When the value of O becomes negative, the result is a sum mixer. Thus the segment of the chart below $T = 1$ represents a sum mixer. The lower part of this segment has been expanded in Figure 3.10. Figure 3.11 is for a difference mixer with a high-side oscillator. In this case it is possible to operate with T below unity. The implication is that the IF is above the signal frequency but below the oscillator frequency. This can be useful to keep the image and IF responses distant from the RF passband and thus reduce the need for tuned filters. Also, the crossovers tend

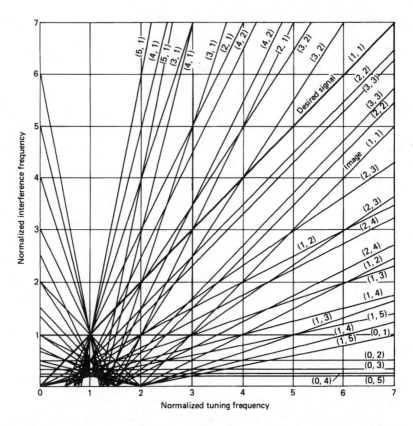

Figure 3.9 Spurious response chart for a difference mixer with a low-side oscillator. Up to sixth-order responses are plotted; (n, m) are the orders of the oscillator and interfering signals, respectively.

to involve higher orders of f_s so that the spurious rejection is often better than where lower orders of f_s cross over. Figure 3.12 is an expansion of the lower left-hand corner of Figure 3.11 so that better judgments can be made in this case.

The difference mixers tend to become free of spurs (to the order plotted) as the low end of the tuning band is moved to the right. Of course, this also moves the high end of the band proportionately more to the right. This causes the parallel spurs to be proportionately closer to the desired signal, and requires improved preselection filtering. It can be seen, for example, that the 0.5 separation of the (2, 2) response is 25 percent of the low end of the band when at $T = 2$, but only 12.5 percent when at $T = 41$. The high-side oscillator arrangement has a lower density of crossovers for a given low-band frequency selection. Similarly, the difference mixer with a high-side oscillator has a lower density of crossovers than the sum

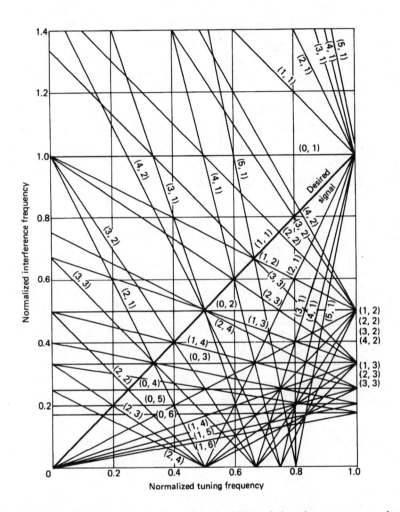

Figure 3.10 Spurious response chart for a sum mixer. Up to sixth-order responses are plotted; (n, m) are the orders of the oscillator and interfering signals, respectively.

mixer and those at lower orders of the oscillator. As a general observation, it appears that the difference mixer with a high-side oscillator provides fewer spurious responses than the other arrangements and should be preferred unless other design factors outweigh this consideration.

3.5.1 D-H Traces

Other methods of plotting spurious response charts are possible. *D-H* traces are an ingenious type of plot proposed by Westwood [3.3]. The two frequencies being mixed are referred to as f_D and f_H for reasons that will become apparent. Which-

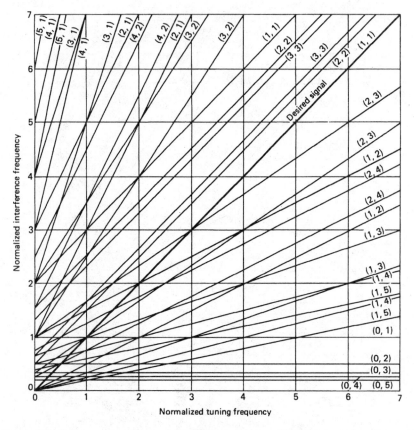

Figure 3.11 Spurious response chart for a difference mixer with a high-side oscillator. Up to sixth-order responses are plotted; (n, m) are the orders of the oscillator and interfering signals, respectively.

ever of the two frequencies is the higher is designated f_D (whether f_o or f_s). The difference between f_D and f_H is designated f. When these frequencies are normalized by dividing by the IF, they are referred to as D, H, and X. The ordinates are made equal to H and the abscissa to X, so that constant H represents a *horizontal* line or trace, and X represents a vertical trace. Because D is a linear combination of H and X, it represents a *diagonal* trace (hence, H and D). Manipulating the various expressions, we find:

$$H = \frac{-NX}{N \pm M} + \frac{1}{N \pm M} \tag{3.31}$$

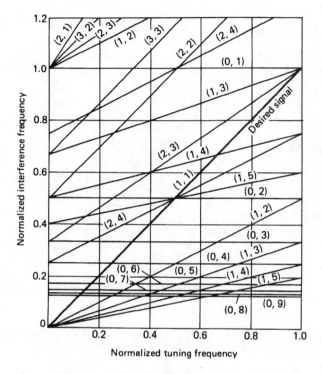

Figure 3.12 An expansion of the lower left-hand segment of Figure 3.11.

where N and M may now be positive or negative integers, including zero. H represents the tuned frequency for a difference mixer with a high-side oscillator, the oscillator frequency for a difference mixer with low-side injection, and the higher of the oscillator and tuned frequency for a sum mixer. The complementary frequency (oscillator or tuned frequency) is determined from the equation $D - H = X$ and is a diagonal line at 45° sloping down to the right from the value of D when $X = 0$.

The various lines defined by Equation 3.31 are, thus, the same for all mixer varieties, and one set of charts—rather than three—can be used to evaluate the location of spurs. Figure 3.13 illustrates these *cross-product* (C-P) charts. They are used in the following manner:

- A potential IF and a mixer type are selected.
- The maximum frequency of the oscillator or tuning is determined.

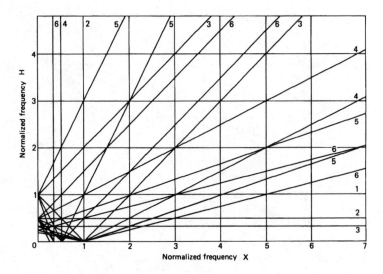

Figure 3.13 Sample *D-H* chart of cross products to the sixth order for maximum X = 7.2.

- The IF is subtracted and the resultant divided by the IF; this is the *H* intercept.
- From this point, the D line is drawn down at a 45° angle to the right.
- The intersection of the *D* trace with the C-P traces indicates the location of spurs.

The *X* axis represents the difference between the oscillator and the interfering signal. The tuned frequency for a difference mixer is represented by the line *X* = 1. For the high-side oscillator, the *H* axis is the same as the *T* axis. For the low-side oscillator, the *H* frequency is the equivalent of the *O* axis, so that unity must be added to the *H* value to get the value of *T*. For a sum mixer, the tuned frequency curve is the diagonal with an *H* intercept of 0.5 and an *X* intercept of 1.0. The *H* axis represents the lower of *T* or *O*. The other may be obtained by subtraction from unity. Westwood [3.3] used a computer plotter to provide a series of charts for 6, 10, and 16 orders of cross products with *X* running from 0 to maxima of 0.18, 0.72, 1.8, 7.2, 18, 72, and 180. An example is reproduced as Figure 3.13 for comparison with the earlier charts.

The nature of the equations defining the location of spurious responses is such that selection of the optimum IF frequency could be programmed readily for computer solution if an appropriate criterion for optimization were established. Alternatively, if the designer wished to make the final selection, a program could be developed so that when the tuning range, mixer class, and proposed IF were entered, a location of all spurs within a predetermined frequency of the tuning frequency up to a specified order could be plotted.

A. Example case

As an example of spur location, consider the frequencies used in Figure 3.1. The input frequency coverage is from 2 to 32 MHz. A low-pass filter with 32-MHz cutoff is used to provide preselection. The first IF is at 75 MHz, and the mixer is a difference type with a high-side oscillator. This results in a normalized tuning range from 0.02667 to 0.42667. Referring to Figure 3.12, we find that the only spurs to the sixth order that fall on the tuning frequency are the harmonics of a signal at the IF from the third order up. At the lower end of the range the (1, 2) product is at its nearest, falling at 0.01333 or 1 MHz at the 2 MHz end of the band. It falls at progressively higher frequencies as the tuning frequency rises. These signals fall well outside the first IF passband and will be removed by the first IF filter.

At the high end of the band, the nearest product is the (2, 4) product. At the low end of the top band (22.8/75 = 0.304), this product is at a frequency of 0.402 (30.15 MHz). Because this is within the passband of the low-pass filter, rejection of this spur depends on the mixer response and the first IF bandwidth. The (2, 4) product at the top of the band occurs at 0.4633 (34.75 MHz), so it has rejection from both the IF and low-pass filter. The subharmonic of the IF at 0.5 (37.5 MHz) is further outside the band, and because it does not change position with tuning, it could be provided with an additional trap in the preselection filtering, if necessary. Thus, the major spur concerns are the IF subharmonics below 0.5 and the sixth-order product in the higher RF bands. The same conclusions can be reached using Figure 3.13 and following the line $X = 1$ from H of 0.02667 to 0.42667.

The spurs resulting from the first IF mixer should also be examined. In this case, the tuned signal is at 75 MHz, the oscillator is at 84 MHz, and the IF is at 9 MHz. This results, again, in a difference mixer with high-side oscillator injection and a T value of 8.3333. This value is off scale in Figure 3.11. However, it is clear that up to the sixth order there will be no crossovers in the vicinity, and the nearest spur is (2, 2), which is 0.5 (4.5 MHz) above the T value. The first IF preselection filter has a bandwidth of 25 kHz, so it should not be difficult to assure adequate filtering of this spur. The only areas of concern then are those associated with the first mixer.

3.6 Selectivity

Because of historical and physical limitations, radio channels are generally assigned on a frequency division basis. Each transmitter is assigned a small contiguous band of frequencies, within which its radiated spectrum must be confined. Transmitters that could interfere with one another are ideally given nonoverlapping channel assignments. However, the demand for spectrum use is so great that at times compromises are made in this ideal situation. From time to time, suggestions have been made of other techniques for spectrum sharing, such as time division or code division. Physical limitations prevent the entire spectrum from being so assigned, but portions of it have been successfully used in special applications.

Despite the usefulness of other spectrum-sharing techniques [3.4], radio receivers in the frequency range we are discussing are likely to use mainly frequency division for the foreseeable future. To operate effectively in the current crowded spectrum environment, a receiver must provide selective circuits that reject adjacent and further separated channel assignments, while passing the desired signal spectrum with low distortion. A reasonably narrow bandwidth is also necessary to minimize the effects of man-made and natural noise outside the channel so as to provide good sensitivity. In general-purpose receivers, it is often desirable to provide a selection of several bandwidths for transmissions with different bandwidth occupancies. Most modern receivers achieve selectivity by providing one or more lumped filter structures, usually near the input of the final IF chain, although distributed selectivity is still used at times. In the following sections, we discuss some common filter characteristics and methods of implementing them.

3.7 Single-Tuned Circuit

The series or parallel combination of an inductance, capacitance, and resistance results in a single resonant circuit. The parallel circuit provides a single pole, is the simplest of filter circuits, and is often used as a tuning or coupling element in RF or IF circuits. As long as the Q of the circuit is high, similar frequency response results from a serial or a parallel circuit (or one that may be tapped to provide impedance match). For a parallel resonant circuit (Figure 3.14), the magnitude of the normalized response may be given by the following:

$$A = (1 + S^2)^{1/2} \qquad\qquad (3.32)$$

where

$$S = Q\left[\frac{f}{f_o} - \frac{f_o}{f}\right] \approx 2\,Q\,\Delta f$$

$$Q = R\left[\frac{C}{L}\right]^{1/2}$$

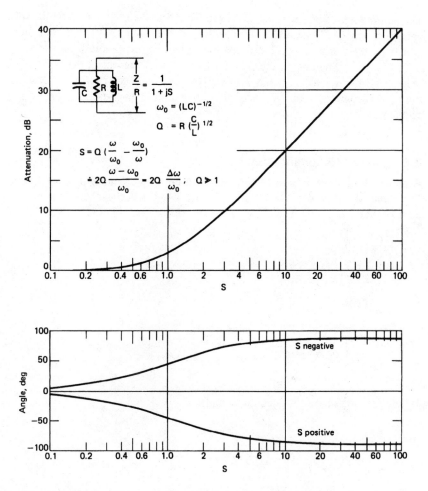

Figure 3.14 Amplitude and phase response of a single resonant circuit.

$$f_o = \frac{1}{2\pi}(L\,C)^{1/2}$$

The phase response is given by:

$$\phi = \tan^{-1} S \qquad\qquad (3.33)$$

The amplitude and phase response are plotted in Figure 3.14. Such circuits, when cascaded, may be tuned synchronously or offset to get particular responses. However, their usual use is for circuit matching or providing limited selectivity. Very often the remainder of the circuit is coupled to the resonator through tapping or through the use of nontuned windings on the coil. Taps may be achieved by multiple capacitors, multiple uncoupled inductors, or by a true tap on a single inductor.

3.8 Coupled Resonant Pairs

Another simple filter frequently used for coupling between amplifiers is a coupled isochronously tuned pair of resonators. Figure 3.15 is a design chart for this circuit arrangement [3.5]. The applicable equations are:

$$U = \frac{[(1 + C^2 - S^2)^2 + 4\,S^2]^{1/2}}{1 + C^2} \tag{3.34}$$

$$\phi = \tan^{-1}\left[\frac{2\,S}{1 + C^2 - S^2}\right] \tag{3.35}$$

Where k = the coefficient of coupling and

$$(1 + C^2) = \left[\frac{2\,Q_1\,Q_2}{Q_1 + Q_2}\right](1 + k^2\,Q_1\,Q_2)$$

$$S = 2\,Q\left[\frac{F_o}{f} - \frac{f}{f_o}\right] \approx \frac{2\,Q f}{f_o}$$

$$Q = \frac{2\,Q_1\,Q_2}{Q_1 + Q_2}$$

The maximally flat condition occurs when $C = 1$. Because above $C = 1$ there are two peaks while below there is one, this condition is called *transitional coupling*. U represents the selectivity, but the output at $S = 0$ varies with k (Figure 3.16). The maximum output occurs when $k^2 = 1/Q_1 Q_2$, which is called the *critical coupling*.

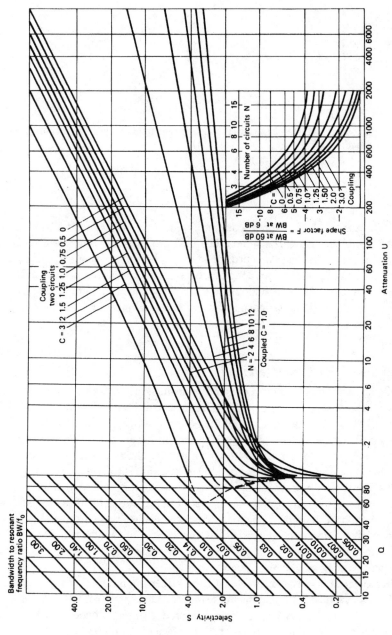

Figure 3.15 Design chart of coupled circuit pairs.

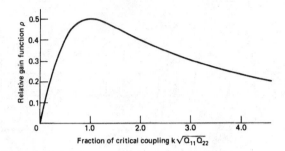

Figure 3.16 Secondary output voltage versus coupling for a tuned coupled circuit pair. The relative gain function is:

$$\rho = \frac{k\sqrt{Q_{11}\, Q_{22}}}{1 + k^2\, Q_{11}\, Q_{22}}$$

If the two Qs are equal, the expressions simplify, and the critical and transitional coupling become the same. Transitional coupling produces the two-pole Butterworth response. The peaked cases correspond to two-pole Chebyshev responses.

Studies have been made of three coupled isochronously tuned circuits. However, with three Qs and two ks as parameters, results tend to become complex. The use of multipole lumped filters designed in accordance with modern filter theory is generally more common.

3.9 Complex Filter Characteristics

When filters require more than two or three resonators, the number of parameters becomes so large that it is necessary to develop special techniques for filter design. The receiver designer normally specifies the performance desired, and the filter is purchased from a manufacturer who specializes in the design of such devices. It is important, however, to be familiar with common filter characteristics. In modern network theory, it is usually assumed that the termination is a constant resistance, although designs may be made with other assumptions for the termination. The networks are designed based on the locations of poles and zeros of the transfer function. A number of families of characteristics are available, often known by the name either of an author who suggested the filter type or of a mathematician who is associated with the characteristic. This makes for some confusion, because some families may be known by several names.

Usually the most important characteristics of a filter are the amplitude response versus frequency (selectivity) and the phase response. The principal interest in the latter results from the fact that it is necessary to reproduce the amplitude and relative phase of the signal transmitted correctly at all significant frequencies in its spectrum to avoid waveform distortion. In some transmissions (speech and music), phase distortion is tolerable. Phase distortion, however, is closely related to delay, and distortion that can cause sufficient delay differences among frequency components can be detected even for audio transmissions. For video transmissions, it is important that the waveform be reproduced with relatively little phase distortion, because relative delays among components can result in poor definition (ghosts and other artifacts). In data transmission, the form of the transient response is more important than perfect reproduction of the original waveform. It is desirable that the step response have a rise time that allows the signaling element to attain its ultimate amplitude before the next element is received and that has minimal ringing to avoid intersymbol interference.

The various characteristics, amplitude, phase, and transient response, are interrelated, because they are completely determined by the poles and zeros of the transfer function. Good transient response with little ringing requires a slow amplitude cutoff and relatively linear phase response. Good waveform reproduction requires constant amplitude response and linear phase response of the transmitted spectrum. On the other hand, the rejection of interference of adjacent channels requires rapid attenuation of frequencies outside the transmitted channel. The physical nature of networks makes a compromise necessary between the in-channel distortion and out-of-channel rejection. For more details on filter design, consult references [3.6] through [3.9].

Some of the more common modern filter families include the following:

- *Butterworth* or *maximally flat amplitude*
- *Chebyshev*
- *Thompson, Bessel*, or *maximally flat group delay*
- *Equiripple linear phase*
- *Transitional*
- *Elliptic* or *Cauer*

The data presented in the following discussion have been taken from [3.8] and [3.9]. In all but the last case, representative curves are given of amplitude response versus normalized frequency. Some curves of group-delay response ($d\theta/d\omega$) versus normalized frequency and responses to impulse and step modulated carriers at midband versus normalized time are also included. For more extensive information see [3.8].

3.9.1 Butterworth Selectivity

Figure 3.17 shows the various responses for the Butterworth, or maximally flat amplitude, response. The poles for this filter type are positioned so that the

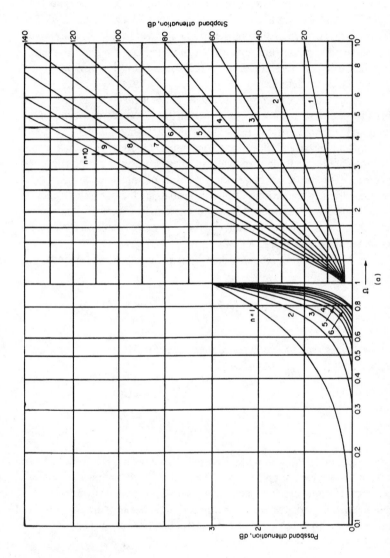

Figure 3.17 Characteristics of Butterworth filters: (a) attenuation, (b) group delay, (c) impulse response, and (d) step response. (*From [3.8]. © 1967 John Wiley and Sons, Inc. Reprinted by permission of the publisher.*)

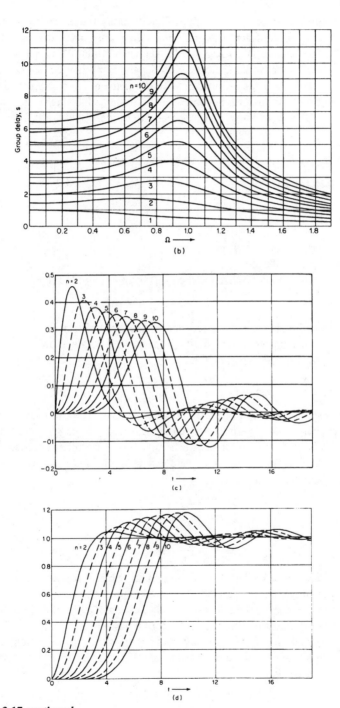

Figure 3.17 *continued*

maximum number of derivatives of amplitude versus frequency are zero at the center frequency of the filter. The more poles there are available in the filter, the more derivatives can be set to zero and the flatter the filter. About halfway to the 3-dB selectivity of these filters, group delay departs from flatness (phase linearity) and rises to a peak near the 3-dB point. The larger the number of poles, the more rapid the amplitude drop-off is beyond the 3-dB point and the higher the deviation of the group delay from flatness. The selectivity of the Butterworth filter may be expressed as:

$$Att = 10 \log(1 + \Omega^{2n}) \tag{3.36}$$

Where:
Att = attenuation expressed in decibels
n = the number of poles
Ω = the bandwidth normalized to the 3-dB bandwidth of the filter

Butterworth multipole filters have substantial ringing, which can result in substantial intersymbol interference for digital signals with data symbol rates that approach or exceed the 3-dB bandwidth.

3.9.2 Chebyshev Selectivity

Figure 3.18 shows the amplitude responses for the Chebyshev, or equal-amplitude ripple, case, when the ripple is 0.5 dB. Other selections ranging from 0.01 to 1 dB of ripple are plotted in the references. The poles for these filters are located so as to provide the equal ripple in the passband and have selectivity that is related to the Chebyshev polynomials as follows:

$$Att = 10 \log(1 + \varepsilon^2 C_n^2 \, \Omega) \tag{3.37}$$

where C_n is the nth-order Chebyshev polynomial, which oscillates between 0 and 1 in the passband, and ε is a parameter selected to provide the desired ripple level. These filters have a more rapid increase in attenuation outside the 3-dB bandwidth than Butterworth filters for the same number of poles. Their group-delay distortion is also higher and there is substantially more ringing. Thus, these filters provide improved adjacent-channel rejection but produce greater intersymbol interference in digital applications.

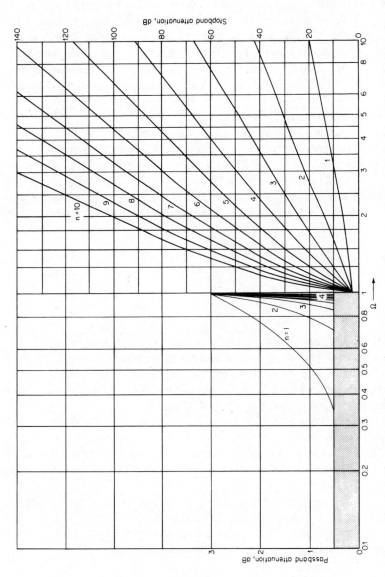

Figure 3.18 Attenuation characteristics of Chebyshev filters with 0.5 dB ripple. *(From [3.8]. © 1967 John Wiley and Sons, Inc. Reprinted by permission of the publisher.)*

3.9.3 Thompson or Bessel Selectivity

The Thompson or Bessel characteristic is obtained by seeking the maximally flat group delay available for a filter with a particular number of poles. The phase delay at any frequency is given by $-\phi/\omega$. Because transmission through a filter invariably results in delay, the phase decreases monotonically with frequency. When expanded as a Taylor series, the first derivative is always negative. The first derivative of the phase measures the rate of change of the phase with frequency, and when the first derivative is multiplied by a small angular frequency change, the result gives the phase change for that small frequency change or the difference in delay. This result is called the *group* or *envelope delay*. If $-d\phi/d\omega$ is constant, there is no change in delay; the phase change is linear. To obtain a maximally flat delay, as many as possible derivatives higher than the first are set to zero. In the Thompson selectivity characteristic, the location of the poles is chosen so that this is the case. The higher the number of poles n, the greater the number of derivatives that may be forced to zero, and the more constant is the group delay. A constant delay transfer function may be expressed as $\exp(-s\tau)$. If the time and complex frequency s are normalized using delay τ, the transfer function T may be expressed as:

$$T(S) = \frac{1}{\exp S} = \frac{1}{\cosh S + \sinh S} \tag{3.38}$$

In the Bessel filter, this is approximated by expanding the hyperbolic functions as continued fractions, truncating them at the appropriate value of n and determining the pole locations from the resulting expressions. Some of the resulting characteristics are shown in Figure 3.19. For the normalized variable Ω up to 2, the attenuation can be approximated by $Att = 3\Omega^2$, but between 2 and the frequency at which the ultimate slope of $20n\Omega$ dB per decade is achieved, the attenuation tends to be higher. The delay is flat and the impulse and step response show no ringing (Figure 3.19b). This type of filter has poorer adjacent-channel response than the previous types discussed, but affords a low level of intersymbol interference in digital transmission.

A related family of filters may be derived from the gaussian transfer function $T(S) = 1/\exp(-\Omega^2)$ by expanding and, truncating the denominator, then locating the poles. The delay curves are not quite so flat, and beyond $\Omega = 2$, the attenuation of the gaussian curves is not so great. They produce similar transient responses.

3.9.4 Equiripple Linear Phase

Just as the Chebyshev (equal-amplitude ripple) shape produces a better adjacent-channel attenuation than a Butterworth shape with the same number of poles, so an equiripple linear phase characteristic produces more adjacent-channel attenuation

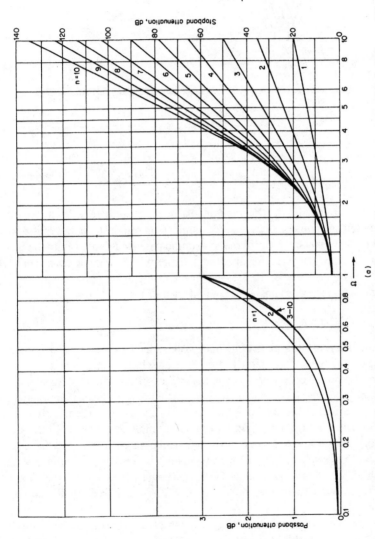

Figure 3.19 Characteristics of Thompson filters: (*a*) attenuation and (*b*) impulse response. (*From [3.8]. © 1967 John Wiley and Sons, Inc. Reprinted by permission of the publisher.*)

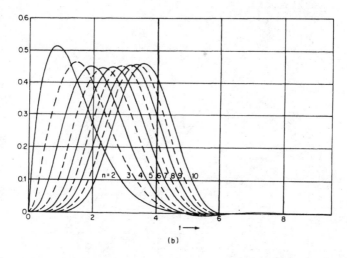

(b)

Figure 3.19 *continued*

than the Thompson shape. The method of locating poles requires successive approximation techniques, but a limited number of response characteristics are available in the references for different maximum passband ripple values ε. Figure 3.20 provides sets of curves for maximum ripple of 0.5°. The adjacent-channel attenuation is higher than for the Thompson shape, the delay shows small ripples, and the transient responses possess a small degree of ringing.

3.9.5 Transitional Filters

Because of the problems of getting both good attenuation and good transient response in a particular filter family, a number of schemes have been devised to achieve a compromise between different families of shapes. One such family presented in [3.8] attempts to maintain a gaussian attenuation shape until attenuation reaches a predetermined level, and drops off thereafter more rapidly than a gaussian or Thompson shape. Figure 3.21 shows amplitude responses for transitional filters that are gaussian to 6 dB. The transient properties are somewhat better than for the Butterworth filter, and the attenuation beyond the transition point is higher. The family that is gaussian to 12 dB has better transient properties, but its attenuation is somewhat poorer than that of Butterworth filters with the same number of poles.

The Butterworth-Thompson family is another transitional family aimed at addressing the same problem. In this case, the poles of the Butterworth shape for a particular n are joined by straight lines in the s plane to the corresponding poles of the Thompson shape. Different transitional families are formed by selecting new pole locations at a fixed fraction m of the distance along these straight lines. For

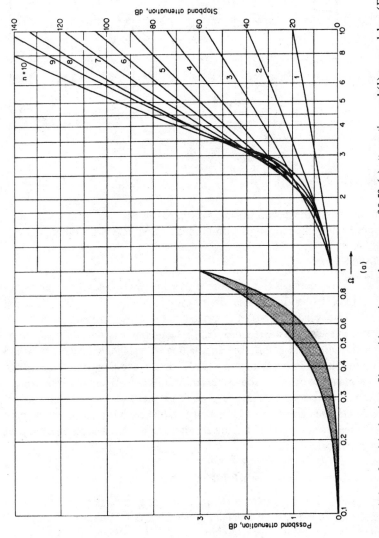

Figure 3.20 Characteristics of equiripple phase filters with maximum phase error of 0.5°: (*a*) attenuation and (*b*) group delay. (*From [3.8].* © *1967 John Wiley and Sons, Inc. Reprinted by permission of the publisher.*)

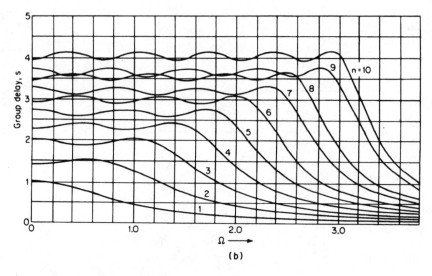

Figure 3.20 *continued*

the case where $m = 0$ the design is Butterworth, for $m = 1$, it is Thompson, and in between these values the properties shift gradually from one to the other.

3.9.6 Elliptic Filters

Elliptic filters, also known as Cauer filters, provide not only poles in the passband, but zeros in the stop band. The singularities are located so as to provide equal ripple (like a Chebyshev shape) in the passband but also to permit equiripple between zeros in the stop band (Figure 3.22). The presence of the stop-band zeros causes a much more rapid cutoff. The attenuation of distant channels is less than for the all-pole filters discussed previously. The phase and transient performance tend to resemble those of Chebyshev filters with similar numbers of poles, but are naturally somewhat poorer. Elliptic filters are used where it is essential to get adjacent-channel attenuation at the lowest cost and where it is unnecessary to pass digital signaling at rates approaching the channel bandwidth. They are also useful for broadband preselector filters.

3.9.7 Special Designs and Phase Equalization

Whenever possible, it is desirable for the receiver designer to select a filter that is a standard device (not a custom unit) and available from a number of suppliers. This results in the most economical design; it is usually possible to find a suitable compromise among the wide variety of filter families available. However, in special cases it may be necessary to incorporate a custom design. If the device can be specified in a form that defines the filter completely and is physically realizable, it is possible to build such a component, but usually at a high cost. One of the

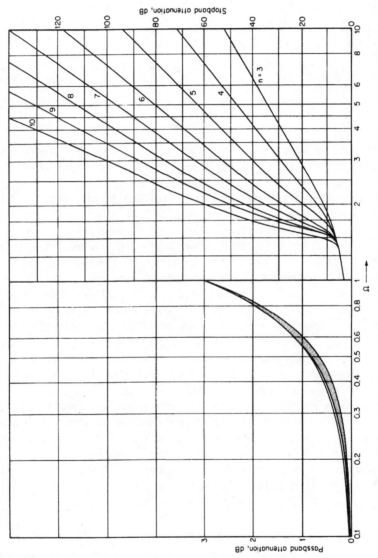

Figure 3.21 Attenuation characteristics of a transitional filter, gaussian to 6 dB. (*From [3.8]. © 1967 John Wiley and Sons, Inc. Reprinted by permission of the publisher.*)

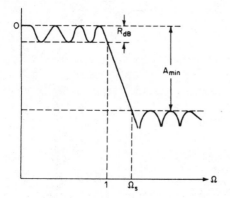

Figure 3.22 Elliptical filter typical amplitude response. (*From [3.9].* *Reprinted by permission of* *McGraw-Hill.*)

parameters that may be important is suppressing one or two specific frequencies outside of the pass band. This can be achieved by a filter that otherwise fits the need, with added zeros at the specific frequencies that must be suppressed. In the trade-off process, however, use of a separate filter (trap) should be considered to provide the zeros.

Another useful technique is phase equalization. Most of the filter families discussed in this chapter are minimum-phase filters (having no zeros in the right half of the s plane). The amplitude characteristics of these filters completely determine the delay characteristics and the transient responses. It was shown previously how sharp cutoff filters tend to produce substantial delay distortion and transient response with significant ringing. Conditions may arise where it is necessary to use a particular amplitude characteristic, but better delay properties are required. This can be achieved by phase equalization, using either a non-minimum phase design or an additional all-pass filter to provide the equalization. Phase equalization can improve both the delay characteristics and the transient response. Linearization of the phase tends to increase the overall delay of transmission, and provides precursive ringing as well as the postcursive ringing common in most of the shapes described. The tendency is for the amplitude of the ringing to be reduced by about half. For data transmission, intersymbol interference now occurs in prior symbols as well as in subsequent symbols. Fortunately, equalization is a problem that the designer must face only seldom; it is advisable to work with a filter expert when such requirements present themselves.

3.10 Filter Design Implementation

Conventional filter design techniques may be implemented using a number of different resonators. The principal available technologies are *inductor-capacitor* (LC) resonators, mechanical resonators, quartz crystal resonators, quartz mono-lithic filters, and ceramic resonators. The classical approach to radio filtering involved cascading single- or dual-resonator filters separated by amplifier stages.

Overall selectivity was provided by this combination of one- or two-pole filters. The disadvantages of this approach were alignment problems and the possibility of IM and overload even in the early IF stages from out-of-band signals. An advantage was that limiting from strong impulsive noise would occur in early stages where the broad bandwidth would reduce the noise energy more than after complete selectivity had been achieved. Another advantage was the relatively low cost of using a large number of essentially identical two-resonator devices. This approach has been largely displaced in modern high-quality radios by the use of multiresonator filters inserted as early as possible in the amplification chain to reduce nonlinear distortion, localize alignment and stability problems to a single assembly, and permit easy attainment of any of a variety of selectivity patterns. The simple single- or dual-resonator pairs are now used mainly for impedance matching between stages or to reduce noise between very broadband cascaded amplifiers.

3.10.1 LC Filters

LC resonators are limited to Q values on the order of a few hundred for reasonable sizes, and in most cases designers must be satisfied with lower Q values. The size of the structures depends strongly on the center frequency, which may range from the audio band to several hundred megahertz. Bandwidth below about 1 percent is not easily obtained. However, broader bandwidths may be obtained more easily than with other resonator types. Skirt selectivity depends on the number of resonators used; ultimate filter rejection can be made higher than 100 dB with careful design. The filter loss depends on the percentage bandwidth required and the resonator Q, and can be expected to be as high as 1 dB per resonator at the most narrow bandwidths. This type of filter does not generally suffer from nonlinearities unless the frequency is so low that very high permeability cores must be used. Frequency stability is limited by the individual components and cannot be expected to achieve much better than 0.1 percent of center frequency under extremes of temperature and aging. Except for front ends that require broad bandwidth filters, *LC* filters have been largely superseded in modern radios.

3.10.2 Electrical Resonators

As frequencies increase into the VHF region, the construction of inductors for use in *LC* resonant circuits becomes more difficult. The *helical resonator* is an effective alternative for the VHF and lower UHF ranges. This type of resonator looks like a shielded coil (see Figure 3.23a). However, it acts as a resonant transmission line section. High Q can be achieved in reasonable sizes (Figure 3.23b). When such resonators are used singly, coupling in and out may be achieved by a tap on the coil, a loop in the cavity near the grounded end (high magnetic field), or a probe near the ungrounded end (high electric field). The length of the coil is somewhat shorter than the predicted open-circuit quarter-wave line because of the end capacity to the shield. A separate adjustable screw or vane may be inserted

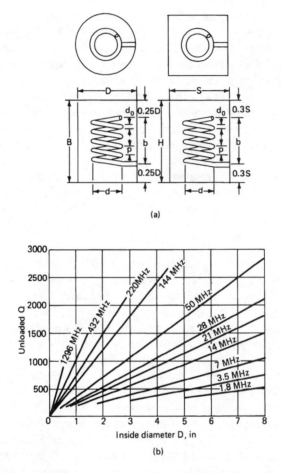

Figure 3.23 Helical resonators: (*a*) round and square shielded types, showing principal dimensions (diameter *D* or side *S* is determined by the desired unloaded *Q*) and (*b*) unloaded *Q* versus shield diameter *D* for bands from 1.8 MHz to 1.3 GHz. (*From [3.10].*)

near the open end of the coil to provide tuning. Multiresonator filters are designed using a cascade of similar resonators, with coupling between them. The coupling may be of the types mentioned previously or may be obtained by locating an aperture in the common shield between two adjacent resonators. At still higher frequencies, coaxial transmission line resonators or resonant cavities are used for filtering (mostly above 1 GHz). The use of *striplines* to provide filtering is another useful technique for these frequency regions.

A. Stripline technology

Stripline typically utilizes a double-sided printed circuit board made of fiberglass. The board is usually 30- to 50-thousandths of an inch thick. The board is uniform over the entire surface and forms an electrical ground plane for the circuit. This ground plane serves as a return for the electrical fields built up on the component side of the board.

The shape and length of each trace of stripline on the component side dictates the impedance and reactance of the trace. The impedance is a function of the width of the trace, its height above the lower surface ground plane, and the dielectric constant of the circuit board material. The length of the trace is another important factor. At microwave frequencies, a quarter-wavelength can be as short as 0.5-in in air. Because all printed circuit boards have a dielectric constant that is greater than the dielectric constant of air, waves are slowed as they travel through the board-trace combination. This effect causes the wavelength on a circuit board to be dependent on the dielectric constant of the material of which the board is made. At a board dielectric constant of 5 to 10 (common with the materials typically used in printed circuit boards), a wavelength may be up to 33 percent shorter than in air. The wider the trace, the lower the RF impedance.

Traces that supply bias and require operating or control voltages are usually made thin so as to present a high RF impedance while maintaining a low dc resistance. Narrow bias and control traces are usually made to be a multiple of a quarter-wavelength at the operating frequency so unwanted RF energy may be easily shunted to ground.

Figure 3.24 shows stripline technology serving several functions in a satellite-based communications system. The circuit includes the following elements:

- A 3-section, low-pass filter
- Quarter-wave line used as half of a transmit-receive switch
- Bias lines to supply operating voltages to a transistor
- Impedance-matching strip segments that convert a high impedance (130 Ω) to 50 Ω
- Coupling lines that connect two circuit sections at one impedance

A wide variety of techniques may be used to synthesize filter and coupling networks in stripline. After an initial choice of one of the components is made, however, only a small number of solutions are practical. While it is apparent that all components must be kept within reasonable physical limits, the most critical parameter is usually the length of the stripline trace. This technique is popular with equipment designers because of the following benefits:

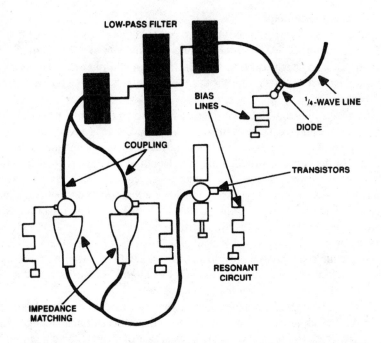

Figure 3.24 A typical application of stripline showing some of the components commonly used.

- **Low cost.** Stripline coupling networks are simply a part of the PC board layout. No assembly time is required during construction of the system.

- **Excellent repeatability.** Variations in dimensions, and therefore perform-ance, are virtually eliminated.

Stripline also has the following drawbacks:

- **Potential for and/or susceptibility to radiation.** Depending on the design, shielding of stripline sections may be necessary to prevent excessive RF emissions or to prevent emissions from nearby sources from coupling into the stripline filter.

- **Repair difficulties.** If a stripline section is damaged, it may be necessary to replace the entire PC board.

3.10.3 Electromechanical Filters

Most of the other resonators used in receiver design are electromechanical, where the resonance of acoustic waves is employed. During a period when quartz resonators were in limited supply, electromechanical filters were constructed from metals, using metal plates or cylinders as the resonant element and wires as the coupling elements. Filters can be machined from a single metal bar of the right diameter. This type of electromechanical filter is limited by the physical size of the resonators to center frequencies between about 60 and 600 kHz. Bandwidths can be obtained from a few tenths of a percent to a maximum of about 10 percent. A disadvantage of these filters is the loss encountered when coupling between the electrical and mechanical modes at input and output. This tends to result in losses of 6 dB or more. Also, spurious resonances can limit the ultimate out-of-band attenuation. Size and weight are somewhat lower, but are generally comparable to *LC* filters. Temperature and aging stability is about 10 times greater than for *LC* filters. Because of their limited frequency range, electromechanical filters have been largely superseded by quartz crystal filters, which have greater stability at comparable price.

3.10.4 Quartz Crystal Resonators

While other piezoelectric materials have been used for filter resonators, quartz crystals have proved most satisfactory. Filters are available from 5 kHz to 100 MHz, and bandwidths from less than 0.01 percent to about 1 percent. (The bandwidth, center frequency, and selectivity curve type are interrelated, so that manufacturers should be consulted as to the availability of specific designs.) Standard filter shapes are available, and with modern computer design techniques, it is possible to obtain custom shapes. Ultimate filter rejection can be greater than 100 dB. Input and output impedances are determined by input and output matching networks in the filters, and typically range from a few tens to a few thousands of ohms. Insertion loss varies from about 1 to 10 dB, depending on filter bandwidth and complexity. While individual crystal resonators have spurious modes, these tend not to overlap in multiresonator filters, so that high ultimate rejection is possible. Nonlinearities can occur in crystal resonators at high input levels, and at sufficiently high input the resonator may even shatter. Normally these problems should not be encountered in a receiver unless it is coupled very closely to a high-power transmitter. Even so, the active devices preceding the filter are likely to fail prior to destruction of the filter. Frequency accuracy can be maintained to about 0.001 percent, although this is relatively costly; somewhat less accuracy is often acceptable. Temperature stability of 0.005 percent is achievable.

3.10.5 Monolithic Crystal Filters

In monolithic crystal filter technology, a number of resonators are constructed on a single quartz substrate, using the *trapped-energy* concept. The principal energy of each resonator is apparently confined primarily to the region between plated electrodes, with a small amount of energy escaping to provide coupling. Usually these filters are constrained to about four resonators, but the filters can be cascaded using electrical coupling circuits if higher-order characteristics are required. The filters are available from 3 to more than 100 MHz, with characteristics generally similar to those of discrete crystal resonator filters, except that the bandwidth is limited to several tenths of a percent. The volume and weight are also much less than those of discrete resonator filters.

3.10.6 Ceramic Filters

Piezoelectric ceramics are also used for filter resonators, primarily to achieve lower cost than quartz. Such filters are comparable in size to monolithic quartz filters but are available over a limited center frequency range (100 to 700 kHz). The cutoff rate, stability, and accuracy are not as good as those of quartz, but are adequate for many applications. Selectivity designs available are more limited than for quartz filters. Bandwidths are 1 to 10 percent. Single- and double-resonator structures are manufactured, and multiple-resonator filters are available that use electrical coupling between sections.

3.10.7 *Resistor-Capacitor* (RC) Active Filters

The advent of high-stability integrated-circuit operational amplifiers, and especially circuits providing multiple operational amplifiers, has made the use of *RC* active filters attractive at low frequencies (up to about 100 kHz). For band-pass applications several approaches are possible [3.9]. For very wide band filters, the cascading of low-pass and high-pass sections may be used. Figure 3.25*a* shows a high-pass section that results in a third-order elliptic section with a cutoff around 300 Hz. Figure 3.25*b* corresponds to a fifth-order 0.5-dB Chebyshev low-pass filter with cutoff at approximately 900 Hz. Cascading these sections (with isolation in between to avoid the impedance changes inherent in direct connection) produces a filter with pass band of 300 to 900 Hz.

The circuits shown in Figure 3.26 produce a pair of poles (and a zero at the origin). By choosing the poles to be complex conjugate, the equivalent of a single resonator is achieved. By cascading such sections, each designed to provide an appropriate pole location, any of the all-pole band-pass filter shapes (Butterworth, Chebyshev, Bessel, etc.) may be achieved. The different configurations have different limitations, among these being the Q that is achievable. The circuits of Figure 3.27 produce greater Q, but with greater complexity. Figure 3.27*a* shows the Q-multiplier circuit, which uses an operational amplifier with feedback to increase the Q of a single-pole circuit. An example of its use with a low-Q *multiple-feedback band-pass* (MFBP) section is shown in Figure 3.27*b*. This technique can also be used to increase the Q of other resonators as long as the gain and phase shift characteristics of the operational amplifier are retained to the resonance frequency.

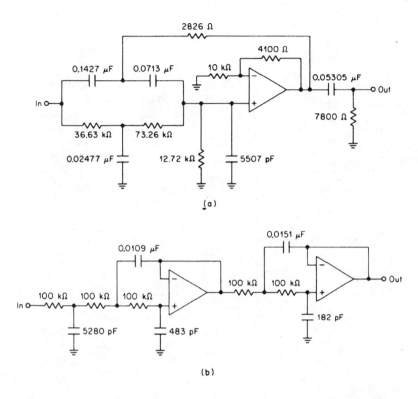

Figure 3.25 Active filter designs: (a) high-pass 3rd-order elliptical section and (b) low-pass 5th-order Chebyshev filter. (*From [3.9]. Reprinted by permission of McGraw-Hill.*)

If an elliptic filter (or other structure) with zeros as well as poles is required, one of the configurations shown in Figure 3.28 may be used to provide a section with conjugate pole and zero pairs, with the positive-frequency zero location either above or below the pole location. The *voltage-controlled voltage-source* (VCVS) configurations (Figure 3.28a and b) use one operational amplifier but many capacitors. The biquad circuit uses only two capacitors but four operational amplifiers. The parameter K of the VCVS circuit is determined from a relationship involving the pole Q and frequency and the zero frequency. Figure 3.29 shows a filter design for a band-pass filter using two VCVS sections and one MFBP section. The design is for a three-pole two-zero filter with a center frequency at 500 Hz, passband of 200 Hz, and stop-band rejection of 35 dB at ±375 Hz.

While active filter sections can be useful for low-frequency use, most low-frequency processing in newer receiver designs uses digital filters. These are discrete-time-sampled filters using sample quantizing. They may be implemented

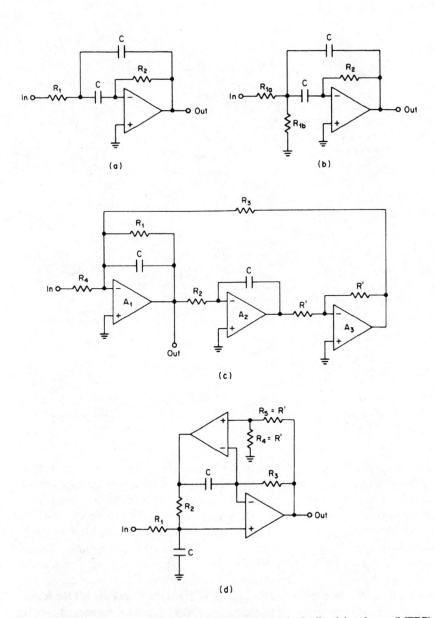

Figure 3.26 All-pole filter configurations: (*a*) basic *multiple-feedback band-pass* (MFBP) circuit (*Q* < 20), (*b*) modified MFBP section, (*c*) biquad circuit (*Q* < 200), and (*d*) dual-amplifier band-pass circuit (*Q* < 150). (*From [3.9]. Reprinted by permission of McGraw-Hill.*)

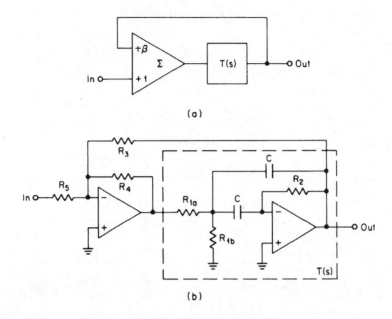

(a)

(b)

Figure 3.27 Q-multiplier circuit: (a) general block diagram and (b) realization using MFBP section. (*From [3.9]. Reprinted by permission of McGraw-Hill.*)

using digital circuits or using microprocessors to achieve the necessary signal processing.

3.11 Time-Sampled Filters

Many modern processing techniques use discrete-time samples of the incoming wave instead of the continuous as-received wave. The sampling theorem states that any band-limited waveform may be reproduced from samples taken at a rate which exceeds twice the highest frequency in the band. Figure 3.30 shows various waveforms and the spectra resulting from regular sampling of a band-limited signal. The sampling duration must be very short compared to the period of the highest frequency in the waveform and, ideally, would be instantaneous. In the time domain, sampling may be thought of as the product of the wave being sampled and a waveform of impulses occurring at the period of the sampling frequency. In the frequency domain, this results in the convolution of the two spectra. The spectrum of the sampling impulse train is a train of impulses in the frequency domain, separated by the sampling frequency. The convolution of this with the band-limited spectrum results in that spectrum being translated by each of the sampling impulses so that the end result is a group of band-limited spectra, all

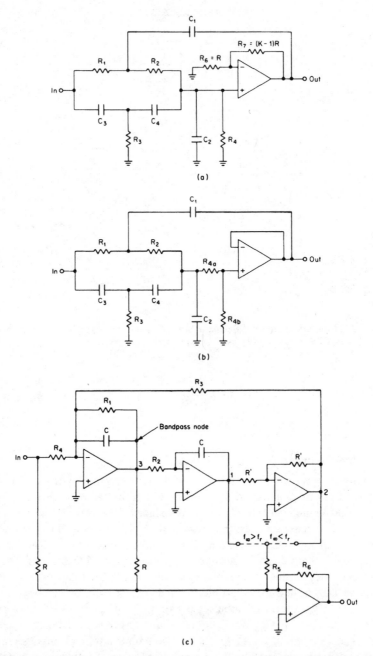

Figure 3.28 Elliptic-function band-pass sections: (*a*) *voltage-controlled voltage source* (VCVS) section for $K < 1$, (*b*) VCBS section for $K > 1$, and (*c*) biquad section. (*From [3.9]. Reprinted by permission of McGraw-Hill.*)

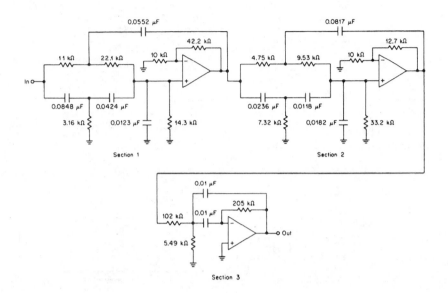

Figure 3.29 Example of an active band-pass filter design. (*From [3.9]. Reprinted by permission of McGraw-Hill.*)

having the same shape, but each displaced from the other by the sampling frequency.

Band-pass spectra do not necessarily have this type of symmetry about their center frequency. A band-pass waveform sampled at a rate that is greater than twice the width of the band-pass spectrum also results in a spectrum with translated replicas for every harmonic of the sampling frequency (also dc and the fundamental). The resulting spectra need have no symmetry about the sampling harmonic, and, indeed, the harmonics should not fall within the replicas. The translated nonzero positive- and negative-frequency replicas may be offset in such a way that the resulting spectra from the convolution have overlapping spectral components that, when summed, bear little resemblance to the original spectrum. Nevertheless it is possible to choose sampling frequencies so that the resulting positive- and negative-frequency convolutions result in symmetrical spectra about the harmonics. If this is done, the low-pass version is the real wave that would result from SSB demodulation of the original wave, and the remainder represents AM of the harmonic frequencies by this wave. This fact can be useful to reduce processing of band-pass signals by sampling them at rates greater than twice the bandwidth, rather than at rates greater than twice the highest frequency.

For processing band-pass waveforms as sampled waveforms, it is convenient to divide them into in-phase and quadrature components, each of which has a

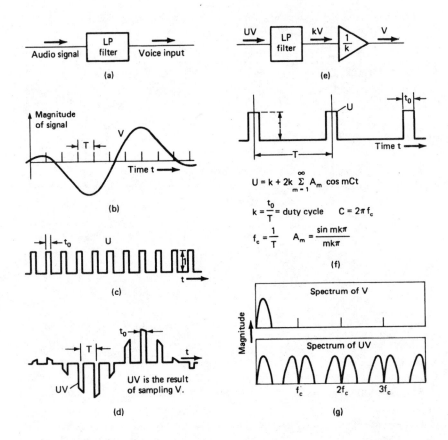

Figure 3.30 Sampled-signal time and spectrum relationships: (*a*) source of voice input, (*b*) typical voice input, (*c*) diagram of *U*, (*d*) diagram of *UV*, (*e*) passing *UV* through a low-pass filter and amplifier to obtain *V*, (*f*) enlarged diagram of unit sampling function *U*, and (*g*) spectrum analysis of *V* and *UV*. (*After [3.11]. Courtesy of Wadsworth Publishing Co.*)

symmetrical amplitude and odd symmetrical phase, and to process these separately. This can be achieved by using filters that separate out the two components into their Hilbert time complements and sampling each, or by sampling with two data streams offset by one-fourth of the carrier frequency of the band-pass spectrum and processing each separately. The same two data streams can be obtained by mixing the wave with sine and cosine waves at the carrier frequency to produce two low-pass waveforms and then sampling them.

The specific time-sampled filters we deal with in this section are essentially low-pass, although this technique can encompass low-frequency band-pass filters. We will assume that where higher-frequency band-pass filtering with such filters needs to be performed, separation into Hilbert components has been accomplished and two similar low-pass structures are used for the processing.

3.11.1   Discrete Fourier and z Transforms

When dealing with nonsampled signals, it is convenient to deal with the time representation, the Fourier transform, and the Laplace transform of the signal. The Fourier transform provides us with the frequency response characteristics of waveforms and permits definition of filters by their amplitude and phase charac-teristics versus frequency. The Laplace transform provides expressions for the response in the complex-frequency plane, and allows us to specify filters by their pole and zero locations. Filters may also be treated in the time domain by consid-ering their responses to the unit impulse and using the convolutional integral. When dealing with discrete-sampled signals we can use similar tools: the *discrete Fourier transform* (DFT), the z transform, and the convolution sum. For more details consult [3.12] through [3.14]. Table 3.1 compares these continuous and discrete tools.

The DFT differs from the continuous Fourier transform in obvious ways. The principal difference in the spectrum is that it is repeated periodically along the frequency axis with a period of $1/T$, the sampling frequency. Because of this, filter designs need consider only one strip in the complex-frequency (s) plane, for example, the area between the line $-\infty - j\pi/T$ to $\infty - j\pi/T$ and the line $-\infty + j\pi/T$ to $\infty + j\pi/T$. It is for this reason that the spectrum of the signal being sampled should be band-limited. Otherwise, after sampling, the periodic spectra would overlap and result in distortion of the signal in subsequent recovery. This type of distortion is known as *aliasing*. It is also for this reason that the z transform is used rather than the discrete Laplace transform in studying the location of singularities in the complex-frequency plane. The transformation $z = \exp(j2\pi fT)$, equivalent to $z = \exp(sT)$, maps all of the periodic strips in the s plane into the z plane. The mapping is such that the portion of the strip to the left of the imaginary axis in the s plane is mapped within the unit circle in the z plane, and the remainder is mapped outside the unit circle. The condition for the stability of circuits, that their poles be in the left half s plane, converts to the condition that poles in the z plane be within the unit circle.

3.11.2   Discrete-Time-Sampled Filters

Because the sampled values $f(nT)$ are each constant, a z transform of a realizable waveform is a polynomial in z. However, the definition of z is such that z^{-1} represents a delay of T in the time domain. These factors facilitate the design of discrete-time-sampled filters through the use of delay lines, and by extension to the design of filters using digital samples and memories to achieve the necessary delay. Figure 3.31 shows some of the configurations that can be used to create discrete-time-sampled filters. Each of the boxes labeled z^{-1} represents a delay of T. Additions are shown by circles about a plus sign, and multiplications are shown by

TABLE 3.1 Comparison of Tools for Continuous and Discrete Signal Processing

	Continuous	Discrete sampled
Time function	$f(t)$ $-\infty < t < \infty$	$f(nT)$ $-\infty < n < \infty$ n = integer; T = sampling period
Real frequency transforms	Fourier transform $F(j\omega) = \displaystyle\int_{-\infty}^{\infty} f(t)e^{-j\omega t}\,dt$ $f(t) = \dfrac{1}{2\pi}\displaystyle\int_{-\infty}^{\infty} F(j\omega)e^{j\omega t}\,d\omega$	Discrete Fourier transform $F(j\omega) = \displaystyle\sum_{n=-\infty}^{\infty} f(nT)e^{-j\omega nT}$ $f(nT) = \dfrac{T}{2\pi}\displaystyle\int_{-\pi/T}^{\pi/T} F(j\omega)e^{j\omega nT}\,d\omega$
Complex frequency plane	Laplace transform $F(s) = \displaystyle\int_{0}^{\infty} f(t)e^{-st}\,dt$ $f(t) = \dfrac{1}{2\pi j}\displaystyle\int_{c-j\infty}^{c+j\infty} F(s)e^{st}\,ds$ $c > 0$	z transform $F(z) = \displaystyle\sum_{n=-\infty}^{\infty} F(nT)z^{-n}$ $f(nT) = \dfrac{1}{2\pi j}\displaystyle\oint_{c2} F(z)z^{n-1}\,dz$ $z = e^{j\omega T}$
Time domain convolution	$g(t) = \displaystyle\int_{-\infty}^{\infty} F(\tau)h(t-\tau)\,d\tau$ For filters, $h(t)$ = impulse response	$g(nT) = \displaystyle\sum_{m=-\infty}^{\infty} f(mT)h(nT-mT)$ For filters, $h(nT)$ = impulse response
Hilbert transforms	$\hat{f}(t) = \dfrac{1}{\pi} P \displaystyle\int_{-\infty}^{\infty} \dfrac{f(\tau)}{t-\tau}\,dT$ $f(t) = -\dfrac{1}{\pi} P \displaystyle\int_{-\infty}^{\infty} \dfrac{\hat{f}(\tau)}{t-\tau}\,dT$	$\hat{f}(nT) = \dfrac{2}{\pi} \displaystyle\sum_{\substack{m=-\infty \\ m \neq n}}^{\infty} f(nT-mT)\,\dfrac{\sin^2(\pi m/2)}{m}$ $f(nT) = -\dfrac{2}{\pi} \displaystyle\sum_{\substack{m=-\infty \\ m \neq n}}^{\infty} \hat{f}(nT-mT)\,\dfrac{\sin^2(\pi m/2)}{m}$

$P\displaystyle\int$ indicates Cauchy's principal value

letters representing coefficients of the numerator and denominator polynomials of the z-domain transfer function.

The a_n values represent the numerator coefficients and jointly produce the zeros of the filters, and the b_n values represent the denominator coefficients (except for the zero-order coefficient, which is normalized to unity) and jointly produce the poles. Because the input samples $x(x)$ of these filters are fed back after various delays, a single impulse fed to the input continues to produce output indefinitely (but gradually attenuated if a stable configuration has been chosen). Such filters are referred to as *infinite impulse response* (IIR) types and correspond to common filters in the continuous time domain. If the various b_n are set equal to zero, the configuration of Figure 3.31c is obtained. This represents a filter with all zeros and no poles. The multipliers are equivalent to the time samples of the unit impulse response of the filter truncated at sample $N-1$. For this reason, this structure is known as a *finite impulse response* (FIR) filter. Many techniques have been devised for the design of such filters, as described in the references.

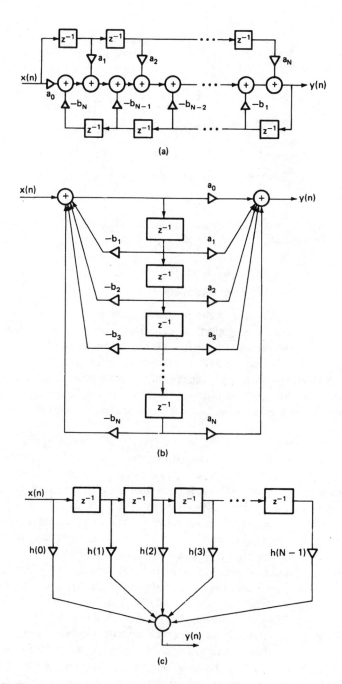

Figure 3.31 Time-sampled filter configurations: (*a*) direct form 1, (*b*) direct form 2, and (*c*) FIR direct form. (*After [3.14].*)

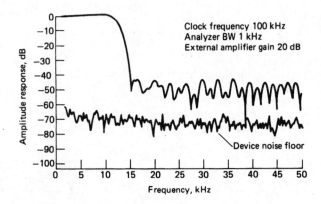

Figure 3.32 Narrow low-pass filter spectral response (Reticon R5602-1). (*After [3.15].* *Courtesy of Reticon Corp.*)

3.11.3 Analog-Sampled Filter Implementations

Time-sampled filters have been constructed by using low-pass filter structures to provide the delays or by using electrical delay line structures (helical transmission lines). However, the principal types that may occasionally be of use for receiver applications are SAW filters and filters using capacitor storage, often called *bucket brigade* devices. These are of considerable interest because it is possible to implement them using microelectronic integrated-circuit techniques. The structures may also be used for other purposes than filtering, where delayed output is useful.

In the bucket brigade types, the input voltage sample is used to charge the first of a series of small capacitors. Between sampling times, circuits are activated to transfer the charge from each capacitor to the next one along the line. Thus, the capacitors constitute a "bucket brigade delay line." By providing readout amplifiers at each stage and attenuators as appropriate, filters of any of the Figure 3.31 configurations can be made. These structures are of particular interest when implemented in integrated circuits. Figure 3.32 shows the response of a 64-stage device offered commercially [3.15] as a low-pass filter (using the Figure 3.31c configuration). Band-pass and chirped filters are also available. The structures can be made to have linear phase response, skirts with greater than 150 dB/ octave roll-off rate, and stop-band rejection of 50 dB. Sampling rates up to 1 MHz are offered, and filter characteristics may be scaled by changing the clock frequency.

SAWs may be excited on piezoelectric substrates. Delayed versions can be picked up from the substrate along the direction of propagation. These devices are not truly discrete-sampled components, because their outputs are a sum of attenuated continuous delayed waveforms. However, the configuration of the filters resembles that in Figure 3.31c, except for the continuity of the output. The

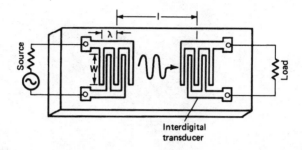

Figure 3.33 Mechanical configuration of a SAW filter using IDTs with uniform-length fingers. (*After [3.16].*)

advantage of SAWs is the reduction of propagation velocity in the acoustic medium and the consequent feasibility of constructing a long delay line in a small space. For example, in single-crystal lithium niobate, which is frequently used, acoustic waves are about five orders of magnitude slower than electromagnetic waves. Thus, a 1-µs delay can be achieved in 1/3 cm of this material, whereas it would require about 1000 ft of coaxial cable.

The surface waves are set up and received on a thin piezoelectric substrate using an *interdigital transducer* (IDT). This consists of thin interleaved conductors deposited on the substrate. Figure 3.33 shows the basic structure. For more information on the construction and theory of operation, consult [3.16] through [3.19]. SAWs are useful for filters over a frequency range from about 10 MHz to 3 GHz and provide bandwidths from about 1 to 25 percent or more of center frequency. Insertion loss is on the order of 10 dB below 500 MHz, but increases above this frequency. The IDT can be weighted by tapering the lengths of overlap of the fingers, as shown in Figure 3.34*a* [3.20]. The response of this filter is shown in Figure 3.34*b*. Had the lengths of the fingers not been tapered, the side lobes of the filter would have been higher, as indicated in the theoretical response shown in Figure 3.34*c*. As filter elements for radios, SAWs should be considered for frequencies above HF and for wide bandwidths, where other filter types are not available.

A. SAW application considerations

Many of the advantages of SAW devices are derived from their basic physical structure. They are inherently rugged and reliable, and—because their operating characteristics are established by photolithographic processes—they do not require tuning adjustment. SAW devices also provide excellent temperature stability. The temperature curve is parabolic and repeatable, from wafer to wafer, to within a few degrees Celsius.

Modern SAW filters use finite impulse response design techniques, similar to those applied to digital filters. The principal design tool is the Fourier transform,

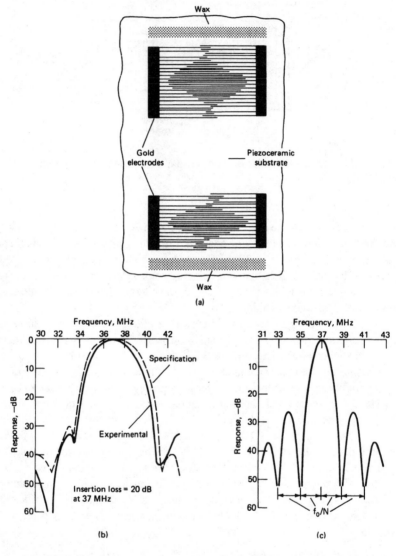

Figure 3.34 SAW filter: (*a*) construction with tapered IDT finger lengths, (*b*) selectivity curve, and (*c*) theoretical selectivity curve for uniform-length fingers. (*After [3.20].*)

which is used to relate the time and frequency responses of the transducers and resultant filter element. Although actual filter synthesis is a complex process, in general, the design engineer derives two impulse responses for the two transducers whose transforms can be combined (in dB) to produce the desired overall frequency response characteristic. These two impulse responses are then etched onto the surface of the metalized piezoelectric substrate.

The *fractional bandwidth* (passband divided by the center frequency) is usually the first and most important parameter to consider when specifying a SAW filter. The fractional bandwidth determines the substrate material, which establishes the temperature stability of the resulting component. Further, it limits the design options available because resonator filter technologies can rarely be used for fractional bandwidths greater than 0.3 percent, and low-loss filter techniques are rarely useful at fractional bandwidths greater than 10 percent. As the fractional bandwidth of a SAW filter increases, the number of interdigital electrodes on the surface of the substrate decreases, requiring higher coupling materials at wider fractional bandwidths.

SAW filters have been used for many years in the IF stages of communications receivers. These devices, however, tended to be physically large and exhibit relatively high insertion loss. Loss of 20 to 35 dB was not uncommon. Newer devices feature reduced insertion loss (2 to 10 dB is typical) and significantly reduced occupied volume. Examples of these devices include:

- *Coupled-resonator* filters
- *Proximity coupled resonator* filters
- *Single-phase unidirectional transducer* filters

The performance of the filter elements in a communications receiver are important no matter what the application. Digital systems, however, are particularly sensitive to filter shortcomings. For SAW devices, *group-delay* variations in the passband can result in pulse ringing on the desired symbol. The result is often *intersymbol interference* (ISI). *Group delay deviation* (GDD) limits on IF filters for digital communications systems, therefore, are significantly tighter than for analog radios.

3.12 Digital Processing Filters

Modern high-quality radio designs are increasingly turning to digital implementation for selectivity, demodulation, and signal processing. As A/D converters have become faster and more accurate, and as integrated digital circuits and microprocessors have become available at low cost, digital techniques have become attractive. Advantages of digital processing include the small size and low cost of the circuits, availability of many filter design techniques, ease of changing filter characteristics under computer control, and absence of costly alignment procedures. Digital processing has progressed from audio circuits into modems, IF filters, and front-ends, leading to a number of commercially-available "all-digital" receivers. Limitations on A/D accuracy and noise, and the speed of such circuits have been the principle design challenges to be overcome in all-digital receivers. Progress, however, continues to be made.

Digital filters are based on discrete sampling concepts, with the added characteristic that the samples are digitized using A/D converters. The A/D converter changes each sample into a group of binary samples that may be stored and on which digital delay, multiplication, and addition can be carried out. Filter circuits may be designed as described for sampled filters in general. Digital storage and processing can easily implement the configurations of Figure 3.31. Because the coefficients and interconnectivities can be stored in digital memories, the filter configurations may be changed readily. The principal disadvantages are limitations on accuracy and stability resulting from the quantization of the samples and coefficients and the rate at which the processes must be carried out (number of operations per second). The latter determines whether a particular process can be performed within a microprocessor or whether separate additional digital circuit elements are required.

A number of basic techniques for the design of digital filters have been devised ([3.13], [3.14], and [3.21]). Three general concepts are used:

- Frequency-domain design
- Complex-frequency-domain singularity selection
- Impulse response (time-domain) design

Because of the close interrelationship between the sampled time series and the z transform, these methods may result in similar filter structures. Frequency-domain designs can use a DFT to convert from the time to the frequency domain, and then modify the resulting spectrum in accordance with a desired frequency-domain response. The resultant can then be restored to the time domain. To provide a continuing time series output, it is necessary to use a "pipeline" DFT process.

An alternative is to use a finite *fast Fourier transform* (FFT) with time padding. The FFT operates on a finite number of time samples but provides conversion between time and frequency domains much more rapidly than a direct DFT performing the same task. Because the number of samples is finite, the conversion back to the time domain is based on a periodic wave, introducing errors in the first and last samples of the group. By padding the input series with sufficient zeros before the first real sample and after the final sample, this influence can be eliminated, and successive groups can be displaced and added to provide a continuous time-sampled output after the filtering process. Where possible, this process requires less digital processing than the continual pipeline DFT conversions.

Using the DFT process, it is also possible to convert the frequency-domain coefficients back to the time domain and then design a filter of the IIR type, depending on the resulting coefficients. This same approach may be used in dealing with the location of singularities in the complex z domain. Once the poles and zeros are located, they can be converted into a ratio of polynomials in z, from which the various filter coefficients can be determined.

The use of the impulse response to define the filter is especially useful for an FIR design (Figure 3.31c) because the coefficients are the samples of this response. The filter differs slightly from an analog filter upon which it might be based because of the finite duration of the impulse. In this case, however, initial requirements should be based on a finite impulse response. This is especially valuable when considering the ISI of digital signals. FIR filters possess a number of advantages. Such filters are always stable and realizable. They can be designed easily to have linear phase characteristics. Round-off noise from the finite quantization can be made small, and these filters are always realizable in the delay line form of Figure 3.31c. Disadvantages of this configuration include the requirement for a large amount of processing when a sharp filter approximation is required, and a linear phase FIR filter may not be compatible with an integral number of samples. Figure 3.35 shows example frequency responses of low-pass FIR filters. Figure 3.36 shows an example of a band-pass filter.

IIR filters cannot have linear phase if they are to satisfy the physical realizability criterion (no response before input). In order to be stable, their poles must lie within the unit circle in the z plane. They are more sensitive to quantization error than FIR filters and may become unstable from it. Generally, IIR filters require less processing for a given filter characteristic than FIR filters. These filters are implemented by the configurations of Figure 3.31a and b. IIR filters can be designed by partial fraction expansion of their z transforms, leading to the use of individual sections involving one or two delays to implement a single pole or a pair of complex pairs of poles, respectively. IIR filters may be designed directly in the z plane, and optimization techniques are available. However, they may also be designed by transformation of analog filter designs. Because the s plane and the z plane do not correspond directly, the transformation cannot maintain all properties of the analog filter. The following four procedures are widely used:

- **Mapping of differentials.** In this technique, the differentials that appear in the differential equations of the analog filter are replaced by finite differences (separated by T). Rational transfer functions in s become rational transfer functions in z. However, filter characteristics are not well preserved.

- **Impulse invariant transformation.** This technique preserves the impulse response by making the samples equal to the continuous response at the same time. There are z-transform equivalents to single-pole and dual complex-conjugate-pole s-plane expressions. The s-plane expression is broken into partial fractions and replaced by the equivalent z-plane partial fractions. The frequency response of the original filter is not preserved.

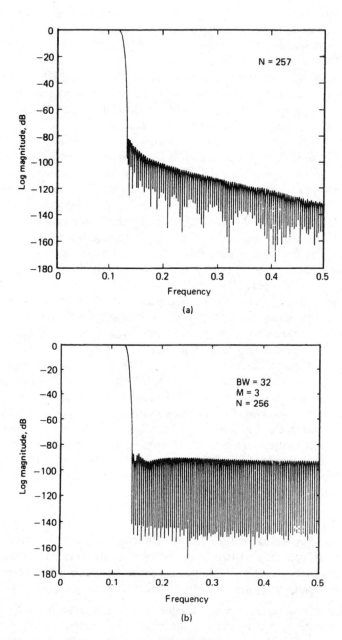

Figure 3.35 Frequency response of low-pass FIR digital filter designs, (*a*) Kaiser windowing design, (*b*) frequency sampling design, and (*c*) optimal (*minimax*) design. (*After [3.12].* *Reprinted by permission of Prentice-Hall, Inc.*)

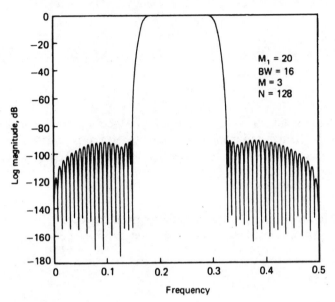

Figure 3.35 *continued*

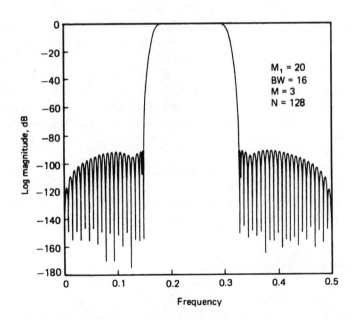

Figure 3.36 Frequency response of a band-pass digital FIR filter using a frequency sampling design. (*After [3.14]. Reprinted by permission of Prentice-Hall, Inc.*)

- **Bilinear transformation.** This technique uses the transformation:

$$s = \frac{2(1 - z^{-1})}{T(1 + z^{-1})}$$

Substitution in the s transfer function yields a z transform to implement the filter. The transformation can be compensated to provide a similar amplitude versus frequency response, but neither phase response nor impulse response of the analog filter is preserved.

- **Matched z transformation.** This technique replaces poles or zeros in the s plane by those in the z plane using the transformation

$$(s + a) = 1 - z^{-1} \exp(-a\,T)$$

The poles are the same as those that result from the impulse invariant response. The zeros, however, differ. In general, use of the matched z transform is preferred over the bilinear transformation.

All of these techniques are useful to simulate an analog filter. When the filter has a sufficiently narrow band, the approximations can be reasonably good. Further, they allow use of the families of filter characteristics that have been devised for analog filters. However, as a general rule for new filter design, it would seem better to commence design in the z plane.

Because of the large number of design types and the difficulty of formulating universal criteria, it is difficult to compare FIR and IIR filters. In a specific case several designs meeting the same end performance objectives should be tried to determine which is easiest and least expensive to produce.

3.13 Frequency Tracking

For many years general-purpose receivers were tuned over their bands using a mechanical control that simultaneously varied the parameters of the antenna coupling, RF interstage, and oscillator resonant circuits. The most common form was the ganged capacitor, which used the rotation of a single shaft to vary the tuning capacitance in each of the stages. Except for home entertainment receivers, which were produced in large numbers, the receivers employed capacitors with

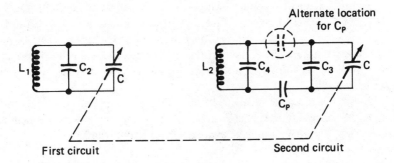

Figure 3.37 Circuits for simple tracking analysis.

essentially identical sections. Because the antenna and interstage circuits often used coils with coupled primaries, each one often different, resonant above or below the frequency band, the effective inductance of the coils varied slightly with the tuning frequency. Also, the LO had a fixed offset equal to the IF either above or below the desired frequency, so the band ratio covered by the LO circuit differed from that covered by the RF and antenna circuits. As a result it was necessary to devise circuits that could tune to the desired frequency with minimum error using the same setting of the variable capacitor element. This was necessary to reduce tracking losses caused by the RF or antenna circuits being off tune, and thus reducing gain. The result was a variation of sensitivity, and of the image and IF rejection ratios.

The modern receiver with synthesizer and up converter can avoid the tracking problem by replacing the variably tuned circuits with switched broad-band filters. However, there are still designs where down converters may be needed and RF tuning may become essential. Therefore, the designer should be aware of the techniques for minimizing losses in tracked circuits. Except in unusual cases, the modern tuning elements are not likely to be mechanical but rather electrically tuned varactors, so that frequency change can be effected rapidly, under computer control. In this discussion, the variable element will be assumed to be a capacitor, but analogous results could be obtained using variable inductor elements and changing the circuits appropriately.

Figure 3.37 shows a simple tuned circuit and a tuned circuit with a fixed series padding capacitor, both tuned by the same value of variable capacitance C. The second circuit covers a smaller range of frequencies than the former, and the frequency of tuning for a given setting of C can be modified by changing C_p and C_3 and/or C_4. Because it was more common for the LO frequency to be above the RF, the second circuit is often referred to as the *oscillator circuit* in the literature. However, this type of circuit may be used whenever one circuit of a group covers a more limited range than another, whether it is used for oscillator, RF, or antenna coupling. If C_p is very large or infinite (direct connection), it is possible to track the circuits at only two frequencies. This can be done by adjusting the total shunt capacitance at the higher frequency and the inductance value at the lower frequency. Such a procedure is known as *two-point tracking* and is useful if the ratio of band coverage is small.

In the superheterodyne receiver, the tuned frequency is determined by that of the LO. The actual difference between the LO frequency and the RF frequency for the circuits of Figure 3.37 may be expressed by the following:

$$4\,\pi^2 f_1^2 = \frac{1}{L_1\,(C_2 + C)} \tag{3.39a}$$

$$4\,\pi^2 f_2^2 = \frac{1}{L_2\,[C_4 + C_p\,(C_3 + C)/(C_p + C_3 + C)]} \tag{3.39b}$$

$$4\,\pi^2 f_2^2 = \frac{1}{L_2\,(C_4 + C_3 + C)} \quad C_p = \infty \tag{3.39c}$$

The frequency difference is $f_2 - f_1$. The points of tracking f_{m1} and f_{m2}, where the difference is zero, will be chosen near the ends of the band. In one form of idealized two-point tracking curve, the points f_{m1} and f_{m2} are selected to produce equal tracking error at the top, bottom, and center of the band. For an assumed quadratic variation, the values of tracking frequency that produce this condition are [3.22]:

$$f_{m1} = 0.854 f_a + 0.146 f_b \tag{3.40a}$$

$$f_{m2} = 0.146 f_b - 0.854 f_a \tag{3.40b}$$

where f_b and f_a are the upper- and lower-end frequencies, respectively.

Generally C has been determined in advance and C_p is infinite or selected. The values of f_a and f_b determine the end points for the oscillator circuit, and f_{m1} and f_{m2} the tracking points for the RF circuit. If these values are substituted in Equation (3.39), relationships among the component values are established that produce the desired tracking curve. As pointed out, only two-point tracking is used when f_a/f_b

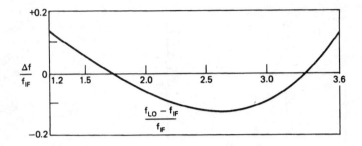

Figure 3.38 An example of two-point tracking for wide bandwidth.

is relatively small (less than 1.5 in most cases). To illustrate the disadvantage of using two-point tracking for wider bands, Figure 3.38 shows a result for wide-band coverage (550 to 1650 kHz, or 3.0 ratio). For a 456 kHz IF, the maximum tracking error is about 60 kHz. If the RF circuit Q were 50, this would represent a tracking loss of 14 dB at a tuning frequency of 1200 kHz. To improve this condition, three-point tracking is used.

In three-point tracking we attempt to select the circuit values so that there is zero tracking error at three points within the tuning range. The values f_a and f_b are established by the band ends of the oscillator circuit. The three tracking frequencies f_1, f_2, and f_3 are chosen to minimize tracking error over the band. As a result, there are five relationships in Equation (3.39) to be fulfilled among the six selectable components. Because negative values of inductance and capacitance are not realizable in passive components, these relationships do not allow completely arbitrary selection of any one of the components. However, some of the components may be arbitrarily selected and the others determined. The results of such calculations under various conditions are summarized in Table 3.2 [3.23].

The various symbols used in Table 3.2 are given here. All frequencies are expressed in megahertz, all inductances in microhenries, and all capacitances in picofarads. The tracking frequencies are f_1, f_2, and f_3. The IF is f_{IF}.

$$a = f_1 + f_2 + f_3$$

$$b^2 = f_1 f_2 + f_1 f_3 + f_2 f_3$$

$$c^3 = f_1 f_2 f_3$$

TABLE 3.2 Summary of Tracking Component Values for Various Conditions (*From [3.23]. Courtesy of Donald S. Bond.*)

Condition	C_P	C_3	C_4	L_2
$C_4 = 0$ or $C_4 \ll C_P$ (usual case)	$C_0 f_0^2\left(\dfrac{1}{n^2} - \dfrac{1}{l^2}\right)$	$\dfrac{C_0 f_0^2}{l^2}$	0	$L_1\dfrac{l^2}{m^2}\left(\dfrac{C_P + C_3}{C_P}\right)$
$C_3 = 0$	$\dfrac{C_0 f_0^2}{n^2}$	0	$\dfrac{C_0 f_0^2}{l^2 - n^2}$	$L_1\dfrac{l^2}{m^2}\left(\dfrac{C_P}{C_P + C_4}\right)$
C_4 known	$A\left(\dfrac{1}{2} + \sqrt{\dfrac{1}{4} + \dfrac{C_4}{A}}\right)$	$\dfrac{C_0 f_0^2}{l^2} - \dfrac{C_P C_4}{C_P + C_4}$	C_4	$L_1\dfrac{l^2}{m^2}\left(\dfrac{C_P + C_3}{C_P + C_4}\right)$
C_3 known	$\dfrac{C_0 f_0^2}{n^2} - C_3$	C_3	$\dfrac{C_P B}{C_P - B}$	$L_1\dfrac{l^2}{m^2}\left(\dfrac{C_P + C_3}{C_P + C_4}\right)$

$$d = a + 2f_{IF}$$

$$l^2 = \frac{b^2 d - c^3}{2f_{IF}}$$

$$m^2 = l^2 + f_{IF}^2 + ad - b^2$$

$$n^2 = \frac{c^3 d + f_{IF}^2 \, l^2}{m^2}$$

C_o = total tuning capacitance of the first circuit, at any frequency f_o

$$L_1 = \frac{25{,}330}{C_o f_o^2}$$

$$A = C_o f_o^2 \, (1/n^2 - 1/l^2)$$

$$B = \frac{C_o f_o^2}{l^2} - C_3$$

The tracking frequencies in three-point tracking may be selected to produce equal tracking error at the ends of the band and at the two intermediate extrema. Under these circumstances [3.22],

$$f_1 = 0.933 f_a + 0.067 f_b \tag{3.41a}$$

$$f_2 = 0.5 f_a + 0.5 f_b \tag{3.41b}$$

$$f_3 = 0.067 f_a + 0.933 f_b \tag{3.41c}$$

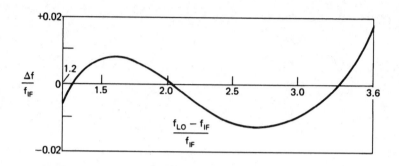

Figure 3.39 An idealized three-point tracking curve.

This results in an idealized cubic curve. For the same example as before, this approach reduces the maximum tracking error to about 5.5 kHz, which for a circuit Q of 50 results in about 0.83-dB tracking loss at 1200 kHz, a considerable improvement over two-point tracking.

The equal tracking error design results in equal tracking loss at each of the maximum error points if the circuit Q increases directly with frequency. This is not necessarily the case. If we examine the case where the Q is constant over the band, we find that the errors in $2\Delta f/f$ rather than Δf should be equal. This produces Equation (3.42) and Figure 3.39 [3.23]:

$$f_1 = 0.98f_a + 0.03f_b \tag{3.42a}$$

$$f_2 = 0.90f_a + 0.27f_b \tag{3.42b}$$

$$f_3 = 0.24f_a + 0.84f_b \tag{3.42c}$$

These values are plotted in Figure 3.40, where $t_1 = f_b/f_a$. The frequencies of error extrema are f_a, f_b, and

$$f_{\Delta1} = 0.93f_a + 0.13f_b \tag{3.43a}$$

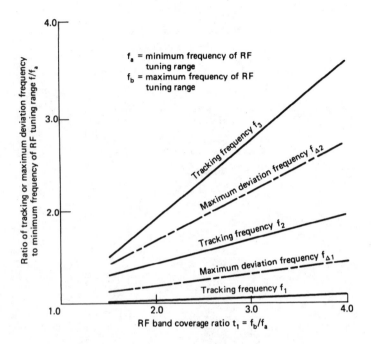

Figure 3.40 Three-point tracking of critical frequencies. (*After [3.23].*)

$$f_{\Delta 2} = 0.64 f_a + 0.53 f_b \tag{3.43b}$$

These values are also plotted in Figure 3.40.

An example of this tracking approach is shown in Figure 3.41, which shows the RF and LO circuits for a 52 to 63 MHz band VHF receiver tuned by varying the voltage of capacitive varactors. The same tuning voltage is used to tune LO and RF amplifiers. In this case we are dealing with a down converter to an IF of 11.5 MHz, and the oscillator is tuned below the RF in this band.

3.14 IF and Image Frequency Rejection

Tuned RF circuits have been devised to increase the rejection of IF or image frequencies to a greater extent than a simple resonator could achieve by itself. The techniques used include locating primary resonances and couplings to provide poorer transfer at the IF or image frequency and tapping the secondary circuit. The technique of using "traps" (transmission zeros) at the IF or image frequency can be used not only with tunable circuits, but with broader band-switched circuits.

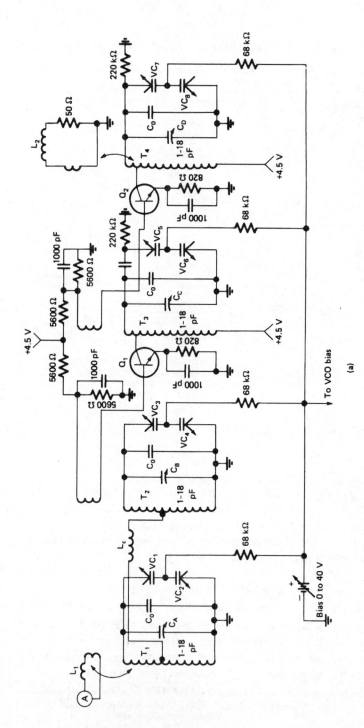

Figure 3.41 Tracked electrically tuned circuits for a VHF-band receiver: (*a*) RF tuner and (*b*) oscillator.

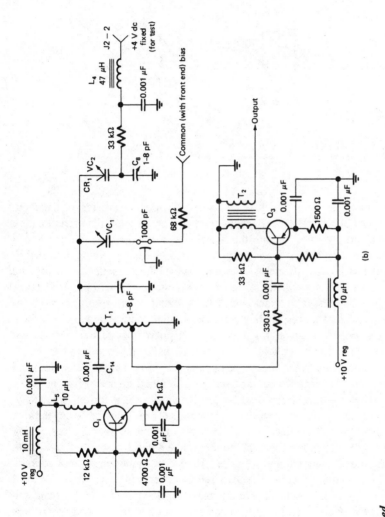

Figure 3.41 *continued*

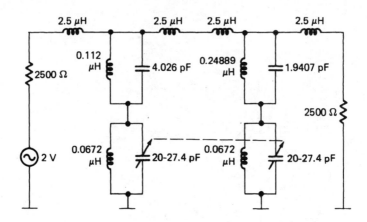

Figure 3.42 Circuit with tracking poles of attenuation.

It is frequently desirable to design receivers for wide dynamic range. The input filters for such receivers need substantial image or local oscillator reradiation suppression, but may not tolerate the losses that a multiresonator filter would produce. A clever technique for such suppression is shown in Figure 3.42 [3.24]. Two coupled tuned circuits are used to tune to the signal frequency. Additional parallel circuits, in series with these, generate two trapping poles, one at the image frequency and the other at the local oscillator frequency. As a result, up to 80-dB image and oscillator suppression relative to the input port are obtainable, as shown in the simulation results given in Figure 3.43. This particular design is for a filter tuning from 117 to 137 MHz, with a 10.7 MHz IF and a high-side local oscillator. Above the tuning frequency, the lower circuits have a capacitive reactance. The resonant frequencies of the upper circuits are such that they present an inductive reactance. The resulting series resonance attenuates the undesired frequency. Proper choice of the rejection circuit inductance and the resonant frequency permits two-point tracking of the rejection frequency over the tuning range. The tracking error is dependent upon the fractional tuning range and the amount of separation of the undesired frequency from the desired frequency. Analogous techniques may be used for series-tuned circuits. The approach is of use for high- and low-side local oscillators when the IF is below the tuning band. For IF above the tuning band, the rejection frequencies are sufficiently separated that fixed-tuned band-pass filters are generally used, and traps are not usually necessary.

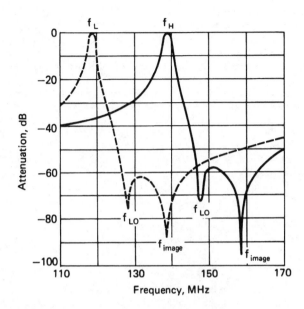

Figure 3.43 Plot of attenuation for the circuit shown in Figure 3.42.

3.15 References

3.1. H. A. Haus et al., "Representation of Noise in Linear Twoports (IRE Subcommittee 7.9 on noise)," *Proc. IRE*, vol. 48, p. 69, Jan. 1960.

3.2. H. T. Friis, "Noise Figures of Radio Receivers," *Proc. IRE*, vol. 32, p. 419, July 1944.

3.3. D. H. Westwood, "D-H Traces, Design Handbook for Frequency Mixing and Spurious Analysis," Internal RCA Rep., Feb. 1969.

3.4. "Authorization of Spread Spectrum and Other Wideband Emissions not Presently Provided for in the FCC Rules and Regulations," General Docket 81-413 of the FCC.

3.5. P. C. Gardiner and J. E. Maynard, "Aids in the Design of Intermediate Frequency Systems," *Proc. IRE*, vol. 32, p. 674, Nov. 1944.

3.6. T. E. Shea, *Transmission Networks and Wave Filters*, Van Nostrand, New York, 1929.

3.7. E. A. Guillemin, *Communication Networks*, vol. 11, *The Classical Theory of Long Lines, Filters and Related Networks*, Wiley, New York, 1935.

3.8. A. I. Zverov, *Handbook of Filter Synthesis*, Wiley, New York, 1967.

3.9. A. B. Williams, *Electronic Filter Design Handbook*, McGraw-Hill, New York, 1981.

3.10. J. R. Fisk, "Helical Resonator Design Techniques," *QST*, July 1976.

3.11. H. S. Black, *Modulation Theory*, Van Nostrand, New York, 1953.

3.12. H. Freeman, *Discrete Time Systems*, Wiley, New York, 1965.

3.13. B. Gold and C. M. Rader, *Digital Processing of Signals*, McGraw Hill, New York, 1969.

3.14. L. R. Rabiner and B. Gold, *Theory and Application of Digital Signal Processing*, Prentice-Hall, Englewood Cliffs, NJ, 1975.

3.15. "Analog Product Summary, Discrete Time Analog Signal Processing Devices," Reticon Corp., Sunnyvale, CA, 1982.

3.16. L. R. Adkins et al., "Surface Acoustic Wave Device Applications and Signal Routing Techniques for VHF and UHF," *Microwave J.*, p. 87, Mar. 1979.
3.17. *IEEE Trans. Microwave Theory and Techniques* (special issue), vol. MTT-17, Nov. 1969.
3.18. J. S. Schoenwald, "Surface Acoustic Waves for the RF Design Engineer," *RF Design*, p. 25, Mar./Apr. 1981.
3.19. H. Matthews, Ed., *Surface Wave Filters*, Wiley-Interscience, New York, 1977.
3.20. R. F. Mitchell et al., "Surface Wave Filters," *Mullard Tech. Commun.*, no. 108, p. 179, Nov. 1970.
3.21. A. V. Oppenheim and R. W. Schaeffer, *Digital Signal Processing*, Prentice-Hall, Englewood Cliffs, NJ, 1975.
3.22. K. R. Sturley, *Radio Receiver Design*, Pt. I: *Radio Frequency Amplification and Detection*, Wiley, New York, 1942.
3.23. D. S. Bond, "Radio Receiver Design," lecture notes of course given in E. S. M. W. T. Program (1942-43), Moore School of Electrical Engineering, University of Pennsylvania, Philadelphia, 2d ed. (1942).
3.24. W. Pohlman, "Variable Band-Pass Filters with Tracking Poles of Attenuation," *Rohde & Schwarz Mitt.* no. 20, p. 224, Nov. 1966.

Additional Suggested Reading

"Communication Receivers for the Year 2000: Part I," *Ham Radio* magazine, Nov. 1981.
"Communication Receivers for the Year 2000: Part II," *Ham Radio* magazine, Dec. 1981.
"Communication Receivers for the Year 2000: Part III," *Ham Radio* magazine, Nov. 1981.
"Communications Receivers Pace Electronics," *Microwaves & RF*, March 1987.
"Crystal Filter Design with Small Computers," *QST*, May 1981.
"Differences Between Tube-Based and Solid State-Based Receiver Systems and Their Evaluation Using CAD," *QEX*, ARRL Experimenter's Exchange, Apr. 1993.
"Eight Ways to Better Radio Receiver Design," *Electronics*, Feb. 20, 1975.
"High Dynamic Range Receiver Input Stages," *Ham Radio* magazine, Oct. 1975.
"How Many Signals Does a Receiver See?," *Ham Radio* magazine, p. 58, June 1977.
"Improved Noise Modeling of GaAs FETS—Using an Enhanced Equivalent Circuit Technique," *Microwave Journal*, p. 87-101, Nov. 1991, and p. 87-95, Dec. 1991.
"Key Components of Modern Receiver Design," *QST*, Mar. and Apr. 1994.
"Recent Advances in Shortwave Receiver Design," *QST*, p. 45, Nov. 1992.
"Recent Developments in Communication Receiver Design to Increase the Dynamic Range," ELECTRO/80, Boston, May 1980.
"Recent Developments in Shortwave Communication Receiver Circuits," National Telecommunications Conference, IEEE, Nov. 1979.
"Testing and Calculating Intermodulation Distortion in Receivers, *QEX*, p.3, Jul. 1994.
"The Design of Wide-Band Amplifier with Large Dynamic Range and Low Noise Figure Using CAD Tools," *IEEE Long Island MTT Symposium Digest*, Crest Hollow Country Club, Woodbury, NY, Apr. 28, 1987.

Chapter 4
Antennas and Antenna Coupling

4.1 Introduction

Selecting the antenna coupling circuit to optimize the NF can be one of the most important choices in receiver design. Over the frequency range with which we are concerned, the wavelength can vary from many miles at the lower frequencies to about 1 ft at 1000 MHz. Antenna efficiency, impedance, bandwidth, and pattern gain are all functions of the relationship of the antenna dimensions relative to the wavelength. Also, the level of atmospheric noise increases directly with the wavelength. To a lesser extent, the excess noise of the input device (amplifier or mixer) tends to increase with frequency.

The antenna may be located some distance from the receiver and connected through a considerable length of transmission line, or it may be a small device mounted directly on the receiver or transceiver housing. Many point-to-point applications are of the former type, because good antenna location can improve reception and reduce the required transmitter power. Receivers that are hand-held or backpacked require integral antenna structures, and vehicular receivers must use antennas of limited size and relatively short lengths of transmission line. For many applications, the impedance characteristics of the receiver antenna are specified by the system designer on the basis of the type of antenna system planned. In such cases, sensitivity is measured using a dummy antenna with these characteristics. In hand-carried applications, the type and size of antenna are often specified, but the implementation is left to the designer. Our discussions of antennas in this chapter deal primarily with such small antennas (those much smaller than a wavelength). Such antennas are characterized by high reactances and low to moderate resistances.

Large antennas are used mainly in point-to-point service and are designed with matching networks to feed their transmission lines. The receiver designer is usually required to work with an impedance that is primarily resistive over the passband. The reactive portion of the impedance seldom becomes larger than the resistive portion. Consequently, in such cases the receiver is usually designed for a resistive dummy antenna. The most common values are 50 or 70 Ω, reflecting common transmission line characteristic impedances. For more detailed discussions of different antenna types, there are many published sources; some representative ones being [4.1] through [4.4].

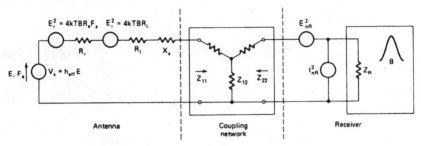

Figure 4.1 The equivalent circuit of an antenna at the receiver terminals.

4.2 Antenna Coupling Network

The function of the antenna coupling network in a receiver is to provide as good a sensitivity as possible, given the remainder of the receiver design. This is different from the antenna coupler role in a transmitter where the objective is to produce maximum radiated power. As indicated in Chapter 3, the condition for maximum power transfer (matched impedances) is not necessarily the condition for minimum NF. At the higher frequencies, where atmospheric noise is low, the antenna may be represented by the equivalent circuit shown in Figure 4.1. The coupling network connects between this circuit and the receiver input device, as shown in Figure 4.2. For the receiver input, the noise model of Figure 3.2b has been used.

The coupling network should have as low a loss as possible, because losses increase the NF. The combination of network and antenna can be represented, using Thévenin's theorem, as a series generator and impedance, feeding the input circuit. If we assume that the coupling is lossless, the contributions of R_a and R_L to the output impedance will be in the same ratio as in the antenna. If we let the impedance transformation be represented by m^2, then the equivalent generator can be represented as mE_a because the available power is unchanged by the lossless network. In Figure 4.2, no output reactance has been shown because, in principle, it can be eliminated by the network. Examining the total noise across the noiseless input impedance of the device and the noise from the antenna alone, at the same point, we find:

$$F = 1 + \frac{E_n^{\,2} + (I_n R_T)^2}{4\,k\,T\,R_T\,B} \tag{4.1}$$

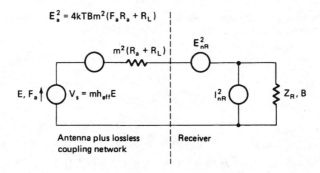

Figure 4.2 The equivalent circuit of an antenna, coupling network, and the receiver input.

where $R_T = m^2 (R_a + R_L)$.

The quantity R_T can be varied by changing m (the network). By differentiating with regard to R_T, we see that F is a minimum when R_T has been adjusted to E_n/I_n and has a value of $1 + E_n I_n / 2kTB$. If there are losses in the coupling network, an optimum NF can be achieved by adjusting the network output impedance to the value E_n/I_n and the input impedance to provide the minimum coupling network loss. When the losses become high, it may prove that a better overall NF is achieved at some value of R_T other than optimum for the input device, if the coupling loss can thereby be reduced.

Note that the antenna itself has an NF greater than unity. The noise resulting from the thermal radiation received by the antenna is that generated by the resistor R_a. The addition of the losses from conductors, dielectrics, transmission lines and other elements, R_L, increases the noise power so that the overall noise is:

$$E_a{}^2 = 4\,k\,T\,B\,(R_a + R_L) \qquad\qquad (4.2)$$

and the noise factor is:

$$F = 1 + \frac{R_L}{R_a} \qquad\qquad (4.3)$$

At frequencies below 30 MHz, atmospheric noise produces an equivalent NF, which is much higher than unity, usually expressed as an equivalent antenna noise factor F_a. In these cases, the overall NF becomes:

$$F = F_a + \frac{R_L}{R_a} \tag{4.4}$$

and the antenna losses, and the importance of the receiver NF in the system, are reduced. If F_a is sufficiently high, a small antenna with very small R_a can often be used despite high antenna and coupling losses.

On the other hand, at much higher frequencies, when highly directional antennas may point at the sky, it is possible that the equivalent noise temperature of R_a is substantially reduced below the standard 300 K normally assumed. In such cases, it is important to minimize all losses and even to use specially cooled input amplifiers to produce maximum sensitivity. This usually occurs for receivers at higher frequencies than are covered in this book.

When a receiver is designed to operate over a small tuning range, it is comparatively straightforward to design networks to couple the antenna. However, when—as in many HF receivers—a wide tuning band (up to 16:1) is covered with the use of a single antenna, the antenna impedance varies widely (Figure 4.3). In this case the antenna coupling network must either be tuned or a number of broadband coupling networks must be switched in and out as the tuning changes. Even when a tuned circuit is used, it is necessary to switch components if the band ratio exceeds 2 or 3:1.

4.3 Coupling Antennas to Tuned Circuits

Until recently, it was customary practice to couple the antenna to a tuned circuit connected to the input amplifier. The tuned element was one of several similar circuits that separated successive amplifiers and the first mixer and could be tuned simultaneously to be resonant at the required RF. Figure 4.4 includes schematic diagrams of several variations of the coupling of an antenna to the tuned circuit. They differ mainly in the details of the coupling. In some cases, several different couplings might be made available in a single receiver to accommodate different antenna structures. Two characteristics of the coupling circuit require attention, the gain and detuning. The voltage gain from the antenna open-circuit generator to the tuned circuit output (including the input impedance of the input device) and the noise factor of the receiver at the input device determine the overall noise factor or sensitivity. The primary circuit to which the antenna is connected is generally resonant above or below the required secondary resonance, and reflects reactance into the secondary circuit so that the tuning differs slightly from that of the other

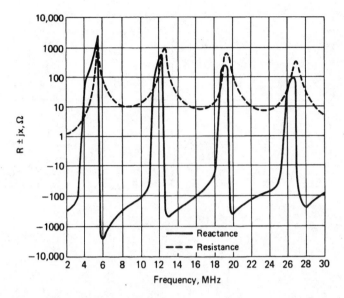

Figure 4.3 Typical impedance variation with frequency for an HF horizontal wire antenna.

circuits, which are tuned simultaneously. This must be taken into account in circuit design. Because advance knowledge of the antenna impedances that may be connected to the receiver are sketchy at best, a small trimmer capacitor is often provided for manual adjustment by the user. Figure 4.5 shows the gain variations of a typical coupling circuit of this type when the primary resonant frequency is below the secondary. Figure 4.6 shows the detuning effects.

With computer control and the need for frequency hopping, the use of mechanical tuning is becoming obsolete. Voltage-tuned capacitors (varactors) or current-tuned saturable magnetic core inductors are used in some applications. Another alternative is switching of broad-band coupling networks. When difference mixers are used, the bandwidth must be restricted to protect against spurious responses. Also, for small antennas with high Q, broad-banding entails excessive losses, especially at higher frequencies where atmospheric noise is low. Most passive broad-banding methods were devised to provide power transfer, which is not as important as noise factor for receiver applications. At frequencies where thermal noise is limiting, the matched solution is sometimes only a few decibels poorer than the optimum noise factor solution, so that broad-band matching techniques can be used. Also if the same antenna is to be used for both transmission and reception in a transceiver, the matching requirement of the transmitter may outweigh the receiver considerations.

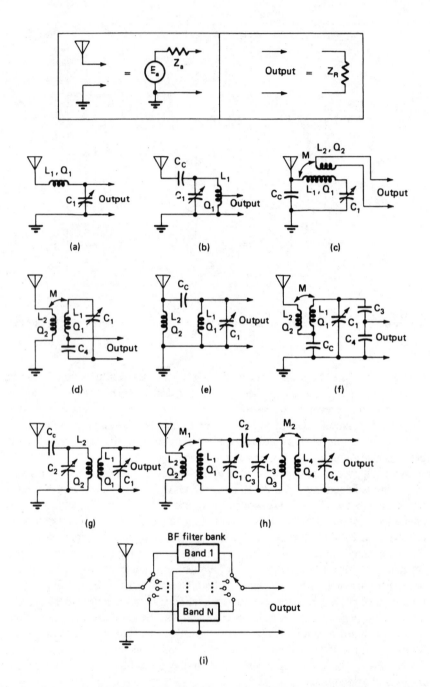

Figure 4.4 Typical circuits used for coupling antennas to tuned resonant circuits.

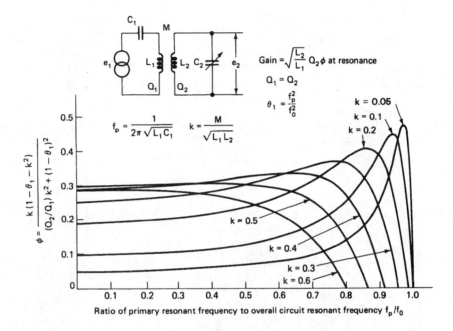

Figure 4.5 Gain characteristics of a coupling circuit with the primary resonant frequency f_p below the secondary at f_0. (*After [3.23]*.)

4.4 Small Antennas

As observed in Figure 4.3, an antenna structure passes through a sequence of series and parallel resonances, similar to a transmission line. When antennas are used at frequencies such that they operate substantially below their first resonance, they are referred to as *small antennas*. There are several reasons for using small antennas:

- At very low frequencies, large antennas are impractical.

- For frequencies where external noise predominates, there is no need for better signal pickup.

- For mobile and backpacked radios, the antennas must remain small enough for easy movement.

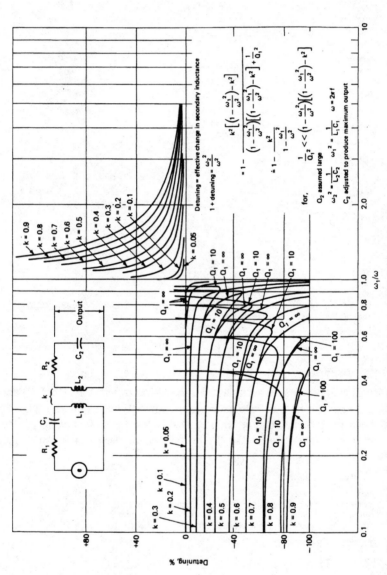

Figure 4.6 Detuning of the secondary resonance by a nonisochronous primary resonance.

When the antenna is to be used with a transmitter as well as a receiver, it is necessary to match it to the transmitter impedance. For mobile radios with substantial frequency coverage, this requires either broad-band matching or an automatic matching network that retunes when the transmitter does. In this case, the receiver may be constrained to use the same network but can use transformers to change the impedance seen by the first active circuit.

The most common short antenna is a vertical whip. For some narrow tuning range applications the whip may be converted to a helical form of similar length so that the extra inductance tunes it to serial resonance. For television receivers, dipoles are used ("rabbit ears") as well as some structures aiming at half-wave resonance at UHF (folded dipoles, bow ties, and circular loops). These antennas are not, strictly speaking, "small antennas." Loops, as small antennas, have been used extensively in portable broadcast receivers, usually in the form of coils wound on ferrite rods, mounted within the plastic case of the receiver. Loops are also used for direction finding receivers because of the sharp nulls in their figure-eight directional patterns. At low frequencies, the loop can also be useful to reduce the near electric field interference produced by frictionally induced voltages, usually known as *precipitation static*. The loop can be shielded from electric fields without shielding it from the electromagnetic radiated fields and thus can provide higher sensitivity than a whip antenna, which cannot be so shielded.

4.4.1 Whip Antennas

For hand-carried sets, whips from about 6 in to 6 ft in length have been used. Generally the shorter the whip, the greater the mobility. Some automotive vehicles have used whips up to 15 ft in length, although, again, shorter sizes are preferred. Usually the longer sizes are used for transceivers where improved transmitter system efficiency is sought. Over this range of lengths, the quarter-wave resonance of a whip for a mobile set may vary from about 15 to 500 MHz. So long as the operating frequency is substantially below this resonance, the whip input impedance appears to be a small capacitance in series with a resistance. Although the radiation resistance of whips tends to be small, losses from the antenna resistance and coupled objects in the vicinity (for example, a person) often cause resistance much higher than the radiation resistance alone.

The radiation resistance of a short vertical whip over a perfect conducting plane is given by:

$$R_r = 40 \, \pi^2 \left(\frac{h}{\lambda}\right)^2 \tag{4.5}$$

where h is the antenna height and λ is the wavelength. Seldom does the mounting surface resemble a plane, let alone one with perfect conduction. However, this

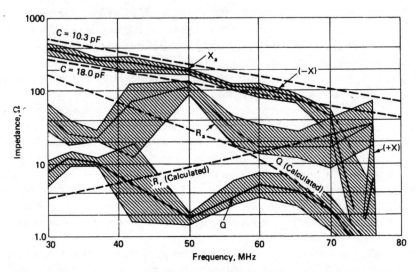

Figure 4.7 Impedance of a short-whip antenna as a function of operating frequency.

equation does provide a rough estimate of the resistance. The open-circuit voltage is the electric field strength multiplied by the antenna height. The capacitance is given by [4.2]:

$$C_a = \frac{24.2h}{\log(2\,h/a) - 0.7353} \tag{4.6}$$

where h and the whip diameter a are measured in meters and C is given in picofarads. The coefficient becomes 0.615 when h is measured in inches. Again, this is only a rough approximation to the real situation. Figure 4.7 gives the results for a 3-ft antenna and compares these with a range of measurements that have been made.

The problem of coupling a short-whip antenna optimally to the first active circuit in a receiver, thus, involves coupling a generator with voltage that varies with frequency, in series with a capacitive reactance and a small resistance that, with its noise temperature, also varies with frequency. This is complicated by the fact that the antenna mounting connector usually introduces a shunt capacitance to ground that can be a substantial fraction of the antenna capacitance. With the values shown in Figure 4.7, the predicted capacitance is 10.3 pF. The reactance for this value is shown in the figure, and the predicted radiation resistance for the short antenna is also plotted. The measured reactance is reasonably close to the calculated value, although a bit lower. The addition of 3 or 4 pF would result in excellent agreement. The measured resistance, however, differs significantly from the radiated resistance and shows a large rise around 50 MHz. Krupka [4.5] suggests that the coupled human represents a resonator. In his tests, the resonance appeared to be around 60 MHz. Without this effect, it appears that the loss resistance might be 30 to 50 Ω at the low frequency, dropping off slowly as the quarter-wave resonance of the whip is approached. Figure 4.7 shows the predicted Q for the whip and the range of measured Q values. At the low end of the band a series tuning circuit would have a bandwidth of about 3 MHz; at 60 MHz it would be about 20 MHz. If the input circuit impedance were matched, the low-frequency bandwidth might be doubled. When such sets had mechanical tuning, it was found that a series tuning inductance in a circuit such as shown in Figure 4.4d, ganged to the tuning circuit, was about the best coupling circuit for use with short whips in this frequency range. With the requirement for quick frequency hopping, switched multiple coupling circuits, such as shown in Figure 4.4e, are a better choice.

The need for carried or backpacked equipment with wide tuning ranges has been primarily a military requirement, principally in the frequency range from 2 to 90 MHz. There are many commercial and industrial applications at higher frequencies, but these are usually confined to the selection of only a few frequencies in a relatively narrow band, so that the serial inductance coupling of Figure 4.4d or direct connection of the whip with a shunt coil for tuning, as shown in Figure 4.4b, prove most useful. In the HF band, whips that are longer than 3 ft are desirable. Between 10 and 30 MHz, such whips show trends similar to those noted previously. Below 10 MHz, much wider ranges of loss resistance are encountered in the measurements, presumably because the longer wavelength permits coupling to a far broader range of the surroundings. The serial inductance tuning, followed by an appropriate resonant circuit step-up, would remain the coupling of choice. However, below 10 MHz, atmospheric noise is sufficiently high that it often limits sensitivity even when coupling is far from optimum. For circuits at such frequencies, active antennas provide the ideal solution for broad-band tuning. Such circuits can also be useful at higher frequencies.

4.4.2 Loop Antennas

The principal uses for loop antennas have been in radio direction finders and in portable broadcast receivers. The loop antenna differs from the monopole in that when the face of the loop is vertical, it responds to the magnetic field rather than the electric field. The first resonance of a loop antenna is a parallel resonance rather than a series resonance. When the dimensions of a loop antenna are small compared to a wavelength, the loop is said to be *small*, and its impedance is an inductance in series with a resistance. This includes the loss resistance and a small radiation resistance. Rather than being omnidirectional in azimuth, like a whip, the loop responds as the cosine of the angle between its face and the direction of arrival of the electromagnetic wave. This is the familiar figure-eight pattern that makes the loop useful for direction finding by providing a sharp null for waves arriving perpendicular to the face. Loops often have multiple turns to increase the effective height and may also have a high permeability core to reduce size.

A single turn loop in air has a low-frequency inductance that is given by:

$$L = 0.01596 D \left[2.303 \log \left(\frac{8D}{d} \right) - 2 \right] \tag{4.7}$$

where D is the diameter of the loop and d is the diameter of the wire in the loop, in inches, and the inductance is given in microhenries. The radiation resistance in ohms is:

$$R_r = 320 \pi^4 \frac{A^2}{\lambda^4} \tag{4.8}$$

where a is the area of the loop and λ is the wavelength, measured in the same units which, when squared, give the units of a. The effective height of the loop is:

$$h_{eff} = 2 \pi \frac{A}{\lambda} \tag{4.9}$$

As the frequency increases so that the dimensions are no longer small, these values change. Figure 4.8 shows the calculated loop impedance at the higher frequencies, and Figure 4.9 gives a comparison of theoretical measurements with experimental data [4.6]. When the loop has N turns, these expressions become:

$$L = 0.01596 D N^2 \left[2.303 \log \left(\frac{8D}{d} \right) - 2 \right] \tag{4.10}$$

$$R_r = 320 \pi^4 \frac{A^2 N^2}{\lambda^4} \tag{4.11}$$

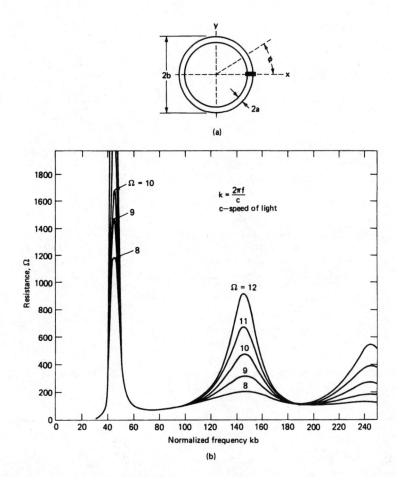

Figure 4.8 Loop antenna characteristics: (*a*) coordinates for a loop antenna, (*b*) calculated impedance of a loop antenna as a function of frequency, and (*c*, next page) reactance as a function of frequency. (*After [4.6].*)

$$H_{eff} = 2 \pi \frac{A N}{\lambda} \tag{4.12}$$

The effect of a ferrite core is to increase inductance, radiation resistance, and effective height. If the loop were simply immersed in a magnetic medium, the inductance and effective height would increase directly as the relative permeability of the medium, and the radiation resistance would increase by the square of the relative permeability. The usual design is to wind a coil on a long thin ferrite

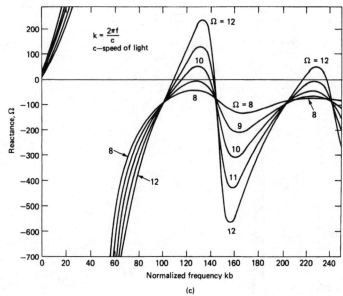

Figure 4.8 *continued*

cylinder. In such a case, the air inductance must first be calculated using one of the standard solenoid formulas, such as Wheeler's [4.7]:

$$L = \frac{R^2 N^2}{9R + 10H}$$

(4.13)

where R and H are the coil radius and length in inches and L is measured in microhenries. The introduction of a ferrite core multiplies the values in Equations 4.12 and 4.13 by an effective permeability

$$\mu_e = \frac{\mu}{1 + D(\mu - 1)}$$

(4.14)

where μ is the relative permeability of the ferrite core and D is a demagnetization factor [4.8] that increases Equation 4.11 by the square of μ_e. Figure 4.10 shows the experimentally determined value of D. It will be noted that even as μ grows

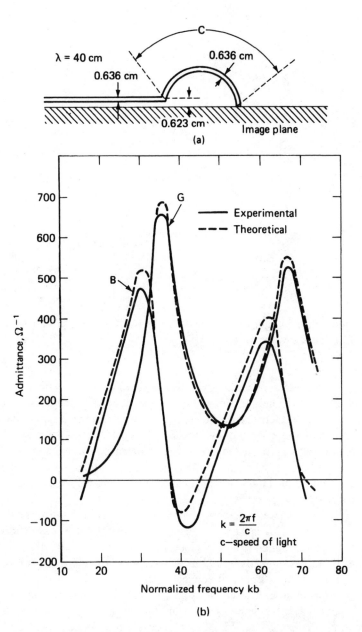

Figure 4.9 A comparison of experimental measurements of a loop antenna with theoretical calculations: (*a*) physical dimensions and (*b*) admittance as a function of frequency. (*After [4.6].*)

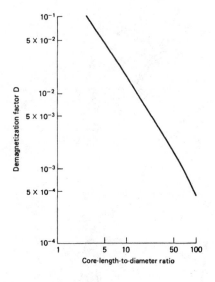

Figure 4.10 A plot of the loop antenna demagnetization factor versus the core-length-to-diameter ratio. (*After [4.8].*)

without limit, the maximum value μ_e can attain is $1/D$. For practical core sizes, this is much less than the value of μ for ferrite.

A. Coupling
Coupling to a loop antenna varies somewhat depending on use. A loop intended for broadcast or other simple communication reception is generally a multiturn device and may have a ferrite core. Such a loop may be tuned by a capacitance and connected directly to the input of the receiver. Figure 4.11 shows some examples of this type of coupling. If the loop has lower inductance than required for proper input impedance, it may be connected in series with an additional inductance for tuning, as shown. If the tuned loop impedance is too high, the receiver input may be tapped down on the circuit.

For applications where the pattern of the loop is important, it must be balanced to ground carefully. This is easier to achieve magnetically than electrically. To prevent capacitive coupling to nearby objects from upsetting the balance, the loop must also be electrostatically shielded. This type of shielding is achieved by enclosing the loop in a grounded conductive tube which has a gap to prevent completion of a circuit parallel to the loop. The loop feed wires are shielded, and the loop input may be fed through a balun whose primary is electrostatically shielded from the secondary. In this way, the entire balanced input circuit is contained within a continuous grounded shield (Figure 4.12). The shielding prevents pickup on the input by the normal antenna mode, the electrical field component of which can distort the pattern and reduce the sharpness of the null. In direction finding applications, a separate whip may be employed to inject a

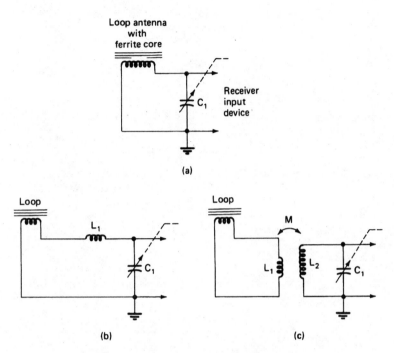

Figure 4.11 Examples of coupling circuits for broadcast reception.

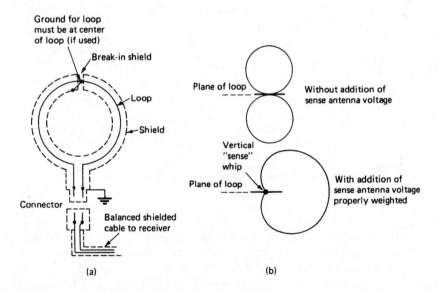

Figure 4.12 Specialized forms of the loop antenna: (*a*) electrostatic shielding of a loop antenna and (*b*) azimuthal pattern change from a controlled "antenna effect."

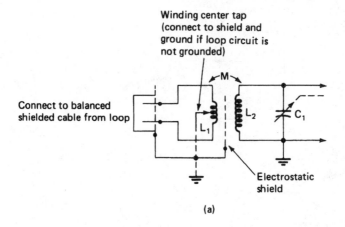

(a)

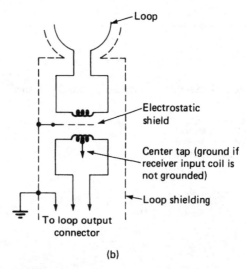

(b)

Figure 4.13 Low impedance coupling of a loop antenna: (*a*) coupling circuit for a low-impedance shielded loop and (*b*) wideband transformer coupling at the loop.

controlled amount of this mode to distort the figure-eight pattern to a cardioid so as to determine the direction of the pickup along the null. In some installations the loop may be located some distance from the receiver. For such a case, a low-reactance loop is desirable so that the cable (which is much shorter than a quarter-wave in these applications) has minimum effect. The loop may be connected directly to the cable, or a separate broad-band transformer may be used (Figure 4.13).

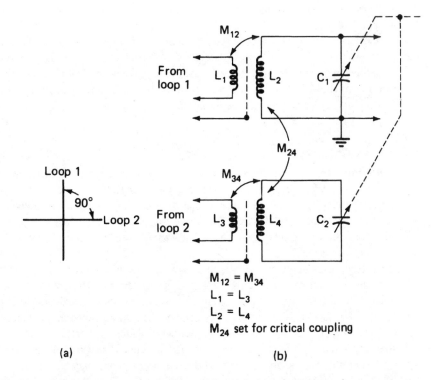

Figure 4.14 A circuit for achieving an omnidirectional pattern from orthogonal loop antennas: (a) the azimuthal location of the orthogonal loop planes and (b) a coupling circuit to produce an omnidirectional pattern from orthogonal loops.

Electrostatic shielding of a loop also reduces the effects of precipitation static, which is the result of electric field components from nearby static discharges. Hence if a loop is to be used for communications in an environment subject to such discharges, it is important that it be shielded.

When two loops are mounted perpendicular to each other, their nulls are orthogonal. If their output voltages are combined in phase, the result is a figure-eight pattern with null at 45°. By controlling the relative input fraction from each component loop and including phase reversal capabilities, it is possible to rotate the null through a complete circle. This technique is used to produce electrically-steerable nulls and automatic direction finders.

If, in this example, the voltages are added with a 90° phase offset, the resultant is a circular pattern. This arrangement can be used in communications applications to permit similar reception from all directions while getting the protection of the loops from precipitation static. One circuit that has been used for achieving such a combination is shown in Figure 4.14. In this case the coupling between the two resonant circuits is set to produce equal levels from both antennas, while achieving the 90° phase shift.

For applications requiring a broad tuning band without mechanical tuning, the design of broad-band networks with adequate performance is more difficult for loops than for open-wire antennas because their Q tends to be much higher, and the contribution of the radiation resistance is lower. When such broad-band designs prove necessary, an active antenna solution may be appropriate.

4.5 Active Antennas

By connecting active devices directly to small receiving antennas, it is possible to achieve similar performance over a wide frequency range. Such circuit and antenna arrangements are referred to as *active*, or *aperiodic*, antennas. Somewhat different arrangements must be used for active transmitting antennas, where the objective is to increase efficiency when using a small antenna. This latter case is not of interest to us here.

Without some form of broad-banding, it would be necessary to use many different antennas for reception over the broad tuning range of 10 kHz to several hundreds of megahertz typically of interest in communications receivers. Using advanced techniques, it is possible to design receivers that can cover substantial parts of this range using computer control. For surveillance applications, multiple antennas and antenna tuning are undesirable. The problems are especially acute at low frequencies where the physical size of the required antenna may be very large. Consider, for example, reception at a frequency of 10 MHz (30 m wavelength) with an antenna having an effective height of 3 m. If the desired signal has a field strength of 10 μV/m, the open-circuit output voltage of the antenna is 30 μV. The antenna impedance may be a resistance of 14 Ω (4-Ω radiation resistance) in series with a 40-pF capacitor (398 Ω). If the antenna were terminated by a 50-Ω resistance, the equivalent circuit would be as shown in Figure 4.15. A quarter-wavelength antenna (7.5 m) might be 40-Ω resistive (36-Ω radiation resistance). In this case, the voltage delivered to the load would be 42 μV (5⁄9 of the open-circuit 75 μV), as compared to 3.7 μV in the first case. In either case, all voltages are similarly reduced, atmospheric and man-made noise as well as the desired signal. Whether the shorter antenna is adequate or whether an aperiodic antenna can be used will depend on whether the mistermination reduces the output signal level below the inherent receiver noise.

A short whip or rod antenna of only a few yards, such as referred to previously, is essentially short-circuited by a 50-Ω termination, and reception may be very poor. The absolute voltage from the receiver is not so much of importance as the S/N. As long as the reduced level of both the signal and noise is sufficiently above the receiver noise threshold, the received S/N will be as good as can be provided by any coupling circuit. Absolute levels of the separate voltages are of no significance. Therefore, it is possible to put an amplifier between an electrically short antenna

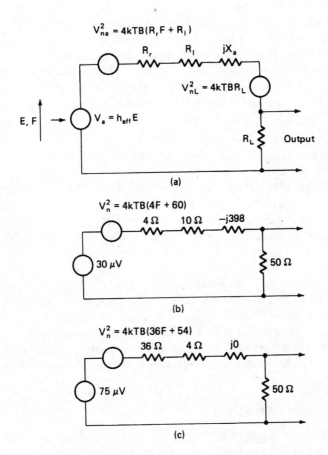

Figure 4.15 The equivalent circuit for noise calculations of antennas with resistive termination: (a) general circuit, (b) 3-ft whip at 10 MHz with a 50-Ω load, and (c) 7.5-ft whip at 10 MHz with a 50-Ω load.

and the receiver as long as the amplifier NF is sufficiently low. If the input impedance of such an amplifier is high enough, the antenna will not be loaded down, and the open-circuit antenna voltage will drive the amplifier.

In the following example a 3-ft-long whip terminated by a noise-free amplifier of high impedance is compared with a quarter-wave antenna at 10 MHz. The field strength of the desired signal is assumed to be 10 μV/m in both cases. The following conditions can be observed:

- **Passive antenna case.** The quarter-wave antenna is 7.5-m long and produces an EMF of 75 μV. The antenna impedance is resistive and somewhat larger than 36 Ω (the radiation resistance). If we assume that the various external noise sources cause an overall noise field in the receiver bandwidth of 1 μV/m, the noise EMF is 7.5 μV. The antenna thermal noise, assuming a 3-kHz bandwidth, is 0.044 μV, so it has no effect on the calculation. The resulting S/N is 20 dB.

- **Active antenna case.** The antenna has an electrical length of about 1 m. The desired signal produces an EMF of 10 μV; the external noise produces 1 μV. The antenna resistance may be as much as 10 or 15 Ω, of which about 0.4 Ω is radiation resistance. The antenna thermal noise is still negligible. The antenna reactance, assuming a 1.5-cm whip diameter, is about 700 Ω. If the amplifier input impedance is much greater than this and it has unity voltage gain and 50-Ω output impedance, the S/N remains 20 dB.

From the foregoing, it is apparent that if an amplifier can be constructed with sufficient gain to compensate for the change in antenna length, the same absolute voltage can be produced by the active or passive antenna. Clearly, the noise-free assumption for the amplifier is the major impediment to achieving equal performance from the active antenna. Thus, the active antenna in its minimum configuration consists of a small passive antenna, typically a whip or dipole, and an integrated amplifying device.

4.5.1 Application Considerations

Let us examine the simple case in which a whip antenna is directly connected to the input gate of a *field effect transistor* (FET). As shown in Figure 4.16, the antenna acts as a source to feed the transistor. An electric field E generates a voltage that can be determined from $V_a = H_{eff}E$. The antenna impedance is determined primarily by the effective capacitance C_a, which may be determined from Equation 4.6, while the transistor has an input capacitance C_r. These two capacitances form a capacitive voltage divider. The signal voltage that drives the transistor is then:

$$V_T = \frac{h_{eff}E}{1 + C_r/C_a} \tag{4.15}$$

For electrically short antennas, the voltage V_t is proportional to E, and nearly independent of frequency. Therefore, the active antenna can operate over an extremely wide bandwidth. The gain-bandwidth product of such a device can be computed from the performance of the FET. At the output, it will reproduce the

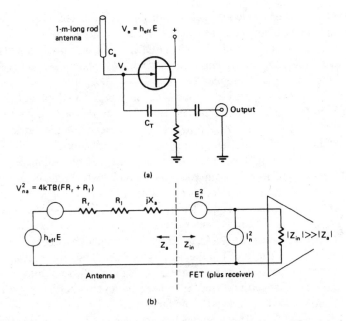

Figure 4.16 An active antenna comprising a short monopole and amplifier: (*a*) circuit and (*b*) equivalent circuit for noise calculations. (*After [4.9].*)

input voltage as long as its cutoff frequency is sufficiently high. Additional reactances may be added intentionally to produce frequency selectivity and thus limit the bandwidth of the active antenna.

The output level is not of primary importance because additional amplifiers can always be added. A more important consideration is the output S/N. If we assume that the active antenna has sufficient gain, the S/N will be determined by it and not the receiver. The only internally generated noise is from the transistor, because the antenna resistance generates negligible thermal noise. In this analysis, there are three components to consider:

- The signal voltage at the operating frequency
- The amplified noise from external sources (man-made, atmospheric, and galactic)
- The transistor noise contribution

If the noise voltage generated by the transistor is sufficiently low, the overall system may achieve as good an S/N as an optimized passive antenna for the same specific frequency.

Consider an active antenna with a 1-m-long rod antenna. The capacitance depends on both the diameter of the rod and the capacitance of the feed connection, but may be taken as about 25 pF. A typical FET has a capacitance of about 5 pF so that at low frequencies, 80 percent of the antenna *electromagnetic field* (EMF) is applied to the FET input. At frequencies of up to 200 MHz, the input series resistive component is small compared to the reactance of 5 pF. The NF of an FET, when fed from a 50-Ω source, can be 3 dB or better. This corresponds to a series noise resistor of 50 Ω or less. The whip, however, is a quarter-wave long at 75 MHz so that above about 30 MHz it can no longer be considered "short." At 200 MHz, the antenna is 0.67 wavelength, and the pattern has begun to be multilobed.

At 30 MHz the radiation resistance is about 4 Ω and losses might be comparable. The NF based on thermal noise alone would approach 9 dB. The level of man-made noise in rural areas produces an equivalent NF of about 25 dB. Under this condition, the effect of the active circuit is to increase it slightly to 25.2 dB. By 75 MHz, the radiation resistance has risen to 36 Ω, the antenna reactance has dropped off to zero, and the losses are probably still in the vicinity of 4 Ω. The noise resistance of the FET is about 50 Ω, and its shunt reactance is greater than 400 Ω. While the voltage division ratio has changed to about 99 percent, the overall NF based on thermal noise is about 3.5 dB. The rural NF from man-made noise is about 15 dB, and the galactic NF is about 7 dB. The overall active antenna NFs resulting from these two noise levels are 15.2 and 8.1 dB, respectively.

These rough estimates indicate the type of performance that can be expected from an active antenna. Because of the lack of detailed information on the variation of the NF with the input impedance, experimental measurements are desirable to determine the actual values attainable. Figure 4.17 compares the NFs for two active antenna types to man-made noise levels based on CCIR report 258-5 [4.10]. The specifications of an HF active antenna are indicated in Table 4.1.

For the same length antenna, the pattern of an active antenna is the same as that of a passive antenna. For the vertical rod antenna, typical elevation patterns are shown in Figure 4.18. The patterns are the same for any azimuth. At HF, the low intensity at high elevation angles leads to a large dead zone, which can be reduced by using horizontal dipoles. The pattern for cross horizontal dipoles with 90° phase difference is shown in Figure 4.19, where the three-dimensional pattern is combined with ray traces at the highest ionospheric reflection angle to indicate the dead zone. The available power from the monopole divided by the matched power of the active antenna amplifier output is designated G_v. The G_v values for two UHF active antennas are shown in Figure 4.20. The ratio of the output voltage from the active antenna to the input field driving the monopole is designated K. Figure 4.21 shows the K values for the two UHF antennas, and Figure 4.22 is a schematic diagram typical of such active antennas.

IM distortion is another source of noise and interference generated in the active device, depending on the level of input signals. Because the objective is a very

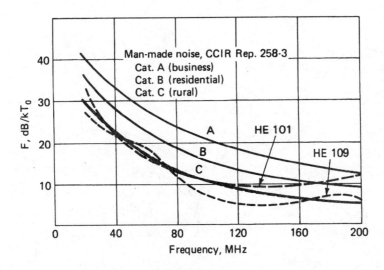

Figure 4.17 A comparison of overall NFs of active antennas with predicted man-made noise levels.

TABLE 4.1 Specifications for Rohde and Schwarz Active Antenna Type HE010 (*Courtesy of Rohde and Schwarz.*)

Frequency range	10 kHz to 80 MHz
Impedance	50 Ω
VSWR	≤2
Conversion factor: field strength to output voltage E/V	0.1 (corresponding to $K \approx 20$ dB)
Intercept point second-order	≥55 dBm
Third-order	≥32 dBm
Cross modulation for cross-modulation products 20 dB down; interfering transmitter modulated at 1 kHz and 30% modulation depth	20 V/m up to 30 MHz; 10 V/m 30 to 80 MHz
Operating temperature range	−40 to +70°C
Storage temperature range	−55 to +85°C
Connectors (two outputs)	Female N type
Supply voltage	18 to 35 V
Current drain	500 mA

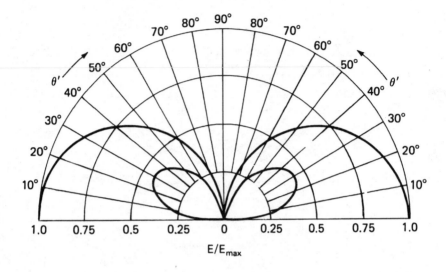

Figure 4.18 Elevation patterns for a vertical monopole. The outer pattern is for a perfect ground; the inner is for dry ground, $\varepsilon = 5$, $\sigma = 0.001 S/m$.

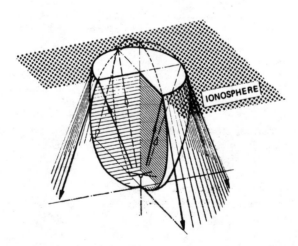

Figure 4.19 The radiation pattern for quadrature-fed crossed horizontal dipoles, showing ray patterns at the highest ionospheric reflection angel, to indicate dead zone. (*Courtesy of News from Rohde and Schwarz.*)

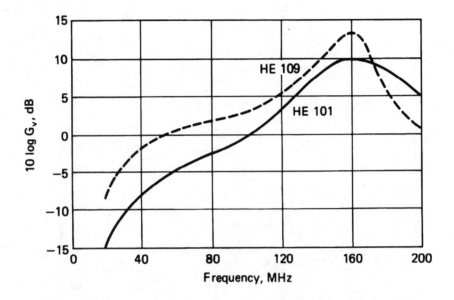

Figure 4.20 Gain G_V for active UHF antennas. (*Courtesy of* News from Rohde and Schwarz.)

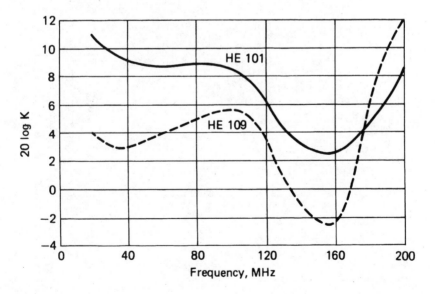

Figure 4.21 Field conversion ratio K for active UHF antennas. (*Courtesy of* News from Rohde and Schwarz.)

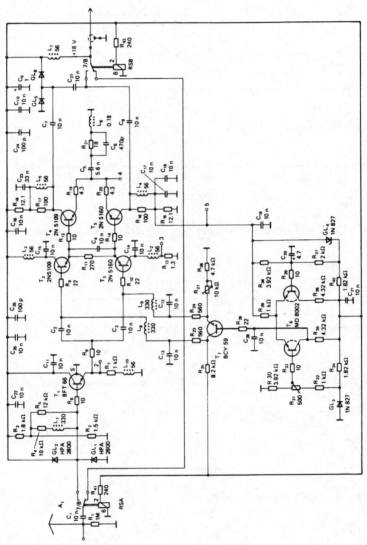

Figure 4.22 Typical schematic diagram of an active antenna. (*Courtesy of* News from Rohde and Schwarz.)

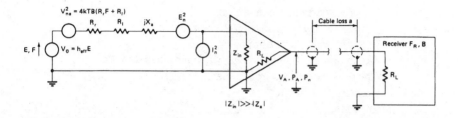

Figure 4.23 Block diagram of a short active antenna connected to a receiver through a cable.

broad-band antenna, many signals may contribute to IM. An active antenna, therefore, is best described by assigning to it:

- A frequency range
- A minimum sensitivity, which is determined by the NF
- A dynamic range, which is determined by the second-, third-, and higher-order IPs
- Polarization (horizontal or vertical)

Next, consider the analysis of the short active antenna. Figure 4.23 is a block diagram indicating the elements of the antenna and its connection into a receiver. The various symbols are defined as follows:

a	Cable loss
B	Receiver bandwidth
E	Received electric field
F	Equivalent noise factor of received noise field
f_A	Noise factor of active antenna
F_R	Receiver noise factor
F_S	System (active antenna plus receiver) noise factor
G_v	Active antenna gain ratio
h_{eff}	Effective radiator height
$IP_{2,3}$	Second- or third-order IP
K	Conversion ratio, $= V_A/E$
P_A	Output power of active antenna into R_L load
$P_{IM\,2,3}$	Output power of second- or third-order IM products

P_n Output noise power from active antenna
R_a Resistance of radiator
R_l Loss resistance of radiator circuit
R_L Amplifier load resistance (50 Ω nominal)
R_r Radiation resistance of radiator
V_A Active antenna output voltage when terminated by R_L
V_0 Open-circuit voltage from radiator, $= h_{eff}E$
X_a Reactance of radiator
Z_a Impedance of radiator, $= R_a + jX_a$

The system noise factor is:

$$F_S = F_A + \frac{(F_R - 1)\,a}{G_v} \tag{4.16}$$

The active antenna noise factor is:

$$F_A = \frac{4\,k\,T\,B(FR_r + R_1) + E_n\,2 + I_n^2\,(R_a^2 + X_a^2)}{4\,k\,T\,B\,R_r} \tag{4.17}$$

To minimize the noise factor, X_a should be adjusted to zero. Because this requires a two-port, the two-port can also be designed to include a transformer so that the resistance R_a, as seen by the amplifier, can be adjusted to minimize the amplifier contribution. Under those conditions the optimum value of R_a, as seen by the amplifier, is E_n/I_n. For FETs, the value of this R_{opt} ranges from 20,000 to 200,000 Ω at 1 MHz and decreases inversely with the frequency. However, for broadband use, which is the objective of the active antenna, the reactance cannot be tuned out, so there is little value in optimizing the resistance, which is much smaller than the reactance as long as the antenna is small.

If we accept the values as they are provided, Equation 4.17 can be rewritten as follows:

$$F_A = F + \frac{R_1}{R_r} + \frac{V_n^2}{4\,k\,T\,B\,R_r} + \frac{I_n^2\,Z_a\,Z_a^*}{4\,k\,T\,B\,R_r} \tag{4.17a}$$

Again, dealing with FETs, at frequencies at least up to 30 MHz, the last term remains negligible, and the next to last one can be simplified to R_n/R_r; R_n varies from about 250 to 1000 Ω, depending on the FET.

In considering the IM components, it is desirable to relate them to the field strength. The amplifier has output IPs as given by:

$$IP_2^{[d]} = 2\,P_a^{[d]} - P_{IM2}^{[d]} \tag{4.18}$$

$$IP_3^{[d]} = \frac{3P_A^{[d]} - P_{IMB}^{[d]}}{2} \tag{4.19}$$

where the superscripts indicate that the power values are measured in dBm. However,

$$P_A = \frac{V_a^2}{R_L} \tag{4.20}$$

But,

$$G_v = \frac{V_A^2/R_L}{V_0^2/4R_a} \tag{4.21}$$

and,

$$K = \frac{V_A}{E} \tag{4.22}$$

Therefore, we can express IP_j in terms of E, the field exciting the antenna, or V_0, the voltage generated by that field as:

$$(j-1)\,IP_j^{[d]} = 10j \log G_v + 20j \log V_0 - 10j \log R_a + 30 - P_{IMj}^{[d]} \tag{4.23}$$

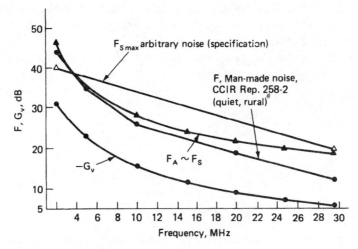

Figure 4.24 Noise performance of an HF active antenna system. (*After [4.9].*)

$$(j-1)\, IP_j^{[d]} = 20j\, \log K + 20j\, \log E - 10j\, \log R_L + 30 - P_{IMj}^{[d]} \qquad (4.24)$$

If P_{IMj} is measured from the noise level, then the IM products produced by two input signals of specified field or input voltage levels are required to be no more than M decibels above the output noise level. Note that $P = F_s kTB$ is the available noise power at the antenna input, and the output noise power is G_v times this value. Therefore,

$$P_{IMj}^{[d]} \leq 10\, \log F_s + 10\, \log G_v + 10\, \log B - 174 + M \qquad (4.25)$$

which may be substituted in Equation 4.23 or Equation 4.24 to determine the required IPs for the amplifier.

Next consider as an example the design of an HF active antenna suitable for shipboard applications. The NF requirements will be taken in accordance with Figure 4.24. The curve marked "F_{Smax} arbitrary noise specification" defines the system NF, except where ambient noise is above this line. For purposes of analysis this noise has been taken as quiet rural man-made noise, although it is well known that shipboard noise is higher than this condition. The antenna is also required not to generate IM products higher than 40 dB above the system's maximum NF caused

TABLE 4.2 Electrical Gain and Data for HF Shipboard Active Antenna as a Function of Frequency

F, MHz	G_v, dB	R_r, Ω	R_l, Ω	F, $\dfrac{dB}{kT0}$	$F_A \sim F_S$, $\dfrac{dB}{kT0}$
2	−31	0.0175	0.0222	44.8	46.4
5	−21.7	0.110	0.208	33.6	36.5
10	−15.5	0.439	0.970	25.2	29.5
15	−12	0.987	2.168	20.3	25.5
20	−10.2	1.755	3.020	16.8	22.7
25	−7.3	2.742	6.785	14.1	20.6
30	−5.5	3.948	10.144	11.8	18.9

by two interfering signals at a field level of 10 V/m. The active antenna design selected uses a 1-m rod and an amplifier with an input FET.

The antenna impedance at 2 MHz (30 pF) is estimated at 2600 Ω, primarily capacitive. At 30 MHz, the reactance is reduced to about 175 Ω, which is still substantially more than the anticipated R_A. Setting the capacitive component of the amplifier FET to 5 pF, the amplifier input voltage is 0.83 times the input field (h_{eff} = 1 m). Therefore, if the amplifier voltage gain is set to 0.6, K will equal 0.5. A fractional value of K reduces the output voltage, so that the high input levels produce lower IM products than for unity or higher K. The input shunt resistance of the FET is extremely high, so that the FET gain is high, despite the low voltage gain. Hence F_s is close to F_A unless F_R is very high. With typical values of E_n and I_n [4.11] for an FET up to 30 MHz and above the antenna impedance values, the I_n term in Equation 4.17a becomes negligible. E_n typically runs between $2 \times 1^{-\theta}$ and 4×10^{-9}, which corresponds to a series noise resistance of 240 to 960 Ω. Then, Equation 4.17 reduces to the following:

$$F_A = F + \frac{R_1}{R_r} + \frac{240}{R_r} \qquad (4.17b)$$

assuming the better noise resistance. F was plotted in Figure 4.24.

R_r may be estimated from Equation 4.5. R_l may be estimated, or it may be calculated from measured values of K and G_v and knowledge of h_{eff}. Values of G_v are shown in Figure 4.24 and are listed in the second column of Table 4.2. Calculated values of R_r are listed in the third column of the table. Since h_{eff} is 1 m, R_l can be calculated for the design K of 0.5; the resulting values are listed in the fourth column of the table. In the fifth column, the values of F from Figure 4.24

are given. Combining these values per Equation 4.17b, the overall NFs shown in the final column and plotted in Figure 4.22 may be determined. Despite the loss G_v, which is as high as 30 dB at the low end, the NF is below the specified maximum from 4 MHz upward. Below 4 MHz the NF is very close to the man-made noise.

The requirement that 10-V/m signals produce IM products no more than 40 dB above the highest noise level sets the maximum IM level to 10 log (F_s) + 10 log G_v + 10 log B – 134 dBm. For a 3-kHz bandwidth, this becomes 10 log (F_s) + 10 log G_v – 99.2 dBm. At 2 MHz, F_s equals 46.4 dB and G_v equals –31 dB, so P_{IM2} ≤ –83.8 dBm. From Equation 4.24, then, IP_2 = –12 + 40 – 34 + 60 + 83.8 = 137.8 dBm. Similarly, IP_3 = 82.4 dBm. These are the IPs required to generate IM at just the specified level. If the amplifier has lower levels of IPs, the specification will not be met. From practical considerations, the 1-dB compression point should be 10 dB above the maximum expected output level, in this case at 37 dBm. This corresponds to 15.8 V at 0.32 A in a 50-Ω load. The operating voltage of this amplifier should be set at 25 V or more. If the input conversion ratio is changed, so that a value smaller than 0.5 is used, the IPs can be reduced.

Suppose that in a practical amplifier, designed as above, IP_2 = 100 dBm and IP_3 = 65 dBm can be attained. Then, the second-order products for the 10 V/m interferers are at a level of –46 dBm and the third-order products are –49 dBm. This contrasts with the –85-dBm limit resulting from the specification. To get the IM products to the level of 40 dB above the noise requires a reduction in field strength of 19.5 dB (1.06 V/m) and 12.3 dB (2.43 V/m), respectively. This means that many more signals can produce substantial IM interference than in the former case.

The dynamic range is usually measured between the sensitivity signal level and the point where IM products are equal to the noise. The lower level is unaffected by a change in the IPs, but the upper level is changed by the decibel change in the IP times the ratio $(j – 1)/j$, where j is the order of the IP. Therefore, the dynamic range is changed identically. In the previous case, the dynamic range defined by second-order products would be reduced by 19.5 dB; that defined by third-order products would be reduced by 12.0 dB.

The foregoing calculations assume a man-made NF of 45 dB at 2 MHz and two 10-V/m interfering carriers generating the IM products. A number of tests in extremely hostile environments have been performed on developmental models of the active antenna described.

The foregoing has been a discussion of some of the characteristics of one variety of active antenna. A more complete summary of such antennas can be found in [4.12].

4.6 References

4.1. H. Jasik, R. C. Johnson, and H. B. Crawford, Eds., *Antenna Engineering Handbook*, 2d ed., McGraw-Hill, New York, 1984.

4.2. S. A. Schelkunoff and H. T. Friis, *Antennas, Theory and Practice*, Wiley, New York, 1952.

4.3. J. D. Kraus, *Antennas*, McGraw-Hill, New York, 1950.

4.4. W. L. Weeks, *Antenna Engineering*, McGraw-Hill, New York, 1968.

4.5. Z. Krupka, "The Effect of the Human Body on Radiation Properties of Small-Sized Communication Systems," *IEEE Trans.*, vol. AP-16, p. 154, Mar. 1968.

4.6. J. E. Storer, "Impedance of Thin-Wire Loop Antennas," *Trans. AIEE*, pt. 1, *Communications and Electronics*, vol. 75, p. 606, Nov. 1956.

4.7. H. A. Wheeler, "Simple Inductance Formulas for Radio Coils," *Proc. IRE*, vol. 16, p. 1398, Oct. 1928.

4.8. E. A. Wolff, *Antenna Analysis*, Wiley, New York, 1966.

4.9. U. L. Rohde, "Active Antennas," *RF Design*, May/June 1981.

4.10. "Man-Made Radio Noise," CCIR Rep. 258-4, in vol. VI, *Propagation in Ionized Media*, ITU, Geneva, 1982.

4.11. C. D. Motchenbacher and F. C. Fitchen, *Low Noise Electronic Design*, Wiley, New York, 1973.

4.12. G. Goubau and F. Schwering, Eds., *Proc. ECOM-ARO Workshop on Electrically Small Antennas* (May 6-7, 1976), U.S. Army Electronics Command, Fort Monmouth, NJ, Oct. 1976. (Available through Defense Technical Information Service).

Additional Suggested Reading

"March Antenna Over 1.5 to 30 MHz Range with Only Two Adjustable Elements," *Electronic Design*, p. 19, Sept 13, 1975.

Chapter 5
Amplifiers and Gain Control

5.1 Introduction

Amplifier circuits are used to increase the level of the very small signals (1 μV or less) to which a receiver must respond so that these signals can be demodulated and produce output of a useful level (on the order of volts). Such circuits may amplify at the received radio frequency, at any of the IFs, or at the lowest frequency to which the signal is transformed. This frequency is generically referred to as *baseband frequency*, but in specific cases, audio frequency (AF), video frequency (VF), or another notation appropriate for the particular application may be used.

Because of the wide range of signals to which a receiver must respond, the input device, at least, must operate over a wide dynamic range (120 dB or more). The input device should be as linear as possible so as to minimize the generation of IM products from the strongest signals that must be handled. Therefore, the number of strong signals should be minimized by restricting the receiver bandwidth at as low a gain level as possible. Thus, the gain should be low prior to the most narrow bandwidth in the receiver. It is not always possible to narrow the bandwidth adequately at RF, and so RF amplifiers are especially subject to many strong interfering signals. If there is more than one RF amplifier, the later ones encounter stronger signal levels, and the first mixer generally encounters the strongest interferers. On the other hand, mixers often have poorer NFs than amplifiers, and input coupling circuits and filters have losses to further increase the NF of the receiver. Consequently, unless the external noise sources produce much higher noise than the receiver NF (which is often the case below 20 MHz), receiver design becomes a compromise between sensitivity and IM levels. At the lower frequencies, it is common practice to avoid RF amplification and use the first mixer as the input device of the receiver. Bandwidth is then substantially restricted by filters in the first IF amplifier section.

When the desired signal is relatively strong, the receiver amplification may raise it to such a level as to cause excessive distortion in the later stages. This may reduce voice intelligibility and recognizability or video picture contrast, or it may increase errors in data systems. We must therefore provide means to reduce the system gain as the strength of the desired signal increases. Gain control can be effected either as an operator function, that is, MGC, or it may be effected automatically as a result of sensing the signal level, namely, AGC. AGC circuits are basically low-frequency feedback circuits. They are needed to maintain a relatively constant output level when the input signal fades frequently. Designing AGC circuits to perform satisfactorily under all expected signal conditions is a major engineering challenge.

5.1.1 Amplifying Devices and Circuits

During the early years of radio communication, amplification was achieved by the use of vacuum tubes. During the 1950s, transistors began to replace tubes because of their lower power consumption, longer life, and smaller size. Current receivers use transistor amplifiers exclusively, either individually or as a combination of transistors in an integrated circuit.

There are a variety of transistors available for use, depending upon the application. There are bipolar transistors in either PNP or NPN configuration and FETs, which may be classified as *junction FET* (JFET), *metallic oxide semiconductor FET* (MOSFET), *vertical MOSFET* (VMOSFET), dual-gate MOSFET, and *gallium arsenide* (GaAs) FET. These differ mostly in the manufacturing process, and new processes are developed regularly to achieve some improvement in performance.

Bipolar transistors may be used in several amplifying configurations, as shown in Table 5.1. Modern transistors have a gain-bandwidth product f_T of 1 to 6 GHz and reach their cutoff frequency at currents between 1 and 50 mA, depending upon the transistor type. These transistors exhibit low NFs, some as low as 1 dB at 500 MHz. Because their gain-bandwidth product is quite high and some are relatively inexpensive, most modern feedback amplifiers use such transistors. Transistors have been developed with special low-distortion characteristics for *cable television* (CATV) applications. Such CATV transistors typically combine low NF, high cutoff frequency, low distortion, and low inherent feedback. For class *A* amplifiers, which provide maximum linearity, dc stability is another important factor.

The gain-bandwidth product of a bipolar transistor is obtained from the base resistance r_{bb}, the diffusion layer capacitance C_D, and the depletion layer capacitance at the input C_E. The depletion layer capacitance depends only on the geometry of the transistor, while emitter diffusion capacitance depends on the direct current at which the transistor is operated. At certain frequencies and certain currents, the emitter diffusion layer becomes inductive and therefore cancels the phase shift, resulting in an input admittance with a very small imaginary part. For switching and power applications, different parameters are of importance. These include saturation voltage, breakdown voltage, current-handling capability, and power dissipation. Special designs are available for such applications.

The JFET has high input and output impedances up to several hundred megahertz and combines low noise with good linearity. These FETs can be used in grounded-source, grounded-gate, and grounded-drain configurations. Table 5.2 shows the characteristics of these basic configurations. They are analogous to the amplifier configurations for the bipolar transistor. The JFET is operated at a negative bias. Positive voltage above about 0.7 V opens the gate-source diode. When the gate-source channel becomes conductive, the impedance breaks down

TABLE 5.1 Basic Amplifier Configurations of Bipolar Transistors

Common emitter	Common base	Common collector

Characteristics of basic configurations			
	Common emitter	Common base	Common collector
Input impedance Z_1	Medium	Low	High
	Z_{1e}	$Z_{1b} \approx \dfrac{Z_{1e}}{h_{fe}}$	$Z_{1c} \approx h_{fe} R_L$
Output impedance Z_2	High	Very high	Low
	Z_{2e}	$Z_{2b} \approx Z_{2e}h_{fe}$	$Z_{2c} \approx \dfrac{Z_{1e} + R_g}{h_{fe}}$
Small-signal current gain	High	< 1	High
	h_{fe}	$h_{fb} \approx \dfrac{h_{fe}}{h_{fe}+1}$	$\gamma \approx h_{fe}+1$
Voltage gain	High	High	< 1
Power gain	Very high	High	Medium
Cutoff frequency	Low	High	Low
	$f_{h_{fe}}$	$f_{h_{fb}} \approx h_{fe}f_{h_{fc}}$	$f_{h_{fc}} \approx f_{h_{fe}}$

and distortion occurs. The transfer characteristic of the FET is defined by the equation:

$$I = I_{DSS}\left(1 - \frac{V_g}{V_p}\right)^2 \tag{5.1}$$

where V_p is the pinch-off voltage at which the transistor ceases to draw any current. The normal operating point for the gate voltage would therefore be roughly at $V_g = V_p/2$. I_{DSS} is the drain saturation current, which is the current observed when zero bias is applied to the transistor. All FETs have a negative temperature coefficient and, therefore, do not exhibit thermal runaway as is observed with bipolar transistors.

The bipolar transistor transfer characteristic is described approximately by the equation:

TABLE 5.2 Basic Amplifier Configurations of FETs

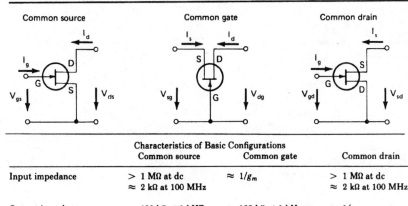

| | Characteristics of Basic Configurations | | |
	Common source	Common gate	Common drain
Input impedance	> 1 MΩ at dc ≈ 2 kΩ at 100 MHz	≈ $1/g_m$	> 1 MΩ at dc ≈ 2 kΩ at 100 MHz
Output impedance	≈ 100 kΩ at 1 kHZ ≈ 1 kΩ at 100 MHz	≈ 100 kΩ at 1 kHz ≈ 10 kΩ at 100 MHz	≈ $1/g_m$
Small-signal current gain	> 1000	≈ 0.99	> 1000
Voltage gain	> 10	> 10	< 1.0
Power gain	≈ 20 dB	≈ 14 dB	≈ 10 dB
Cutoff frequency	$g_m/2\pi C_{gs}$	$g_m/2\pi C_{ds}$	$g_m/2\pi C_{gd}$

$$I = I_0 \exp\left(\frac{V_0}{V_T}\right) \tag{5.2}$$

This exponential transfer characteristic, for even small variations in input voltage, produces a drastic change in the direct current. Because the first derivative of the transfer characteristic is also exponential, the small-signal transfer function of the bipolar transistor is highly nonlinear. In contrast, the FET has a small-signal transfer function that is linear, as can be seen by differentiating its transfer characteristic. The transconductance g_m is directly proportional to the voltage applied to the gate.

The MOSFET has an insulation layer between the gate and the source-drain channel, and therefore has an extremely high impedance at dc. Several thousand megohms have been measured. JFETs have somewhat better NFs than MOSFETs. This is apparently caused by the input zener diode usually included in the manufacture of the MOSFET to protect the gate against static charges that could destroy the transistor. Otherwise there is very little difference between the parameters of JFETs and MOSFETs.

For higher-power applications, transistors are manufactured with several channels in parallel. The Siliconix U320 or Crystalonics CP640 and CP643 are typical examples of this family. These are medium-power, low-distortion devices for linear amplification. For still higher power applications, VMOSFETs have been developed. Their drain saturation voltage is very low because the r_{on} resistance is kept small. These transistors can be operated at 25 to 50 V at fairly large rates of dissipation. Being FETs, they have a transfer characteristic that follows a square law almost precisely. These VMOSFETs can be operated at several watts output at RF with low IM distortion products.

A first approximation to the gain-bandwidth product of a FET is:

$$f_{T\max} = \frac{g_m}{2 \pi C_{gs}} \tag{5.3}$$

Thus, the cutoff frequency varies directly with the transconductance g_m and inversely with the gate-source capacitance. A typical JFET, such as the 2N4416, which has found widespread use, has $f_{T\max} = 10\ mS/10\pi\ pF = 318MHz$. A VMOSFET, in comparison, has a g_m of 200 mS and an input capacitance of 50 pF, resulting in a gain-bandwidth product of 637 MHz.

One might be tempted to assume that FETs offer a significant advantage over bipolar transistors. The bipolar transistor, however, has an input impedance of about 50 Ω at a frequency of 100 MHz, while the input impedance of a FET in a grounded-source configuration is basically that of a capacitor. In order to provide proper matching, feedback circuits or matching networks must be designed to provide proper 50 Ω termination over a wide band. Because of this need for additional circuitry, it is difficult to build wide-band amplifiers with FETs in high-impedance configurations. Wide-band FET amplifiers are typically designed using the transistor in a grounded-gate configuration, in which the FET loses some of its low-noise advantage.

Feedback from output to input, through the internal capacitive coupling of a device, is referred to as the *Miller effect*. In order to reduce this effect in FETs, one package is offered containing two MOSFETs in a cascode circuit. The output of the first FET is terminated by the source of the second transistor, the gate of which is grounded. Therefore, the feedback capacitance from drain to source of the first transistor has no influence because the drain point has very low impedance. The grounded gate of the second FET has very low input capacitance. Therefore, its feedback capacitance C_{sd} has little effect on the operation. A dual-gate FET, with the second gate grounded, accomplishes much the same effect, using a single source-drain channel.

For applications above 1 GHz, GaAs FETs have been developed. The carrier mobility of GaAs is much higher than that of silicon, so that for the same geometry, a significantly higher cutoff frequency is possible. Moreover, using modern technology, GaAs FETs can be made smaller than other technologies, such as bipolar transistors. At frequencies above 1 GHz, GaAs FETs have better noise and IM distortion than bipolar transistors. The advantage of the FET versus the bipolar transistor, in this frequency range, changes from time to time as technology is improved or new processes are devised.

The emergence of the 1.9-GHz personal communications market has intensified development of GaAs technology for *low-noise amplifier* (LNA) applications. While 2-GHz LNAs are—of course—possible with silicon technology, the improved NF performance of GaAs provides designers with greater operating margins. GaAs-based LNA devices have been produced that integrate bias and impedance matching, eliminating perhaps 10 or more components required to implement a comparable discrete design. For portable wireless applications, such devices can be integrated with switching functions and other related mobile radio circuits within the same physical package.

Another technology that shows promise for microwave receiver low-noise amplifiers and related circuits is the *silicon-germanium heterojunction bipolar transistor* (SiGe HBT). SiGe HBTs are designed for low voltage operation and low power consumption. ICs fabricated from SiGe offer the potential of operating above 10 GHz, with the cost and integration advantages of silicon devices. This technology uses *bandgap engineering* in the silicon device to decrease the base bandgap, which increases electron injection into the base. This enables base doping to be increased without a concomitant sacrifice in current gain. The increased base doping permits a lower base resistance to be obtained in SiGe than in Si for the same current gain. Furthermore, the graded Ge content in the base region induces a *drift field* in the base, which decreases the base transit time. The end result is an increase in the transition frequency of the device and significantly better performance in microwave circuits.

5.2 Representation of Linear Two-Ports

All active amplifiers, regardless of internal configuration, can be described by using a linear two-port model. Table 5.3 shows the relationship between input and output for the hybrid (*h*) parameters, which are generally used for the audio range and specified as real values. Table 5.4 shows the relationship between the *h* parameters in common-base and common-emitter bipolar configurations. The same relationships apply for FETs if the word "base" (*b*) is exchanged with "gate" (*g*), and "emitter" (*e*) is exchanged with "source" (*s*). The subscript for "collector" (*c*) should also be exchanged with "drain" (*d*).

For higher-frequency applications, between 1 and 100 MHz, it has become customary to use the admittance (*y*) parameters. Table 5.5 describes transistors by *y* parameters in the RF range. Once the *y* parameters in emitter or source configu-

TABLE 5.3 Relationships Among _h_ Parameters

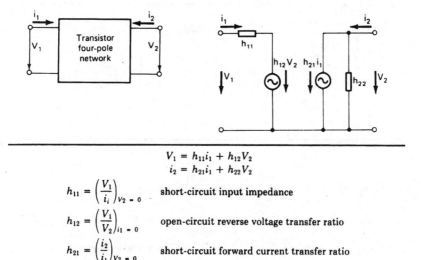

$$V_1 = h_{11}i_1 + h_{12}V_2$$
$$i_2 = h_{21}i_1 + h_{22}V_2$$

$h_{11} = \left(\dfrac{V_1}{i_i}\right)_{V_2\,=\,0}$ short-circuit input impedance

$h_{12} = \left(\dfrac{V_1}{V_2}\right)_{i_1\,=\,0}$ open-circuit reverse voltage transfer ratio

$h_{21} = \left(\dfrac{i_2}{i_1}\right)_{V_2\,=\,0}$ short-circuit forward current transfer ratio

$h_{22} = \left(\dfrac{i_2}{V_2}\right)_{i_1\,=\,0}$ open-circuit output admittance

rations are known, they can be calculated for the common-base, common-emitter, common-gate, or common-source configurations:

$$h_{11} = \frac{1}{y_{11}} \qquad\qquad h_{21} = \frac{y_{21}}{y_{11}}$$

TABLE 5.4 Relationships Among _h_ Parameters for Common-Base and Common-Emitter Configurations

$$\begin{pmatrix} h_{11b} & h_{12b} \\ h_{21b} & h_{22b} \end{pmatrix} = \frac{1}{1 + h_{21e} - h_{12e} + \Delta h_e}\begin{pmatrix} h_{11e} & -(h_{12e} - \Delta h_e) \\ -(h_{21e} + \Delta h_e) & h_{22e} \end{pmatrix}$$

$$\begin{pmatrix} h_{11b} & h_{12b} \\ h_{21b} & h_{22b} \end{pmatrix} \approx \frac{1}{1 + h_{21e}}\begin{pmatrix} h_{11e} & -(h_{12e} - \Delta h_e) \\ -h_{21e} & h_{22e} \end{pmatrix}$$

$$\begin{pmatrix} h_{11e} & h_{12e} \\ h_{21e} & h_{22e} \end{pmatrix} = \frac{1}{1 + h_{21b} - h_{12b} + \Delta h_b}\begin{pmatrix} h_{11b} & -(h_{12b} - \Delta h_b) \\ -(h_{21b} + \Delta h_b) & h_{22b} \end{pmatrix}$$

$$\begin{pmatrix} h_{11e} & h_{12e} \\ h_{21e} & h_{22e} \end{pmatrix} \approx \frac{1}{1 + h_{21b}}\begin{pmatrix} h_{11b} & -(h_{12b} - \Delta h_b) \\ -h_{21b} & h_{22b} \end{pmatrix}$$

$$\Delta h = h_{11}h_{22} - h_{12}h_{21}$$

TABLE 5.5 Relationships Among *y* Parameters

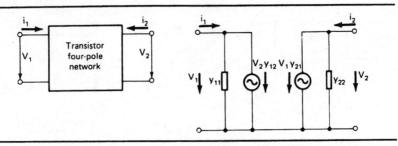

$$i_1 = y_{11}V_1 + y_{12}V_2$$
$$i_2 = y_{21}V_1 + y_{22}V_2$$

$y_{11} = g_{11} + jb_{11} = \left(\dfrac{i_1}{V_1}\right)_{V_2=0}$ short-circuit input admittance

$y_{12} = g_{12} + jb_{12} = \left(\dfrac{i_1}{V_2}\right)_{V_1=0}$ short-circuit reverse transfer admittance

$y_{21} = g_{21} + jb_{21} = \left(\dfrac{i_2}{V_1}\right)_{V_2=0}$ short-circuit forward transfer admittance

$y_{22} = g_{22} + jb_{22} = \left(\dfrac{i_2}{V_2}\right)_{V_1=0}$ short-circuit output admittance

$$h_{12} = \frac{-y_{12}}{y_{11}} \qquad\qquad\qquad h_{22} = \frac{\Delta y}{y_{11}} \qquad\qquad (5.4)$$

$$\Delta y = y_{11}y_{22} - y_{12}y_{21} = \frac{h_{22}}{h_{11}}$$

Similarly,

$$y_{11} = \frac{1}{h_{11}} \qquad\qquad\qquad y_{21} = \frac{h_{21}}{h_{11}}$$

$$y_{12} = \frac{-h_{12}}{h_{11}} \qquad\qquad\qquad y_{22} = \frac{\Delta h}{h_{11}} \qquad\qquad (5.5)$$

$$\Delta h = h_{11}h_{22} - h_{12}h_{21} = \frac{y_{22}}{y_{11}}$$

The power gain can be calculated from either the h or the y parameters,

$$G_p \equiv \frac{P_2}{P_1} = \frac{|G_v|^2 G_L}{G_1} = \frac{h_{21}^2 R_L}{(1 + h_{22} R_L)(h_{11} + \Delta h \, R_L)}$$

(5.6)

$$G_p = \frac{|y_{21}|^2 G_L}{\text{Re}\,[(Y_L \, \Delta \, y_{11} + \Delta y)(Y_L^* + y_{22}^*)]}$$

The power gain referred to the generator available gain (the data sheet specification G_p is generally based on this definition) is given by:

$$G_p \equiv \frac{P_2}{P_{G\,opt}} = \frac{4\,h_{21}^2 \, R_G \, R_L}{[(1 + h_{22} R_L)(h_{11} + R_G) - h_{12} h_{21} R_L]^2}$$

(5.7)

$$G_p = \frac{4\,|y_{21}|^2 \, G_G \, G_L}{[(Y_G + y_{11})(Y_L + y_{22}) - y_{12} y_{21}]^2}$$

With matching at the input and output, i.e., with $Z_G = Z_1^*$ and $Z_L = Z_2^*$, an ideal power gain can be achieved, but only if the following stability conditions are observed:

$$1 - Re\left\{\frac{y_{12} y_{21}}{g_{11} g_{22}}\right\} - \frac{1}{4}\left[Im\left\{\frac{y_{12} y_{21}}{g_{11} g_{21}}\right\}\right]^2 > 0;\ \Delta h > 0 \ (\text{every } h \text{ real})$$

(5.8)

If there were no feedback, i. e., $h_{12} = 0$, $y_{12} = 0$, the optimum gain $G_{p\ opt}$ would become:

$$G_{p\,opt} = \frac{h_{21}^2}{4\,h_{11}\,h_{22}} = \frac{|y_{21}|^2}{4\,g_{11}\,g_{22}}$$

(5.9)

TABLE 5.6 S Parameters for Linear Two-Ports

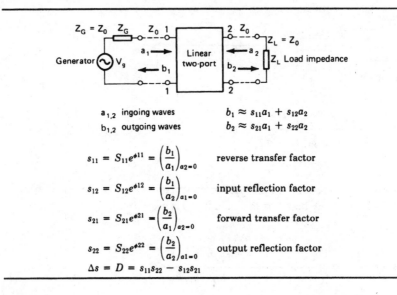

$a_{1,2}$ ingoing waves

$b_{1,2}$ outgoing waves

$b_1 \approx s_{11}a_1 + s_{12}a_2$

$b_2 \approx s_{21}a_1 + s_{22}a_2$

$$s_{11} = S_{11}e^{\phi 11} = \left(\frac{b_1}{a_1}\right)_{a2=0} \qquad \text{reverse transfer factor}$$

$$s_{12} = S_{12}e^{\phi 12} = \left(\frac{b_1}{a_2}\right)_{a1=0} \qquad \text{input reflection factor}$$

$$s_{21} = S_{21}e^{\phi 21} = \left(\frac{b_2}{a_1}\right)_{a2=0} \qquad \text{forward transfer factor}$$

$$s_{22} = S_{22}e^{\phi 22} = \left(\frac{b_2}{a_2}\right)_{a1=0} \qquad \text{output reflection factor}$$

$$\Delta s = D = s_{11}s_{22} - s_{12}s_{21}$$

In the case of neutralization, the values h_{ik}^2 and y_{ik}^2 modified by the neutralization four-pole network must be used.

In the frequency range above 1 GHz it has become customary to use the scattering (S) parameters. These are specified as complex values referred to a characteristic impedance Z_0. Table 5.6 shows the relationships and Table 5.7 lists the input and output reflection factors, voltage gain, power gain, stability factors, unilateral power gain, and ideal power gain in terms of the S parameters.

The stability factor of a transistor stage can be calculated from the feedback. The input admittance is given by:

$$Y_1 = y_{11} - \frac{y_{12}y_{21}}{y_{22} + Y_L} \tag{5.10}$$

If y_{12} is increased from a very small value, the input admittance can become zero, or the real part can become negative. If a tuned circuit were connected in parallel with y_{11} or if Y_L were a tuned circuit, the system would become unstable and oscillate.

TABLE 5.7 Performance Characteristics in Terms of S Parameters

Input reflection factor at any output Z_L	$s'_{11} = s_{11} + \dfrac{s_{12}s_{21}\Gamma_L}{1 - s_{22}\Gamma_L}$
Output reflection factor at any output Z_G	$s'_{22} = s_{22} + \dfrac{s_{12}s_{21}\Gamma_G}{1 - s_{11}\Gamma_G}$
Reflection factors of Z_G, Z_L referred to Z_0	Γ_G, Γ_L
Voltage gain at any output Z_G, Z_L	$G_v = \dfrac{V_2}{V_1} = \dfrac{s_{21}(1 + \Gamma_L)}{(1 - s_{22}\Gamma_L)(1 + s_{11})}$
Power gain	$G_p = \dfrac{P_2}{P_1} = \dfrac{|s_{21}|^2(1 - |\Gamma_L|^2)}{(1 - |s_{11}|^2) + |\Gamma_L|^2(|s_{22}|^2 - |D|^2) - 2\,\mathrm{Re}\,(\Gamma_L N)}$
Power gain referred to the generator performance available	$G_P = \dfrac{P_2}{P_{G\ \mathrm{opt}}} = \dfrac{|s_{21}|^2(1 - |\Gamma_G|^2)(1 - |\Gamma_L|^2)}{|(1 - s_{11}\Gamma_G)(1 - s_{22}\Gamma_L) - s_{12}s_{21}\Gamma_{LG}|^2}$
Stability factor	$K = \dfrac{1 + |D|^2 - |s_{11}|^2 - |s_{22}|^2}{2\,|s_{12}s_{21}|}$
Maximum power gain available ($K > 1$)	$G_{p\ \mathrm{max}} = \dfrac{s_{21}}{s_{12}}(K + \sqrt{K^2 - 1})$ for positive h_{FE1} where $T = \dfrac{s_{21}}{s_{12}}(K - \sqrt{K^2 - 1})$ for negative h_{FE1} $\Gamma_G = M^* \dfrac{h_{FE1} \pm \sqrt{h_{FE1}^2 - 4M^2}}{2|M|^2}$ $M = s_{11} - Ds^*_{22}$ $\Gamma_L = N^* \dfrac{h_{FE2} \pm \sqrt{h_{FE2}^2 - 4|N|^2}}{2|N|^2}$ $N = s_{22} - Ds^*_{11}$ $h_{FE1} = 1 + \left|s_{11}\right|^2 - \left|s_{22}\right|^2 - \left|D\right|^2$ $h_{FE2} = 1 + \left|s_{22}\right|^2 - \left|s_{11}\right|^2 - \left|D\right|^2$
Unilateral power gain ($s_{12} = 0$)	$G_p v = G_{p0}G_{p1}G_{p2}$ $G_{p0} = |s_{21}|^2;$ $G_{p1} = \dfrac{1 - |\Gamma_G|^2}{|1 - s_{11}\Gamma_G|^2};$ $G_{p2} = \dfrac{1 - |\Gamma_L|^2}{|1 - s_{22}\Gamma_L|^2}$
Ideal unilateral power gain ($s_{12} = 0$; $\Gamma_G = s^*_{11}$; $\Gamma_L = s^*_{22}$)	$G_{p\ \mathrm{opt}} = G_{p0}G_{p\ \mathrm{max}}G_{p2\ \mathrm{max}} = \dfrac{|s_{21}|^2}{(1 - |s_{11}|^2)(1 - |s_{22}|)^2}$

Courtesy of Siemens AG.

Table 5.8 shows the relationships between the S and y parameters. If the h parameters are known, Table 5.9 gives the relationships between them and the S parameters. In most cases, the S and y parameters are provided by the transistor manufacturer's data sheet. Sometimes these values may not be available for certain frequencies or certain dc operating points. The h, y, or S parameters can be calculated from the equivalent circuit of a transistor. In the case of a bipolar

TABLE 5.8 Relationships Between S and y Parameters

$s_{11} = \dfrac{(1 - y'_{11})(1 + y'_{22}) + y'_{12}y'_{21}}{(1 + y'_{11})(1 + y'_{22}) - y'_{12}y'_{21}}$	$y'_{11} = \dfrac{(1 - s_{11})(1 + s_{22}) + s_{12}s_{21}}{(1 + s_{11})(1 + s_{22}) - s_{12}s_{21}}$
$s_{12} = \dfrac{-2y'_{12}}{(1 + y'_{11})(1 + y'_{22}) - y'_{12}y'_{21}}$	$y'_{12} = \dfrac{-2s_{12}}{(1 + s_{11})(1 + s_{22}) - s_{12}s_{21}}$
$s_{21} = \dfrac{-2y'_{21}}{(1 + y'_{11})(1 + y'_{22}) - y'_{12}y'_{21}}$	$y'_{21} = \dfrac{-2s_{21}}{(1 + s_{11})(1 + s_{22}) - s_{12}s_{21}}$
$s_{22} = \dfrac{(1 + y'_{11})(1 - y'_{22}) + y'_{12}y'_{21}}{(1 + y'_{11})(1 + y'_{22}) - y'_{12}y'_{21}}$	$y'_{22} = \dfrac{(1 + s_{11})(1 - s_{22}) + s_{12}s_{21}}{(1 + s_{11})(1 + s_{22}) - s_{12}s_{21}}$

y parameters are standardized to Z_0.
Actual values are $y_{ik} = y'_{ik}/Z_0$; $i, k = 1,2$.

TABLE 5.9 Relationships Between S and h Parameters

$s_{11} = \dfrac{(h'_{11} - 1)(h'_{22} + 1) - h'_{12}h'_{21}}{(h'_{11} + 1)(h'_{22} + 1) - h'_{12}h'_{21}}$	$h'_{11} = \dfrac{(1 + s_{11})(1 + s_{22}) - s_{12}s_{21}}{(1 - s_{11})(1 + s_{22}) + s_{12}s_{21}}$
$s_{12} = \dfrac{2h'_{12}}{(h'_{11} + 1)(h'_{22} + 1) - h'_{12}h'_{21}}$	$h'_{12} = \dfrac{2s_{12}}{(1 - s_{11})(1 + s_{22}) + s_{12}s_{21}}$
$s_{21} = \dfrac{-2h'_{21}}{(h'_{11} + 1)(h'_{22} + 1) - h'_{12}h'_{21}}$	$h'_{21} = \dfrac{-2s_{21}}{(1 - s_{11})(1 + s_{22}) + s_{12}s_{21}}$
$s_{22} = \dfrac{(1 + h'_{11})(1 - h'_{22}) + h'_{12}h'_{21}}{(1 + h'_{11})(1 + h'_{22}) - h'_{12}h'_{21}}$	$h'_{22} = \dfrac{(1 - s_{11})(1 - s_{22}) - s_{12}s_{21}}{(1 - s_{11})(1 + s_{22}) + s_{12}s_{21}}$

h parameters are standardized to Z_0.
Actual values h are: $h_{11} = h'_{11}Z_0$, $h_{12} = h'_{12}$, $h_{21} = h'_{21}$, $h_{22} = h'_{22}/Z_0$.

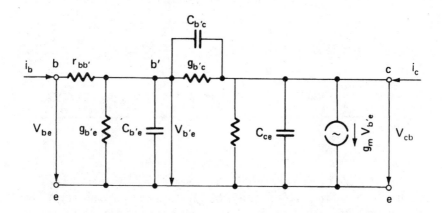

Figure 5.1 Bipolar transistor π-equivalent circuit.

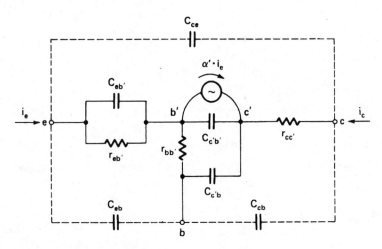

Figure 5.2 Bipolar transistor T-equivalent circuit.

transistor we can use Figure 5.1, recommended by Giacoletto [5.1], to determine the y parameters. For higher frequencies, the T-equivalent circuit shown in Figure 5.2 is recommended.

All of these calculations are independent of whether the transistors are of the bipolar or the field effect variety. However, FETs have slightly different equivalent circuits. Figure 5.3 shows the equivalent circuit for a JFET. This can also be applied to the MOSFET, VMOSFET, and GaAs FET. The dual-gate MOSFET, because of its different design, has a different equivalent circuit, as shown in Figure 5.4.

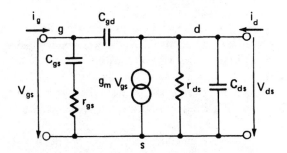

Figure 5.3 The equivalent circuit of a JFET.

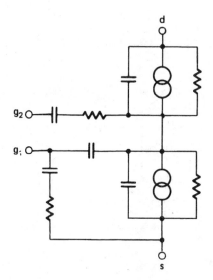

Figure 5.4 The equivalent circuit of a dual-gate MOSFET.

5.3 Noise in Linear Two-Ports with Reactive Elements

In Chapter 4, we discussed noise factor calculations, when the input is a generator with resistive internal impedance and where the various coupling or tuning circuits introduced only resistive components. The noiseless input impedance of the amplifier was also tacitly considered resistive, although that impedance does not affect the noise factor of the stage. We will now consider a more complete representation of a stage that is especially applicable at the higher frequencies. First we note that in using Figure 3.2b, we assumed that the noise voltage and current were independent random variables. (In estimating noise power, terms involving their product were assumed to be zero.) Because the devices we use are quite complex and feedback may occur within the equivalent two-port, this is not necessarily true. Our first correction, then, will be to assume that there may be a correlation between these variables, which will be represented by a correlation coefficient C, which may vary between ± 1.

While in Figure 4.1 we showed the possibility of generator and circuit reactances, in our simplified diagram of Figure 4.2 we assumed that they had been so adjusted that the net impedances were all resistive. Our second correction shall be to retain such reactances in the circuit. Even though it might be better to have net reactances of zero, it is not always possible to provide broad-band matching circuits that will do this for us. With these simplifications, we may replace Figure 3.2b with Figure 5.5.

Using the Thévenin representation of the generator plus extra noise sources, we convert the current source I_n to a voltage source $I_n Z_g$. Having done this, we find that our mean-square voltage includes five components, V_g^2, $4kTBR_g$, E_n^2,

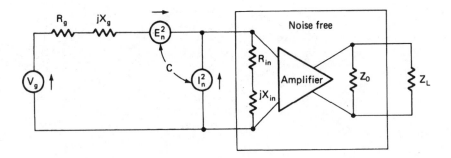

Figure 5.5 Representation of an amplifier with correlation of noise sources.

$I_n^2 Z_g Z_g^*$, and $C E_n I_n Z_g$. It is still assumed that the generator and the noise sources are uncorrelated. The resultant noise factor may be expressed as:

$$F = 1 + \frac{E_n^2 + I_n^2 (R_g^2 + X_g^2) + C E_n I_n (R_g + j X_g)}{4 \, k \, T \, B \, R_g} \tag{5.11}$$

We are left with a complex number, which is not a proper noise factor, so we must collect the real and imaginary parts separately and take the square root of the sum of their squares to get a proper noise factor. The resultant terms in the sum of the squares of the real and imaginary parts that involve X_g are:

$$S = I_n^4 X_g^4 + X_g^2 \, [C^2 \, E_n^2 \, I_n^2 + 2 \, I_n^2 \, (4 \, k \, T \, B \, R_g + E_n^2 + I_n^2 \, R_g^2 + C E_n I_n R_g)]$$

This will clearly be minimized by minimizing X_g, provided that the term in brackets is positive. The only thing that might make this expression negative would be a sufficiently large negative value of C. However, $|C|$ is at most unity, and this would make the last three terms in brackets $(E_n - I_n R_g)^2$, which must be positive. Consequently the noise factor is minimized by making X_g zero, as in the prior case. If this is done, Equation (5.11) becomes:

$$F = 1 + \frac{E_n^2 + I_n^2 \, R_g^2 + C E_n I_n R_g}{4 \, k \, T \, B \, R_g} \tag{5.12}$$

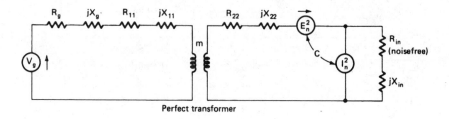

Figure 5.6 Schematic diagram of a circuit with a coupling network between the source and a noisy amplifier.

The optimum value of R_g is found by differentiating by it and setting the result to zero. As in the prior case, $R_{g\,opt} = E_n/I_n$. Thus, the correlation has not changed the optimum selection of R_g (assuming $X_g = 0$), and the minimum noise factor now becomes:

$$F_{opt} = 1 + \frac{E_n\,I_n\,(2 + C)}{4\,k\,T\,B} \tag{5.13}$$

This differs from the value obtained in Section 3.2 only by the addition of C, which was previously assumed to be zero. Depending on the value of C, then, the optimum NF may vary between $1 + E_nI_n/4kTB$ and $1 + 3E_nI_n/4kTB$.

Because the generator impedance is generally not directly at our disposal, any attempt at optimization must use a matching circuit. If the matching circuit were an ideal transformer, it would introduce an impedance transformation proportional to the square of its turns ratio m^2 without introducing loss or reactance. A separate reactance could be used to tune X_g to zero, and m could be adjusted to provide an optimum noise factor. Unfortunately, real coupling circuits have finite loss and impedance, as indicated in Figure 5.6. Because of the added losses, the overall circuit optimum noise factor may not require the same conditions as presented previously, nor will the optimum be so good. While it would be possible to deal with the overall circuit in Figure 5.6 and derive conditions for optimum operation, such a step will not be pursued because of the fact that many of the parameters cannot be varied independently. In the real world, variation in m may result in variations of the other circuit values. Moreover, E_n, I_n, and C are generally not available in the manufacturers' data sheets.

Consequently, in practice, the optimum result or a result as close to it as practical is achieved experimentally. It is important to remember to tune out reactances or at least to keep them much lower than the noise resistances. It is also important to remember that the optimum noise design is not necessarily the optimum power match for the input generator. Commonly, manufacturers provide data on the NF achieved with a particular matching circuit from a particular generator resistance under specified operating conditions. It can be assumed that the match provided by the particular circuit is near optimum, since manufacturers are interested in showing their products in the best light.

In the foregoing, we have ignored the input impedance of the amplifier because it does not affect the amplifier NF. While this is literally true, the relationship of the input impedance to the impedance of the source can affect the gain of the amplifier. We remember, however, that the effect of the noise generated in later stages depends upon the gain of the stage. A low NF in one stage is of little value if there is not sufficient gain in the stage to make its noise dominant over that generated by later stages. If there is a reactive component of the input impedance of a stage, it will reduce the current supplied to the input resistance and, consequently, reduce the input power to the stage. In essence, this reduces the gain of the preceding stage, so that its noise is reduced relative to that being generated by the stage under consideration. This results in a poorer overall NF. It is, therefore, desirable to use the input reactance of the amplifier as part of the reactance used to tune out the source reactance.

If we consider a FET, at 200 MHz we might have values of $E_n/\sqrt{B} = 2\times10^{-9}$ and $I_n/\sqrt{B} = 4\times10^{-12}$, leading to $R_{g\ opt} = 500\ \Omega$. To the extent that we can tune out the reactance and eliminate losses, the optimum noise factor of the amplifier, from Equation (5.13), becomes $1 + 0.503\ (2 + C)$. This represents a range of NFs from 1.8 to 4.0 dB. With $C = 0$, the value is 3.0 dB. These values are comparable to those listed by manufacturers. In the case of a bipolar transistor, extrapolating data from [4.11] on the 2N4124, we find that for 200 MHz, $E_n/\sqrt{B} = 1.64\times10^{-9}$ and $I_n/\sqrt{B} = 3.41\times10^{-11}$, leading to $R_{g\ opt} = 48.1\ \Omega$ and $F_{opt} = 1 + 3.52\ (2 + C)$. This represents a NF range of 6.55 to 10.63 dB. These values are rather higher than the typical 4.5 dB value for a VHF transistor, and 1.3 to 2 dB at this frequency for transistors designed for microwave use. We note that the 360-MHz f_t of the 2N4124 is substantially below the 600- to 1000-MHz f_t of VHF types, and the 2 GHz or above f_t of microwave devices.

5.4 Wide-Band Amplifiers

We next consider a common-emitter stage, as shown in Figure 5.7. Maximum gain is obtained if the collector impedance is raised to the maximum level at which the amplifier remains stable because the voltage gain is $G_v = -y_{21}R_L$. In this type of stage, there is a polarity inversion between input and output. The current gain β decreases by 3 dB at the β cutoff frequency f_β, for example, 30 MHz. This, in turn,

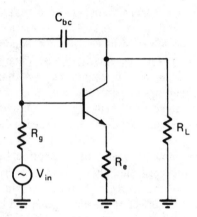

Figure 5.7 Schematic diagram of a common-emitter amplifier stage.

reduces the input impedance and decreases the stage gain as the frequency increases further. In addition, the collector-base feedback capacitance C_{CB} can further reduce the input impedance and can ultimately cause instability. The increase of input capacitance resulting from the voltage gain and feedback capacitance is called the *Miller effect*. The Miller effect limits the bandwidth of the amplifier.

The single common-emitter stage can be analyzed using the equivalent circuit shown in Figure 5.7. The resulting input impedance, output impedance, and voltage gain are plotted in Figure 5.8a, while Figure 5.8b, in comparison, plots the same parameters for a common-base stage. The short-circuit current gain α for the common-base configuration is much less frequency-dependent than the short-circuit current gain β for the earlier configuration. If we compare the gain-bandwidth products of the two circuits, we note that the common-base circuit, while having less gain, can be operated to higher frequencies than the common-emitter circuit.

To overcome this problem in the common-emitter stage, circuits have been developed using two or more transistors to eliminate the effect of the Miller capacitance and early reduction in β. An example is the differential amplifier shown in Figure 5.9, which combines an emitter-follower circuit with a grounded-base circuit. The emitter-follower stage guarantees a high input impedance, in contrast to a common-base input stage, and the cutoff frequency of the emitter-follower stage is substantially higher than that of the common-emitter. For all practical purposes, we can assume that the emitter-follower and grounded-base stages have the same cutoff frequency. Such a differential stage combines medium input impedance with extremely low feedback and is, therefore, suitable as a wide-band amplifier.

Another circuit that can be used successfully is the cascode arrangement. This circuit consists of a common-emitter stage whose output provides the input to a common-base stage. Since the output of the first transistor practically operates into

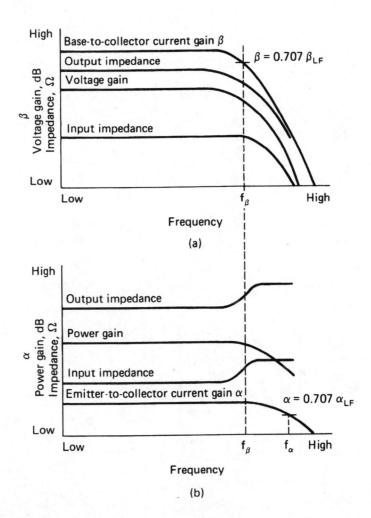

Figure 5.8 Performance curves: (*a*) common-emitter configuration and (*b*) common-base configuration.

a short circuit, this circuit combines the low feedback of the common-base stage with the medium input impedance of the common-emitter stage. The cascode arrangement has a somewhat better NF than the differential amplifier.

In integrated circuits, a combination of the two techniques is frequently used. Figure 5.10 shows the schematic diagram of a wide-band amplifier (MC1590 made by Motorola). Here the differential amplifier and the cascode arrangement are combined. The advantages of the differential amplifier in this particular case are

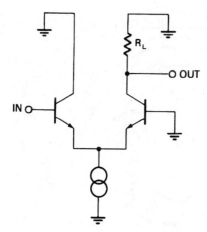

Figure 5.9 Schematic diagram of a differential amplifier circuit.

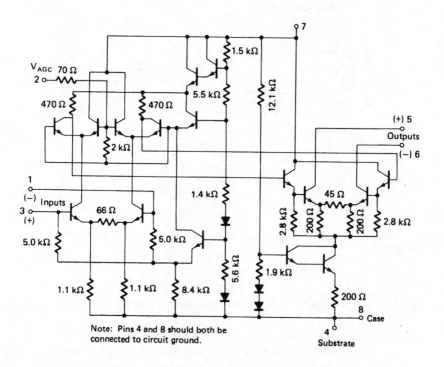

Figure 5.10 Schematic diagram of the MC1590 wide-band amplifier. (*Courtesy of Motorola.*)

TABLE 5.10 Parameters of Differential and Cascode Amplifiers

	Differential pair compared to common-emitter stage with twice the dc bias current of each transistor of the differential pair	Cascode connection compared to common-emitter stage with the same dc bias current
y_{11}	1/4	1
y_{12}	1/30 to 1/200	1/200 to 1/2000
y_{21}	1/4	1
y_{22}*	1 to 1/3	1 to 1/3

* For $\omega C_{b'e} r_{bb} \ll 1$, $y_{22} \approx p C_{b'c}(1 + g_m r_{bb'})$, while for the two configurations with common-base output stages it is approximately $p C_{b'c}$. (Ideally the cascode case should have an even smaller y_{22}; in practice, parasitic terms tend to keep it from being much smaller.) ($p = j\omega$)

thermal stability and the possibility of applying AGC. Table 5.10 provides the information necessary to calculate the gain of the differential amplifier and the cascode amplifier. Table 5.11 indicates typical matched gains and NFs for these configurations as compared to a single unit.

With respect to the high-frequency parameters, it is interesting to compare the input and output admittances of a single transistor, a differential amplifier, and a cascode amplifier using the same type of transistor. For very high isolation, sometimes a cascode arrangement with three transistors is used. The dual-gate MOSFET is based on the cascode principle. In our previous discussions, the question was raised whether bipolar transistors or FETs are of greater use at high frequencies. The gain-bandwidth product by itself is not sufficient to answer this question. The feedback component y_{12} in the equivalent circuit for FETs is typically 1 pF and is higher than what is found in bipolar transistors. Special transistors have been developed that have extremely low feedback capacitance by using an internal Faraday shield. Typical versions are the BF173 for non-AGC stages in IF amplifiers and the BF167 for AGC amplifiers. The feedback capacitance of these transistors is in the vicinity of 0.12 pF so that they can be used without neutralization. Dual-gate MOSFETs provide still lower feedback capacitance, 0.02 to 0.03 pF.

The drawback of FETs and, specifically, VMOS stages is that the input capacitance can be very high. VMOS transistors operated at 150 MHz with a 12-V supply

TABLE 5.11 Comparison of Matched Gains and NFs

	Single unit	Cascode connection	Differential pair
Gain	41.1 dB	44.5 dB	39 dB
Noise figure	4 dB	6 dB	7 dB

can produce 10-W output. However, for these devices the input impedance is 100-pF capacitance in series with a few ohms. It is possible to develop a wide-band matching circuit for this input, but the design requires a sufficiently high voltage to be generated into this very low impedance. As a result, the usable gain is much less than predicted by the theoretical gain-bandwidth product. For narrow-band operations, which are not suitable in many power amplifier applications, stable gains of 20 dB at 150 MHz can be obtained. The input capacitance of low-power FETs intended for receiver amplifiers is substantially lower (4 to 8 pF), but the input susceptance is sufficiently low that a similar broadbanding problem exists.

For example, the input Q of the 3N200, obtained from a representative data sheet, is 13.6 at 100 MHz, reducing to 2.7 at 500 MHz. The input Q of a typical grounded-emitter bipolar transistor intended for use in this frequency range would be about one-third of these values. It would, however, require neutralization to realize its full gain capability. Confirming these indications, experience has shown that wide-band operation is much easier to achieve with bipolar transistors than with FETs.

Wide-band amplifiers are typically used to increase the signal level with the highest possible reverse isolation. Input and output impedance cannot be modified, and only the effects of cutoff frequency can be compensated. In the following section, it is shown that properly designed feedback amplifiers allow adjustment of the impedances (within limits), and the use of feedback techniques can produce improved amplifier linearity.

5.5 Amplifiers with Feedback

The wide-band amplifiers discussed previously achieved their bandwidth through the clever combination of two or more transistors. This allowed compensation of the Miller effect, and circuits such as the cascode arrangement are widely used as wide-band amplifiers in antenna distribution systems, for example. Another technique that results in increased bandwidth is the use of negative feedback. In the feedback amplifier, a signal from the output is applied to the input of the amplifier with a reversal of phase. This reduces distortion introduced by the amplifier and makes the amplifier less dependent upon transistor parameters. At the same time, it reduces the gain of the amplifier and, depending on the particular feedback circuit, can change and stabilize input and output impedances.

In discussing feedback amplifiers, we distinguish between three classes:

- Single-stage resistive feedback
- Single-stage transformer feedback
- Multistage and multimode feedback

Before discussing specific feedback designs, however, we shall review the general effects of negative feedback on gain stability and noise factor.

5.5.1 Gain Stability

Because the individual transistor stages have a gain-bandwidth factor that depends on the device configuration and operating point, uniformity of gain over a wide bandwidth is achieved by reducing the overall gain. The net gain of a feedback amplifier can be expressed as:

$$A = \frac{A_0}{1 - F A_0} \tag{5.14}$$

where F is the feedback factor, which is adjusted to be essentially negative real in the frequency band of interest. When this is so, $A < A_0$. When $FA_0 \gg 1$, then A reduces to $-1/F$. In practice, A_0 may decrease with frequency and may shift in phase, and F may also be a complex number with amplitude and phase changing with frequency. To maintain constant gain over a wide band, with small dependence on transistor parameters, the magnitude of A_0 must remain large and F must remain close to a negative real constant value. Outside of the band where these conditions exist, the feedback stability criteria must be maintained. For example, the roots of the denominator in Equation (5.14) must have negative real parts, or the locus in the Argand diagram of the second term in the denominator must satisfy Nyquist's criterion [5.2].

Modern transistors have a drift field in the base-emitter junction, generated in the manufacturing process, which produces excess phase shift at the output. To maintain stability for feedback, it is necessary to compensate for this excess phase shift. In a simple voltage divider used for feedback, such excess phase shift cannot be easily compensated. For complex feedback systems, such as multistage amplifiers with both transformer and RC feedback, additional all-pass networks are required to correct for excessive phase shift.

5.5.2 Noise Considerations

If noise is considered a form of distortion introduced in the amplifier, similar to the nonlinear effects, we might expect feedback to improve the S/N of the system. In practice, this does not occur. The input noise sources of the amplifier are not changed by the feedback, so that the amplified S/N at the output remains the same. The feedback reduces both signal and noise amplification in the same ratio. Noise and other distortion products generated later in the amplifier are reduced by about the same amount. This implies, however, that the total output S/N should remain about the same whether or not feedback is applied.

The additional components necessary to produce the feedback add noise, so that the overall noise factor of the circuit may be somewhat poorer. More importantly, with feedback connected, the gain of the amplifier is reduced, so that the effect of noise from subsequent circuits on overall NF will increase. Countering this trend, feedback can change the input impedance of the circuit so that it may be possible to produce higher gain from the prior circuit, thereby tending to reduce the effect of the feedback amplifier's noise contribution.

If resistive feedback is used, especially emitter degeneration, the high-frequency NF is increased. It can be observed easily that simple feedback, such as those found in RF circuits, produces a substantially improved dynamic range and, simultaneously, a poorer NF than the circuit without feedback. While the noise degradation in simple RC feedback can be explained mathematically, it is difficult to forecast the actual resultant NF. It is more useful to determine NF experimentally.

Where it is essential to minimize NF degradation, a technique called *noiseless feedback* may be used. Noiseless feedback is based on the concept of transforming the output load resistance to the input in such a way as to provide the necessary feedback without introducing additional thermal noise. As a result, the NFs of such systems are minimally changed. The transformers may have losses in the vicinity of 1 dB or less, which can change the NF by this amount.

The NF of a stage is determined by the various parameters in the equivalent circuit. Depending on the type of feedback (positive or negative), the input impedance can be increased or decreased. If the equivalent noise resistor R_n remains unchanged while the input impedance is increased, the overall NF will decrease. If the feedback method changes both equivalent noise resistor and input impedance similarly, then the NF may remain unchanged. This is the usual effect, because the amplified noise as well as the amplified signal are fed back in the process. The feedback is more likely to have an effect upon the NF because of the change in input impedance and amplifier gain, so that the noise in prior and subsequent circuits may play a larger or smaller part in determining the overall NF.

Feedback can change both the resistive and the reactive parts of the input impedance, as well as other parameters. Therefore, it is possible to find a combination where the feedback by itself cancels the imaginary part and changes the input impedance in such a way that the overall NF is improved. Such "noise matching" can be achieved by an emitter-base feedback circuit. This is the only circuit where power, noise matching, and minimum reflection can be achieved simultaneously.

The influence of feedback is best understood by studying an example. We will examine a case where an input filter is used between the input signal generator and the first transistor, and where feedback can modify impedances and other design parameters. The example starts with an amplifier that has been designed initially neglecting circuit losses and potential transistor feedback. The basic circuit schematic is shown in Figure 5.11a. The transistor input admittance y_{11}, and its estimated added R_n have been considered, but the effect of the feedback admittance (Miller effect) has been ignored. The transistor and generator impedances have been

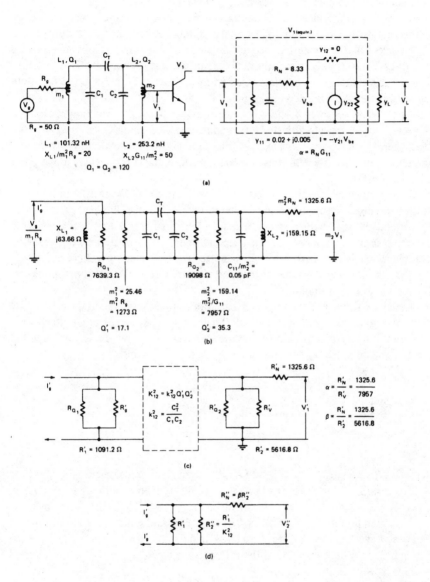

Figure 5.11 Simple feedback example: (a) schematic diagram, (b) equivalent circuit at 100 MHz, (c) equivalent circuit at resonance, and (d) equivalent circuit referred to the input circuit.

stepped up to produce initial operating Q values of 20 and 50, as shown. These values of Q will produce a 3-dB bandwidth about the 100-MHz center frequency in the vicinity of 5 MHz, with transitional coupling between the tuned coupled pair.

To select and evaluate the design parameters more fully, it is assumed that a circuit Q of 120 (rather than infinite Q) exists and that the overall NF of the stage is to be determined for coupling adjusted for (A) optimum NF and (B) optimum power transfer. As indicated in Figure 5.11b, the effective operating Q values have now been reduced to 17.1 and 35.3, respectively. Next, the coupling coefficient $k_{12} = C_T/\sqrt{C_1 C_2}$ must be adjusted to provide the desired conditions. It is well known that when $k_{12}\sqrt{Q_1 Q_2}$ is equal to unity, maximum power is transferred between the circuits. It is convenient to measure coupling in units of this value, so we write $K_{12} = k_{12}\sqrt{Q_1 Q_2}$.

At resonance the reactance in both circuits is tuned out. The resistance reflected from the second circuit to the first can be shown to be $R_2'' = R_1'/K_{12}^2$, where R_1' is the effective total shunt resistance in the first circuit, and R_2'' is the reflected effective shunt resistance from the second circuit. R_1' is made up of the parallel combination of circuit loss resistance and effective generator input shunt resistance; R_2'' has the same proportions of loss and transistor effective shunt resistances as the circuit shown in Figure 5.11c.

The noise factor of the circuit is the relationship of the square of the ratio of the noise-to-signal voltage at V_2'', with all the noise sources considered, to the square of the ratio of the generator open-circuit noise-to-signal voltage. Referring to the simplified equivalent circuit shown in Figure 5.11d, we find:

$$F = \frac{R_g'}{R_1'}\left[\frac{1 + K_{12}^2}{K_{12}^2} + \frac{\beta\,(1 + K_{12}^2)^2}{K_{12}^2} \right] \tag{5.15}$$

with $R_N'/R_V' = \alpha$, $\alpha R_N'/R_2' = \beta$, and $K_{12}^2 = R_1'/R_2''$ (as indicated previously).

To optimize F for variations in K_{12}, we set its derivative with regard to K_{12}^4 in the previous equation equal to zero and find $K_{12}^2 = (1 + \beta)/\beta$. This leads to $K_{12} = 1.513$, and in turn to $k_{12} = 0.0616$, $R_g'/R_1' = 1.1666$, and $F = 2.98$, or NF = 4.74 dB. For best power transfer, $K_{12} = 1$, k_{12} 0.0407, and $F = 3.43$, or NF = 5.36 dB. Because NF is based on the generator as the reference, it includes the losses in the tuned circuits as well as the transistor NF.

Figure 5.12 shows the selectivity curves for the two filters with different coupling factors. The coupling that produces the higher NF provides narrower selectivity. The coupling capacitor may be determined as $C_t = k_{12}\sqrt{C_1 C_2}$ where C_1 and C_2, the capacitances required to tune the two coupled coils to resonance at the carrier frequency, are 25 and 10 pF, respectively. Thus, in case A, $C_{12A} = 0.97$ pF and in case B, $C_{12B} = 0.64$ pF. Because of the difficulty of controlling such small capacitances, it would be better to convert from the π arrangement of capacitors to

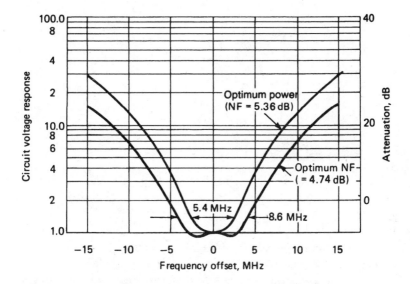

Figure 5.12 Selectivity curves for different coupling factors and feedback.

a T arrangement. The main tuning capacitors must be adjusted to compensate for the coupling capacitor, the reflected reactance from the transistor, the coil distributed capacitance, and any other strays.

Let us now assume that because of feedback through the base-collector junction (Miller effect), the input admittance is altered so that the input conductive component leads to a shunt resistance of 300 Ω instead of 50 Ω. We will also assume that the noise resistor stays the same, although—in fact—it is likely to change as well. We must change our tap so that the impedance step-up is decreased to produce the equivalent loading in the second circuit. This means that $m_2{}^2$ becomes 26.5 instead of 159.1. The equivalent noise resistor at the secondary is reduced in this same ratio to 220.7 Ω. The new value of α becomes 0.0277; that of β, 0.0393. This results in $K_{12} = 2.27$, $F = 1.73$ and NF = 2.38 dB. In the matched case, the NF is also improved to 4.01 dB. In both cases, this represents more than 1 dB of improvement; for optimum F, the bandwidth is further widened.

5.5.3 Types of Feedback

If a single transistor stage, as shown in Figure 5.13, is operated in small-signal condition and the effect of the Miller capacitance is not neglected, we can distinguish between two types of distortion: voltage distortion and current distortion. The current distortion is the result of the transfer function of the device. This transformation of the input voltage to an output current is nonlinear. The output current multiplied by the output load impedance becomes the output voltage.

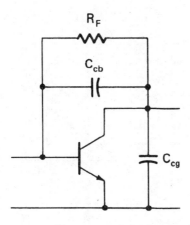

Figure 5.13 Schematic diagram of a single transistor stage, showing collector-base feedback.

The voltage distortion is observed because the output of the transistor has two semiconductor capacitances that are voltage-dependent. The feedback capacitance C_{cb} and the output capacitance C_{ce} both vary with the output voltage. If this voltage reaches levels of several volts, substantial variations of the capacitances occur. This, as well as modulation of the output collector-base junction, results in the nonlinear distortion called *voltage distortion*.

It should be noted that current distortion can only be compensated by current feedback, and voltage distortion by voltage feedback. This can best be shown by measuring the IM distortion under two-tone test conditions in a CATV transistor, such as the 2N5179, as a function of direct current. Figure 5.14 shows that the IM distortion products become smaller as the direct current is increased, with the drive

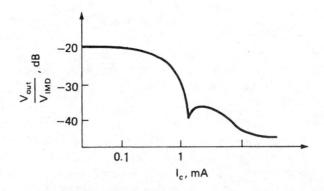

Figure 5.14 Variation of IM products with dc level for constant small-signal drive.

level constant. This is because the exponential transfer characteristic of the base-emitter diode is more nearly linearized.

As the direct current is increased, NF deteriorates slightly, and optimum NF and IM do not occur at the same operating point. This particular effect corresponds to the feedback circuit in Figure 5.15a, where there is an input voltage V_{in} and resistors R_g and R_e in series with the generator and the emitter, respectively. The presence of the unbypassed resistor in the emitter circuit increases the input and output impedances of the transistor and decreases the IM distortion products. If we analyze the same figure, we will notice that the input and output impedances are also changed as a function of the feedback resistance R_F. However, as long as the dynamic input impedance generated by considering R_F is not reduced below R_g, IM distortion products generated by current distortions are not compensated.

These are the two most important feedback types, and combinations of them are in common use. In practice we find the following feedback systems:

- Voltage series or voltage ratio feedback
- Current series or transimpedance feedback
- Voltage shunt or admittance feedback
- Current shunt or current ratio feedback

Based on the particular feedback technique, the input impedance may be increased or decreased. In some cases, both feedback systems are used simultaneously. The input impedance, then, can be set to whatever value is required, while still reducing the distortion products. Each case must be analyzed individually. Figure 5.15 shows a number of simple feedback circuits at frequencies low enough that the internal Miller feedback is negligible. Table 5.12 summarizes their gain and impedance characteristics.

5.5.4 Mixed Feedback Circuits

Purely resistive feedback amplifiers have the disadvantage that their noise performance is poorer than that of the transistor itself under optimum conditions. Mixed feedback allows wider flexibility since the input impedance, output impedance, and gain can be set more or less independently. We now examine one design example in detail. The interested reader can employ the same techniques used in this example to analyze other circuits. In such circuits, we rely heavily on the use of ferrite core transformers. It is important that these transformers have minimum stray inductance and that the bandwidth ratio $B = f_{max}/f_{min} = 1/s$ be as large as possible. Ratios of more than 200 are possible.

Figure 5.16 shows the circuit of the amplifier using voltage feedback. The following equations apply:

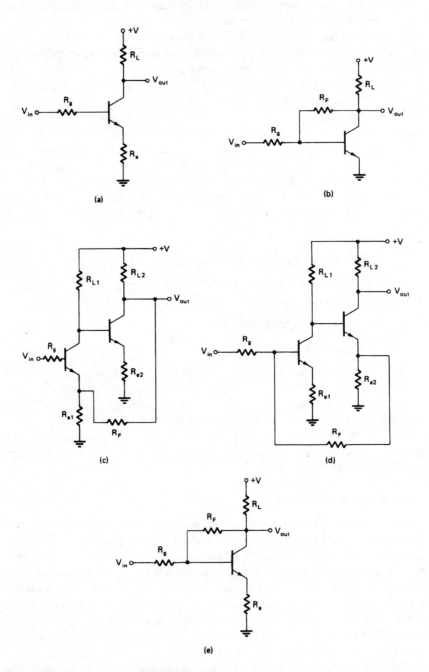

Figure 5.15 Simple transistor feedback circuits.

TABLE 5.12 Characteristics of Transistor Feedback Circuits Shown in Figure 5.15

	Fig. 5.15a	Fig. 5.15b	Fig. 5.15c	Fig. 5.15d	Fig. 5.15e
Voltage gain A_v			$\dfrac{A_0}{1 + FA_0}$		
Input impedance R_{in}	$R_0(1 + FA_0) = R_g + r_{be} + R_e(1+\beta)$	$R_g + \dfrac{R_F}{1 + (R'/R_p')FA_0}$	$(R_g + r_{d1})(1 + FA_0)$	$R_g + \dfrac{R_F'}{1 + FA_0}$	$R_g + \dfrac{R_F'}{1 + (R'/R_F')FA_0}$
Output impedance R_{out}	R_L	$\dfrac{R_L'}{1 + FA_0}$	$\dfrac{R_{L2}}{1 + FA_0}$	R_{L2}	$\dfrac{R_L'}{1 + FA_0}$
Open-loop voltage gain A_0	$\dfrac{-\beta R_L}{R_g + r_{be} + R_e}$	$\dfrac{-\beta R_L' R'}{r_{be}R_g}$	$A_1 \cdot A_2$	$A_1 \cdot A_2 \cdot \dfrac{R'}{R_g}$	$\dfrac{-\beta R_L R'}{r_{be} + (\beta + 1)R_e}$
Feedback factor F	$\dfrac{R_e}{R_L}$	$-\dfrac{R_g}{R_F}$	$\dfrac{R_{e1}}{R_{e1} + R_F}$	$\dfrac{R_{e2} R_g}{R_{L2} R_F}$	$-\dfrac{R_g}{R_F}$
Other		$R^1 = \dfrac{r_{be}R_g R_F}{r_{be}(R_g + R_F) + R_g R_F}$ $R_L^1 = \dfrac{R_L R_F}{R_L + R_F}$	$A_1 = \dfrac{-\beta_1 R_{L1}}{R_g + r_{d1}}$ $A_2 = \dfrac{-\beta_2 R_{L2}}{R_{L1} + r_{d2}}$ $r_{d1} = r_{be1} + R_{e1}(1 + \beta_1)$ $r_{d2} = r_{be1} + R_{e2}(1 + \beta_2)$ $R_{L2}^1 = \dfrac{R_{L2}(R_F + R_{e1})}{R_{L2} + R_F + R_{e1}}$ $R_{e1}^1 = \dfrac{R_{e1}R_F}{R_{e1} + R_F}$	$A_1 = \dfrac{-\beta R_{L1}}{r_{d1}}$ $A_2 = \dfrac{-\beta R_{L2}}{R_{L1} + r_{d2}}$ $r_{d1} = r_{be1} + (1 + \beta_1)R_{e1}$ $r_{d2} = r_{be2} + (1 + \beta_2)R_e$ $R_{e2}^1 = \dfrac{R_{e2}R_F}{R_{e2} + R_F}$ $R^1 = \dfrac{r_{be}R_g R_F}{r_{be}(R_g + R_F) + R_g R_F}$ $R_F^1 = \dfrac{R_F r_d}{R_F + r_d}$	$R_L^1 = \dfrac{R_L R_F}{R_L + R_F}$ $R^1 = \dfrac{r_d R_e R_F}{r_d(R_g + R_F) + R_e R_F}$ $r_d = r_{be} + (1 + \beta)R_E$

Transistor Approximation: $h_{11} = r_{be}$; $h_{21} = \beta$; $h_{12} = 0$; $h_{22} = 0$.

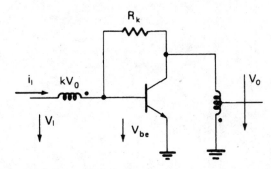

Figure 5.16 Schematic diagram of an amplifier using voltage feedback.

$$V_I = k V_0 + V_{be} \tag{5.16}$$

$$i_I = \frac{V_{cb}}{R_k} + V_{be} y_{11} \tag{5.17}$$

$$Z_I \equiv \frac{V_I}{i_I} = \frac{k V_0 + V_{be}}{V_{cb}/R_k + V_{be} y_{11}} \tag{5.18}$$

For an open-loop voltage gain $A_0 \equiv V_0/V_{be} \geq 10$ and an operating frequency $f \leq f_T/10$, so that $V_{be} y_{11}$ is negligible, the following simplifications are possible:

$$A \equiv \frac{V_0}{V_I} = \frac{1}{k} \tag{5.19}$$

$$Z_I = k R_k \tag{5.20}$$

As an example, let us consider a circuit with $R_k = 200 \ \Omega$ and $k = 0.2$, using a transistor type 2N5109 at an i_c of 80 mA. At this operating point, approximately,

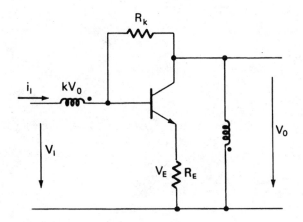

Figure 5.17 Transistor amplifier with current feedback.

$g_m = 1.5$ S, $R_{ce} = 200\ \Omega$, and $f_T = 1400$ MHz. This leads to $A_0 = g_m\ [(R_{ce}R_k/R_{ce} + R_k)] \approx 150$. Therefore, the approximation holds when $f < 1400/10 = 140$ MHz. Thus, $A = 5$ and $Z_I = 40\ \Omega$.

We now introduce current feedback, which results in the new schematic diagram shown in Figure 5.17. For this circuit, we can write the following equations:

$$V_I = k\ V_0 + V_{be} + V_E \tag{5.21}$$

$$i_I = V_{be}\,y_{11} + \frac{V_{cb}}{R_k} \tag{5.22}$$

$$Z_I = \frac{k\ V_0 + V_{be} + V_E}{V_{be}\,y_{11} + V_{cb}/R_k} \tag{5.23}$$

$$V_E = I_e\ R_E \tag{5.24}$$

$$i_e \approx i_c \text{ with } f \leq f_T/10 \tag{5.25}$$

$$i_c = \frac{V_0}{R_L} + \frac{V_{cb}}{R_k} \tag{5.26}$$

or, after some rearranging,

$$i_C = \frac{V_0 \, (R_k/R_L + 1)}{R_k + R_E} \tag{5.27}$$

If we assume V_{be} is small enough to be ignored, we have for the input impedance:

$$Z_I = R_k \frac{k + (R_E/R_L) \, (R_k + R_L)/(R_E + R_k)}{1 - (R_E/R_L) \, (R_k - R_L)/(R_E + R_k)} \tag{5.28}$$

We may write this as:

$$Z_I = \frac{R_k \, (k + C)}{1 + C} \tag{5.28a}$$

with,

$$C = \frac{R_E}{R_L} \frac{R_k + R_L}{R_E + R_k} \tag{5.29}$$

Also, we finally obtain the formula for the voltage gain,

$$A = \frac{1}{k + C} \tag{5.30}$$

The lower cutoff frequency of the circuit is determined by the main inductor of the transformer. An experimental circuit of this kind was measured to have 50-Ω input impedance between 1 and 150 MHz, with a VSWR of 1.3. The VSWR to 200 MHz was 1.5. The noise factor up to 40 MHz was 2.5, increasing to 7.5 at 200 MHz. Two signals at a level of +6 dBm generate two spurious signals 60 dB below the normal reference signal.

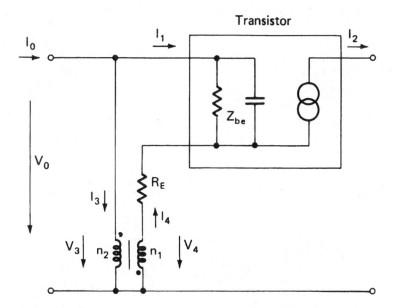

Figure 5.18 Schematic diagram of a base-emitter feedback amplifier.

5.5.5 Base-Emitter Feedback Circuit

The base-emitter feedback circuit is a configuration that has seen moderate use, particularly for conditions where the lowest possible NF, highest possible gain, and minimum standing wave ratio are required simultaneously. Experience with transistor amplifiers has shown that simultaneous satisfaction of these three conditions is not possible with most feedback circuits.

The emitter-base feedback circuit is probably the only circuit in which such performance can be achieved. Further, use of a bipolar transistor in this circuit leads to excellent large-signal handling performance. An FET should not be used because its square-law characteristic generates a second harmonic of the source current that is transferred to the gate by the transformer. This results in the generation of the products $f_1 \pm f_2$ and $f_2 \pm f_1$. Because of the exponential form of the bipolar transistor, the generation of these products is much smaller. Figure 5.18 shows the schematic diagram of this circuit.

The transformer relationships are the turns ratio, $u = n_2/n_1$, $V_4 = - V_3/u$, and $I_4 = -I_3 u$. The transistor relationships are $I_2 = -I_1\beta \exp{(-jf/f_T)}$ and $Y_{be} \equiv 1/Z_{be} = S_0/\beta + j\omega S_0/2\pi f_T$, where $S_0 = I_E/V_T$. Other basic relationships in the circuit are $I_4 = I_2 - I_1$, $V_0 = V_3$, $V_3 = I_1 Z_{be} - I_4 R_E + V_4$, and $I_0 = I_1 + I_3$. From the foregoing, the following intermediate relationships are derived:

$$I_4 = -I_1 \left[1 + \beta \exp\left(-jf/f_T\right)\right] \tag{5.31}$$

$$V_0 = I_1 Z_{be} + u I_3 R_E - \frac{V_0}{u} \tag{5.32}$$

$$I_3 = \frac{-I_4}{u} = \frac{I_1 \left[1 + \beta \exp\left(-jf/f_T\right)\right]}{u} \tag{5.33}$$

It will be convenient to let $\beta \exp\left(-jf/f_T\right) = D$ in the equations that follow:

$$V_0 \left(1 + \frac{1}{u}\right) = I_1 \left[Z_{be} + R_E \left(1 + D\right)\right] \tag{5.34}$$

$$I_0 = I_1 \left(1 + \frac{1+D}{u}\right) \tag{5.35}$$

The input impedance can be calculated from $Z_I = V_0/I_0$, resulting in:

$$Z_I \left(1 + \frac{1}{u}\right) = \frac{Z_{be} + R_E \left(1 + D\right)}{1 + \left(1 + D\right)/u} \tag{5.36}$$

Generally $\beta \gg 1$, and if $\beta f < f_T$ as well, we can approximate D as $\beta \left(1 - jf/f_T\right)$. Then, we have:

$$Z_I \left(1 + \frac{1}{u}\right) = \frac{Z_{be} + R_E D}{1 + D/u} \tag{5.37}$$

The real part of D is β, and $\beta/u \gg 1$ will also be assumed.

$$Z_I \left(1 + \frac{1}{u}\right) = \frac{u Z_{be}}{D} + u R_E \tag{5.38}$$

Recalling the definition of Z_{be}, we may write:

$$Z_I = \frac{u^2}{u+1} \left[\frac{1}{D\ S_0\ (1/\beta + jf/f_T)} + R_E \right]$$

$$Z_I = \frac{u^2}{u+1} \left[R_E + \frac{1}{S_0\ (1 - jf/f_T)\ (1 + j\beta f/f_T)} \right]$$

$$(5.39)$$

We have already assumed $\beta f/f_T < 1$, which implies $f/f_T \ll 1$. Then, the parenthetical terms in the denominator become $1 + j(\beta - 1)f/f_T$. The imaginary term is less than 1. If it is sufficiently less to be neglected, we finally arrive at the following:

$$Z_I = \frac{u^2}{1+u} \left(R_E + \frac{1}{S_0} \right)$$

$$(5.40)$$

Therefore, the input impedance is the transformed sum of the reciprocal mutual conductance and the resistive portion of the emitter feedback resistor.

As an example, we select the values $n_2 = 4$, $n_1 = 2$, $\beta = 20$, $f = 40$ MHz, $f_T = 1.66$ GHz, $R_E = 7\ \Omega$, and $S_0 = 0.385S$. This gives us $u = 2$, $\beta f = 800 < 1660 = f_T$, and

$$Z_I = \frac{2^2}{3} \left(7 + \frac{1}{0.385} \right) = 1.333(7 + 2.6) + 12.8\Omega$$

The complete expression for the input impedance of this circuit was calculated for various frequencies, using a computer program, with results given in Table 5.13.

5.6 Gain Control of Amplifiers

The large dynamic range of signals that must be handled by most receivers requires gain adjustment to prevent overload or IM of the stages and to adjust the demodulator input level for optimum operation. A simple method of gain control would involve the use of a variable attenuator between the input and the first active stage. Such an attenuator, however, would decrease the signal level, but it would also reduce the S/N of any but the weakest acceptable signal. Most users are willing to tolerate an S/N of 10 to 20 dB for weak signals, but expect an S/N of 40 dB or more for stronger signals.

TABLE 5.13 Calculated Input Impedance of Base-Emitter Feedback Circuit in the Example Circuit

Frequency, MHz	Input impedance, Ω	Real	Imaginary
10	11.87	11.87	-0.12
46.4	11.85	11.84	-0.57
100	11.81	11.74	-1.22
464.4	10.66	9.54	-4.78
1000	7.61	4.88	-5.80

Therefore, gain control is generally distributed over a number of stages, so that the gain in later stages (the IF amplifiers) is reduced first, and the gain in earlier stages (RF and first IF) is reduced only for signal levels sufficiently high to assure a large S/N. In modern radios, where RF gain tends to be small, this may mean switching in an attenuator at RF only for sufficiently high signal levels. Variable gain control for the later stages can operate from low signal levels. Variable-gain amplifiers are controlled electrically, and when attenuators are used in receivers, they are often operated electrically either by variable voltages for continuous attenuators or by electric switches (relays or diodes) for fixed or stepped attenuators. Even if attenuators are operated electrically, the operator sometimes needs direct control of the gain. This may be made available through a variable resistor or by allowing the operator to signal the control computer, which then sets the voltage using one or more *digital-to-analog* (D/A) converters. Control should be smooth and cause a generally logarithmic variation (linear decibel) with the input variable. In most instances, because of fading, AGC is used to measure the signal level into the demodulator and to keep that level in the required range by a feedback control circuit.

The simplest method of gain control is to design one or more of the amplifier stages to change gain in response to a control voltage. In tube radios, gain was changed by changing the amplifier's operating point. It was found necessary to design special tubes for such control in order to avoid excessive IM distortion. Similarly, transistor amplifiers require special circuits or devices for amplifier stage gain control. One circuit arrangement for this application uses one gate of a dual-gate FET as the gain control device while the signal is applied to the second gate. In this way the g_m of the device is varied with minimum change in the operating point of the signal gate. Figure 5.19 shows the schematic diagram of this arrangement, using a 3N200, along with the change in gain with the control voltage on the second gate. Because a dual-gate FET is the equivalent of a cascode connection of two FETs, that circuit works similarly. A cascode arrangement of bipolar transistors may also be used.

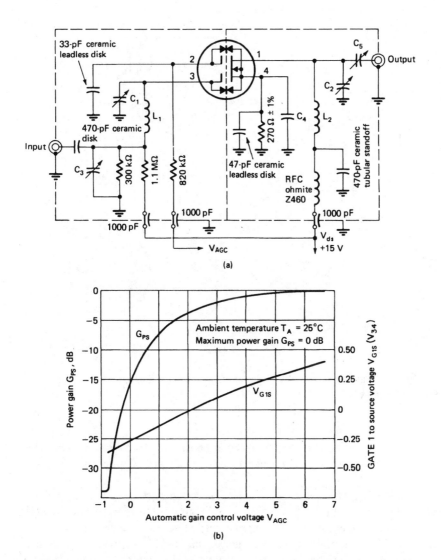

Figure 5.19 Gain-controller amplifier: (*a*) schematic diagram and (*b*) curve of gain versus control voltage.

A common bipolar circuit for gain control is a differential pair of common-emitter amplifiers whose emitters are supplied through a separate common-emitter stage. The gain-control voltage is applied to the base of the latter stage while the signal is applied to one (or, if balanced, both) of the bases of the differential pair.

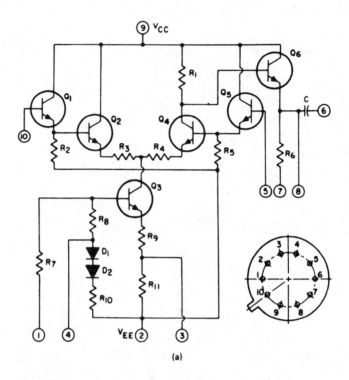

(a)

Figure 5.20 Gain-controlled IC amplifier: (*a*) schematic diagram and (*b*, next page) gain-control curve.

This arrangement has been implemented in linear integrated circuits. Figure 5.20 shows the schematic diagram of such a gain-controllable amplifier stage, the RCA CA3002, and its control curves.

A PIN diode attenuator can provide the low-distortion gain control that is especially important prior to the first mixer. Figure 5.21 shows such a circuit for the HF band. Its control curve has approximately linear decibel variation over most of its 60-dB range. The π-type attenuator circuit is used to provide a good match between 50-Ω terminations over the control range. The minimum useful frequency for a PIN diode attenuator varies inversely with the minority carrier lifetime. For commonly available diodes, the low end of the HF band is near this limit.

Other devices without a low-frequency limitation have been used for low-distortion gain control, including *positive temperature coefficient* (PTC) resistors controlled by a dc level to heat them and photoconductive devices controlled by the level of impinging light. A receiver normally uses several stages of controlled gain to produce the total range of control required.

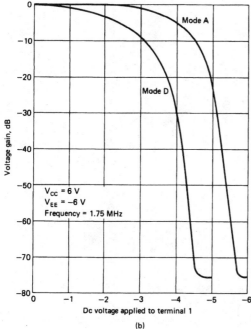

(b)

Figure 5.20 *continued*

5.6.1 AGC

The narrow-band signals encountered in receiver design may be modulated in amplitude or frequency, or both simultaneously. The baseband output depends on the modulation index and the signal level at the demodulator input. An AGC circuit in the receiver provides a substantially constant signal level to the demodulator independent of the input signal level. In the previous paragraphs we have discussed devices that can maintain linear performance over a wide range of programmable gain levels. The AGC provides to one or more of such devices the external control signals necessary to maintain the constant signal level required by the demodulator.

To help understand the operation of the AGC action, let us examine the block diagram in Figure 5.22. An input voltage, which may lie between 1 μV and 1 V, is fed to the input amplifier. The envelope of this voltage is detected at the input to the detector. This voltage is processed to produce the control voltages for variable-gain devices, which reduce the input to the amplifier and the gain within the amplifier. As we try to maintain constant output voltage with varying input voltage, we are dealing with a nonlinear system, which can be described accurately by nonlinear differential equations. The literature does not provide a complete mathe-

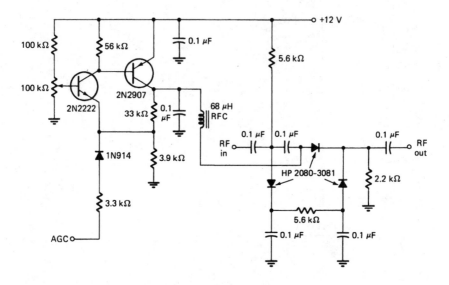

Figure 5.21 Schematic diagram of a PIN diode attenuator.

matical analysis of such a system. We will, instead, use a linearized model to deal with the problem. A more complete result, if required, may be achieved through computer simulation of the nonlinear system.

Let us assume for the moment that the amplifier has a control range of 120 dB. This implies that we have at least two gain-controlled devices with 60 dB control range, because few devices are available—either discrete or integrated—with more than 60-dB gain range. For the sake of simplicity, we will not assume a NF for the system, although we will assume that in the output bandwidth, the output noise

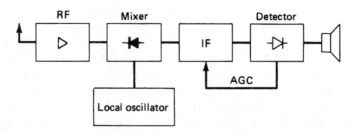

Figure 5.22 Simplified block diagram of a communications receiver with AGC.

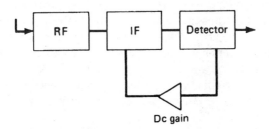

Figure 5.23 Block diagram of a receiver with amplified AGC.

voltage is 100-mV rms and the output impedance is low. The 1-µV input signal would generate a 1-V output signal (with 20-dB S/N). For further simplification, we assume that the amplifier is linear with constant gain for an output voltage below 1 V. For a control voltage between 1 and 10 V, we assume that the gain is reduced (generally linearly in decibels with the voltage) by 120 dB to unity gain. The AGC detector, which generates the control voltage, is assumed to have 1-V threshold, above which its output responds linearly to its input.

Neglecting the dynamics of the situation, we would find that with no signal voltage present at the input, the AGC detector receives 0.1-V rms, which is below threshold and thus generates no AGC control voltage. As the voltage at the AGC detector rises to 1 V (input of 1 µV), the control begins. As the output voltage rises to 10 V, the amplifier gain is reduced to 0 dB (input of 10 V). We assume that this would occur in a relatively linear manner. Thus, we have an output variation of 10:1 (20 dB) for an input variation of 120 dB. For most professional receiver applications, such performance would be considered rather poor.

In a good receiver design, the AGC onset may well be at a 20-dB S/N, as in our example, at 1 µV, but would maintain the output within 6 dB or less for the 120-dB change in input. To reduce the output level swing from 20 to 6 dB, an increased gain reduction sensitivity is required. Thus, if the gain change of 120 dB can be achieved from 1 to 2 V (rather than 1 to 10 V), the desired improvement results. This might be achieved by using more sensitive gain change devices or by using more devices in the amplifier. Another alternative is to provide an amplifier for the AGC voltage. An amplification of five times produces the desired effect. The block diagram of Figure 5.23 shows such a configuration. If we assume that the control time constants are determined primarily by the detector circuit and the additional amplifier has a wider bandwidth than the detector, then the attack and decay times will be shortened by the amount of the postamplification (five times in the example).

In a high-quality receiver, the AGC detector operates essentially in a linear mode, the dc output being proportional to the RF input voltage. It is possible for a diode to operate as a square-law detector when the RF voltage is small compared to the typical levels of one to several volts for the linear detection region of typical germanium, silicon, or hot-carrier diodes. In the square-law region, the diode is barely biased *on* by its junction bias, and its response is nonlinear. The detection sensitivity is higher in this region of operation. Such operation occurs typically when the design does not provide adequate IF gain so that voltage levels as low as 0.1 V may need to be rectified. A dc control is typically superimposed in the circuit to bias the diode slightly in order to provide higher efficiency and lower threshold than otherwise possible. The effect can be observed in inexpensive AM pocket radios, where the same diode is used for both AM detection and AGC. As the signal increases, the output S/N improves up to a certain point, then deteriorates, and finally improves again. This "holdback" area has to do with bias changes of these devices running in the square-law region, and is tolerated only to avoid providing more IF gain, which would increase the cost.

5.6.2 AGC Response Time

The AGC system has a delay in its response to changes in input. This means that the AGC control voltage holds constant for a short time after a change in signal level and then follows the change to compensate for the level change. In practice, it is not desirable for an AGC to have too fast a reaction time. In such a case, any static pulse, ignition noise, or other impulsive interference with very fast rise time would be detected by the AGC detector and would desensitize the receiver for a hold time required to discharge the AGC filter capacitors.

For many years, communications receivers used attack times between 1 and 5 ms. For CW operation and in thunder storms, this proves too fast and hangs up the receiver. The fastest attack time that is possible depends upon the filtering of the detector, the response of the amplifiers, the IF selectivity, and the IF itself. For SSB reception, some receivers derive the AGC from the baseband signal. Rather than use a high-level IF amplifier, such receivers may use a product detector to convert the signal to baseband at a level of about 10 mV. The resultant signal is then amplified and rectified to develop the AGC control voltage. If an IF-derived AGC is used, the lowest practical frequency is about 30 kHz. The minimum attack time would require about one cycle of the IF, or 33 µs. If the lowest audio frequency generated were 50 Hz, the audio-generated AGC attack time might be extended to 20 ms. Care would need to be taken not to use the audio-derived system for control of an RF carrier (at or near zero beat). On the whole, it appears that a baseband-derived AGC design should be avoided in high-performance receivers.

Selection of the proper AGC time constant is a subjective decision. Most receiver manufacturers now set the attack time between 20 and 50 ms and resort to additional means to combat short-term overload of the system. An excellent method of achieving this objective is to run the second stage of the amplifier in such a mode that it will clip at 6 dB above nominal output level.

For example, let us assume that our previous case is implemented, whereby an AGC amplifier with a gain of 5 allows a variation of the dc control voltage between

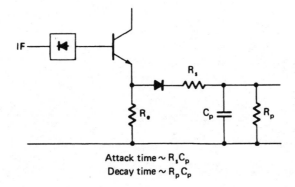

Attack time ~ $R_s C_p$
Decay time ~ $R_p C_p$

Figure 5.24 Schematic diagram of an AGC system with independent attack and decay time constants.

1 and 2 V. The clipping level would be set at twice the higher value, or 4 V. This can be achieved either by proper biasing of the amplifier or by using symmetrical diodes at the output. In some designs, the Plessey SL613 logarithmic amplifier has proved useful, because it prevents the output from exceeding a certain value. Such a circuit arrangement prevents audio amplifier overload during the attack time of the receiver, and fast static crashes will not block the receiver as a result of a fast AGC response time.

This discussion of AGC times applies primarily to CW and SSB reception, where there is no carrier to serve for AGC control. We do not want the background noise to rise between transmissions, so a dual-time-constant system is desirable. Figure 5.24 shows a dual-time-constant system in which the attack time is determined by values R_s and C_p, while the decay time is dependent on R_p and C_p. The diode prevents a discharge of the capacitor from the source. This exponential decay is not sufficient in some cases, and an independent setting of three time constants—attack, hold, and decay—is desirable. Figure 5.25 shows a three-time-constant circuit where all of these values can be established independently.

Attack and decay times are typically defined as the time it takes to get within a certain percentage of the final value after a signal appears or disappears. It turns out, however, that the loop gain, which determines the loop band-width, is dependent upon the actual gain reduction. Therefore, the attack and decay times should be defined for the highest-gain reduction or maximum input voltage. In most cases, the receiver designer will find that for the first 60 dB of increase in amplitude, up to 1 mV, the AGC will behave well. However, for the next 20 dB, the AGC may become unstable and oscillate. There are several causes for such instabilities. Assuming the case of the simple AGC circuit, which we will analyze later, where there are no delay or dead times, we have to deal with the phase shifts of the various amplifiers, and therefore instabilities can occur. In addition, the capacitor C_p in our

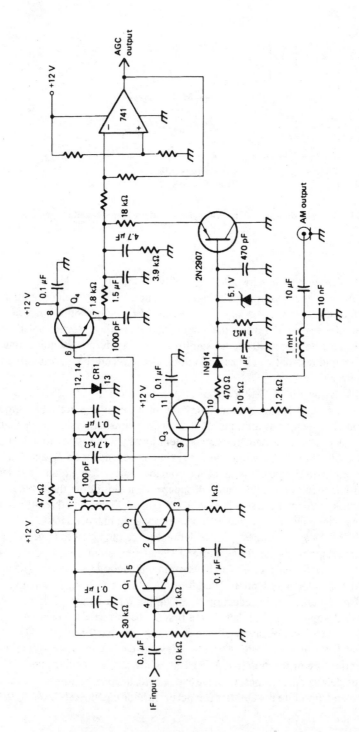

Figure 5.25 Schematic diagram of an AGC system with independent attack, hold, and decay time constants.

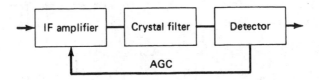

Figure 5.26 Block diagram of an IF amplifier with AGC and a crystal filter.

previous example has to be charged, so the current source has to be able to supply enough current for the charge. In many cases, the dc bias of the transistor stage that charges the capacitor is wrongly adjusted and therefore cannot follow. It is important to understand that the driving source has to be capable of providing proper currents.

In the case of an AM signal, the AGC cannot be made faster than the lowest modulation frequency. In a broadcast receiver, 50 Hz or 20 ms is too small a margin, and a 60- to 100-ms attack time should be preferred. If the AGC time constant is made too fast, the modulation frequency response will be changed and distortion may occur.

5.6.3 Effect of IF Filter Delays

A selective filter introduces not only frequency selectivity but also delay, the amount depending on the specific design. Most IF filters use crystal resonators for the selective elements, but delays result from any filter type. Figure 5.26 is a block diagram of an IF amplifier that incorporates a crystal filter prior to the AGC detector. The purpose of the filter is to limit the noise bandwidth of the circuit. If it is assumed that the amplifier comprises two wide-band stages (such as the Plessey SL612), the noise bandwidth is 30-MHz wide. For the AM or AGC detector, this would produce an extremely poor S/N. The introduction of a single- or dual-pole monolithic crystal filter will limit the noise bandwidth to about ±5 kHz, thus improving the S/N substantially. The filter, especially if it has a flat top and sharp cutoff characteristic (like a Chebyshev filter), may introduce substantial delay, ranging from a few microseconds to as much as 50 ms. Some mechanical filters, resonant at low frequencies, such as 30 kHz, have extremely steep shirts, and also can have delays of 50 to 100 ms. With such a delay, the AGC detector produces a gain-control voltage responding to the signal at a substantially earlier time. If the AGC attack time is smaller than the delay, such delays can cause AGC instabilities. It is important, therefore, to make sure that the AGC attack time is longer than possible delays or else to avoid delays in the system so as to use a short attack time. The delay varies across the filter band, the most critical points being between the –3 and –10 dB points of the selectivity curve. At these points, extreme delays occur and the AGC is most vulnerable.

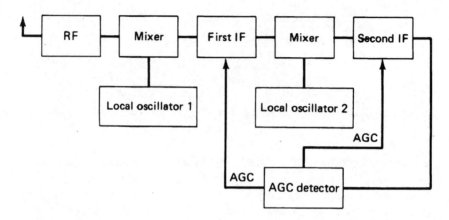

Figure 5.27 Block diagram of a dual-conversion superheterodyne receiver with AGC applied to both IF amplifiers.

It is common practice to design an AGC so as to avoid the excessive delays of crystal filters. By reorganizing the previous example and using separate signal and AGC detectors, it is possible to use the high-delay crystal filter in the AM or SSB detector path, and use a broader filter with smaller delay in the AGC loop.

Some designers feel that there are merits to applying AGC to both the first and the second IF amplifiers of a dual-conversion receiver, such as shown in Figure 5.27. In this situation, we can again experience the delay introduced by the selective filters. If there are multiple bandwidths, the delay will vary with the bandwidth, being most pronounced for the most narrow filter (CW operation). This system can be further complicated if there is an RF amplifier and the AGC is extended to it. Because of the potentially longer delay through the longer path, such a configuration can become difficult. The AGC loop will tend to be very sluggish, with long attack time to provide AGC stability under all circumstances.

5.6.4 Analysis of AGC Loops

The AGC system is basically a feedback amplifier or servo system, as shown in Figure 5.28. A number of authors have provided treatments of the AGC loop ([5.3] to [5.8]). The gain-control curve is essentially nonlinear, so in most cases, linearized treatments of the loop are given. In [5.5], an exponential control curve of the gain is assumed, which leads to a direct analysis without linearization. Our treatment will mostly follow reference [5.6].

The static performance of the AGC system is an important characteristic that shows how successful the design is in maintaining a constant output for varying input voltage. Such curves may be drawn easily if the control characteristic of the

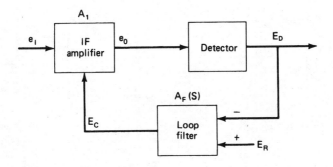

Figure 5.28 Block diagram of an AGC loop.

amplifier and the dc transmission of the AGC loop are known. Figure 5.29 shows a typical amplifier control characteristic, in this case with a range of nearly 100 dB. If we assume the AGC detector has 100 percent efficiency, the dc control voltage is equal to the output voltage. For each output amplitude level, we may compute the control voltage produced, and from Figure 5.29 we can determine the gain. The output divided by the gain determines the input voltage. Figure 5.30 shows two such curves as solid curves. In the lower curve the control voltage is assumed to be equal to the output voltage, while in the upper curve the control voltage is assumed

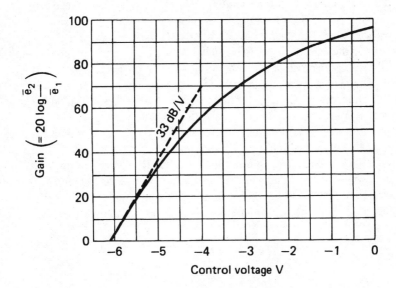

Figure 5.29 Representative gain-control characteristic curves.

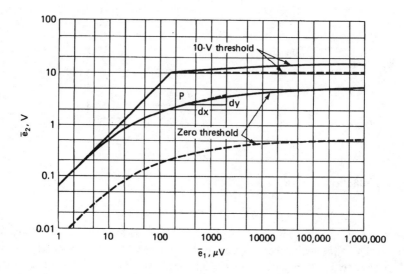

Figure 5.30 AGC regulation characteristics.

to be zero until the output exceeds 10 V, whereafter the control voltage is equal to the output voltage less 10 V. The advantage of using a delay voltage in the control circuit is obvious.

The advantage of adding an amplifier in the AGC path may be seen in the dashed curves of Figure 5.30, where the control voltage is amplified ten times. Where no delay is used, the result is simply reduction of the output voltage ten times, without modifying the control shape. In the delayed case, only the difference is amplified, and the result is a much flatter AGC curve. If the amplification were provided prior to the delay voltage, the result would be to drop the onset of AGC to an output level of 1 V. The curve is parallel to the delayed curve without amplification but provides control over an extra octave.

As with all feedback systems, care must be taken to design the loop to avoid oscillation. Also, the AGC loop has a closed-loop gain characteristic, which is essentially low-pass. The values must be selected to minimize reduction of the desired amplitude modulation of the signal. Clearly, the loop response must be slow compared to signal modulation, yet it should be comparable to fading. To deal with a linearized loop, the following equations can be written. First, assume a linear detector:

$$E_D = k\, e_0 - E_d \qquad\qquad \frac{d\, E_D}{d\, e_0} = k \qquad\qquad (5.41)$$

where E_d is the diode voltage drop. Second, for a square-law detector,

$$E_D = k\, e_0^2 \qquad\qquad \frac{d\, e_D}{d\, e_0} = 2\, k\, e_0 \qquad\qquad (5.42)$$

The relationship between input and output voltages is given by,

$$e_0 = A_I\, e_I \qquad\qquad (5.43)$$

where A_I is the amplifier voltage gain. If we differentiate this relative to control voltage E_c, and divide by itself we obtain the derivative of the logarithms,

$$\frac{d(\ln e_o)}{d\, E_C} + \frac{d(\ln A_I)}{d\, E_C} \qquad\qquad (5.44a)$$

$$\frac{d(\ln e_0)}{d\, E_C} = \frac{d(\ln e_I)}{d\, E_C} + K_n \qquad\qquad (5.44b)$$

where K_n is the amplifier gain-control constant. We define $1/K_D$ as the logarithmic derivative of e_O with regard to E_D and obtain:

$$\frac{1}{K_D} = \frac{d\, e_0}{e_0\, d\, E_D} = \frac{d(\ln e_0)}{d\, E_D}; \qquad K_D = \frac{d\, E_D}{d(\ln e_0)} \qquad\qquad (5.45)$$

$$\frac{d\, E_D}{K_D\, d\, E_C} = \frac{d(\ln e_I)}{d\, E_C} + K_n \qquad\qquad (5.44c)$$

Let us now refer to Figure 5.31; using the standard terminology for the closed control loop, we obtain:

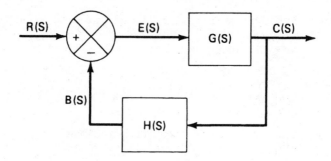

Figure 5.31 AGC block diagram using standard control-loop terminology.

$$M(S) = \frac{C(S)}{R(S)} = \frac{G(S)}{1 + G(S)\,H(S)} \qquad (5.46)$$

From Figure 5.28, $M(S) = dE_D/d\,[\ln (E_I)]$; $A_f(S) = -dE_c/dE_D$. From Equation (5.44), $-[K_D A_f(S)]^{-1} = -[M(S)\,A_f(S)]^{-1} + K_N$. Therefore, $G(S) = K_D$ and $H(S) = K_n A_F(S)$. Also, the amplifier gain is usually measured in decibels per volt rather than in nepers per volt. If we let K_I be the sensitivity in decibels per volt and $K_c = 0.11513$ Np/dB, then $K_n = K_c K_I$. The open loop transfer function is:

$$\frac{B(S)}{E(S)} = G(S)\,H(S) = K_D\,K_c\,K_I\,A_F(S) \qquad (5.47)$$

The loop error transfer function is:

$$\frac{E(S)}{R(S)} = \frac{1}{1 + K_D\,K_c\,K_I\,A_F(S)} \qquad (5.48)$$

A unit step function input is used in evaluating the transient response. The attack time of the AGC loop is the time required for the resulting error $e(t)$ to fall to a specified value, usually 0.37 or 0.1.

The most common type of loop filter is the simple integrator shown in Figure 5.32. If we substitute $A_F(S) = A_1/S$ in Equation (5.48), we find the error response:

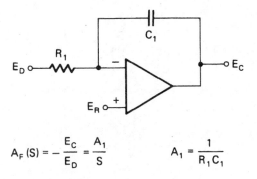

$$A_F(S) = -\frac{E_C}{E_D} = \frac{A_1}{S} \qquad\qquad A_1 = \frac{1}{R_1 C_1}$$

Figure 5.32 Simple integrator filter circuit.

$$\frac{E(S)}{R(S)} = \frac{S}{S + K_v} \tag{5.49}$$

where $K_v = K_D K_c K_I A_1$. To calculate the error response to a step input, we set $R(S) = 1/S$, yielding $E(S) = 1/(S + K_v)$. The inverse Laplace transform yields a time error response $e(t) = \exp(-K_v t)$, a simple exponential decay with time constant equal to K_v.

When the loop filter is a simple integrator, if the input voltage is an AM sinusoid, with modulation or index m and frequency f_A, the detector output can be shown to be [5.5]:

$$E_D = E_R \left[\frac{1 + m \sin(2\pi f_A t)}{1 + \beta\, m \sin(2\pi f_A t + \theta)} \right] \tag{5.50}$$

where $\beta = [1 + (2\pi f_A)/K_v^2]^{-1/2}$ and $\theta = -\tan^{-1}(2\pi f_A/K_v)$. The quantities β and θ are easily identified as the magnitude and phase angle of the unity feedback closed-loop gain. From Equation (5.50), it can be seen that at low frequency, the detector output is a dc level, $E_D = E_R$, and at high frequency the output is the undistorted modulation. The distortion that the denominator of Equation (5.50) describes is dependent on the modulation index. We arbitrarily define the tolerable distortion for maximum modulation index ($m = 1$) and $\beta < 0.2$. This allows us to set a low-frequency cutoff $\beta^2 = 1/26$, or $2\pi f_c = 5K_v$. Because the loop gain will have a similar effect on the modulation, whatever the type of loop filter, we may use the same expression to define the cutoff frequency even when the filter is more complex than a simple integrator.

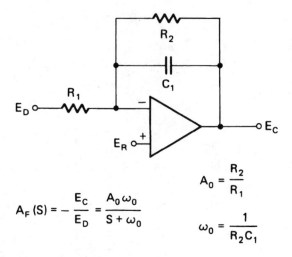

$$A_0 = \frac{R_2}{R_1}$$

$$A_F\,(S) = -\frac{E_C}{E_D} = \frac{A_0\,\omega_0}{S + \omega_0}$$

$$\omega_0 = \frac{1}{R_2 C_1}$$

Figure 5.33 Loop filter with finite dc gain.

For a filter with finite dc gain, such as shown in Figure 5.33,

$$G(S)\,H(S) = K_D\,K_c\,K_I\,A_0 \left[\frac{2\,\pi\,f_0}{S + 2\,\pi\,f_0}\right] \qquad (5.51)$$

Substituting the dc loop gain or positional error constant $K_P = K_D K_c K_I A_0$ in Equation (5.51), we obtain:

$$G(S)\,H(S) = \frac{K_p\,2\,\pi\,f_0}{S + 2\,\pi\,f_0} = \frac{N(S)}{D(S)} \qquad (5.52)$$

$$\frac{E(S)}{R(S)} = \frac{D(S)}{N(S) + D(S)} = \frac{S + 2\,\pi\,f_0}{S + (1 + K\,P)2\,\pi\,f_0} \qquad (5.53)$$

For a unit step function input,

$$E(S) = \frac{S + 2\pi f_0}{S[S + (1 + K P)2\pi f_0]} \tag{5.54}$$

This yields, by using the inverse transform pair,

$$e(t) = \frac{1}{1 + K_P} + \frac{K_P \exp[-(1 + K_P)2\pi f_0 t]}{1 + K_P} \tag{5.55}$$

The unity-gain closed-loop transfer is:

$$\frac{B(S)}{R(S)} = \frac{N(S)}{N(S) + D(S)} = \frac{K_P 2\pi f_0}{S + (1 + K_P)2\pi f_0} \tag{5.56}$$

Setting this equal to 1/26, as above, and solving for cutoff, we obtain:

$$f_c = f_0(25 K_P^2 - 2 K_P - 1)^{1/2} \tag{5.57}$$

For $K_P \gg 1$, this becomes $f_c \approx 5 K_P f_0$.

5.6.5 Dual-Loop AGC

In applications where NF is very important, high-gain preamplifiers may be used preceding the first mixer. Figure 5.34 shows a block diagram where a preamplifier with variable gain is used. As in the case of a microwave receiver, where such a configuration would find its most likely use, the selectivity prior to the second IF filter would be relatively wide. This will be assumed here. The design is somewhat more complex in this case than where a preamplifier is not used. For operation over a wide dynamic range, it is necessary to apply control to the preamplifier to prevent strong input signals from overloading the following stages. On the other hand, reduction of the preamplifier gain increases the NF of the system. Low NF is the reason for using the preamplifier stage in the first place. To solve these difficulties, it is necessary to delay application of AGC to the preamplifier until the gain in the later stages has been sufficiently reduced by the input level. Thus, as shown in Figure 5.34, two threshold references are provided, the initial one for the IF stages, and the second one, which prevents reduction of preamplifier gain until there is an adequate S/N at the output. Figure 5.35 shows a loop filter for this

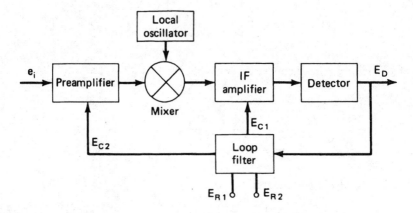

Figure 5.34 Block diagram of a receiver including a controlled preamplifier.

configuration, and Figure 5.36 shows a reconfiguration as a standard dual-control loop.

A loop filter of this type determines the closed-loop parameters, and when f_1 is properly chosen, the bandwidth changes very little when the second threshold is reached. Improper choice of the parameters can cause an undesired change of bandwidth and gain at the second threshold, which could cause loop instabilities. The open-loop gain of the system when both thresholds are exceeded is given by:

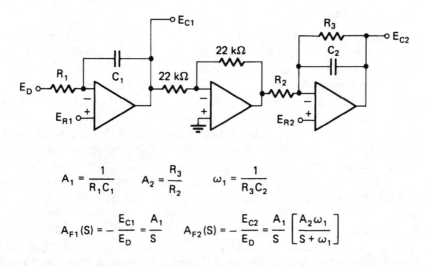

$$A_1 = \frac{1}{R_1 C_1} \qquad A_2 = \frac{R_3}{R_2} \qquad \omega_1 = \frac{1}{R_3 C_2}$$

$$A_{F1}(S) = -\frac{E_{C1}}{E_D} = \frac{A_1}{S} \qquad A_{F2}(S) = -\frac{E_{C2}}{E_D} = \frac{A_1}{S}\left[\frac{A_2 \omega_1}{S + \omega_1}\right]$$

Figure 5.35 Loop filter for a dual AGC loop.

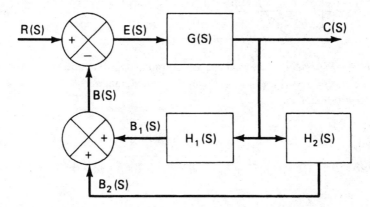

Figure 5.36 Dual-loop AGC system redrawn as a standard control loop.

$$\frac{B(S)}{E(S)} = G(S)\,[H_1(S) + H_2(S)]$$

$$\frac{B(S)}{E(S)} = \frac{K_D\,K_c\,K_{I1}\,A_1}{S} + \frac{K_D\,K_c\,K_{I2}\,A_1\,A_2\,2\,\pi f_1}{S(S + 2\,\pi f_1)} \qquad (5.58)$$

$$\frac{B(S)}{E(S)} = \frac{K_D\,K_c\,K_{I1}\,A_1}{S}\left[1 + \frac{K_{I2}\,A_2\,2\,\pi f_1}{K_{I1}(S + 2\,\pi f_1)}\right]$$

With $K_1 = K_D K_c K_{I1} A_1$ and $K_2 = A_2 K_{I2}/K_{I1}$, we obtain:

$$\frac{B(S)}{E(S)} = \frac{K_1}{S}\left(1 + \frac{K_2\,2\,\pi f_1}{S + 2\,\pi f_1}\right) \qquad (5.59)$$

Now let us also introduce a delay or dead time factor τ. With this included, we obtain:

$$\frac{B(S)}{E(S)} = \frac{K_1}{S}\left[1 + \frac{K_2\,2\,\pi f_1\exp(-s\tau)}{S+2\,\pi f_1}\right] \tag{5.60}$$

For sufficiently small $s\tau$, we may use a polynomial approximation of the exponential, $\exp(-s\tau) = 1 + a_1(s\tau) + a_2(s\tau)^2$, $a_1 = -0.9664$, and $a_2 = 0.3536$. The closed-loop error function is:

$$\frac{E(S)}{R(S)} = \frac{S(S+2\,\pi f_1)}{S^2 + S(2\,\pi f_1 + K_1) + K_1\,2\,\pi f_1 + K_1\,K_2\,2\,\pi f_1\exp(-s\tau)} \tag{5.61}$$

For a step change, the error voltage is:

$$E(S) = \frac{S+2\,\pi f_1}{X(S^2 + SY/X + Z/X)} \tag{5.62}$$

where,

$$X = 1 + K_1\,K_2\,2\,\pi f_1\,a_2\,T^2$$

$$Y = 2\,\pi f_1 + K_1 + K_1\,K_2\,2\,\pi f_1\,a_1\,T$$

$$Z = K_1\,2\,\pi f_1(1 + K_2)$$

For the purposes of examination, assume the following values:

$$2\pi f_n = [Z/X]^{1/2}$$
$$\zeta = Y/4\pi f_n$$
$$a = \zeta 2\pi f_n$$
$$b = 2\pi f_n[\zeta - (\zeta^2 - 1)^{1/2}]$$
$$c = 2\pi f_n[\zeta + (\zeta^2 - 1)^{1/2}]$$
$$f_0 = f_n(1 - \zeta^2)^{1/2}$$

Deleting the *dead time* for a moment (the dead time is included by multiplying all expressions of $e(t)$ with $1/X$), we obtain:

$$\left[\cos 2\,\pi f_0\, t + \frac{(2\,\pi f_1 - a)\sin 2\,\pi f_0\, t}{2\,\pi f_0}\right]\exp(-at)\ \zeta < 1$$

$$e(t) = [1 + (2\,\pi f_1 - a)t]\,\exp(-at)\ \zeta = 1$$

$$\frac{(2\,\pi f_1 - b)\exp(-bt) - (2\,\pi f_1 - c)\exp(-ct)}{c - b}\ \zeta > 1$$

For $K_2 = 0$ (or $f_1 = 0$), $e(t)$ is a simple exponential decay, $\exp(-k_1 T)$. Using Equation (5.57) to solve for f_c, we find:

$$2\,\pi f_c = [d + (d^2 + 400\,\pi^4 f_n^4)^{1/2}]^{1/2}$$

where

$$d = 13K_1^2 + (1 - 2\,\zeta^2)\,4\,\pi^2 f_n^2$$

Design of the AGC loop filter consists of selecting values of K_1, K_2, and f_1 and then determining the filter components that will result in these values. Below the second threshold, the loop is described by Equations (5.48) through (5.50) and the associated relationships. Thus, K_1 is determined by the cutoff frequency or response time. In most cases, it is determined simply by $K_1 = 2\pi f_c/5$.

K_2 is determined by the relative gain reductions of the IF amplifier and preamplifier required above the second threshold. K_2 is the ratio of preamplifier gain reduction to IF amplifier gain reduction, in decibels, produced by a small change in the control voltage E_{c1}. Because $K_1 = dA_1\,(dB)/dE_C$, at direct current, $A_2 = dE_{C2}/dE_{C1}$, $K_2 = (dE_{C2}/dE_{C1})\,(dA_{I2}/dE_{C2})\,(dE_{C1}/dA_{I2})$ or $K_2 = dA_{I2}(dB)/dA_{I1}(dB)$. For given values of K_1 and K_2, f_1 adjusts the damping factor. In most cases, approximately critical damping ($\zeta = 1$) is desirable. For this condition,

$$2 \pi f_1 = K_1 \left\{ (1 + 2 K_2) - [(1 + 2 K_2)^2 - 1]^{1/2} \right\} \tag{5.63}$$

For example, assume that a receiver requires 75-dB AGC range. The IF gain is to be reduced 25 dB before the second threshold is reached. Above this threshold, the IF amplifier gain must be reduced an additional 20 dB and the preamplifier gain must be reduced an additional 30 dB. The required cutoff frequency is 250 Hz. Then:

$$K_1 = \frac{2 \pi 250}{5} = 314.2$$

$$K_2 = \frac{30}{20} = 1.5$$

$$f_1 = 50[4 - (16 - 1)^{1/2}] = 39.903$$

A Bode plot of the gain of this loop is shown in Figure 5.37. Typical values for K_D, K_{I1}, and K_{I2} are 2.0 V/Nep, 10 dB/V, and 5 dB/V, respectively. Using these values, the filter in Figure 5.35 can be calculated as follows:

$$A_1 = \frac{K_1}{K_D K_c K_{I1}} = \frac{314.16}{2.0 \times 0.1153 \times 10} = 136.43$$

Selecting $C_1 = C_2 = 0.1 \, \mu F$, we find:

$$R_1 = \frac{1}{A_1 C_1} = 7.329 \times 10^4 \approx 75 \, k\Omega$$

$$A_2 = \frac{K_{I1} K_2}{K_{I2}} = 3.0$$

$$R_3 = \frac{1}{2 \pi f_1 C_2} = 2.506 \times 10^5 \approx 240 \, k\Omega$$

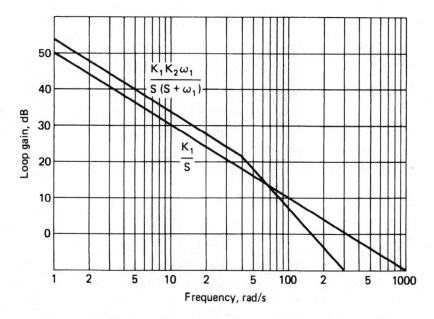

Figure 5.37 Bode plot of AGC loop gain. (*After [5.6].*)

$$R_2 = \frac{R_3}{A_2} = 8.353 \times 10^4 \approx 82\ k\Omega$$

5.6.6 Digital Switching AGC

A digital gain control systems is shown in Figure 5.38. Multiple switches are configured as required to provide the proper overall gain control. The implementation shown provides resolution of 1 dB over a 71-dB range. The switching actually occurs between a high-gain amplifier and a low-gain amplifier in each stage.

5.7 Integrated IF Amplifier and AGC Systems

Large-scale integration of circuit elements is the key to affordable receivers today. The implementation of radio systems on one or more LSI or VLSI devices has led to reduced costs, improved performance, and increased functions and features. The device shown in Figure 5.39 is representative of this trend. The IC (Analog

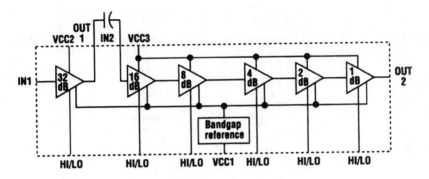

Figure 5.38 IF amplifier with digital gain control. (*After [5.9].*)

Devices AD607) provides linear IF amplification and demodulation for the *global system for mobile* (GSM) and the *digital communications system* (DCS1800) standards. The device consists of a wide dynamic range double-balanced mixer followed by an IF amplifier strip, which provides more than 80 dB of gain-control and an on-chip *I/Q* demodulator. Gain of the IF stage can be automatically controlled using an on-chip peak detector to derive the AGC voltage or manually controlled using an external voltage source. An indication of received signal strength is provided from the AGC detector at the *received signal strength indicator* (RSSI) terminal. Note that the AGC acts on both the IF amplifiers (in this case, a three-stage system) and the mixer. Implementation of such a device on a single piece of silicon provides close control of important operating parameters such as drift, parasitics, and noise.

5.8 Digital IF Processing

Processing of the IF signal in the digital domain, rather than through conventional analog means, opens a range of new possibilities in terms of performance and features. Meeting the criteria for operating frequency and dynamic range, however, is a challenge for digital designers. A practical application can be found in *undersampling* the signal at a frequency that meets Nyquist's criterion with respect to its *bandwidth*, rather than its *frequency* [5.10]. This technique allows the designer to eliminate one or more IF stages, producing a circuit that consumes less power and is less expensive to implement than the conventional approach.

If the bandpass signal is positioned so that it is attenuated by the required amount by the time it crosses integer multiples of the Nyquist frequency and folds back into the band of interest, the band of interest will alias to baseband without destructive interference. Furthermore, if the lower frequency edge of the passband is the edge closest to an odd multiple of the Nyquist frequency, the resulting spectrum is

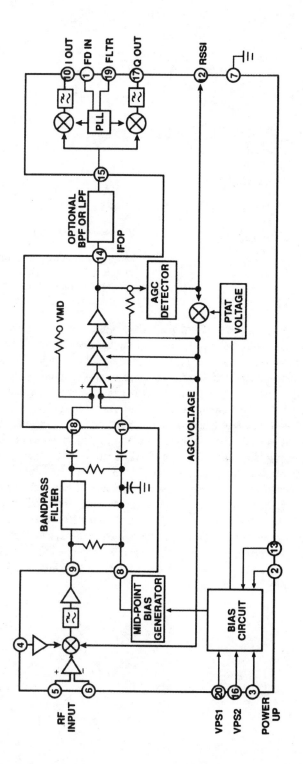

Figure 5.39 Block diagram of the AD607 receiver system integrated circuit. (*Courtesy of Analog Devices.*)

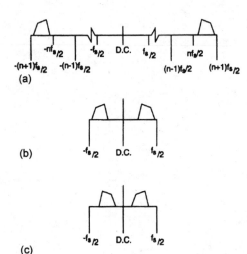

(a)

(b)

(c)

Figure 5.40 Subsampling of an input signal: (*a*) original spectrum of the band-limited signal, (*b*) result of subsampling for *n* even, and *c*) result of subsampling for *n* odd. (*After [5.10].*)

reversed. This "reversal" takes the form of aliasing frequencies in the lower portion of the bandpass signal into the upper portion of the baseband sampled result, and frequencies in the upper portion of the bandpass signal aliasing into the lower portion of the baseband sampled result. This concept is illustrated in Figure 5.40. Subsampling shifts the majority of the performance burden of the digitizer stage to the *sample-and-hold* (S&H) circuit, which is an easier design objective, given current A/D converter devices. Under such a scheme, the A/D converter operates at a rate consistent with the bandwidth of the band of interest, while the S&H operates with a bandwidth consistent with the IF location of the band of interest. This increases the allowable conversion time for the ADC.

Subsampling performs a part of the down-conversion task, which can result in the elimination of the second IF stage (in some cases). Such a receiver is illustrated in Figure 5.41. Conversion from the 70-MHz first IF to the 10.7-MHz second IF is handled by a digital circuit consisting of a mixer, a *numerically controlled oscillator* (NCO) and digital filters. Other functions typically implemented in a digital IF system include fine frequency tuning, channel selection filtering, and phase and frequency tracking of the carrier.

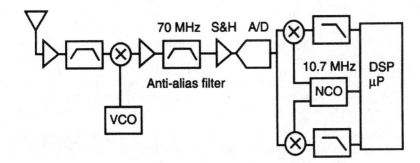

Figure 5.41 Basic receiver architecture using subsampling. (*After [5.10].*)

5.9 References

5.1. Giacoletto, "Study of P-N-N Alloy Junction Transistor from D-C through Medium Frequencies," *RCA Rev.*, vol. 15, p. 506, Dec. 1954.

5.2. Nyquist, "Regeneration Theory," *Bell Sys. Tech. J.*, vol. 11, p. 126, Jan. 1932.

5.3. Oliver, "Automatic Volume Control as a Feedback Problem," *Proc. IRE*, vol. 36, p. 466, Apr. 1948.

5.4. Victor and M. H. Brockman, "The Application of Linear Servo Theory to the Design of AGC Loops," *Proc. IRE*, vol. 48, p. 234, Feb. 1960.

5.5. Ohlson, "Exact Dynamics of Automatic-Gain Control," *IEEE Trans.*, vol. COM-22, p. 72, Jan. 1974.

5.6. Porter, "AGC Loop Control Design Using Control System Theory," *RF Design*, p. 27, June 1980.

5.7. Mercy, "A Review of Automatic Gain Control Theory," *Radio Electron. Eng.*, vol. 51, p. 579, Nov./Dec. 1981.

5.8. C. Kermarrec, et. al, "New Application Opportunities for SiGe HBTs," *Microwave Journal*, pp. 22-35, Oct. 1994.

5.9. W. Pratt, "Process IF Signals with Amplifiers and Quad Demodulators," *Microwaves & RF*, pp. 109-116, December 1993.

5.10. C. Olmstead and M. Petrowski, "Digital IF Filtering," *RF Design*, pp. 30-40, September 1994.

Additional Suggested Reading

"Accurate Noise Simulation of Microwave Amplifiers Using CAD," *Microwave Journal*, Dec. 1988.

"An Accurate Expression for the Noise Reistance R_n of a Bipolar Transistor for Use with the Hawkins Noise Model," *IEEE Microwave and Guided Wave Letters*, vol. 3, no. 2, p. 35, Feb. 1993.

"CAD Packages Improve Circuit-Optimization Methods," *MSN & CT*, May 1985.

"Designing a Matched Low Noise Amplifier Using CAD Tools," Microwave Journal, Oct. 1986.

"Eight Ways to Better Radio Receiver Design," *Electronics*, Feb. 20, 1975.

"Harmonic Balance Method Handles Non-Linear Microwave CAD Problems," *Microwave Journal*, Oct. 1987.

"IF Amplifier Design," *Ham Radio* magazine, Mar. 1977.

"Key Components of Modern Receiver Design, Part I," *QST*, p. 29, May 1994.

"Key Components of Modern Receiver Design, Part II," *QST*, p. 27, June 1994.

"Key Components of Modern Receiver Design, Part III," *QST*, p. 43, July 1994.

"Low-Noise Source Uses Heterojunction Bipolar Transistor," *Microwaves & RF*, Feb. 1989.

"Microwave Harmonica 7.0, A Circuit Simulator for Microwave and Wireless Applications," *RF Design*, Aug. 1966.

"New Nonlinear Noise Model for MESFETS Including MM-Wave Application," *INMMC '90 Digest*, First International Workshop of the West German IEEE MTT/AP Joint Chapter on Integrated Nonlinear Microwave and Millimeter Circuits, Duisburg University, Duisburg, West Germany, Oct 3, 1990.

"Overview of the State-of-the-Art of Modeling the Dynamic Range of VHF, Microwave, and Millimeter Receiver Systems Using Computer-Aided Design," VHF Conference, Weinheim, Germany, Sept 19, 1992

"State-of-the-Art in Nonlinear Microwave CAD Software," *Mikrowellen & HF* magazine, vol. 17, no. 4, p. 252, 1991.

"Wideband Amplifier Summary," *Ham Radio* magazine, Nov. 1979.

Chapter 6

Mixers

6.1 Introduction

Modern communications receivers are almost invariably of superheterodyne design. Depending on the application, there may be one, two, or occasionally three frequency conversions. The circuit in which a frequency conversion is performed is usually referred to as a *mixer*, although the term *converter* is also used. In older literature, the first or second detector is often used to designate the first or second mixer. The demodulator circuit in this case is usually referred to as the *n*th detector, where *n* is one more than the number of frequency conversions. In the mixer circuit, the RF signal and an LO signal are acted upon by the nonlinear properties of a device or devices to produce a third frequency, referred to as an IF, or the *n*th IF when there is more than one mixing process. The IF is selected by a filter from among the various frequencies generated, and higher-order products may produce various spurious responses, as described in an earlier chapter.

Because of the large number of signals received by the antenna, it is customary to use preselection filtering to limit the potential candidates for producing spurious responses. The narrower the preselection filter, the better the performance of the receiver in regard to spurious responses and other IM products. However, narrow filters tend to have relatively large losses that increase the NF of the receiver, and for receivers designed for covering a wide range of frequencies, the preselector filters must be either tunable or switchable. The NF can be improved by providing one or more RF amplifiers among the preselection circuits. The RF amplifier compensates for the filter loss and, at the same time, generally has a better NF than the mixer. It provides additional opportunities for the generation of IM products and increases RF signal levels at the mixer input, which can cause poorer spurious response performance. As pointed out in Chapter 3, receiver design requires compromises among a number of performance parameters.

Below 30 MHz, communications receivers are now being built without RF preamplifiers, and the antenna signal is fed directly to the mixer stage. In this frequency range, the man-made and atmospheric noise received by the antenna has a higher level than a modern low-NF receiver generates internally. Until recently it was customary to build receivers in the range below 30 MHz with an NF less than 10 dB, but more modern designs have tended to use values between 10 and 14 dB. Above 30 MHz, receiver noise is more significant and lower NFs are desirable. NFs of 4 to 6 dB are common, and occasionally values as low as 2 dB are encountered. For lower NFs, special cooled amplifiers are required. The mixer is located in the signal chain prior to the narrow filtering of the first IF and is affected by many signals of considerable amplitude. Its proper selection is very important in the design of a communications receiver.

Ideally a mixer should accept the signal and LO inputs and produce an output having only one frequency (sum or difference) at the output, with signal modulation precisely transferred to this IF. Actual mixers produce the desired IF but also many undesired outputs that must be suitably dealt with in the design.

Any device with nonlinear transfer characteristics can act as a mixer. Cases have been reported where antennas built of different alloys and metals having loose connections produced nonlinear distortion and acted as diode mixers. The same has been reported when structures having different metals corroded and were located in strong RF fields. The resultant IM produced interference with nearby receivers; this process has been called the "rusty bolt effect." In this chapter, we will examine three classes of mixers:

- Passive mixers, which use diodes as the mixing elements
- Active mixers, which employ gain devices, such as bipolar transistors or FETs
- Switching mixers, where the LO amplitude is either much greater than required by the device or is rectangular, so that the mixing elements are essentially switched on and off by the LO

6.2 Passive Mixers

Passive mixers have typically been built using thermionic diodes, germanium diodes, and silicon diodes. The development of hot carrier diodes, however, has resulted in significant improvement in the performance of passive mixers. Figure 6.1 shows the schematic diagram of a frequently used doubly balanced mixer circuit. To optimize such a mixer, it is important to have a perfect match among the diodes and transformers. The manufacturing process for hot carrier diodes has provided the low tolerances that make them substantially better than other available diode types. The use of transmission line transformers and modern ferrites with low-leakage inductance has also contributed substantially to increased operating bandwidth of passive mixers.

A single diode can be used to build a mixer. Such an arrangement is not very satisfactory, however, because the RF and LO frequencies—as well as their harmonics and other odd and even mixing products—all appear at the output. As a result, there are a large number of spurious products that are difficult to remove. Moreover, there is no isolation of the LO and its harmonics from the input circuit so that an RF amplifier is required to reduce oscillator radiation from the antenna. The double balanced mixer with balanced diodes and transformers cancels the even

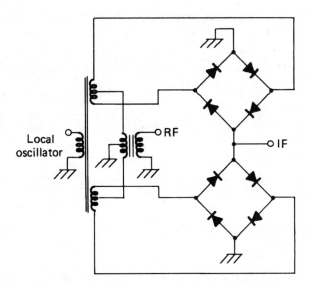

Figure 6.1 Schematic diagram of a double balanced mixer.

harmonics of both RF and LO frequencies and provides isolation among the various ports. Therefore, change of termination has less influence on the mixer performance than with circuits without such balance and isolation. However, this statement is not true for nonlinear products from a single terminal. If two RF signals with frequencies f_1 and f_2 are applied to the input of the mixer, the third-order products $2f_1 \pm f_2$ and $2f_2 \pm f_1$, which can be generated, are extremely sensitive to termination. It can be shown that for any type of mixer, a nonresistive termination results in a reflection of energy at the output so that the RF currents no longer cancel. The third-order intercept point of the mixer is directly related to the quality of termination at the mixer output.

The double balanced mixer has very little spurious response. Table 6.1 shows typical spurious responses of a high-level double balanced mixer. The mixing products are referenced in dB below the desired output or 0 level at f_{IF}. This performance can be typically obtained with f_{LO} and f_{RF} at approximately 100 MHz, f_{LO} at + 17 dBm, and f_{RF} at 0 dBm, using broadband resistive terminations at all ports.

Let us now consider the basic theory of mixers. Mixing is achieved by the application of two signals to a nonlinear device. Depending upon the particular device, the nonlinear characteristic may differ. However, it can generally be expressed in the form:

TABLE 6.1 Typical Spurious Responses of High-Level Double Balanced Mixer

RF input signal harmonics		f_{LO}	$2f_{LO}$	$3f_{LO}$	$4f_{LO}$	$5f_{LO}$	$6f_{LO}$	$7f_{LO}$	$8f_{LO}$
$8f_{RF}$	100	100	100	100	100	100	100	100	100
$7f_{RF}$	100	97	102	95	100	100	100	90	100
$6f_{RF}$	100	92	97	95	100	100	95	100	100
$5f_{RF}$	90	84	86	72	92	70	95	70	92
$4f_{RF}$	90	84	97	86	97	90	100	90	92
$3f_{RF}$	75	63	66	72	72	58	86	58	80
$2f_{RF}$	70	72	72	70	82	62	75	75	100
f_{RF}	60	0	35	15	37	37	45	40	50
		60	60	70	72	72	62	70	70

$$I = K(V + v_1 + v_2)^n \tag{6.1}$$

The exponent n is not necessarily integral, V may be a dc offset voltage, and the signal voltages v_1 and v_2 may be expressed as $v_1 = V_1 \sin(\omega_1 t)$ and $v_2 = V_2 \sin(\omega_2 t)$.

When $n = 2$, Equation (6.1) may then be written as:

$$I = K\left[V + V_1 \sin(\omega_1 t) + V_2 \sin(\omega_2 t)\right]^2 \tag{6.2}$$

This assumes the use of a device with a square-law characteristic. A different exponent will result in the generation of other mixing products, but this is not relevant for a basic understanding of the process. Expanding Equation (6.2),

$$I = K[V^2 + V_1^2 \sin^2(\omega_1 t) + V_2^2 \sin^2(\omega_2 t) + 2VV_1 \sin(\omega_1 t)$$
$$+ 2VV_2 \sin(\omega_2 t) + 2V_2 V_1 \sin(\omega_2 t) \sin(\omega_1 t)] \tag{6.2a}$$

The output comprises a direct current and a number of alternating current contributions. We are only interested in that portion of the current which generates the IF; so, if we neglect those terms that do not include both V_1 and V_2, we may write:

$$I_{IF} = 2KV_1 V_2 \sin(\omega_1 t) \sin(\omega_2 t)$$

$$I_{IF} = KV_2 V_1 \left\{\cos\left[(\omega_2 - \omega_1)t\right] - \cos\left[(\omega_2 + \omega_1)t\right]\right\} \tag{6.3}$$

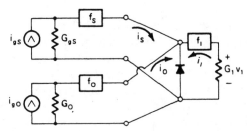

Figure 6.2 Representation of a mixer circuit.

This means that at the output, we have the sum and difference signals available, and the one of interest may be selected by the IF filter.

A more complete analysis covering both strong and weak signals is given by Perlow [6.1]. We outline this procedure next. The semiconductor diode current is related to the input voltage by:

$$i = I_{sat}[\exp(av) - 1] \tag{6.4}$$

where I_{sat} is the reverse saturation current. We may expand this equation into the series,

$$i = I_{sat}\left[av + \frac{(av)^2}{2!} + \frac{(av)^3}{3!} + \ldots \frac{(av)^n}{n!} + \ldots \right] \tag{6.5}$$

The desired voltages across the terminal are those resulting from the input, LO, and IF output signals (Figure 6.2). If the selective filter circuits have high impedance over sufficiently narrow bandwidths, the voltage resulting from currents at other frequencies generated within the diode will be negligible. We write the diode terminal voltage as:

$$v = V_s \cos(2\pi f_s t + \theta_S) + V_O \cos(2\pi f_O t + \theta_O) + V_1 \cos(2\pi f_I t + \theta_I) \tag{6.6}$$

where the subscripts S, O, and I refer to the input signal, LO, and IF outputs, respectively.

The output current may be written by substituting Equation (6.6) into Equation (6.5). The resultant can be modified, using the usual trigonometric identities for the various products of sinusoids. If n is the highest expansion term used, the process produces currents at the nth harmonics of the input frequencies as well as IM products of all frequencies $jf_S \pm kf_O \pm lf_I$, where j, k, and l are positive integers (including zero) whose sum is less than or equal to n. Because $f_I = f_O - f_S$, there are a number of components that fall at these three frequencies. They may be summed to provide the current that flows in the various loads of Figure 6.2. When divided by the voltage at each of these frequencies, the current gives the effective conductance of the diode, in parallel with the various load conductances. The currents at other frequencies flow freely in the external circuits, producing negligible voltage. They do not affect the power relationships, which exist only among signal, LO, and IF outputs.

In the case of the square-law device, where $n = 2$, the conversion voltage gain may be expressed as:

$$\frac{V_I}{V_S} = \frac{A_0}{1 + (A_1 \, V_S)^2} \tag{6.7}$$

where:

$$A_0 = \frac{a^2 \, I_{sat} I_O}{2(a \, I_{sat} + G_L)(a \, I_{sat} + G_O)}$$

$$A_1^2 = \frac{(a^2 \, I_{sat} I_O)^2}{(a \, I_{sat} + G_L)(a \, I_{sat} + G_O)}$$

The gain is thus a function of the signal level. For small V_S the gain is A_0, but as V_S increases, the gain decreases as $1/V_S^2$. The output voltage initially rises with increasing input voltage until it reaches a peak of $A_0/2A_1$ when V_S is $1/A_1$. After this saturation, the output voltage decreases with increasing signal voltage. The levels of gain and saturation are dependent on the diode parameters, the loads, and the level of LO current delivered to the diode.

When higher-order terms are considered, the conversion gain retains the same form as Equation (6.7); however, the expressions for A_0 and A_1 become more complex, varying with all of the three input voltage levels [6.2]. The conductances presented by the diode are no longer simply aI_{sat} at all three frequencies but vary also with the various voltage levels. For optimum power transfer, the internal and external signal and IF output conductances must be matched. For minimum LO power requirement, the source and diode impedance must also be matched.

To provide a high level of saturation, it is essential that the oscillator power to the diode be high. The minimum loss between signal and IF in a receiver is needed especially at low signal levels, where maximum sensitivity must be retained. Consequently, for receiver design we often have signal and IF power near zero. This produces the small-signal conductances from the diode,

$$G_{Ss} = G_{Is} = aI_{sat}[I_0^2\,(a\,V_O) - I_1^2\,(a\,V_O)]^{1/2} \tag{6.8a}$$

$$G_{Os} \equiv \frac{2\,I_{sat}I_1(a\,V_O)}{V_O} \tag{6.8b}$$

where $I_0(aV_O)$ and $I_1(aV_O)$ are modified Bessel functions of the first kind.

The source and load conductances are equal and depend only on the diode parameters and the LO voltage. The LO conductance is also a function of these same parameters. The LO level must be selected to provide sufficient power to avoid saturation at the high end of the dynamic range of the receiver. From the LO level, signal and IF conductances are determined, and the filters and loads are designed in accordance with Equations (6.8a and b) to provide optimum match at the low end of the dynamic range. We may choose as an example a silicon diode with $a = 38$ V^{-1} and $I_{sat} = 10^{-14}$ A. For 0.5-W LO drive, we can estimate $V_O^2 G_{OS} = 0.5$. With our other assumptions, this yields $aV_O I_1\,(aV_O) = 9.5 \times 10^{14}$. From tables, we find $aV_O = 33.7$, $V_O = 0.889$ V, $G_{Os} = 0.636$ S, and $G_{Ss} = G_{Ls} = 1.88$ S. With 10-mW drive, $V_O = 0.784$ V, $G_{Os} = 0.01626$ S, and $G_{Ss} = G_{Ls} = 0.04477$ S. The conductances increase with increasing drive.

Using the square-law form of the expression, it is possible to develop expressions for IM distortion ratios. Because the same general form of expression holds for the complete analysis, although the coefficients vary with signal level, it is reasonable to assume that the distortion would show similar variations, but with some deviation from the simpler curves. For the second-order case the maximum output power, at $V_s = 1/A_I$, turns out to be one-fourth of the LO power. It is convenient to measure the output power level as a fraction of this maximum power P_{Imax}. Then, in the square-law case, the mth in-band IM product IMR resulting from two equal input signals may be shown to have the value:

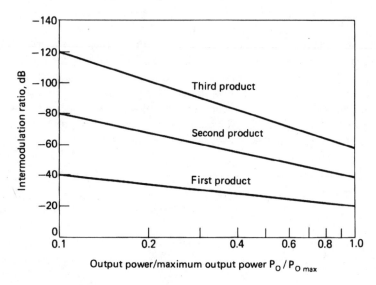

Figure 6.3 IM distortion ratios. (*After [6.3].*)

$$IMR_m = m\left[20\log\left(\frac{P_I}{P_{Imax}}\right) - 19.5\right] \tag{6.9}$$

where the result is expressed in decibels, and $2m + 1$ is the usual order assigned to the product. Figure 6.3 shows plots of these curves for $m = 1, 2,$ and 3 (third-, fifth-, and seventh-order IM). Figure 6.4 shows a comparison of measured third-order IM ($m = 1$) for two different mixer types, each with several levels of LO power. The maximum deviation from the theory is 4 dB.

The gain saturation effects can thus be used to predict the nonlinear effects encountered in mixers [6.3]. Similar predictions can be made for other types of mixers and for amplifiers with gain saturation. Such effects include distortions such as IM distortion (discussed previously), triple-beat distortion, cross modulation, AM-to-PM conversion, and hum modulation.

Passive mixers can be described as low-, medium-, or high-level mixers, depending on the diodes used and the number of diodes in the ring. Figures 6.1 and 6.5 show several arrangements of high-level double balanced mixers. The arrangement with two quads has the advantage of higher LO suppression, but is also more expensive. In addition to these types, a number of other special-purpose passive mixers are available (References [6.4] to [6.10]).

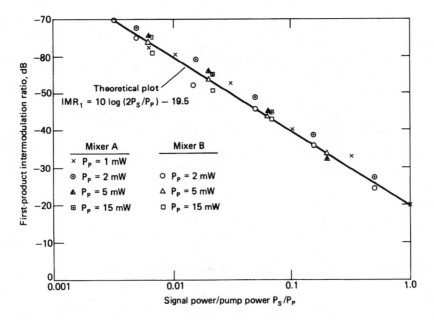

Figure 6.4 Experimental IM ratio measurements. (*After [6.3]*.)

A. SSB mixer

An SSB mixer is capable of delivering an IF output that includes only one sideband of the translated RF signal. Figure 6.6 shows the schematic diagram of such a mixer, which provides the USB at port A and LSB at port B.

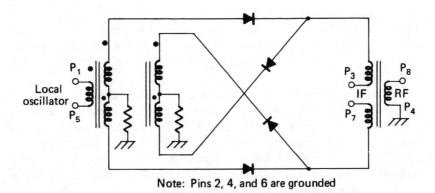

Note: Pins 2, 4, and 6 are grounded

Figure 6.5 Double balanced mixer using a single-diode quad. (*After [6.22]*.)

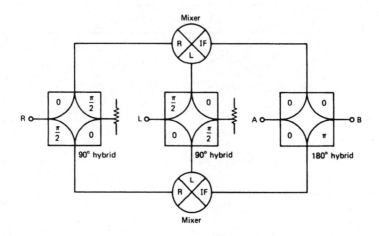

Figure 6.6 Schematic diagram of an SSB mixer. (*Courtesy of Adams Russell, Anzac Division.*)

B. Image-rejection mixer

An LO frequency of 75 MHz and RF of 25 MHz would produce an IF difference frequency of 50 MHz. Similarly an image frequency at 125 MHz at the mixer RF port would produce the same 50-MHz difference frequency. The image-rejection mixer shown in Figure 6.7 is another form of SSB mixer and produces the IF difference frequency at port *C* from an RF signal that is lower in frequency than the LO, while rejecting the same difference frequency from an RF signal higher than the LO frequency.

C. Termination insensitive mixer

While the phrase *termination insensitive* is somewhat misleading, the circuit shown in Figure 6.8 results in a mixer design that allows a fairly high VSWR at the output without the third-order IM distortion being significantly affected by port mismatches.

It has been mentioned previously that a double balanced mixer, unless it is termination insensitive, is extremely sensitive to nonresistive termination. This is because the transmission line transformers do not operate properly when they are not properly terminated, and the reflected power generates high voltage across the diodes. This effect results in much higher distortion levels than in a properly terminated transformer. It is sometimes difficult to provide proper termination for the mixer. However, this can be achieved by using a grounded-gate FET circuit, such as shown in Figure 6.9, or by a combination of a diplexer with a feedback amplifier, such as shown in Figure 6.10.

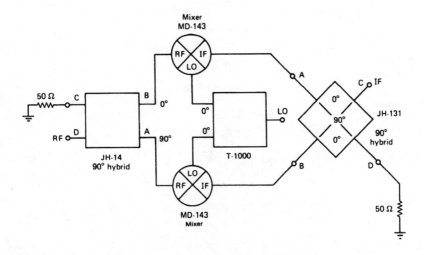

Figure 6.7 Schematic diagram of an image-rejection mixer. (*Courtesy of Adams Russell, Anzac Division.*)

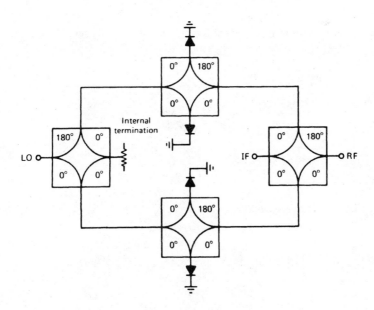

Figure 6.8 Schematic diagram of a termination-insensitive mixer. (*Courtesy of Adams Russell, Anzac Division.*)

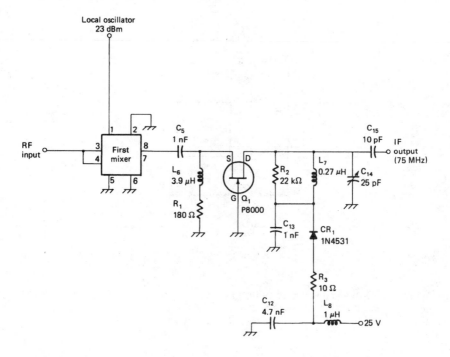

Figure 6.9 Provision of resistive termination by use of a grounded-gate FET.

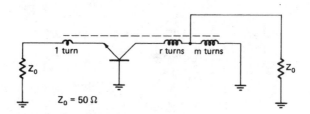

Figure 6.10 Schematic diagram of a feedback amplifier.

The impedance of the diplexer circuit can be expressed as:

$$Z^{-1} = \frac{j\omega C_1}{1 + j\omega C_1 (R + j\omega L_1)} + \frac{1 - \omega^2 L_2 C_2}{R(1 - \omega^2 L_2 C_2) + j\omega L_2} \qquad (6.10)$$

It is desired that $Z = R$. Therefore,

$$R^2 = \frac{L_2(1 - \omega^2 L_1 C_1)}{C_1(1 - \omega^2 L_2 C_2)} \tag{6.11}$$

Because both tuned circuits should resonate at the same frequency, this condition becomes:

$$R^2 = \frac{L_2}{C_1} = \frac{L_1}{C_2} \tag{6.12}$$

The bandwidth of the tuned circuit determines the value of $Q = f_s B$, where B is the bandwidth and f_s is the resonant frequency. Since $Q = 2\pi f_s L_1 / R$, these relationships result in the following design equations:

$$L_1 = \frac{R}{2\pi B} \tag{6.13a}$$

$$L_2 = \frac{B R}{2\pi f_s^2} \tag{6.13b}$$

$$C_1 = \frac{B}{2\pi f_s^2 R} \tag{6.13c}$$

$$C_2 = \frac{1}{2\pi B R} \tag{6.13d}$$

Let us consider the following example. The IF following a mixer is 9 MHz, and the IF bandwidth is 300 kHz. The double balanced mixer should be terminated in a 50-Ω resistor. Thus, $Q = 9.0/0.3 = 30$, and $L_1 = 50/2\pi\ 0.3 = 26.5\ \mu H$, $C_1 = 11.8$ pF, $L_2 = 29.5$ nH, and $C_2 = 10.6$ nF. Because L_2 has such a small value, a suitable capacitor for C_2 must be chosen to avoid excessive lead inductance.

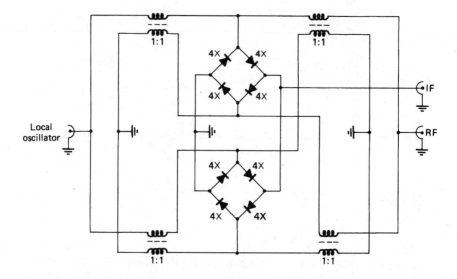

Figure 6.11 Double balanced mixer circuit for low local oscillator drive. Each diode symbol represents four diodes connected in series.

The large-signal handling capacity of passive double balanced mixers has increased tremendously over the past decade. The dynamic range is directly proportional to the number of diodes or diode rings used, as well as the LO drive. For a selected LO to IF port isolation, there is usually a tradeoff between IM distortion performance and feedthrough. In some applications, the absolute level of the IF feedthrough is restricted. Hence, it may be necessary to trade off among various performance criteria.

A very interesting mixer circuit is shown in Figure 6.11. While requiring only 17 dBm of LO drive, it has about 60-dB isolation and an IP at the input of about 15 dBm. This circuit has been used in the designs of several HF receivers.

6.3 Active Mixers

Active mixers were prevalent for many years, and many references treat their design and analysis (References [6.11] to [6.22]). The simplest active mixer is an FET or bipolar transistor with LO and RF signals applied to the gate-source or base-emitter junction. This unbalanced mixer has the same drawbacks as the simple diode mixer and is not recommended for high-performance operation. The next step in performance improvement is the use of a dual-gate FET or cascode bipolar arrangement with the LO and RF signals applied to different gates (bases).

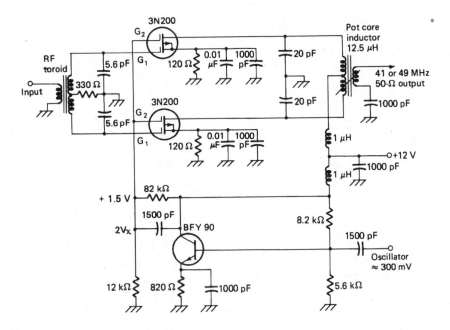

Figure 6.12 Push-pull dual-gate FET balanced mixer. (*After [6.22].*)

The balanced transistor arrangement of Figure 5.20*a* can also be used as a mixer with the LO applied to the base of Q_3 and the signal applied to the bases of Q_1 and/or Q_5.

There are a number of balanced mixer designs that provide reasonably good performance, as indicated in Figures 6.12 through 6.15 and outlined here:

- *Push-pull balanced FET mixer* (Figure 6.12). This circuit uses two dual-gate FETs with a push-pull arrangement between the first gates and the IF output, while the oscillator is injected in parallel on the second gates.

- *Double balanced FET mixer* (Figure 6.13). Four JFETs are arranged in a double balanced quad, with the RF signal being injected on the sources and the LO on the gates.

- *Bipolar mixer array* (Figure 6.14). This circuit provides a push-pull type arrangement similar to that in Figure 6.12, except that the device is bipolar. (The arrangement shown uses the Plessey SL6440C.)

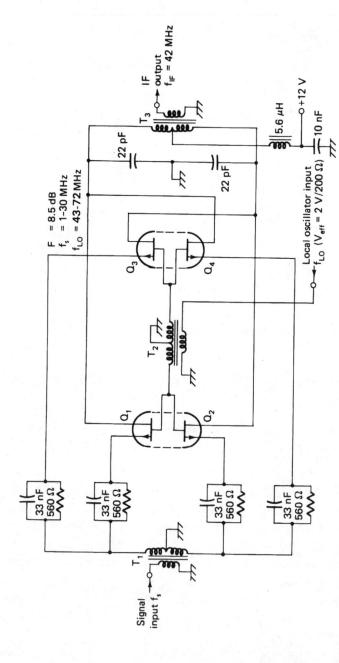

Figure 6.13 Double balanced JFET mixer circuit. (*After [6.22].*)

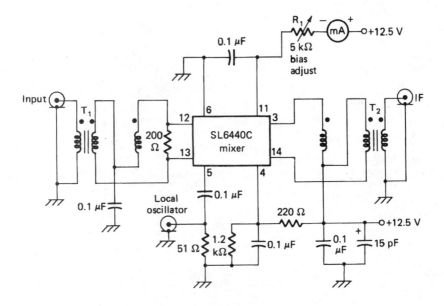

Figure 6.14 Balanced active mixer using a bipolar array. (*After [6.22].*)

- *VMOS balanced mixer* (Figure 6.15). VMOSFETs are capable of handling very high power and have been used in the arrangement shown, which again resembles in general configuration the mixer in Figure 6.12.

Active mixers have gain and are sensitive to mismatch conditions. If operated at high levels, the collector or drain voltage can become so high that the base-collector or gate-drain junction can open during a cycle and cause severe distortion. One advantage of the active mixer is that it requires lower LO drive. However, in designs such as the high-input FET, special circuits must be used to generate sufficiently high voltage at fairly high impedance. This can be difficult. The FET between gate and source shows only a capacitive and no resistive termination. Sometimes, therefore, circuits must be designed to operate into a resistive termination of 50 Ω, for example. This, then, requires power that the FET itself does not require.

A class of active mixers that is of special interest at the higher frequencies uses varactor diodes in an up-converter configuration (References [6.23] to [6.27]).

These mixers use the power from the oscillator (pump) to vary the capacitance of the varactor diodes. When used in an up-converter configuration, a gain is obtained in the ratio of output (IF) power to input (RF) power. Excellent IM and spurious response performance is possible from these mixers. However, for systems

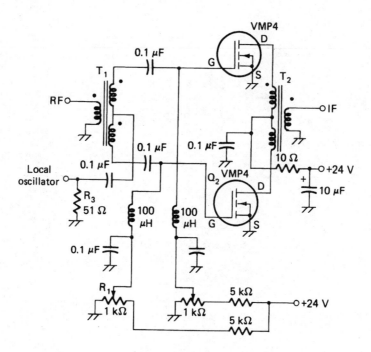

Figure 6.15 VMOS balanced mixer circuit. (*After [6.22].*)

covering a wide RF band, the termination and drive variation problems are substantial.

6.4 Switching Mixers

It is possible to overdrive any active or passive mixer by using very high LO drive, or to use a rectangular LO waveform. This switches the diode or transistor stages on and off. Provided that the devices are sufficiently rapid, they should be able to follow the oscillator drive. Such circuits have been used in the past ([6.28] and [6.29]). However, it has been found that the harmonic content of the output causes unnecessary difficulties, so the technique must be used with care.

A more satisfactory approach is the use of FETs as switches in a passive configuration (References [6.30] to [6.32]). Such a circuit is shown in Figure 6.16. It has been reported that for 1-V RF inputs (+13 dBm), the third-order IM distortion products are –83 dBm, or 100 dB down. This corresponds to a third-order IP of +70 dBm, but such a performance can only be achieved in narrow-band configurations. In a wide-band configuration, an IP of 40 to 42 dBm is attainable. The isolation

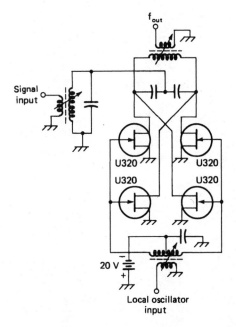

Figure 6.16 Switching mixer circuit after Squires patent 3383601. (*After [6.22].*)

between oscillator and signal ports is about 60 dB, and about 40-dB isolation is provided to the IF signal.

The type E1700 HF receiver (AEG Telefunken) makes use of all the techniques mentioned in this chapter. Figure 6.17 shows the block diagram of the mixer. The mixer consists of a quad switch arrangement of four transistors (SD210). At the output, phase shifters are used to split the energy components and feed them through crystal filters before subsequently recombining them. By this method, selectivity is added at the output, but the termination problem is avoided. Figure 6.18 shows the mixer termination with diplexer, hybrid power splitter, and phase shifter. The two outputs go to the crystal filters.

6.5 IC-Based Mixers

Mixing circuits integrated into an IC package usually comprise an RF input section, which provides voltage-to-current conversion, and a two- or four-transistor current-mode switching core, which introduces an abrupt sign change into the signal path between the RF input and the IF output, controlled by the LO [6.34]. A typical IC mixer is illustrated in Figure 6.19. The advantages of such a device include the following:

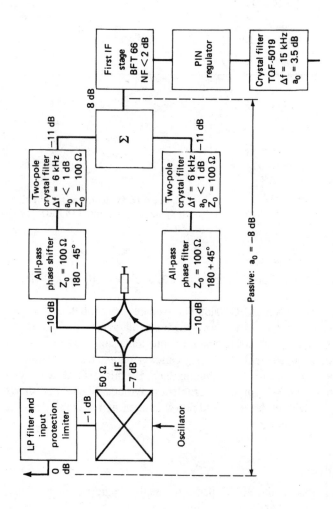

Figure 6.17 Block diagram of the mixer circuit in the E1700 HF receiver (AEG Telefunken). (*After [6.33].*)

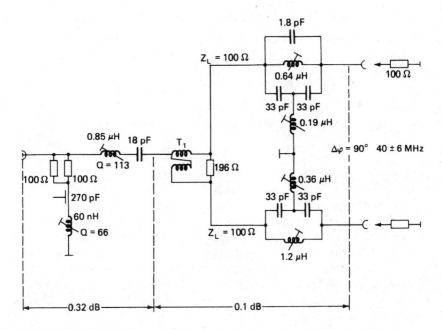

Figure 6.18 Mixer termination circuit of the system shown in Figure 6.17. (*After [6.33].*)

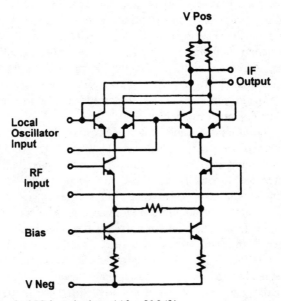

Figure 6.19 A typical IC-based mixer. (*After [6.34].*)

- Conversion gains of 10 to 20 dB can be achieved
- Low LO input power requirement
- Modest gain control is available for AGC purposes
- Excellent isolation between ports in achieved
- Termination sensitivity at the IF port is greatly reduced
- Simplified integration with other circuit and system components

The primary disadvantages of an IC-based mixer over a passive mixer are somewhat reduced dynamic range and, typically, lower maximum operating frequency. Power consumption of an IC mixer is roughly proportional to the signal handling capability; the wider the dynamic range, the higher the required power. The 1-dB *compression point* (the level at which the output is 1 dB below the ideal value extrapolated from the small-signal response) is a key parameter in this regard as is the *third-order intercept* (the extrapolated input level at which the amplitude of the fundamental and the third-order products in the IF output are equal). Because the voltage-to-current converter section, which operates in a Class A mode, typically uses current-sources and resistors to define the maximum RF input, low noise operation demands the use of low-value resistors (in the range of 50 Ω is convenient) to achieve a low NF. This requires large bias currents to support the peak input signals without introducing significant *intermodulation distortion* (IMD). Bias currents of 20 to 30 mA are not uncommon.

A typical high-performance mixer optimized for use in critical applications is shown in Figure 6.20. The device (Analog Devices AD831) is intended for applications where a combination of high third-order intercept (+30 dB) and low NF (12 dB) are required. The IC is usable at RF inputs of up to 800 MHz, and includes a wide-bandwidth (100 MHz) low-noise IF amplifier, which can be configured to provide gain in the system. Filtering may also be achieved through proper selection of external components.

A related trend is the inclusion of an LNA within the IC mixer package. A schematic diagram of one such device [6.35] is shown in Figure 6.21. As illustrated in Figure 6.21*a*, shunt feedback is used to lower the input impedance of the common-emitter stage to establish a nominal 50 Ω input match without significant NF degradation. As with LNA design using discrete components, interdependency exists between a number of amplifier performance parameters. Balancing those values to meet the design objectives of the receiver is usually a challenge for the design engineer. With an IC-based design, of course, that work is done up-front, and the receiver designer need only be concerned with how to optimize the device for a particular application. A simple power-down feature is implemented in the LNA through the use of a large PMOS switch (M_1 in the diagram). In the power-down mode, the switch is open and the supply current drawn by the amplifier

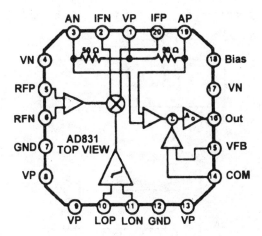

Figure 6.20 Generalized schematic of the Analog Devices AD831 mixer. (*After [6.34].*)

is determined by the leakage current of the switch, typically less than 1 μA. In the power-on mode, the switch is closed to provide bias to the amplifier.

The mixer circuit of the LNA-mixer IC is shown in Figure 6.21b. The circuit is based on a *Gilbert cell* (described in the next section), and although a circuit of this type inherently operates in a differential manner, the mixer has been configured with single-ended inputs and outputs. This is the most convenient mode of operation for most applications. Single-ended inputs and outputs avoid the need for balun devices, which add to the complexity of the receiver. As in the LNA section, power-down operation is accomplished through the use of MOS switches.

6.5.1 Gilbert Cell Mixer

The Gilbert cell has become a popular topology for RF mixer design in monolithic ICs. Although it was designed to be an analog four-quadrant multiplier for small signals, it can also be used in a switching mode for mixing. Because the Gilbert cell is based on coupled differential amplifiers, it can offer high gain, wide bandwidth, and low power consumption, and is amenable to monolithic IC implementation. BJT-based Gilbert cells, however, offer inferior linearity (IP3) due to the highly nonlinear nature of BJTs when designed for voltage-controlled applications. MOS and MESFET Gilbert cells have demonstrated better linearity due to the typical quadratic dependence on input voltage.

One of the difficulties of implementing a Gilbert cell mixer results from the use of differential amplifiers in the cell itself. Virtually all RF signals are single-ended,

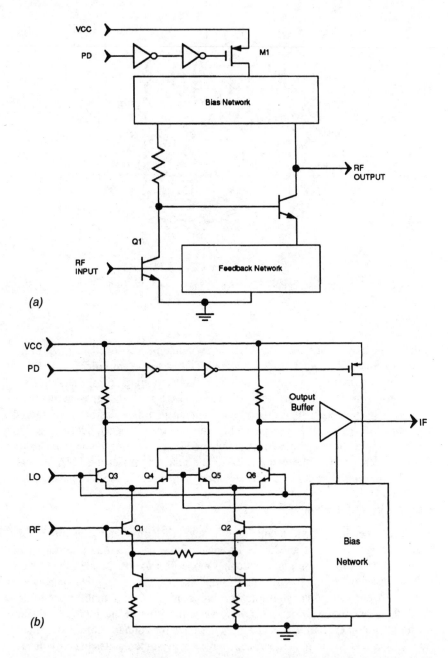

Figure 6.21 Simplified schematic diagram of an LNA-mixer IC: (*a*) LNA circuit and (*b*) mixer circuit. (*After [6.35.]*)

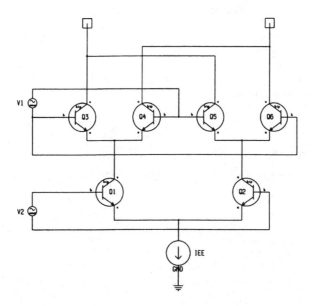

Figure 6.22 The ideal Gilbert cell mixer. (*After [6.36].*)

and obtaining a differential signal is difficult. The use of high-frequency transformers, as used in diode mixers, largely negates the advantages of realizing the Gilbert cell monolithically. High-frequency transformers significantly add to cost and size, and also reduce manufacturability.

The core ideal Gilbert cell is shown in Figure 6.22, with input voltages V_1 and V_2. Transistors Q_1 and Q_2 form one differential amplifier with I_{EE}. Transistors Q_3/Q_4 and Q_5/Q_6 form the coupled differential amplifiers with Q_1 and Q_2, respectively. All transistors are considered identical. To develop the mathematics of the analog multiplier, we start with a basic differential pair and consider the base-emitter voltage V_{be} of a transistor as a function of its collector current I_c, as the following:

$$V_{be}(Q_1) = V_t \ln \frac{I_c(Q_1)}{I_{sat}} \tag{6.14}$$

Usually, I_c is written in terms of V_{be}, but here we write the inverse form neglecting the -1 term in the usual equation. Because $V_2 = V_{be}(Q_1) - V_{be}(Q_2)$, we can form Equation 6.14 to obtain the I_c ratios in a differential amplifier. Then, by writing $I_{EE} = I_e(Q_1) + I_e(Q_2) \ I_c(Q_1) + I_c(Q_2)$ and substituting,

$$\frac{I_c(Q_1)}{I_c(Q_2)} = \exp\left\{\frac{V_2}{V_t}\right\} \tag{6.15}$$

the following is derived:

$$I_c(Q_1) = \frac{I_{EE}}{1 + \exp\left\{-\dfrac{V_2}{V_t}\right\}} \tag{6.16}$$

$$I_c(Q_2) = \frac{I_{EE}}{1 + \exp\left\{\dfrac{V_2}{V_t}\right\}} \tag{6.17}$$

Now, considering the complete cell, ΔI is defined as:

$$\Delta I = I_c(Q_3) + I_c(Q_5) - I_c(Q_4) - I_c(Q_6) \tag{6.18}$$

The I_c values in Equation 6.18 are developed in a manner similar to the single differential pair described previously. Using $\tanh(x) = (e^x - e^{-x})/(e^x + e^{-x})$, the following equation can be derived:

$$\Delta I = I_{EE}\left[\tanh\left(\frac{V_1}{2V_t}\right)\right]\left[\tanh\left\{\frac{V_2}{2V_t}\right\}\right] \tag{6.19}$$

When V_1 and V_2 are small, $\tanh(x) \equiv x$ and ΔI is proportional to $V_1 \times V_2$.

In mixer applications, the LO is often applied as V_1, although it can also be applied as V_2 and uses less LO power. The transistors driven by the LO are essentially switched from cutoff to near saturation and act as a chopper, much like a diode mixer. The IF signal can be amplified by the transistors and is collected as ΔI, usually as a voltage drop across resistors attached to the collectors.

Although the ideal Gilbert cell is useful for developing the theory of operation, it is difficult and costly to realize at RF. In order to develop a Gilbert cell mixer monolithically, certain trade-offs need to be made. The most significant trade-off is the ideal differential driver of the RF and LO ports. Rather than using costly transformers, there are two options that can be applied to obtain an effective differential driver: The first uses a pre-driver stage to convert a single-ended line into a balanced line, and the second uses a combination of *common emitter* (CE) inverting stages and *common base* (CB) noninverting stages to realize a pseudodifferentially driven Gilbert cell. The second approach is often used since it is a simpler design and uses fewer components. This is the approach presented here.

As shown in the schematic of Figure 6.22, the LO and RF signals are applied to the CE inverting stages and the CB noninverting stages of the cell. Because the CE and CB stages have different gain and not exactly 180° phase difference at 900 MHz, the mixer does not operate ideally. Because the output of the cell is a current difference (ΔI) in the collector arms of the upper transistors, the output impedance is high. The ΔI is converted into a voltage by the drop across the biasing resistor. Because the output impedance is high, an emitter follower stage is used to buffer the IF and drive a low 50-Ω impedance load.

6.5.2 Gilbert Cell Performance Analysis

Using harmonic-balance analysis a comparison can be made between the ideal Gilbert cell performance and a "realizable" cell. Mixer analysis can be performed in two modes:

- Analyze the circuit using a general two-tone analysis where the LO and RF signals can be of any power level

- Analyze the circuit with only the LO signal having finite power, assume the RF signal has negligible power, and apply frequency conversion methods to determine mixer conversion gain

The latter method was used in this analysis because it is faster to compute. The first method is useful to determine mixer compression.

The *conversion gain* (CG) of the nonideal cell is about 3 dB lower than the ideal cell, but now requires –4 dBm LO power to reach maximum CG as compared to –14 dBm for the ideal cell. (See Figure 6.23.) At the maximum, the CGs are identical, primarily due to the IF buffer that was added. Because the IF buffer is an emitter follower, it does not provide voltage gain but does provide power gain, and CG is a measure of power ratios.

The mixer NF of the non-ideal cell is about 10 dB higher than the ideal cell and also needs higher LO power to reach a minimum. At the minimum, the difference in NF decreases to about 3 dB. This difference is primarily due to the lower CG of the cell itself; providing additional gain through the IF buffer does little to improve the NF.

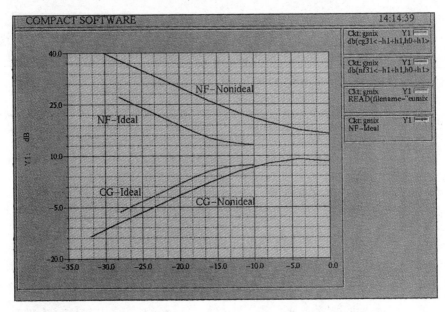

Figure 6.23 NF and conversion gain as a function of LO power. (*After [6.36]*.)

One simple way to improve the NF of the mixer is by preceding it with an LNA. Because a complex matching circuit would take up valuable real estate on the IC dice and would make the mixer highly frequency dependent, a simple low noise transistor was chosen instead. Although this trade-off does not yield optimum NF, it does keep the LNA-mixer suitable for a wide frequency range. External circuitry can be added off-chip if desired.

A schematic diagram of the LNA is shown in Figure 6.24a. The gain and NF of the device are shown in Figure 6.24b; the transistor provides 12 dB of gain and a 2 dB NF at 900 MHz. In Figure 6.25, a comparison of CG and NF is made with the mixer alone and with the LNA connected to the RF input. The LO power is swept at LO = 900 MHz and the RF = 945 MHz. As expected, the CG is improved by 12 dB with the LNA attached, and the maximum occurs at the same LO power. The minimum NF of the LNA/mixer is now 9 dB at an LO power of –5 dBm, slightly lower than the LO power for maximum CG. This NF is approximately 0.8 dB higher than that calculated using the cascaded noise factor formula:

$$F = 1 + F_1 + \frac{F_2}{G_1} \qquad (6.20)$$

where:

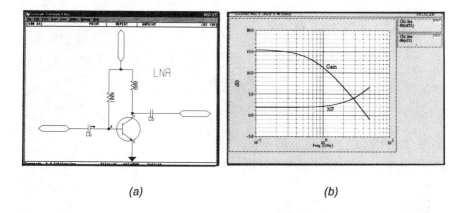

(a)   (b)

Figure 6.24 Low noise amplifier: (*a*) schematic diagram and (*b*) plot of gain and NF. (*After [6.36].*)

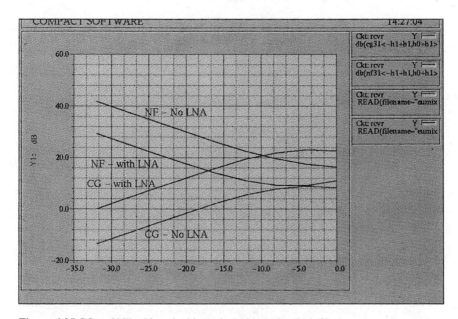

Figure 6.25 CG and NF with and without the LNA. (*After [6.36].*)

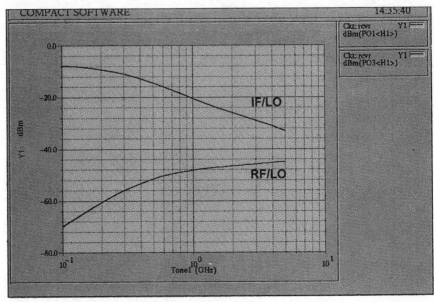

Figure 6.26 Isolation analysis of the RF and IF ports. (*After [6.36].*)

F_1 = LNA noise factor = 2 dB = 1.58
F_2 = mixer noise factor = 18 dB = 63.1
G_1 = LNA power gain = 12 dB = 15.9
F = 6.56
NF = 8.2 dB

Because the input impedance of the mixer is not an ideal 50-Ω, and the LO is leaking into the LNA output because the mixer is not well balanced, the NF of the LNA-mixer cascade does not follow the noise factor formula. These results are only possible through general circuit analysis.

From the same harmonic-balance analysis, the LO leakage out of the RF and IF ports can be characterized, as illustrated in the plot of Figure 6.26. The LNA-mixer cascade exhibits small LO leakage out of the RF port due to the small feedback in the parasitics of the LNA transistor. More significant leakage will probably occur through package parasitics that can be characterized separately.

The leakage into the IF port is quite significant because of the imperfect balancing of the CE and CB stages. In the ideal Gilbert cell, the rejection is very high at the LO frequency, but at 2 LO, significant power exits the IF port. In either case, a low-pass filter is needed to keep stray LO out of the IF stages. This is often built into the IF amplifier.

By holding the LO power fixed at -10 dBm and sweeping the LO and RF sources while holding the IF fixed, the frequency characteristics can be determined (Figure

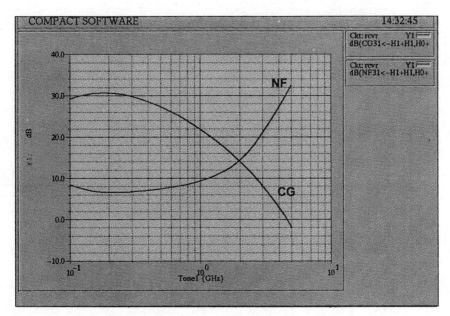

Figure 6.27 Plot of CG and NF for the example circuit. *After [6.36].*)

6.27). The low-frequency rolloff is determined by the 50-pF capacitors used for coupling and CB ac ground. The high-frequency rolloff is primarily determined by the transistor parasitics and can be tailored through careful device size selection for higher frequency ranges. The transistor used in this example is a scaled version of an MRF941, which has an f_t of 8 GHz. More modern BJTs have an f_t of 15 to 20 GHz and can significantly improve high-frequency mixer performance.

Because most wireless communications applications are narrowband (3 to 5 percent) and have even smaller channel bandwidths, the frequency dependence is of little consequence. However, frequency compensation circuits can easily be realized off-chip for wider bandwidth applications.

Using three-tone harmonic-balance analysis, IM mechanisms in mixers can be performed. By using one tone for the LO excitation and applying two closely spaced RF tones, IM products can arise around the IF. By sweeping the RF powers, the IF IM product power can be calculated and plotted with the IF power, as shown in Figure 6.28. The extrapolation of these to an intersection point yields the IP3 figure-of-merit for the mixer. The LNA-mixer combination produces an IP3 of –6 dB referenced to the output.

Three-tone analysis generates a large number of spectral components that must be accounted for during the harmonic balance analysis. The typical case of using three LO harmonics and third-order IM produces 87 spectral components. This large number of spectral components taxes the capacity of modern workstations in

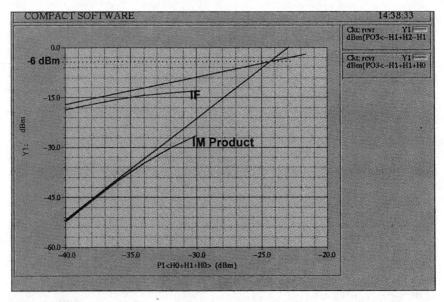

Figure 6.28 IP3 analysis of the example circuit. (*After [6.36].*)

both memory requirements and *central processing unit* (CPU) time. It has been found that reducing the number of spectral components has significant effects on IP3 accuracy and is not a viable method of reducing the problem size[6.36]. A better approach involves the use of *sparse matrix* techniques to help reduce the time and memory needed to analyze a three-tone problem.

The frequency spectrum that results from three-tone analysis is shown in Figure 6.29. Around each LO harmonic and at the baseband, an *intermodulation spectrum* (IMS) is placed. Typically, at the upper LO sideband, the RF_1 and RF_2 are placed as the most significant components. The IF_1 and IF_2 form the most significant components of the baseband IMS. Each IMS contains 12 spectral components. If three LO harmonics are used, then the total number of components is $12 + 3 \times (12 + 1 + 12) = 87$.

A. Supply voltage effects

One of the important characterizations of handheld wireless design is the dependency of key parameters on bias voltage as the battery degrades. By sweeping the bias voltage, CG, NF, and IP3 can be determined. As shown in Figure 6.30, the optimal point occurs near 3 V and degrades as the bias voltages increases or decreases. At higher voltages, the fixed LO power is insufficient to drive the mixer into switching, so the CG degrades, as does the NF and IP3. At lower voltages, the devices are saturating and CG again degrades. Analyses such as these assist in

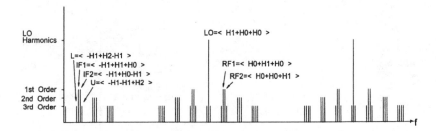

Figure 6.29 Three-tone mixer spectrum. (*After [6.36].*)

determining the useful range of the subsystem, and its effect on the entire system can be assessed through the system-level analysis.

6.6 Wide Dynamic Range Converters

The superheterodyne receiver systems discussed so far in this book have been intended for reception of a single service at a time. In that regard, they can be described as *narrow band*. A *wide-band receiver*, on the other hand, allows many megahertz of signals to pass to the demodulation stage, where the individual signals or services are sorted out. Figure 6.31 shows a block diagram of the typical wide-band receiver architecture [6.37]. Note that the LO operates at a fixed

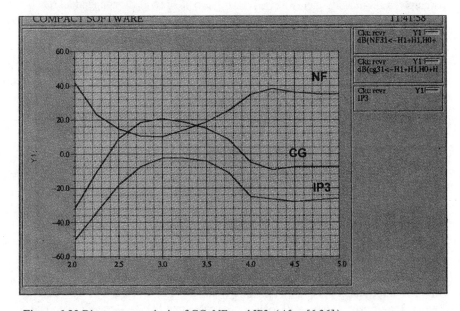

Figure 6.30 Bias sweep analysis of CG, NF, and IP3. (*After [6.36].*)

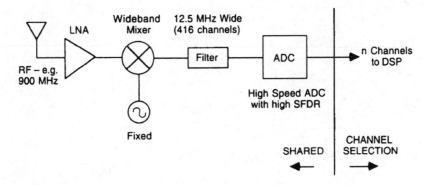

Figure 6.31 Basic architecture of a wide-band receiver. (*After [6.37].*)

frequency, rather than the variable configuration commonly associated with a superheterodyne receiver. Initial selectivity is determined by the wide-band mixer and broad bandwidth filter. The work of decoding is performed by a high-speed A/D converter.

A typical wide-band radio of this type may process a signal bandwidth of 5 to 25 MHz simultaneously, a scheme frequently referred to as *block conversion*. Block conversion provides significant economies for the end-user in certain applications, including cellular radio base stations and wide-band frequency scanning radios. For example, a single wide-band receiver can replace 48 individual independent receivers in a cellular base station environment. The final frequency selection process is performed in one or more digital processing stages using channelizers to select and filter the desired signal(s) from the wide-band input. This data is then passed to a DSP system, usually implemented on a single VLSI chip. Because the DSP is programmable, virtually any demodulation scheme may be used (AM, SSB, QAM, FM, etc.). Filtering of the signal may also be performed by the DSP circuit, offering unique filter shapes that would be difficult (or even impossible) to achieve in hardware. Systems incorporating such techniques are commonly referred to as *software radios*.

The technical requirements of wide-band receivers are stringent. A dynamic range of 90 dB or more is not uncommon between the strongest and weakest signals to be received by the radio. The *spurious free dynamic range* (SFDR) provides a measure of the analog front end and A/D converter performance as a unit; SFDR values are typically in the range of 95 to 100 dB for a high-quality radio system. SFDR is useful because it provides a measurement of performance as the input signal approaches the noise floor of the receiver. This yields an indication of overall receiver SNR or the BER of a digital receiver. Figure 6.32 plots SFDR as a function of input amplitude A_{in}. Note that the SFDR actually improves as the signal level is reduced to approximately 10 dB below full scale input (0 dB). Depending upon the

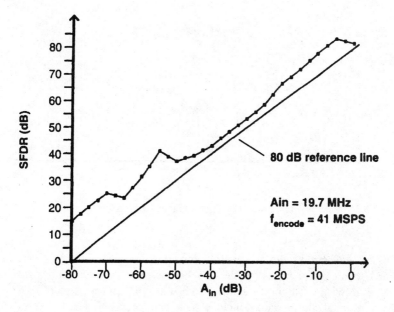

Figure 6.32 SFDR as a function of A/D converter input amplitude. (*After [6.37].*)

situation, this improvement may be greater than the loss of signal range and actually provides more dynamic range despite the reduction in input signal amplitude. The full scale degradation is typically the result of nonlinearities associated with the static transfer function near the full scale point of the ADC. Slew rate limitations of the track-and-hold circuit—an integral part of the ADC—may also contribute to reduced performance near the full scale point.

Dithering may be used to reduce A/D converter nonlinearities into the effective noise floor by forcing the converter to use different parts of its range each time it samples a given analog input value. Dithering is implemented by digitally generating a *pseudo-random number* (PRN) and applying it to a D/A converter, the output of which is summed with the analog input signal to be processed. After processing, the PRN is digitally subtracted from the output of the ADC, the end result being a randomization of the nonlinearities of the converter. A by-product of this process is a reduction in the spectral content generated by the A/D converter by repetitively exercising the same nonlinearity.

The wide input frequency band and wide dynamic range of the input signals place significant requirements on the front end and the converter for low IMD. IMD measurements on an A/D converter are made with multiple tones; 16, 24, or even more tones are common, depending upon the intended service. Spurious signals produced as a result of nonlinearities in the front end can, if at a sufficiently high level, override weaker desired signals located at the same or nearby frequencies.

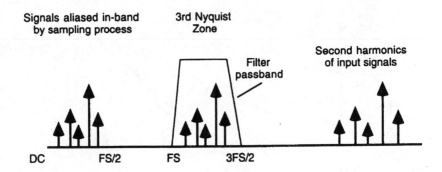

Figure 6.33 Upper Nyquist zone sampling technique. (*After [6.37].*)

Many wide-band radio designs mix down the RF spectrum to baseband using wide dynamic range, high IP mixers. In this case, the converter sample rate must be at least twice the highest frequency to be received, thus satisfying the Nyquist rate. Additional considerations include the type of modulation used in the system. For digitally modulated data, the A/D converter should sample at an integer multiple of the data rate; for analog modulation, sample rates are used that are a multiple of the channel bandwidth.

An alternative approach to baseband sampling involves sampling at the second or third *Nyquist zone*. As illustrated in Figure 6.33, the first Nyquist zone stretches from dc to one-half the sampling rate ($f_s/2$). The second Nyquist zone is the range from $F_s/2$ to F_s and the third Nyquist zone is the range from F_s to $3F_s/2$. Sampling at the second or third Nyquist zone reduces the harmonic-suppression requirements of previous amplification stages because filtering is significantly easier when the conversion is shifted above the first Nyquist zone.

Some of the RF-to-baseband conversion techniques covered in this section may also be applied to narrow-band radio systems to provide improved performance or reduced cost. For example, an ADC can be configured to act as an IF sampler to perform the last mix-down. Using the ADC in an undersampling mode eliminates the requirement for the final mixer and reduces the need for filtering as well. Once digitized, filtering can be performed by a DSP using FIR techniques, potentially improving receiver performance and lowering manufacturing costs.

6.6.1 Process Gain

Process gain refers to the improvement in SNR in a digital radio system by various numerical operations. In any digitization process, the faster the input signal is sampled, the lower the *apparent* noise floor. The total integrated noise remains constant, but it is spread out over more frequencies of the sample, per the following equation [6.37]:

$$N_f = 6.02 \times B + 18 + 10 \log\left(\frac{f_s}{2}\right) \qquad (6.21)$$

where N_f is the the noise floor. Every time the sample rate is doubled, the effective noise floor will improve by 3 dB. This improvement, while significant, can be hard to realize cost effectively. Larger processing gains may, however, be achieved in the digital filtering stage(s) of the receiver. Current DSP designs are quite effective in lowering the effective SNR at the channel output through creative manipulation of the digital bit stream.

6.7 Mixer Design Considerations

Several nonideal effects of mixers and oscillators can work to degrade the effective SNR of the receiver and contribute to adjacent channel interference. The origin of these effects lies in the inherent nonlinearity of the semiconductor devices used to create the components and in the sources of noise inherent in every lossy- or semiconductor-based component. Typically, an effort is made to linearize the operation of the devices to improve component performance at the expense of power consumption or other design parameters. The corresponding figures-of-merit used to characterize these nonideal effects include mixer IP3 and IP5, mixer NF, and oscillator phase noise specifications at one or more specific offset frequencies.

IM in the mixer tends to have two severe effects. One is distortion of the desired strong signal through generation of new spectral components within the channel bandwidth and just outside the channel bandwidth (*spectral regrowth*). The second effect, which is of primary concern, is the growth of spectral products within the channel bandwidth due to the presence of a strong modulated carrier and one or more modulated or unmodulated carriers in nearby channels. The IM products of these carriers fall within the bandwidth of the desired channel and cannot be differentiated from the desired modulated signal. This often occurs when mobile stations are operating in proximity to each other or there may be a base station operating in multiple nearby channels. IM products can be produced through third- or fifth-order products, which are significant at higher signal power levels. IP3 and IP5 represent figures-of-merit for IM distortion, but actual IM product powers may be higher than the linear 3:1 and 5:1 power ratios assumed in the IP3 and IP5 numbers.

Mixer (and amplifier) NF contributes directly to receiver SNR degradation in a straightforward manner. Oscillator phase noise can also be viewed as SNR degradation because the noise sideband injected by the LO into the mixer translates down to the IF and appears as an added noise component.

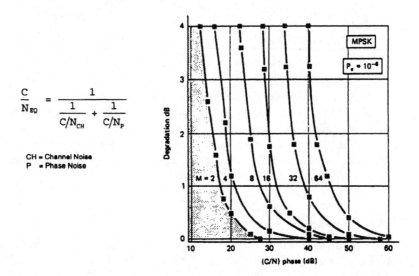

$$\frac{C}{N_{EQ}} = \frac{1}{\dfrac{1}{C/N_{CH}} + \dfrac{1}{C/N_{P}}}$$

CH = Channel Noise
P = Phase Noise

Figure 6.34 BER degradation as a function of LO C/N for various modulation formats. (*After [6.38].*)

Figure 6.34 shows the degradation of BER as a function of the *carrier-to-noise ratio* (C/N) of the LO for various *m*-ary MPSK modulation formats (the reference BER is 10^6). As the phase noise of the oscillator increases (C/N decreases), the BER deteriorates. As you would expect, *m*-ary formats with larger number of states are highly sensitive compared to formats with less states. Although large *m*-ary systems can achieve high spectral efficiency, high quality components must be used. Ultimately, these effects work to deteriorate the BER of the receiver, and it is important to understand trade-offs in component performance with cost, reliability, and BER degradation.

Besides stand-alone mixer considerations, the evaluation of the complete demodulator should consider effects such as unbalanced mixers (both amplitude and phase imbalance) and phase errors of the $\pi/4$ phase delay element. Modulation formats that contain information in the amplitude of the signal (e.g., *quadrature phase-shift-keying* (QPSK) modulation) will suffer from amplitude imbalance. Therefore, demodulator design with discrete mixers must consider imbalance over the bandwidth of operation. Virtually all popular modulation formats are sensitive to phase imbalances and errors in the phase delay element over the bandwidth of operation. For example, a 3° error in the phase delay element will degrade the suppression of the sidebands and carrier in the demodulator to −25 dB (from an ideal demodulator).

Other considerations in the selection of mixer technologies include trade-offs in conversion gain, linearity, dc power, LO power, cost, and manufacturability.

TABLE 6.2 Comparison of the Typical Performance of a Passive and Active Mixer
(Operating Frequency = 900 MHz)

Parameter	Passive	Active
Conversion gain	−10 dB	+10 dB to +20 dB
IP3 (input)	+15 dBm	−20 dBm
DC power	0	15 mW
LO power	+10 dBm	−7 dBm
1 dB comp. (input)	+3 dBm	−10 dBm
Size (mm)	10 × 20 × 10	5 × 4 × 1.75 (SO-8)
Technology	hybrid	Si RF IC

Mixers with conversion can reduce the gain requirements of the LNA and trade off gain, noise, and linearity. Active mixers use precious dc power in portable applications but greatly reduce the LO power requirements (by 10 to 15 dB), therefore allowing a less power-hungry LO to be utilized.

Typical differences between passive and active mixers (using BJT technology) are shown in Table 6.2. As discussed previously, the cg of active mixers can help offset LNA gain and allow designers to select an LNA design for optimum noise performance. The linearity of passive and active mixers differs considerably when referenced at the mixer input (as most manufacturers specify). However, when scaled by the CG, the IP3s referenced to the output differ by 5 to 15 dB, with passive technology still winning. The small amount of dc power used by active mixers is more than compensated when the low LO powers are considered. Active mixers typically have low compression points, which reduce the dynamic range of the receiver.

The mechanisms used for the computation of mixer noise are shown in Figure 6.35. The LO excitation modulates the nonlinear devices and their noise contributions. Through frequency conversion, these noise sources, along with the thermal noise and LO injected noise, are converted to the IF. The frequency conversion of the noise sources is pictorially described in Figure 6.36. The noise sources at a frequency deviation from all LO harmonics are accounted for in the analysis. Network analysis is used to compute the contribution of the frequency converted noise power to the IF port.

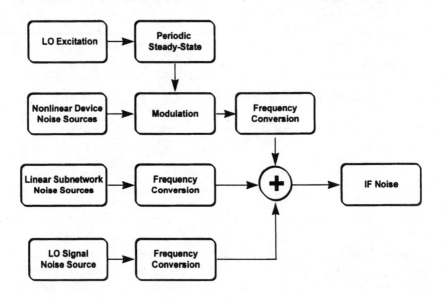

Figure 6.35 The elements of mixer noise. (*After [6.38].*)

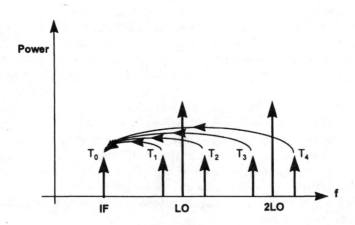

Figure 6.36 Mixer noise sources. The noise at each sideband contributes to the output noise at the IF through frequency conversion. (*After [6.38].*)

6.8 REFERENCES

6.1. S. M. Perlow, "Intermodulation Distortion in Resistive Mixers," *RCA Review.*, vol. 35, p. 25, Mar. 1974.

6.2. P. Torrione and S. Yuan, "Multiple Input Large Signal Mixer Analysis," *RCA Rev.*, vol. 26, p. 276, June 1965.

6.3. S. M. Perlow, "Third-Order Distortion in Amplifiers and Mixers," *RCA Rev.*, vol. 37, p. 257, June 1976.

6.4. H. Bley, "Eigenschaften und Bemessung von Ringmodulatoren," *NTZ*, vol. 13, pp. 129, 196, 1960.

6.5. R. M. Mouw and S. M. Fukuchi, "Broadband Double Balanced Mixer Modulators," *Microwave J.*, vol. 12, p. 131, Mar. 1961 and p. 71, May 1961.

6.6. J. G. Gardiner, "The Relationship between Cross-Modulation and Intermodulation Distortions in the Double Balanced Mixer," *Proc. IEEE*, vol. 56, p. 2069, Nov. 1968.

6.7. J. G. Gardiner, "An Intermodulation Phenomenon in the Ring Modulator," *Radio Electron. Eng.*, vol. 39, p. 193, 1970.

6.8. B. L. J. Kulesza, "General Theory of a Lattice Mixer," *Proc. IEE*, vol. 118, p. 864, July 1971.

6.9. J. G. Gardiner, "Local-Oscillator-Circuit Optimization for Minimum Distortion in Double-Balanced Modulators," *Proc. IEE*, vol. 119, p. 1251, Sept. 1972.

6.10. H. P. Walker, "Sources of Intermodulation in Diode Ring Mixers," *Radio Electron. Eng.*, vol. 46, p. 247, 1976.

6.11. E. W. Herold, "Superheterodyne Converter System Considerations in Television Systems," *RCA Rev.*, vol. 4, p. 324, Jan. 1940.

6.12. E. W. Herold, "Operation of Frequency Converters and Mixers for Superheterodyne Reception," *Proc. IRE*, vol. 30, p. 84, Feb. 1942.

6.13. R. V. Pound, *Microwave Mixers*, M.I.T. Rad. Lab. ser., vol. 16, McGraw-Hill, New York, 1948.

6.14. D. G. Tucker, *Modulators and Frequency Changers,* MacDonald Co., London, 1953.

6.15. R. G. Meyer, "Noise in Transistor Mixers at Low Frequencies," *Proc. IEE*, vol. 114, p. 611, 1967.

6.16. R. G. Meyer, "Signal Processes in Transistor Mixer Circuits at High Frequencies," *Proc. IEE*, vol. 114, p. 1604, 1967.

6.17. J. S. Vogel, "Non-Linear Distortion and Mixing Processes in FETs," *Proc. IEEE*, vol. 55, p. 2109, Dec. 1967.

6.18. J. M. Gerstlauer, "Kreuzmodulation in Feldeffekttransistoren," *Int. elektron. Rundsch.*, vol. 24, no. 8, p. 199, 1970.

6.19. M. Göller, "Berechnung nichtlinearer Verzerrungen in multiplikativen Transistormischern," *Nachrichtentechnik*, vol. 24, no. 1, p. 31 and no. 2, p. 65, 1974.

6.20. U. L. Rohde, "High Dynamic Range Active Double Balanced Mixers," *Ham Radio*, p. 90, Nov. 1977.

6.21. D. DeMaw and G. Collins, "Modern Receiver Mixers for High Dynamic Range," *QST*, p. 19, Jan. 1981.

6.22. U. L. Rohde, "Performance Capability of Active Mixers," *Prof. Program Rec.*, WESCON '81, p. 24; also, *Ham Radio*, p. 30, March 1982, and p. 38, April 1982.

6.23. L. Becker and R. L. Ernst, "Nonlinear-Admittance Mixers," *RCA Rev.*, vol. 25, p. 662, Dec. 1964.

6.24. S. M. Perlow and B. S. Perlman, "A Large Signal Analysis Leading to Intermodulation Distortion Prediction in Abrupt Junction Varactor Upconverters," *IEEE Trans.*, vol. MTT-13, p. 820, Nov. 1965.

6.25. J. M. Manley and H. E. Rowe, "Some General Properties of Nonlinear Elements—Part I," *Proc. IRE*, vol. 44, p. 904, July 1956.

6.26. C. F. Edwards, "Frequency Conversion by Means of a Nonlinear Admittance," *Bell Sys. Tech. J.*, vol. 20, p. 1403, Nov. 1956.

6.27. P. Penfield and R. Rafuse, *Varactor Applications*, M.I.T. Press, Cambridge, Mass., 1962.

6.28. A. M. Yousif and J. G. Gardiner, "Multi-Frequency Analysis of Switching Diode Modulators under High-Level Signal Conditions," *Radio Electron. Eng.*, vol. 41, p. 17, Jan.1971.

6.29. J. G. Gardiner, "The Signal Handling Capacity of the Square-Wave Switched Ring Modulator," *Radio Electron. Eng.*, vol. 41, p. 465, Oct. 1971.

6.30. R. P. Rafuse, "Symmetric MOSFET Mixers of High Dynamic Range," *Dig. Tech. Papers, Int. Solid-State Circuits Conf.*, (Philadelphia, Pa., 1968), p. 122.

6.31. R. H. Brader, R. H. Dawson, and C. T. Shelton, "Electronically Controlled High Dynamic Range Tuner," Final Rep. ECOM-0104-4, Ft. Monmouth, N.J., June 1971.

6.32. R. H. Dawson, L. A. Jacobs, and R. H. Brader, "MOS Devices for Linear Systems," *RCA Eng.*, vol. 17, p. 21, Oct./Nov. 1971.

6.33. M. Martin, "Verbesserung des Dynamikbereichs von Kurzwellen-Empfängern," *Nach. Elektronik*, vol. 35, no. 12, 1981.

6.34. B. Gilbert, E. Brunner, T. Kelly, and B. Clarke, "Analog Signal Processing: Mixers and AGC Amplifiers," *Proc. Wireless Symposium*, p. 16, San Jose, CA, January 1993.

6.35. D. Bien, "A 2 GHz BiCMOS Low-Noise Amplifer and Mixer," *Proc. Wireless Symposium*, p. 56, San Jose, CA, January 1993.

6.36. J. Gerber, "GSM Circuit Design Considerations, Part 1: Mixer Design," Compact Software, Patterson, NJ, p. 12, September 1995.

6.37. B. Brannon, "Using Wide Dynamic Range Converters for Wide Band Radios," *RF Design*, p. 50, May, 1995.

6.38. J. Gerber, "GSM Circuit Design Considerations, Part 1: Mixer Design," Compact Software, Patterson, NJ, p. 1, September 1995.

Additional Suggested Reading

"Eight Ways to Better Radio Receiver Design," *Electronics*, Feb. 20, 1975.

"Performance Capability of Active Mixers," WESCON/81, San Francisco, Sept. 1981.

Chapter 7

Frequency Control and Local Oscillators

7.1 Introduction

Communications receivers are seldom single-channel devices, but more often cover wide frequency ranges. In the superheterodyne receiver, this is accomplished by mixing the input signal with an LO signal. The LO source must meet a variety of requirements, including:

- It must have high spectral purity.
- It must be agile so that it can move rapidly (jump) between frequencies in a timeframe that can be as short as a few microseconds.
- The increments in which frequencies may be selected must be small.

Frequency resolution between 1 and 100 Hz is generally adequate below 30 MHz; however, there are a number of systems which provide 0.001-Hz steps. At higher frequencies, resolution is generally greater than 1 kHz.

In most modern receivers, such a frequency source is typically a synthesizer that generates all individual frequencies as needed over the required frequency band. The modern synthesizer provides stable phase-coherent outputs. The frequencies are derived from a master standard, which can be a high-precision crystal oscillator, a secondary atomic standard (such as a rubidium gas cell), or a primary standard using a cesium atomic beam. The following characteristics must be specified for the synthesizer:

- Frequency range
- Frequency resolution
- Frequency indication
- Maximum frequency error
- Settling time

- Reference frequency
- Output power
- Harmonic distortion
- SSB phase noise
- Discrete spurs (spurious frequencies)
- Wide-band noise
- Control interface
- Power consumption
- Mechanical size
- Environmental conditions

Free-running tunable oscillators, once used in radio receivers, have generally been replaced in modern communications receivers because of their lack of precision and stability. Fixed-tuned crystal-controlled oscillators are still used in second- and third-oscillator applications in multiconversion superheterodyne receivers that do not require single-reference precision. Oscillators used in synthesizers have variable tuning capability, which may be voltage-controlled, generally by varactor diodes.

Synthesizer designs have used mixing from multiple crystal sources and mixing of signals derived from a single source through frequency multiplication and division. Synthesizers may be "direct" and use the product of multiple mixing and filtering or "indirect" and use a PLL locked to the direct output to provide reduced spurious signals. There are a number of publications describing these and other techniques, the classic texts being [7.1] through [7.3]. Most modern communications receivers operating below 1 GHz use single- or multiple-loop digital PLL synthesizers, although for some applications, direct digital waveform synthesis may be used. The material in this chapter considers only such synthesizers and is based primarily on [7.4]. This reference contains a much more detailed treatment of digital PLL synthesizers and provides many additional references. The briefer treatment in [7.5] may also be of interest to the reader. In this chapter, we treat synthesizers before reviewing oscillator design.

7.2 PLL Synthesizer Principles

Figure 7.1 shows the block diagram of a single-loop synthesizer. Unless special techniques are used, such as the *fractional division-by-N* principle, the step size or channel frequency is equal to the reference frequency. When describing frequency synthesizers mathematically, we usually use a linearized model. Because most

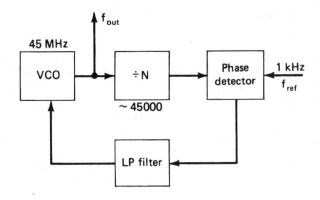

Figure 7.1 Block diagram of a single-loop synthesizer. (*After [7.4].*)

effects occurring in the phase detector are highly nonlinear, only the so-called piecewise linear treatment allows adequate approximation.

We assume that the VCO in Figure 7.1 is tunable over the frequency range from 41 to 51 MHz. Its output is divided to the reference frequency in a programmable divider (i. e., divide by N), whose output is fed to one of the inputs of the phase-frequency detector and compared with the reference frequency supplied to the other input. The loop filter at the output of the phase detector suppresses reference frequency components, while also serving as an integrator. The dc control voltage at the output of the loop filter pulls the *voltage-controlled oscillator* (VCO) until the divided frequency and phase equal those of the reference. In this simple example, with the divider set to 45,000 and the reference to 1 kHz, the VCO is controlled to a frequency of 45 MHz. A fixed division of the frequency standard output produces the reference frequency of appropriate step size. Frequency standards are typically operated at 1, 5, or 10 MHz to take advantage of high crystal stability. A 5 MHz frequency standard would be divided by 5000 in the example. The operating range of the PLL is determined by the maximum operating frequency of the programmable divider, by its division range ratio, and by the tuning range of the VCO.

The PLL is nonlinear because the phase detector is nonlinear. However, it can be accurately approximated by a linear model when the loop is in lock. The response, when the loop is closed, may be expressed as:

$$\frac{\theta_c(s)}{\theta_r(s)} \equiv B(s) = \frac{forward\ gain}{1 + open-loop\ gain}$$

$$= \frac{G(s)}{1 + G(s)/N} \tag{7.1}$$

where $G(s) = G_1(s)G_2(s)F(s)/s$, and θ_c and θ_r are the phases of the controlled oscillator and the reference, respectively.

When the loop is locked, it is assumed that the phase detector output voltage is proportional to the difference in phase between its inputs, that is,

$$V_\theta = K_\theta(\theta_r - \theta_i) \tag{7.2}$$

where V_θ is the output voltage of the phase detector, and θ_r and θ_i are the phases of the reference signal and the divided VCO signal, respectively. K_θ is the phase detector gain factor and has the dimensions of volts per radian. It is also assumed that the VCO can be modeled as a linear device whose output frequency differs from its free-running frequency by an increment of frequency

$$2\pi\,\delta f = K_0\,V_c \tag{7.3}$$

where V_c is the voltage of the output of the low-pass filter and K_0 is the VCO gain factor with the dimensions of radians per second per volt. Because frequency is the time derivative of phase, the VCO operation can be described as:

$$2\pi\,\delta f \equiv \frac{d\theta_c}{dt} = K_0 V_c \tag{7.4}$$

With these assumptions the PLL may be represented by the linear model shown in Figure 7.2. The linear transfer function relating $\theta_c(s)$ and $\theta_r(s)$ is:

$$B(s) = \frac{\theta_c(s)}{\theta_r(s)} = \frac{K_\theta\,K_0\,F(s)/s}{1 + K_\theta\,K_0\,F(s)/Ns} \tag{7.5}$$

The forward gain is:

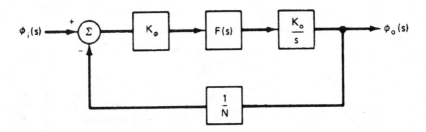

Figure 7.2 Block diagram of a linearized model of a PLL. (*From [7.4]. Reprinted with permission.*)

$$G(s) = \frac{K_\theta K_0 F(s)}{s} \qquad (7.6)$$

and the open-loop gain is:

$$G(s)\,H(s) = \frac{K_\theta K_0 F(s)}{Ns} \qquad (7.7)$$

which leads to the transfer formula of Equation (7.1).

There are various choices of filter response $F(s)$. Because the VCO by itself is an integrator, we can use a simple RC filter following the phase detector. This arrangement is called a *type 1 filter*. Because the components used, together with feedthrough capacitors and other stray effects, can cause excess phase shift, it is necessary to ensure that stability criteria are satisfied. If the gain of a passive loop is too small to provide adequate drift stability of the output phase, especially if a high division ratio is used, the best solution to this problem is the use of an active amplifier as an integrator. In most frequency synthesizers, the active filter-integrator approach is preferred to the passive one. Some frequency synthesizer chips have a single-ended output. In such cases, the use of an additional integrator requires some precautions.

Figure 7.3 shows the passive RC filter for the second-order loop typically used in PLL synthesizers (References [7.6] to [7.8]). The transfer characteristic of the filter is:

$$\frac{V_O(s)}{V_I(s)} = \frac{1 + s\,\tau_2}{1 + s(\tau_1 + \tau_2)} \qquad (7.8)$$

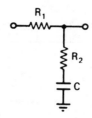

Figure 7.3 Schematic diagram of a typical passive RC filter. (*From [7.4]. Reprinted with permission.*)

where $\tau_1 = R_1 C$ and $\tau_2 = R_2 C$.

Figure 7.4 shows the schematic for the active filter for the second-order loop. Its transfer characteristic is:

$$\frac{V_0(s)}{V_I(s)} = \frac{1 + s\,\tau_2}{s\,\tau_1} \tag{7.9}$$

where $\tau_1 = R_1 C$ and $\tau_2 = R_2 C$.

If only one active integrator is used, we have a *type 1 PLL*. If two integrators are used, as in building an active filter, we have a *type 2 second-order loop*. Here, "second-order" refers to the denominator polynomial of the transfer function. If we insert a simple low-pass filter such as the one shown in Figure 7.3, but with $R_2 = 0$, we obtain:

$$F(s) = \frac{1}{1 + s\tau} \tag{7.10}$$

If we let $K = K_0 K_\theta / N$, the transfer function $B(s)$ becomes:

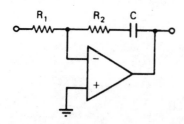

Figure 7.4 Schematic diagram of an active filter for a second-order loop. (*From [7.4]. Reprinted with permission.*)

$$B(s) = \frac{N}{s^2/\omega_n^2 + 2\zeta s/\omega_n + 1} \qquad (7.11)$$

where $\omega_n = \sqrt{K/\tau}$ and $2\zeta = \omega_n/K = \sqrt{1/K\tau}$. Here ζ is the damping factor of the loop and ω_n is the natural frequency.

The frequency response of the second-order transfer function is determined by ζ. For $\zeta = 0.707$, the transfer function becomes the second-order maximally flat, or Butterworth, response. For values of $\zeta < 0.707$ the gain exhibits peaking in the frequency domain. The maximum value of the frequency response can be found by setting the derivative of its magnitude to zero. The frequency at which the maximum occurs is:

$$\omega_p = \omega_n \sqrt{1 - 2\zeta^2} \qquad (7.12)$$

The 3-dB bandwidth B is found to be:

$$B = f_n \left[1 - 2\zeta^2 + (2 - 4\zeta^2 + 4\zeta^4)^{1/2}\right]^{1/2} \qquad (7.13)$$

where $f_n = \omega_n/2\pi$.

The time required for the output to rise from 10 to 90 percent of its final value is the rise time t_r. It is approximately related to the system bandwidth by the relation:

$$t_r = \frac{2.2}{B}$$

The RC time constant of this simple filter determines both the natural loop frequency and the damping factor ζ. In order to improve the performance of the filter, we need more flexibility. When the series resistor R_2 is not zero, we obtain the original RC filter of Figure 7.3. The transfer function of this filter is:

$$B(s) = \frac{N[s\,\omega_n\,(2\zeta - \omega_n/K) + \omega_n^2]}{s^2 + 2\zeta\omega_n s + \omega_n^2} \qquad (7.14)$$

where $\omega_n = \sqrt{K/\tau}$ and $2\zeta = (1 + K\tau_2)/\sqrt{K\tau}$ and τ is written for $\tau_1 + \tau_2$.

The determination of the 3-dB bandwidth of this general type 1 second-order loop is somewhat more complex than the earlier computation, but after calculation, we obtain:

$$B = f_n \left[a + (a^2 + 1)^{1/2} \right]^{1/2} \tag{7.15}$$

where we have written:

$$a = 2\,\zeta^2 + 1 - \frac{\omega_n \,(4\,\zeta - \omega_n/K)}{K} \tag{7.16}$$

The noise bandwidth of the type 1 second-order loop is:

$$B_n = \pi f_n \left(\zeta + \frac{1}{4\zeta} \right) \tag{7.17}$$

In the case of the active filter, where we have two integrators, the closed-loop transfer function of the type 2 second-order PLL with a perfect integrator is:

$$B(s) = \frac{N\,(2\,\zeta\,\omega_n\,s + \omega_n^2)}{s^2 + 2\,\zeta\,\omega_n\,s + \omega_n^2} \tag{7.18}$$

where $\omega_n = (KR_2/\tau_2 R_1)^{1/2}$, $2\zeta = (K\tau_2 R_2/R_1)^{1/2}$, and $K = K_\theta K_0/N$, as usual. The 3-dB bandwidth of the type 2 second-order filter is:

$$B = f_n \left\{ 2\,\zeta^2 + 1 + [(2\,\zeta^2 + 1)^2 + 1]^{1/2} \right\}^{1/2} \tag{7.19}$$

and the noise bandwidth is:

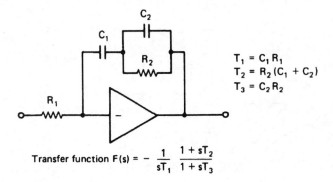

$$T_1 = C_1 R_1$$
$$T_2 = R_2(C_1 + C_2)$$
$$T_3 = C_2 R_2$$

Transfer function $F(s) = -\dfrac{1}{sT_1} \dfrac{1 + sT_2}{1 + sT_3}$

Figure 7.5 Schematic diagram of an active filter for a third-order loop. (*From [7.4]. Reprinted with permission.*)

$$B_n = \frac{(K R_2/R_1) + (1/\tau_2)}{4} \tag{7.20}$$

The type 2 third-order loop is defined by the active integrator, as shown in Figure 7.5. The additional capacitor across the second resistor increases suppression of the reference frequency. The advantage of the higher-order loop is that, for the same loop bandwidth, it offers more reference frequency suppression than the second-order loop. Conversely, for the same suppression it offers a faster lock in time. More details are given in [7.4].

7.2.1 Transient Response

The Laplace transform can be used to calculate the response of the PLL to a change in frequency. Figure 7.6 shows the normalized output response of the type 1 second-order loop, and Figure 7.7 shows the normalized output response of the type 2 second-order loop. We determine from both functions that a damping ratio of 0.707 will produce a peak overshoot of less than 10 percent for the type 1 second-order loop and of less than 20 percent for the type 2 second-order loop when $\omega_n t \geq 4.5$. The settling time is therefore determined to be $t_s = 4.5/\omega_n$. For the actual design of synthesizer loops, the reader should refer to a more detailed reference, such as [7.4].

7.3 Loop Components

The single-loop synthesizer consists of a VCO, digital divider, phase-frequency detector, and reference. In addition, some auxiliary components are required, such

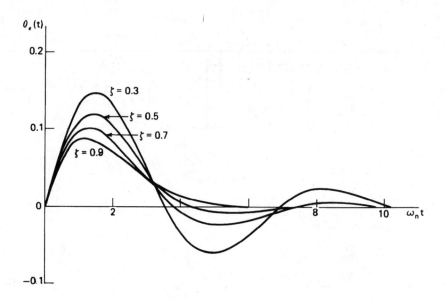

Figure 7.6 The error response of a type 1 second-order PLL to unit-step change in frequency for various damping ratios ζ with K constant. The steady-state error $2\zeta/\omega = 1/K$. (*After [7.5].*)

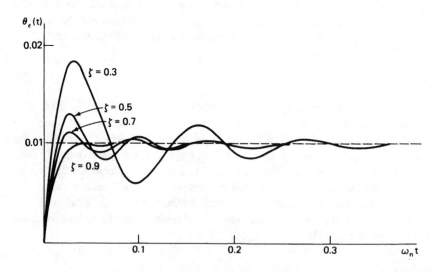

Figure 7.7 The error response of a type 2 second-order PLL to unit-step change in frequency for various damping ratios ζ with ω_n constant. The steady-state error is zero. (*After [7.5].*)

as limiting amplifiers and loop filters. In this section we discuss frequency dividers and phase-frequency detectors, which are the integrated circuits of most interest.

7.3.1 Frequency Dividers

Frequency dividers are built using *transistor-transistor logic* (TTL), complementary MOS (CMOS), and low-power *emitter-coupled logic* (ECL) integrated-circuit technologies. Dividers are commonly of two varieties—*synchronous counters* and *asynchronous counters*. A type 7490 divide-by-10 ripple counter is a classic device of the latter type. The counter produces an output square wave with 50 percent duty cycle. The frequency limitation of the ripple counter depends on the propagation delay of the device, but is higher than that of the synchronous counter. A synchronous counter, such as type 74196, requires internal resetting, a process involving a time loss, which results in a reduction of the maximum operating frequency.

If a frequency counter must operate at a frequency above 10 to 30 MHz, a prescaler is usually required. There are two forms of prescalers: variable-ratio divide-by-$k/(k + 1)$ and fixed-ratio divide-by-k. Fixed-ratio dividers are available in ECL. Their power consumption is rather high, and they are used as ripple counters, preceding a synchronous counter. A single-loop synthesizer loses resolution by the amount of prescaling. The term *prescaling* is used in the sense of a predivider that is nonsynchronous with the rest of the chain.

Figure 7.8 shows a block diagram of the Motorola MC12012 dual-modulus prescaler. By external programming, this ECL divider can be made to divide in various $k/(k + 1)$ ratios. If a clock pulse from the synchronous divider chain is used to reset this prescaler to satisfy the relationship:

$$f_{ref} = \frac{f_{osc}}{k\,A + (k+1)\,B} \tag{7.21}$$

During one full reference cycle, the prescaler can be considered a synchronous counter. With such a system, it is possible to increase the maximum frequency to at least 400 MHz without losing resolution.

In Figure 7.8, the three flip-flops are wired so that a change in coding input can change the division ratio among the values 2, 5, 6, 10, 11, and 12. *Dual-modulus* prescalers are also known as *swallow counters*. From a chain of pulses, one or two are *blocked*, or *swallowed*, at appropriate times to provide the changed counting. In addition to the MC12012, common swallow counters include the 95H90 (350 MHz) and 11C90 (520 MHz), made by Fairchild, and the SP8692 (200 MHz, 14 mA, and 5/6), SP8691 (200 MHz, 14 mA, and 8/9), SP8690 (200 MHz, 14 mA, and 10/11), and SP8786 (1300 MHz, 85 mA, and 20/22), made by Plessey. The division of a swallow counter is controlled by two inputs. The counter divides by

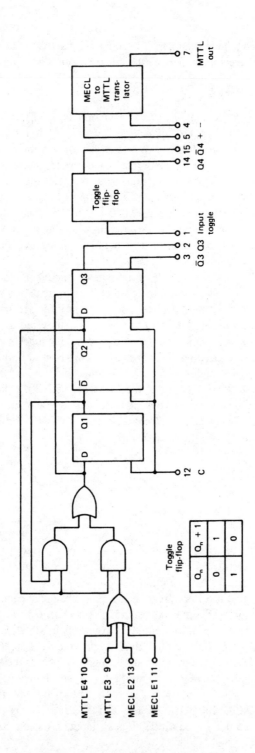

Figure 7.8 Block diagram of a divide-by-10/11 dual-modulus prescaler (Motorola MC12012). (*Courtesy of Motorola.*)

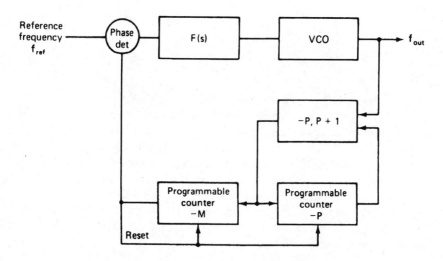

Figure 7.9 Block diagram of a PLL system using a dual-modulus prescaler. (*From [7.4].* *Used with permission.*)

k when either input is in the high state and by $k + 1$ when both are in the low state. The 10/11 division ratio lets us build fully programmable dividers to 500 MHz. The principle of switching the count permits high-frequency prescaling to occur without reduction in the comparison reference frequency. The disadvantage of this technique is that a fully programmable divider is required to control the division ratios, and there is a minimum limit on the possible division ratio. This is not a serious problem in practice.

Figure 7.9 is a block diagram showing the use of a swallow counter of division ratios $P/(P + 1)$. In our example, these are set to 10/11. The A counter counts the units, and the M counter counts the tens. The mode of operation depends on the type of programmable counter used, but the system might operate as follows. If the number loaded into the A counter is greater than zero, then the $P/(P + 1)$ divider is set to divide by $P + 1$ at the start of the cycle. The output from the $P/(P + 1)$ divider serves as a clock for both A and M. When A is full, it ceases counting and sets the $P/(P + 1)$ divider into the P counting mode. Only M is then clocked; when it is full, it resets both the A and M counters and the cycle repeats. The divider chain, therefore, divides by:

$$(M - A) P + A(P + 1) = MP + A \tag{7.22}$$

and

$$f_{osc} = (MP + A)f_{ref} \tag{7.23}$$

If A is incremented by 1, the output changes by f_{ref}. Thus, the channel spacing is equal to f_{ref}. This is the channel spacing that would be obtained with a fully programmable divider operating at the same frequency as the $P/(P + 1)$ divider.

For this system to work, the A counter must underflow before the M counter does, otherwise the $P/(P + 1)$ counter would remain permanently in the $P + 1$ count mode. There is, therefore, a minimum division ratio M_{min}, below which the $P/(P + 1)$ system will not function. The A counter must be capable of counting all numbers up to and including $P - 1$ if every division ratio is to be possible. Thus $A_{max} = P - 1$. Then $M_{min} = P$, because $M > A$. The divider chain divides by $MP + A$; therefore, the minimum system division ratio is:

$$D_{min} = M_{min}(P + A_{min}) = P(P + 0) = P^2 \tag{7.24}$$

Using a 10/11 ratio, the minimum division ratio is 100.

In the system shown in Figure 7.9, the fully programmable counter A must be quite fast. With a 350-MHz clock to the 10/11 divider, only about 23 ns is available for counter A to control the 10/11 divider. To reduce cost, it would be desirable to use TTL logic for the fully programmable counter. However, when the delays through the ECL and TTL translators are considered, very little time remains for the fully programmable counter. The 10/11 function can be extended easily, using external logic, to give an $N/(N + 1)$ counter with a longer control time for a given input frequency. Figure 7.10 shows a 20/21 system, for which the time available for control is typically 87 ns at 200 MHz and 44 ns at 350 MHz. To understand the operation of such circuits, we remember that both control leads must be low for the count to be 11; if either is high, the count is 10. One of the leads is retained for external control, the other is controlled by the added flip-flops (and gates, when needed). In Figure 7.10 the Q lead of the flip-flop is low at the start of operation. If the external control lead is also low, then, at count 11, the counter output flips the Q lead to high. The control lead PE_2 to the counter is now high, but after a count of 10, the flip-flop is again actuated, and the Q lead is brought back to low. Thus, as long as the external control lead PE_1 is low, the flip-flop Q lead goes low after every count of 21. When PE_1 is high, the divider always operates the flip-flop after 10 counts, so the count between Q-lead lows is 20.

The circuit of Figure 7.11, showing a 40/41 system, operates analogously, except that a count of 4 is now needed to restore the Q leads of both flip-flops to the low

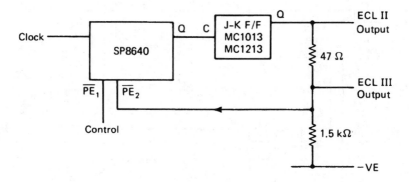

Figure 7.10 Block diagram of a 20/21 counting system for increasing control time. (*From [7.4]. Reprinted with permission.*)

condition so that the gate restores $\overline{PE}_2$ to the low condition. The cycle is now one 11 count followed by three 10 counts for a total of 41 when $\overline{PE}_1$ is low, but four 10 counts for a total of 40 when $\overline{PE}_1$ is high. The time available to control the count is approximately 180 ns at 200 MHz and 95 ns at 350 MHz. This technique can be extended to a count of 80/81 or further, which would allow the control to be

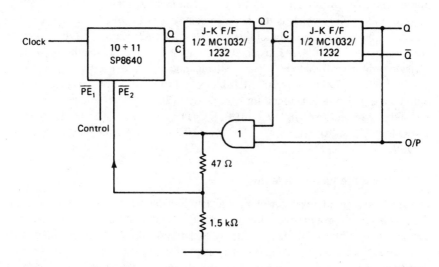

Figure 7.11 Block diagram of a 40/41 counting system for increasing control time. (*From [7.4]. Reprinted with permission.*)

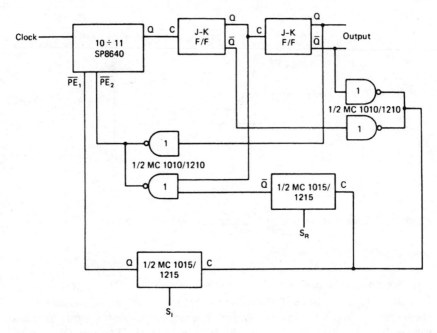

Figure 7.12 Block diagram of a 40/41/42 three-modulus counter. (*From [7.4]. Reprinted with permission.*)

implemented in CMOS. However, the extension increases the minimum division ratio to 6400 (80^2), which is too large for many synthesizers. The ratio can be reduced to 3200 by controlling for three counts, 80/81/82. Similarly, a 40/41 count can be extended to 40/41/42, as shown in Figure 7.12, to reduce the minimum division ratio from 1600 to 800. The available time to control the 40/41/42 is a full 40-clock pulse—200 ns for a 200 MHz clock or 110 ns for a 350 MHz clock. The minimum division ratio is found for s_1 high to be $(20 \times 40) + (0 \times 41) + (0 \times 42) = 800$; for s_1 low and s_2 high, $(19 \times 40) + (1 \times 41) + (0 \times 42) = 801$; and for both s_1 and s_2 low, $(19 \times 40) + (0 \times 41) + (1 \times 42) = 802$. More information can be found in [7.9].

7.3.2 Phase-Frequency Detector

The simplest form of phase detector is the double-balanced mixer shown in Figure 7.13. This circuit provides very low dc output, so that a postamplifier is required. The noise contribution of the amplifier is sometimes a limiting factor. Balanced-mixer phase detectors have a maximum range of π rad. The circuit is frequency-sensitive, so that presteering is required until the limited pull-in range of the circuit is reached. Modern frequency detectors typically use edge-triggered JK master-

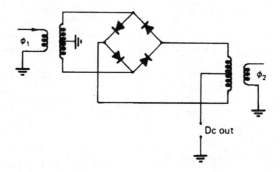

Figure 7.13 Schematic diagram of a double-balanced mixer as a phase detector.

slave flip-flops, which have an operating range from -2π to 2π rad, but there are difficulties at zero phase error because of phase jitter and finite resolution. Figure 7.14 shows the arrangement of such a flip-flop. The best performance is obtained from the so-called tristate phase-frequency comparator, which requires a charge-pump-type integrator at the output. An early version of this was the MC4044 detector. Similar systems were available in CMOS, such as the CD4046.

More modern synthesizer circuits, such as the MC145156, have an improved phase detector, which avoids the dead zone at zero phase difference. A type of antibacklash circuit is incorporated to avoid loop instabilities. At zero correction, a charge pump would not supply any correcting voltage. Because the gain is determined by the amount of energy supplied by the charge pump, zero correction would result in zero gain. What occurs in practice is that the gain can drop 20 dB and heavy phase jitter can occur. Correction of this situation requires an increase in leakage with consequent reduction in reference suppression or the use of the modern detector circuit with the antibacklash circuit, as shown in Figure 7.15. For a perfect tristate phase-frequency comparator, reference frequency suppression would be infinite. Typical values of real circuits fall between 40 and 60 dB. This type of detector is available in chips using CMOS, TTL, or ECL.

Another useful detector is the sample-and-hold detector. It is, however, only phase-sensitive. Figure 7.16 illustrates the principle of the sample-and-hold detector. The input signal, a very linear sawtooth, is sampled with a sufficiently narrow pulse and the voltage is then stored in a minimum-leakage capacitor. The voltage across this capacitor is updated for each sample, and if the sample pulse is sufficiently narrow, excellent reference suppression is possible. In many cases, two such samplers are put in cascade, as shown in Figure 7.17. The sampler has a delay, dependent upon the time constants chosen, which reduces the phase margin in the feedback loop.

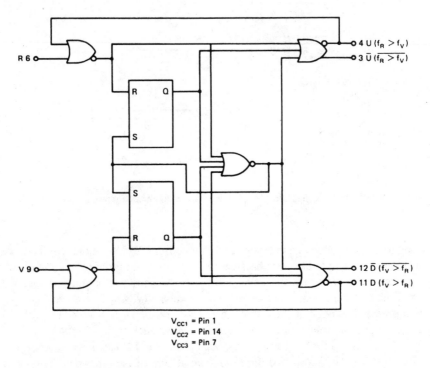

$V_{CC1} = \text{Pin 1}$
$V_{CC2} = \text{Pin 14}$
$V_{CC3} = \text{Pin 7}$

Figure 7.14 Block diagram of a phase-frequency comparator, the Motorola MC12040. (*Courtesy of Motorola, Inc.*)

7.4 Loop Optimization

Previously we considered the performance a of simple loop. We now examine the performance of an optimized loop. This optimization is aimed toward minimum noise, fastest acquisition, maximum reference frequency suppression, and minimum number of parts. The stability of the resulting design must also be demonstrated.

Figure 7.18 shows the relevant portions of a frequency synthesizer, comprising the Motorola MC145152 LSI frequency synthesizer chip, an external 10/11 counter SP8690, the loop filter, and amplifiers. The VCO is not shown in the figure. The 10/11 counter keeps the maximum operating frequency of the MC145152 below 16.8 MHz. The MC145152 has two divider chains, one for the reference divider and one for the programmable divider. The reference frequency is set to 25 kHz by the connections to pins 4, 5, and 6. Outputs 7 and 8 from the synthesizer integrated circuit are used to drive the CA3160 operational amplifier, which acts as an integrator and loop filter. The input is symmetrical, and the 47-kΩ resistor, together with the 390-pF capacitor, form a spike suppression filter.

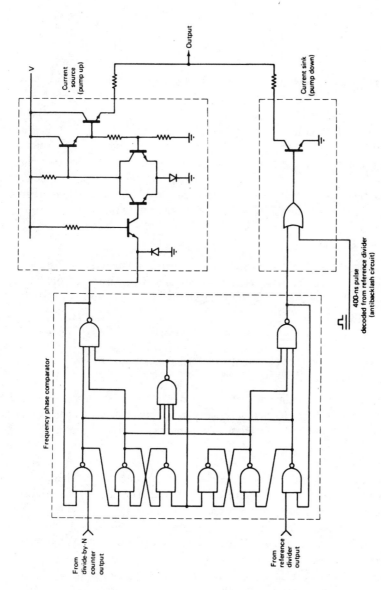

Figure 7.15 A tristate detector with antibacklash circuit included. (*From [7.4]. Reprinted with permission.*)

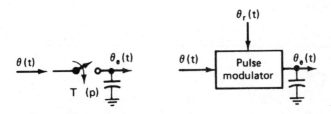

Figure 7.16 Simplified diagram showing the principle of a sample-and-hold phase detector.

The type 2 fifth-order loop includes an active integrator and an additional second-order low-pass filter. Because of the additional low-pass filter, the reference frequency suppression is substantially higher than could be provided by a simple filter and increases at harmonics of the reference frequency. Such an elaborate system as the fifth-order loop cannot be calculated easily, so calculations are made iteratively. A criterion during iteration is the loop stability.

7.4.1 Loop Stability

A PLL system can be analyzed for stability using various criteria. The easiest approach, which can be used with loops of any order, involves plotting the open-loop gain, phase, and frequency responses (the Bode plot). Two basic requirements must be satisfied:

- The gain must fall below unity (0 dB) and must remain so before the phase shift exceeds 180°.
- To maintain stability over the range of changes of parameters, adequate margins must be selected: a phase margin at the gain crossover frequency and a gain margin at the phase crossover frequency.

The damping factor ζ cannot be defined for high-order loops. Therefore, it is recommended that the phase margin should be used as a criterion. The settling time can be shown to be minimum for a phase margin of 45° at the gain crossover frequency. To maintain stability safely, the gain margin should be 10 dB or more at the phase crossover frequency. These considerations provide two requirements, which are basically valid for loops of all kinds.

Figure 7.19 shows the Bode plot provided by a CAD program for the optimized loop. From this we can see that the phase margin at the gain crossover frequency is 45° and the gain margin at the phase crossover frequency is about 25 dB. With such a phase margin, we have a stable loop and can allow adequate tolerances. The Bode plot also allows us to determine the reference frequency suppression of the

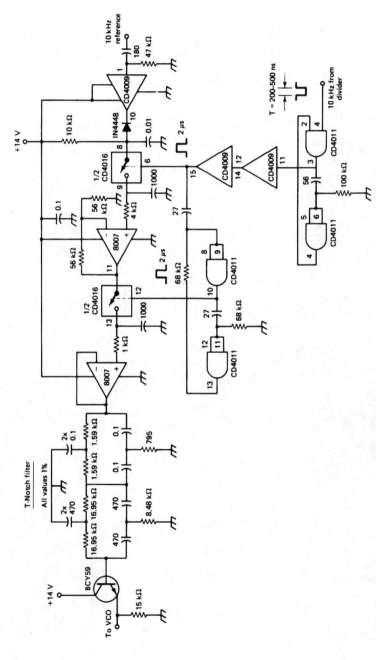

Figure 7.17 Dual sample-and-hold comparator with additional filtering. *(From [7.4]. Reprinted with permission.)*

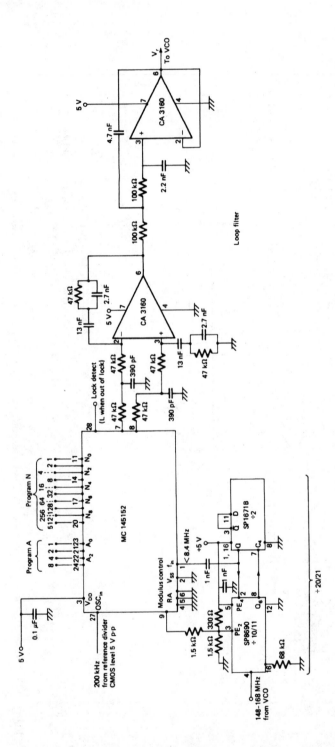

Figure 7.18 Schematic diagram of an optimized type 2 fifth-order loop.

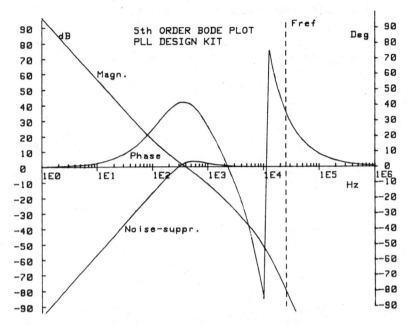

Figure 7.19 Bode plot of the optimized loop of Figure 7.18.

loop filter, which, in this case, is approximately 75 dB. Additional suppression is obtained from the inherent reference frequency suppression of the balanced integrator, as mentioned before. Thus, 75 dB is the minimum reference frequency suppression provided by the loop. The sideband noise of the VCO is improved below 350 Hz, as can be seen from the noise suppression curve. This curve, which assumes no noise contribution from the reference, gives the noise improvement of the VCO output as a function of offset frequency. There is a 3-dB overshoot at about 500 Hz, which is a typical indication of a higher-order loop. A printout of frequency versus time for this synthesizer indicates a settling time of 5 ms for the loop frequency error to be within a small fraction of the channel bandwidth.

Most apparent type 2 second-order loops are, in reality, higher-order loops because of additional low-pass action from active devices. The resulting stray phase shifts in such devices as operational amplifiers and feedthrough capacitors must be taken into consideration, especially in loops with higher frequency cutoffs.

7.4.2 Programming the Counters

Because the LSI circuit used as a programmable counter has only a limited number of pins available, a programming technique must be used that keeps the number required for this function small. A number of modern frequency divider chips have been optimized for use with a microprocessor bus input. However, the one se-

TABLE 7.1 Loop-Divider Program Inputs N and A for Use with Figure 7.18

Channel no.	N	A
0	592	0
1	592	1
2	592	2
.	.	.
.	.	.
.	.	.
9	592	9
10	593	0
.	.	.
.	.	.
19	593	9
20	594	0
.	.	.
.	.	.
798	671	8
799	671	9
800	672	0

lected in Figure 7.18 works well with an external *read-only memory* (ROM). The frequency is controlled by the loop dividers designated N and A in the figure. The frequency range of 148 to 168 MHz is covered in 801 channels with 25-kHz separation. A VCO frequency of 148 MHz corresponds to channel 0, while channel 800 corresponds to 168 MHz. The division ratio thus ranges from a minimum $N_{min} = 5920$ to a maximum $N_{max} = 6720$. The loop divider's total division ratio is related to the binary numbers at the N and A inputs as $N_{tot} = NP + A$, where P is the lower dividing ratio of the prescaler. The resultant programming data are shown in Table 7.1, from which we can derive the following scheme: N is composed of the sum of 592 and the two more significant digits of the channel number, and A corresponds to the least significant digit of the channel number.

For example, let us select channel 573. $N = 592 + 57 = 49$, $A = 3$, $N_{tot} = (649 \times 10) + 3 = 6493$, and $f_{vco} = N_{tot}(25 \times 10^3) = 162.325$ MHz. We can verify this by noting that channel 573 is $573(25 \times 10^3) = 14.325$ MHz higher than channel 0 at 148 MHz, and $148 + 14.325 = 162.325$ MHz.

A circuit implementation for this scheme is provided in Figure 7.20. Three *binary-coded decimal* (BCD) switches are used to provide the human interface. Because the *least significant digit* (LSD) does not exceed 9, the BCD code is the same as the binary code, and the switch output can be applied directly to program input A. The program input N is also encoded in binary form, which above 9 differs

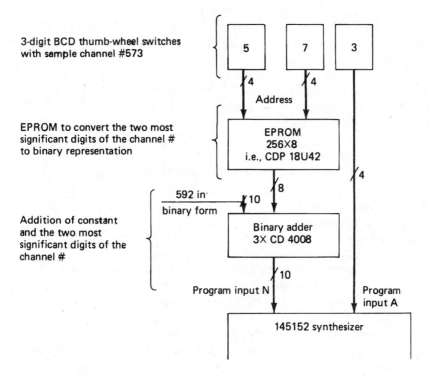

3-digit BCD thumb-wheel switches
with sample channel #573

5   7   3

4   4

Address

EPROM to convert the two most
significant digits of the channel #
to binary representation

EPROM
256X8
i.e., CDP 18U42

4

592 in·
binary form

10   8

Addition of constant
and the two most
significant digits of the
channel #

Binary adder
3X CD 4008

10

Program input N

Program
input A

145152 synthesizer

Figure 7.20 Frequency control circuit for Figure 7.18.

from the BCD numbers generated. Translation to binary is provided by addressing
the EPROM memory as if the two BCD numbers were a binary address and at that
address, storing the binary number corresponding to the two *most significant digits*
(MSD) of the address.

The minimum space required to program the EPROM is 81 × 8 bits, since the
channel number ranges from 0 to 800, yielding 00 to 80 for the two MSDs. With
the selected addressing method, however, the EPROM requires an 8-bit address.
For this number of address lines, available devices will probably be organized as
256 × 8 bits. Only 81 of the 256 addresses will be used. These addresses are
composed of the binary 8-bit word combining the two MSD BCD values for the
channel number. For example, the decimal number 57 would be encoded 01010111.
At this address location, corresponding to binary address 87, the binary representation of 57, which is 00111001, must be stored. In other words, if the two MSD
digits of the channel number are 57, then applying them (in BCD format) to the
EPROM address lines will place the 8-bit binary representation of 57 in the binary
adder input, where the hard-wired summing constant 592 (also in binary repre-

sentation) is added. The sum appearing at the adder output is applied to the N counter inputs of the MC145152 synthesizer block. The LSD of the channel number, ranging from 1 to 9, is applied directly to the A inputs, as mentioned previously. If the synthesizer range were limited to 800 channels (0 to 799), or from 148 to 167.975 MHz, only three lines would be required for the MSD of the channel number (the MSD being limited to 7). This would allow reduction in size of the EPROM to 128×8 bits when using this addressing technique. Other methods of conversion of the address might also be applied to permit using a smaller EPROM.

7.5 Noise and Performance Analysis Using CAD Tools

One of the key measurements of oscillator operation is the S/N of the circuit at certain offsets from the carrier. Ideally, the output of the oscillator would consist of only a single frequency signal, but in reality there is energy—noise—at nearby frequencies as well. Although oscillator amplitude variations will result in side-band energy, most of this noise is phase noise, the main sources of which are the flicker noise contribution of the (semiconductor) active device, the NF of the device under large-signal conditions, and the filtering effect of the resonator.

The actual phase noise of an oscillator is a composite of near- and far-carrier noise. This noise is most easily viewed in terms of its spectral power density, which is the power contained in a 1-Hz bandwidth at a particular frequency. Measuring the spectral power density of an oscillator results in a curve such as that of Figure 7.21. Rather than all of the power being concentrated at the oscillator frequency, some is distributed at frequencies around the oscillator frequency.

Because noise is a form of instability, it is useful to characterize frequency stability in the time domain, too. Short-term stability extends between a very small fraction of a second to 1 second, and occasionally up to 1 minute, although normally values for the stability over 1 second and 1 minute will be about the same. Long-term stability, or aging, is a frequency change typically expressed in terms of parts in 10^{-10} or 10^{-11} per day. Again, this information is in the time domain; in the frequency domain, we use terms such as *random walk*, *flicker*, and *wide phase noise* to describe the slope of the spectral density curve.

The Fourier frequency, usually labeled f_m, is at times called the *sideband frequency, offset frequency, modulation frequency,* or *baseband frequency*. In this chapter, we will refer to it as the offset frequency and will describe the S/R of an oscillator at a given offset from the center frequency. This way of describing oscillator noise—the spectral power density—is the most common way of characterizing phase noise, probably because this is precisely what is seen on a spectrum analyzer when the AM noise contribution is insignificant. Note that the spectrum analyzer display—and the oscillator noise—is symmetrical around the center frequency. Curves such as that of Figure 7.21, therefore, only show one side of the noise spectrum.

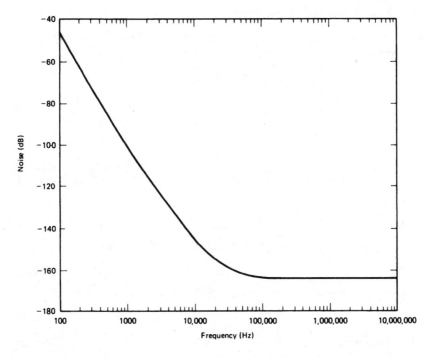

Figure 7.21 Typical noise sideband curve of a free-running oscillator (Rohde and Schwarz signal generator SMDU).

Looking at sideband noise in a 1-Hz bandwidth leads to the definition of $L(f_m)$: the ratio of the single-sideband noise power in a 1-Hz bandwidth, f_m hertz away from the carrier frequency, to the total signal power (as plotted in Figure 7.21).

These unwanted frequency components are referred to as *oscillator noise*. The oscillator output $S(t)$ can be expressed by the equation:

$$S(t) = A(t) \cos [\omega_o t + \theta(t)] \tag{7.25}$$

where $A(t)$ describes the amplitude variation as a function of time and $\theta(t)$ is the phase variation, or phase noise. A well-designed, high-quality oscillator is very amplitude stable, so $A(t)$ can be considered constant. Thus, for a constant-amplitude signal, all oscillator noise is due to $\theta(t)$. Leeson has developed a linear model that describes the origins of phase noise in oscillators, and because it closely fits experimental data, the model is widely used in describing the phase noise of oscillators and frequency synthesizers [7.10].

7.5.1 A Model for SSB Phase Noise

A carrier signal of amplitude V that is frequency modulated by a sine wave of frequency f_m can be represented by the equation:

$$S(t) = V \cos\left(\omega_o t + \frac{\Delta f}{f_m} \sin \omega_m t \right) \qquad (7.26)$$

where Δf is the peak frequency deviation and $\theta_p = \Delta f / f_m$ is the peak phase deviation, often referred to as the *modulation index*, β. Equation 7.26 can be expanded as:

$$S(t) = V \left[\cos(\omega_o t) \cos(\theta_p \sin \omega_m t) - \sin \omega_o t \sin(\theta_p \sin \omega_m t) \right] \qquad (7.27)$$

If the peak phase deviation is much less than 1, i. e., $\theta_p \ll 1$, then:

$$\cos(\theta_p \sin \omega_m t) \approx 1 \qquad (7.28)$$

and

$$\sin(\theta_p \sin \omega_m t) \approx \theta_p \sin \omega_m t \qquad (7.29)$$

Thus, for ($\theta_p \ll 1$), the signal $S(t)$ is approximately equal to:

$$S(t) = V \left[\cos(\omega_o t) - \sin \omega_o t (\theta_p \sin \omega_m t) \right] \qquad (7.30)$$

$$S(t) = V \left\{ \cos(\omega_o t) - \frac{\omega_p}{2} \left[\cos(\omega_o + \omega_m) t - \cos(\omega_o - \omega_m) t \right] \right\} \qquad (7.31)$$

That is, when the peak phase deviation is small, the phase deviation results in frequency components on each side of the carrier of amplitude $\theta_p/2$. This frequency distribution of a narrowband FM signal is useful for interpreting an oscillator's power spectral density resulting from phase noise. The phase noise in a 1-Hz bandwidth has a noise-power to carrier-power ratio of:

$$L(f_m) = \left(\frac{V_n}{V}\right)^2 = \frac{\theta_p^2}{4} = \frac{\theta_{rms}^2}{2} \tag{7.32}$$

The total noise is the noise in both sidebands and will be denoted by S_θ. That is,

$$S_\theta = 2\frac{\theta_{rms}^2}{2} = \theta_{rms}^2 = 2\,L(f_m) \tag{7.33}$$

With this interpretation of the noise power, the noise can now be described in terms of its origin, as illustrated in Figure 7.22.

7.5.2 Spectral Density of Frequency Fluctuations

Stability measurements using frequency comparators give the *spectral density of frequency fluctuations*,

$$S_{\Delta f}(f_m) = \Delta f_{rms}^2 \tag{7.34}$$

To relate the spectral density of frequency fluctuations to the spectral density of phase noise, recall that,

$$\Delta f(t) = \frac{1}{2\pi}\frac{d\,\Delta\theta(t)}{d\,t} \tag{7.35}$$

Transformed into the frequency domain, this gives

$$\Delta f(f_m) = f_m\,\Delta\theta(f_m) \tag{7.36}$$

$$S_{\Delta f}(f_m) = \Delta f_{rms}^2(f_m) = f_m^2\,S_{\Delta\theta}(f_m) = 2\,f_m^2\,L(f_m) \tag{7.37}$$

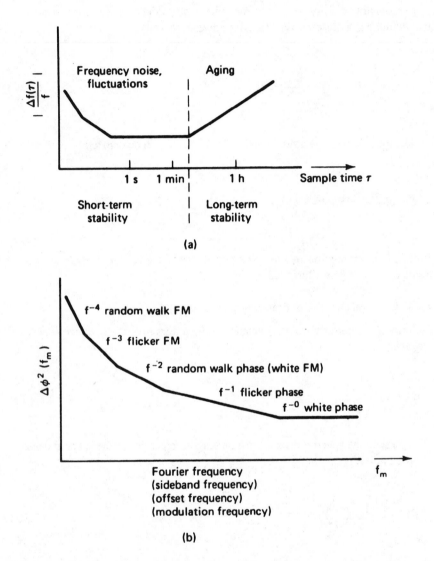

Figure 7.22 Characterization of the sources of noise sidebands: (*a*) time domain and (*b*) frequency domain.

The *National Institute of Standards and Technology* (NIST, formerly the National Bureau of Standards) has proposed to standardize the definition of the spectral density of fractional frequency fluctuations [7.11]. The instantaneous frequency deviation is normalized to the carrier frequency f_0 as follows:

$$y(t) = \frac{\Delta f(t)}{f_o} \tag{7.38}$$

$$S_y(f_m) = \frac{1}{f_o^2} S_{\Delta f}(f_m) = \frac{f_m^2}{f_o^2} S_{\Delta\theta}(f_m) = \frac{2 f_m^2}{f_o^2} L(f_m) \tag{7.39}$$

Characterizing fractional frequency fluctuations allows better comparison of sources with different carrier frequencies.

7.5.3 Residual FM Related to L(*fm*)

Residual FM, the total rms frequency deviation within a specified bandwidth, is another common way to specify the frequency stability of an oscillator. Commonly used bandwidths are 50 Hz to 3 kHz, 300 Hz to 3 kHz, and 20 Hz to 15 kHz. The following equation applies:

$$\Delta f_{res} = \sqrt{2} \sqrt{\int_a^b L(f_m) f_m^2 \, df_m} \tag{7.40}$$

Table 7.2 correlates Δf_{res} and $L(f_m)$ for specific slopes of $L(f_m)$ and L at 1 kHz = -100 dBc.

7.5.4 Allan Variance Related to L(*fm*)

For many applications, such as high-stability crystal oscillators or Doppler radar systems, it is more useful to describe frequency stability in the time domain. The time-domain characterization is based on the sample variance of fractional frequency fluctuations. Averaging differences of consecutive sample pairs with no dead time in between yields the *Allan variance*, $\sigma_y^2(\tau)$, which is the proposed standard measure of frequency stability, as follows:

$$\sigma_y^2(\tau) \frac{1}{2(M-1)} \sum_{k=1}^{M-1} \left(\bar{y}_{k+1} - \bar{y}_k \right)^2 \tag{7.41}$$

$\bar{y}_k$ is the average fractional frequency difference of the kth sample measured over the sample time.

TABLE 7.2 Correlation of Δf_{res} and $L(f_m)$ for Specific Slopes of $L(f_m)$ and L

$\mathcal{L}^a$ at 1 kHz (dBc)	Slope of $\mathcal{L}(f_m)$ Exponent	dB/oct	Residual FM Δf_{res} 50 Hz to 3 kHz	300 Hz to 3 kHz	20 Hz to 15 kHz
−100	0	0	1.34	1.34	15.0
−100	−1	−3	0.95	0.94	4.74
−100	−2	−6	0.77	0.73	1.73
−100	−3	−9	0.90	0.68	1.15

^{a}For any $\mathcal{L}$ at 1 kHz different to −100 dBc, multiply Δf_{res} of the table by

$$\text{antilog} \frac{100 - |\mathcal{L} \text{ at 1 kHz/dBc}|}{20}$$

The table does not take into account any microphonic or spurious sidebands.

Example: $\mathcal{L}$ at 1 kHz $= -88$ dBc, slope -9 dB. For bandwidth 20 Hz to 15 kHz:

$$\Delta f_{res} = 1.15 \text{ Hz} \times \text{antilog} \frac{100 - 88}{20} = 4.6 \text{ Hz}$$

Conversions from frequency- to time-domain data and vice versa are possible but tedious. The power spectrum, $L(f_m)$, needs to be approximated by integer slopes of 0, −1, −2, −3, and −4. Then conversion formulas (Table 7.3) can be applied. A good description of this procedure is given in [7.11] and [7.12]. A noise-conversion nomograph is given in Table 7.4. The example shown is self explanatory.

As most of these relationships, for reasons of convenience, are expressed in decibels rather than absolute values, the following formulas are commonly used:

$$L(f_m) = 10 \log_{10} \left(\frac{\Delta f_{peak}}{2 f_m} \right)^2 \tag{7.42}$$

$$L(f_m) = 10 \log_{10} \left(\frac{\Delta f_{rms}}{\sqrt{2} f_m} \right)^2 \tag{7.43}$$

$$L(f_m) = 20 \log_{10} \left(\frac{\Delta f_{rms}}{\sqrt{2} f_m} \right) \tag{7.44}$$

TABLE 7.3 Frequency Stability Conversion Equations

	Slope of $\sigma_y^2(\tau)$	$\sigma_y(\tau) =$	$\mathcal{L}(f) =$	Slope of $\mathcal{L}(f)$
White phase	-2	$\dfrac{\sqrt{\mathcal{L}(f)f_h}}{2.565f_o}\tau^{-1}$	$\dfrac{[\sigma_y(\tau)\tau f_o(2.565)]^2}{f_h}f^0$	0
Flicker phase	-1.9	$\dfrac{\sqrt{\mathcal{L}(f)[2.184+\ln(f_h\tau)]}}{2.565f_o}\tau^{-1}$	$\dfrac{[\sigma_y(\tau)\tau f_o(2.565)]^2}{2.184+\ln(f_h\tau)}f^{-1}$	-1
White frequency	-1	$\dfrac{\sqrt{\mathcal{L}(f)f^2}}{f_o}\tau^{-1/2}$	$[\sigma_y(\tau)\tau^{1/2}f_o]^2f^{-2}$	-2
Flicker frequency	0	$\dfrac{1.665\sqrt{\mathcal{L}(f)f^3}}{f_o}\tau^0$	$0.361[\sigma_y(\tau)f_o]^2f^{-3}$	-3
Random walk frequency	$+1$	$\dfrac{3.63\sqrt{\mathcal{L}(f)f^4}}{f_o}\tau^{1/2}$	$[(0.276)\sigma_y(\tau)\tau^{-1/2}f_o]^2f^{-4}$	-4

ᵃτ = measurement time, $y = \Delta f_o/f_o$, f_o = carrier, f = sideband frequency, f_h = measurement system bandwidth.

TABLE 7.4 Noise-Conversion Nomograph: Relationship among Modulating Frequency, Power Spectral Density of Phase, Modulation Index, Sideband-to-Carrier Ratio, and Frequency Deviation

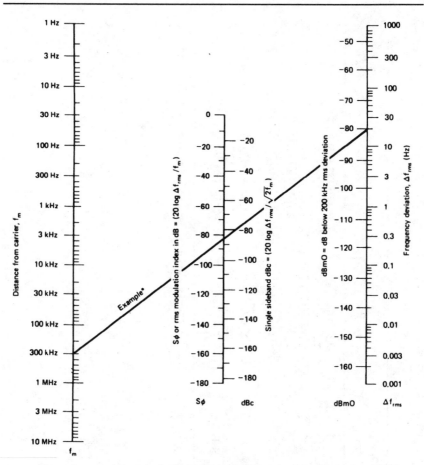

ªUse consistent measurement bandwidth. Example: 20-Hz deviation in a 1-kHz band at 300 kHz from carrier = single-sideband dBc of −87 dB in a 1-kHz band.

$$L(f_m) = 20 \log_{10} \frac{\theta_d}{2} \qquad (7.45)$$

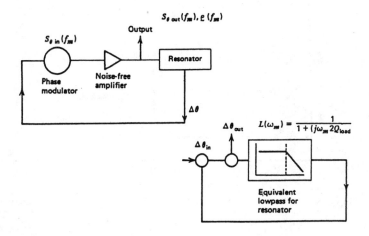

Figure 7.23 Block diagram of the feedback model of an oscillator consisting of a phase modulator, noise-free amplifier, and resonator.

7.5.5 Calculation of Oscillator Noise for Linear Model

We will assume that the oscillator is composed of a linear amplifier with gain A and a high-Q resonant circuit, as illustrated in block diagram form in Figure 7.23. The gain of the resonant circuit has been normalized to unity at the resonant frequency f_0. The amplifier gain must be unity or greater in order for the circuit to oscillate. Let S represent the amplifier noise referred to the amplifier input. The white noise N per unit bandwidth at the amplifier input is given by

$$N = N_i + N_a = F\,k\,T \tag{7.46}$$

where N_t is the thermal noise, N_a is the noise contributed by the amplifier, and F is the amplifier noise factor. Therefore, the ratio of noise power per unit bandwidth to signal power P_s is FkT/P_s, which is a component of S_θ. In addition, amplifiers generate an additional *flicker*, or $1/f$ phase noise, about the carrier frequency because of carrier density fluctuations in the base resistance. A plot of S, shown in Figure 7.24, demonstrates that for frequencies near the carrier, S_θ has a $1/f$ spectrum. At high frequencies, the spectrum is flat and equal to FkT/P_s the thermal noise floor. The frequency f_c below which the spectrum has a $1/f$ shape depends on the characteristics of the individual amplifiers. For the circuit of Figure 7.23 with positive feedback and $A = 1$, the closed-loop steady-state transfer function between the amplifier output and the amplifier input is given by:

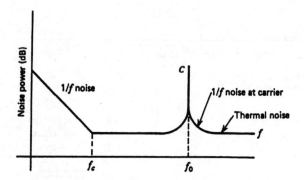

Figure 7.24 Noise power versus frequency of a transistor amplifier with an input signal applied.

$$B(j\omega) = \frac{1}{1 - H(j\omega)} \tag{7.47}$$

where

$$H(j\omega) = \frac{1}{1 + jQ\left(\dfrac{\omega}{\omega_o} - \dfrac{\omega_o}{\omega}\right)} \tag{7.48}$$

Because $H(j\omega)$ is a high-Q filter, and we are interested in describing the noise power distribution about the center frequency, ω_o, $H(j\omega)$ can be replaced by its low-pass equivalent,

$$H_L(j\omega) = \frac{1}{1 + j\omega/\omega_L} \tag{7.49}$$

where

$$\omega_L = \frac{\omega_o}{2Q} \tag{7.50}$$

is the low-pass equivalent bandwidth. The noise spectral density $S_o(\omega)$ at the output of a filter with a transfer function $G(j\omega)$, in terms of the spectral density of the input noise $S_i(\omega)$, is given by:

$$S_o(\omega) = S_i(\omega) |G(\omega)|^2 \tag{7.51}$$

Thus, the equivalent phase noise of the closed-loop system is:

$$S_o = S_\theta \frac{1}{|1 - H(\omega)|^2} \tag{7.52}$$

$$S_o = \frac{S_\theta}{\left|1 - \dfrac{1}{1 + j\omega/\omega_L}\right| \left|1 - \dfrac{1}{1 - j\omega/\omega_L}\right|} \tag{7.53}$$

$$S_o = \frac{S_\theta (1 + \omega^2/\omega_L^2)}{\omega^2/\omega_L^2} \tag{7.54}$$

$$S_o = S_\theta \left(1 + \frac{\omega_L^2}{\omega^2}\right) \tag{7.55}$$

which can be written using Equations 7.32, 7.46, and 7.50 as:

$$S_\sigma(\omega) = \frac{F\,k\,T}{P}\left(1 + \frac{\omega_o^2}{4\,Q^2\,\omega^2}\right) \tag{7.56}$$

which is the expression proposed by Leeson [7.10] for describing the noise at the output of an oscillator.

As mentioned previously, there are various types of noise, and so far Leeson's model does not contain an allowance for the flicker noise. We will, therefore, modify our equation into:

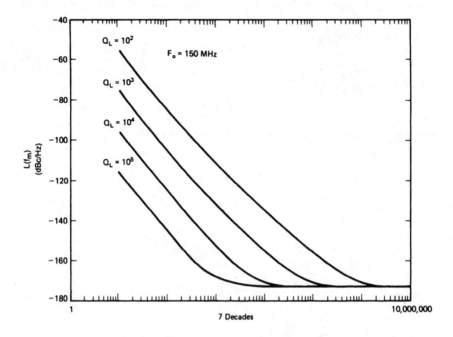

Figure 7.25 Noise sideband of an oscillator at 150 MHz as a function of the loaded Q of the resonator.

$$L(f_m) = \frac{1}{2}\left[1 + \frac{1}{\omega_m^2}\left(\frac{\omega_o}{2\,Q_{load}}\right)^2\right]\frac{F\,k\,T}{P_{sav}}\left(1 + \frac{f_c}{f_m}\right)$$ (7.57)

Equation 7.57 describes the phase noise at the output of the amplifier of the oscillator.

Figure 7.25 shows the difference in oscillator noise as a function of the Q of the resistor. In accordance with [7.13], we will expand Leeson's equation further for an actual oscillator to show how the noise performance can be optimized. Loaded Q can be expressed as

$$Q_{load} = \frac{\omega_o\,W_e}{P_{diss,\,total}} = \frac{\omega_o\,W_e}{P_{in} + P_{res} + P_{sig}}$$ (7.58)

$$Q_{load} = \frac{reactive\ power}{total\ dissipated\ power} \qquad\qquad (7.59)$$

where W_e is the reactive energy stored in L and C,

$$W_e = \frac{1}{2} C V^2 \qquad\qquad\qquad P_{res} = \frac{\omega_o W_e}{Q_{unl}}$$

$$L(f_m) = \frac{1}{2}\left[1 + \frac{\omega_o^2}{4\,\omega_m^2}\left(\frac{P_{in}}{\omega_o W_e} + \frac{1}{Q_{unl}} + \frac{P_{sig}}{\omega_o W_e}\right)^2\right]\left(1 + \frac{\omega_c}{\omega_m}\right)\frac{F k T_o}{P_{sav}} \qquad (7.60)$$

where Q_{unl} = the resonator Q.

This equation is significant because it contains most of the causes of phase noise in oscillators. To minimize the phase noise, the following design rules apply:

1. Maximize the unloaded Q.

2. Maximize the reactive energy by means of a high RF voltage across the resonator, and obtain a low LC ratio. The limits are set by the breakdown voltages of the active devices and the tuning diodes, and by the requirement to avoid a forward-bias condition of the tuning diodes.

3. Avoid saturation at all cost, and try to have either limiting or AGC without degradation of Q. Isolate the tuned circuit from the limiter or AGC circuit. Use antiparallel tuning diode connections to avoid forward bias.

4. Choose an active device with the lowest available NF. At this writing, the best bipolar transistor is the Siemens BFQ81, and the lowest-noise field-effect transistors are U310 and 2N5397, up to 500 MHz. The NF of interest is the NF obtained at the actual impedance at which the device is operated. When using field-effect transistors, it is preferable to deal with the equivalent noise voltage and noise currents rather than with the NF because they are independent of source impedance. The NF improves as the ratio between source impedance and equivalent noise resistance increases. In addition, in a tuning circuit, the source impedance changes drastically as a function of the offset frequency, and this effect has to be considered.

5. Phase perturbation can be minimized by using high-impedance devices such as field-effect transistors where the signal-to-noise ratio of the signal voltage relative

to the equivalent noise voltage can be made very high. This also indicates that in the case of a limiter, the limited voltage should be as high as possible.

6. Choose an active device with low flicker noise. The effect of flicker noise can also be reduced by RF feedback. An unbypassed emitter resistor of 10 to 30 Ω in a bipolar circuit can improve the flicker noise by as much as 40 dB. In a later example we will study such an oscillator. The proper bias point of the active device is important, and precautions should be taken to prevent modulation of the input and output dynamic capacitance of the active device, which will cause amplitude-to-phase conversion and therefore introduce noise.

7. The energy should be coupled loosely from the resonator rather than from another portion of the active device, so that the resonator limits the bandwidth. A crystal oscillator using this principle is described later in this chapter.

Equation 7.60 assumes that the phase perturbation and the flicker effect are the limiting factors, as practical use of such oscillators requires that an isolation amplifier be used.

In the event that the energy is taken directly from the resonator and the oscillator power can be increased above 0 dBm, the S/N can be increased above the theoretical limit of −174 dB, due to the low-pass filter effect of the tuned resonator. However, this is mainly a theoretical assumption that does not represent the real world. In a real system, this noise performance cannot be obtained. In an oscillator stage, even a total noise floor of −170 dB is rarely achieved.

What other influences do we have that cause the noise performance to degrade? So far, we have assumed that the Q of the tuned circuit is really determined only by the LC network and the loading effect of the transistor. In synthesizer applications, however, we find it necessary to add a tuning diode. The tuning diode has a substantially lower Q than that of a mica capacitor or even a ceramic capacitor. As a result of this, the noise sidebands change as a function of the additional loss. This is best expressed by adjusting the value for the loaded Q in Equation 7.57.

There seems to be no precise mathematical way of predetermining the noise influence of a tuning diode, but the following approximation of the Q that results from adding the tuning diode to the resonator seems to give proper results:

$$\frac{1}{Q_{T\,load}} = \frac{1}{Q_{load}} + \frac{1}{Q_{diode}} \qquad (7.61)$$

The loading effect of the tuning diode is the result of losses, and these losses can be described by a resistor parallel to the tuned circuit. The tuning diode is specified to have a cutoff frequency, f_{max}, that is determined from the loss resistor and the value of the junction capacitance as a function of voltage (i. e., measured at 3 V).

This means that the voltage determines the Q and, consequently, the noise bandwidth.

It is possible to define an equivalent noise R_{aeq} that, inserted in Nyquist's equation yields:

$$V_n = \sqrt{4\,K\,T_o\,R\,\Delta f} \qquad (7.62)$$

where $KT = 4.2 \times 10^{-21}$ at about $300°K$, R is the equivalent noise resistor, and f is the bandwidth, determines a noise voltage across the tuning diode. Practical values of R_{aeq} for carefully selected tuning diodes are in the vicinity of $1000\ \Omega$ to $50\ k\Omega$. For $R_{aeq} = 10\ k\Omega$, we obtain:

$$V_n = \sqrt{4 \times 4.2 \times 10^{-21} \times 10{,}000} \qquad (7.63)$$

The resulting noise voltage is $1.265 \times 10^{-8}\ V\,\sqrt{Hz}$.

This noise voltage, generated from the tuning diode, is now multiplied by the VCO gain, resulting in the rms frequency deviation

$$(\Delta f_{rms}) = K_o \times (1.265 \times 10^{-8}\ V)\ \text{in a 1-Hz bandwidth} \qquad (7.64)$$

In order to translate this into the equivalent peak phase deviation,

$$\theta_d = \frac{K_o\,\sqrt{2}}{f_m}\,(1.265 \times 10^{-8})\ \text{rad in a 1-Hz bandwidth} \qquad (7.65)$$

or, for a typical oscillator gain of 100 kHz/V,

$$\theta_d = \frac{0.00179}{f_m}\ \text{rad in a 1-Hz bandwidth}$$

For f_m = 25 kHz (typical spacing for adjacent-channel measurements for FM mobile radios), θ_c = 7.17 × 10^{-8}. This can be converted now into the SSB S/N as:

$$L(f_m) = 20 \log_{10} \frac{\theta_c}{2} = -149 \text{ dBc/Hz} \tag{7.66}$$

This is the value typically achieved in the Rohde & Schwarz SMDU generator or with the Hewlett-Packard 8640 signal generator, and is considered state-of-the-art for a free-running oscillator. It should be noted that these two signal generators use slightly different tuned circuits; the Rohde & Schwarz generator uses a helical resonator, whereas the Hewlett-Packard generator uses an electrically shortened quarter-wavelength cavity. Both generators are mechanically pretuned, and a tuning diode with a gain of about 100 kHz/V is used for frequency-modulation purposes or for the *automatic frequency control* (AFC) input. It is apparent that, because of the nonlinearity of the tuning diode, the gain is different for low dc voltages than for high dc voltages. The impact of this is that the noise varies within the tuning range.

In many applications, VCOs are required to span large frequency ranges. This can be accomplished either solely with a tuning voltage, or by using switching diodes to select ranges via different tuning diodes or fixed reactances.

In low-energy-consuming circuits, the VCO frequently is divided into a coarse-tuning section and a fine-tuning section, both using tuning diodes. In the coarse-tuning range, this results in very high gains, such as 1 to 10 MHz/V, for the diodes. The noise contribution of those diodes is therefore very high and can hardly be compensated by the loop. For low-noise applications, which automatically mean higher power consumption, the use of switching diodes rather than high-gain tuning diodes is unavoidable.

Let us now examine some test results. If we go back to Equation 7.57 and Figure 7.25, they show the noise sideband performance as a function of Q, where the top curve of the figure, with Q_L = 100, represents a somewhat poor oscillator, and the lower curve, with Q_L = 100,000 probably represents a crystal oscillator where the unloaded Q of the crystal was in the vicinity of 3 × 10^6.

Figure 7.26 shows the influence of flicker noise. Corner frequencies of from 10 Hz to 10 kHz have been analyzed, and it becomes apparent that at around 1 kHz, the influence of flicker noise is fairly dramatic, whereas its influence at 20 kHz from the carrier is not significant. Finally, Figure 7.27 shows the influence of the tuning diodes on a high-Q oscillator. Curve A uses a lightly coupled tuning diode with a K_o of 10 kHz/V; the lower curve is the noise performance without any diode. As a result, the two curves are almost identical, which can be seen from the somewhat smeared form of the graph. Curve B shows the influence of a tuning diode at 100 kHz/V and represents a value of −143 dBc/Hz, up from −155 dBc/Hz

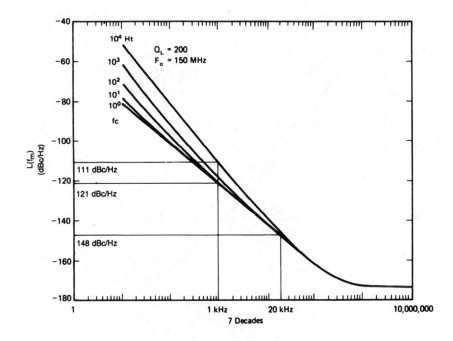

Figure 7.26 Noise sideband as a function of flicker frequency.

and already showing some deterioration. Curve C shows the noise if the tuning diode operates at a 1 MHz/V VCO gain, and the noise sideband at 25 kHz has now deteriorated to −123 dBc/Hz. These curves speak for themselves.

It is of interest to compare various oscillators. Figure 7.28 shows the noise performance of a 10-MHz crystal oscillator, 40-MHz LC oscillator, the HP 864C cavity-tuned oscillator at 500 MHz, the 310- to 640-MHz switched reactance oscillator of the HP 8662, and a 2- to 6-GHz YIG oscillator operating at 6 GHz.

The foregoing theory of the linear approach has clearly indicated that the key factors for success in building a low-noise oscillator are:

- Using a high-Q resonator, referring to the loaded Q
- Obtaining the highest possible output from the oscillator while maintaining the smallest possible large-signal NF of the transistor
- Minimizing the flicker noise and/or AM-to-PM conversion

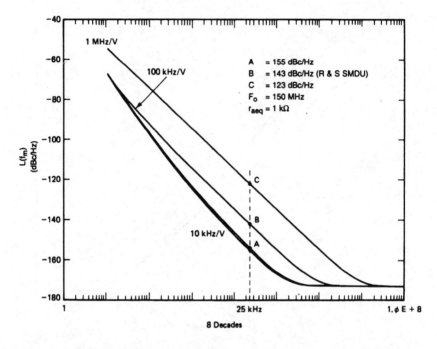

Figure 7.27 Noise sideband performance of an oscillator at 150 MHz, showing the influence of various tuning diodes and increased noise as a function of modulation sensitivity.

Figure 7.29 shows the carrier-to-phase-noise ratio versus the frequency offset from the carrier for low- and high-Q resonators. It is necessary to look at this plot to understand that the breakpoint for the corner frequency of the flicker frequency can move around as the Q changes. Remember also that Q is a function of loading the resonator with the transistor, and therefore the highest open-loop gain should be used.

Oscillators are frequently multiplied up to the actual desired output frequency. For microwave applications, frequencies between 2 and 10 GHz are often required. Figure 7.30 shows a phase noise comparison of YIG and varactor-tuned oscillators normalized to a center frequency of 6 GHz. This represents the state-of-the-art in tunable oscillators; of course, crystal oscillators, oscillators with ceramic resonators, and dielectric resonators have much better performance.

7.5.6 Nonlinear Model

The usual active device in an oscillator is either a bipolar transistor or a member of the FET family. Depending on the frequency range, the FETs used for oscillators are either N junction FETs, MOS devices, or GaAs FETs. In the case of bipolar

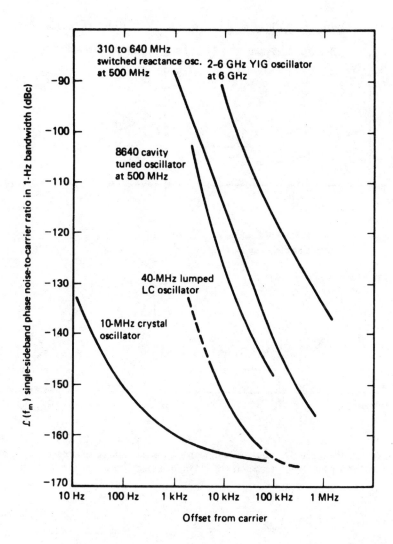

Figure 7.28 Comparison of noise sideband performance of a crystal oscillator, *LC* oscillator, cavity-tuned oscillator, switched reactance oscillator, and YIG oscillator.

transistors, the model of choice for nonlinear analysis is the Gummel-Poon bipolar transistor model, typically used in a modified version.

For the FET family, several nonlinear models are available. In the case of the GaAs FET, popular models include:

- The Curtice-Edenberg

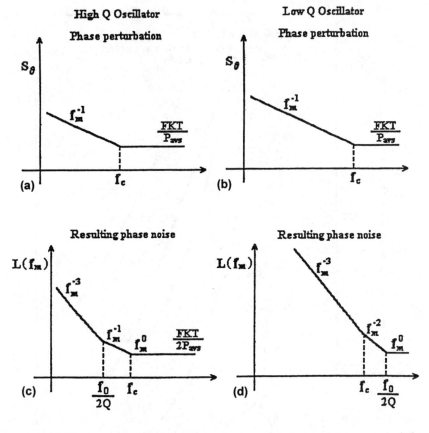

Figure 7.29 The spectral phase noise of the carrier-to-noise ratio versus frequency deviation from the carrier for low-Q resonators (a) and (c) and high-Q resonators (b) and (d).

- Raytheon Stutz
- TOM
- Materk

For MOS devices, we choose a MOS level 3 Spice type model. These models are typically found in nonlinear circuit simulator programs such as Microwave Harmonica (Compact Software). These active devices are modeled by a variety of nonlinear elements such as diodes and voltage- and/or current-dependent capacitances, which are the major form of nonlinearities. The device changes as a function of bias, and therefore the NF at each bias point is different. At high current levels, the noise, of course, will increase significantly. This requires the

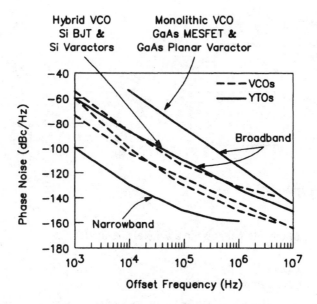

Figure 7.30 Phase noise comparison of YIG and varactor-tuned oscillators normalized to a center frequency of 6 GHz.

generation of a bias-dependent noise model, which is available for applications such as mixers and oscillators.

7.5.7 General Concept of Noise Contribution

In the evaluation of a noisy two-port network, it is important to know the amount of noise added to a signal passing through a network, represented as:

$$\frac{S_{in}}{N_{in}} \rightarrow \text{Network} \rightarrow \frac{S_{out}}{N_{out}} \tag{7.67}$$

An important parameter for expressing this characteristic is the noise factor (or NF).

$$\text{Noise Factor} = F = \frac{S_{in}/N_{in}}{S_{out}/N_{out}} \tag{7.68}$$

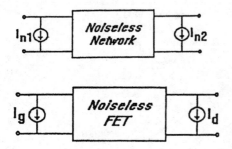

Figure 7.31 Noiseless FET with noise sources at the input and output.

$$NF = 10 \log (F) \tag{7.69}$$

The sources of the internal noise in a general circuit are:

- Noise from linear elements
- Thermal noise related to the admittance of the elements

A noise network can be treated as a noiseless network with an equivalent noise current source at each external port, as shown in Figure 7.31. The correlation of the noise current sources of a linear network is related to the Y matrix of this network as:

$$C_n(\omega) = \frac{1}{\pi} K_B T \delta \omega [Y(\omega) + Y^*(\omega)] \tag{7.70}$$

The intrinsic noise sources of an active device (e. g., MESFET or BJT) can be separated into a noiseless FET with noise sources at the input and the output as shown in Figure 7.32. The intrinsic noise model can be expressed by the four measured parameters:

F_{min} = Minimum NF
R_n = Equivalent normalized noise resistance = $r_n/50$
M_{Go} = Magnitude of the optimal noise reflection coefficient
P_{Go} = Phase of the optimal noise reflection coefficient

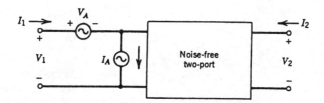

Figure 7.32 The intrinsic noise sources of an active device can be separated into a noiseless FET with noise sources at the input and output.

From these four parameters, the Van der Ziel noise model of the GaAs FET, as an example, can be derived as:

$$C_n(\omega) = \frac{2}{\pi} K_B \, T \, \delta \, \omega \begin{bmatrix} \dfrac{\omega^2 \, C_{gs}^{\;2}}{g_m} R & -j \, \omega \, C_{gs} \, \sqrt{P \, R \, C} \\[2ex] j \, \omega \, C_{gs} \, \sqrt{P \, R \, C} & g_m \, P \end{bmatrix} \tag{7.71}$$

This conversion for all active devices such as FETs and bipolar transistors has been implemented in linear simulators. (Super-Compact and Microwave Harmonica from Compact Software are two such simulators.) In addition, we have to add the flicker noise contribution of an active device. Figure 7.33 shows the flicker noise as a function of frequency.

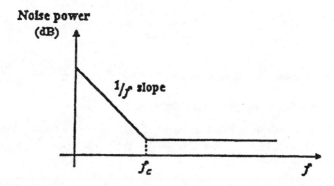

Figure 7.33 Noise power as a function of frequency. The major parameter used to describe the flicker noise is f_c, the corner frequency.

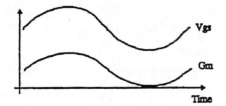

Figure 7.34 The variation of parameters as a function of LO power. The voltages and currents of devices are determined by the harmonic balance calculations.

7.5.8 Large Signal Condition of the Active Device

We now look at the noise model of an active device when pumped by an LO. The noise sources and equivalent circuit model parameters are modulated by the LO. Figure 7.34 shows the variation of some important parameters as a function of the LO power.

The noise correlation matrix of the device is now modulated by the LO. This means variation in the noise correlation component and nonlinear parameters such as:

$$R, P, C, g_m, C_{Gs}, \ldots = F (V_{Gs}, V_{ds}) \tag{7.72}$$

In addition, for FETs, as an example, the flicker noise is modulated by the drain current using the following equation:

$$\left\langle | I_f |^2 \right\rangle = 2 K_B T \delta \omega Q \frac{|I_D|^\beta}{f^a} \tag{7.73}$$

For the bipolar model we use a similar approach.

7.5.9 Phase Noise Spectrum of the Oscillator

For calculation of the phase noise spectrum of the oscillator, a numerical approach is considered to provide nonlinear analysis. It can be shown that the phase noise is composed of two parts:

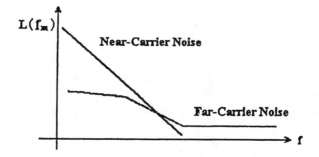

Figure 7.35 The components of oscillator noise. Noise consists of the near- and far-carrier noise.

- **Near-carrier noise.** This consists of contributions from the perturbation of the oscillating frequency caused by the noise sources at each sideband frequency. This part is the major noise source at near-carrier frequencies.
- **Far-carrier noise.** This consists of contributions from each sideband noise source through sideband-to-sideband transfer functions. This part is similar to a mixer noise calculation and is the major noise source at frequencies far from the carrier.

Figure 7.35 illustrates how the oscillator noise consists of the near- and far-carrier noise.

7.5.10 NF of the Mixing Circuit

In order to calculate the NF, we assume that the pumped active device acts like a mixer circuit. We need to calculate the total internal noise of the circuit at the "IF" frequency. This is a two-step process:

1. Using a nonlinear simulator, we perform a harmonic balance analysis to determine the steady state of the mixer. Figure 7.36 shows the mixing arrangement. The harmonic balance calculation of the nonlinear simulator determines the Fourier coefficients of the voltages and currents of the circuit. Any receiver configuration (e.g., LNAs, IF, or AMP) may be considered.

2. We next calculate the transfer functions of the sideband signals to the IF-band signal. Figure 7.37 shows a summary of noise sources mixed to the IF. The noise at each sideband frequency contributes to the noise at the IF through frequency conversion. Figure 7.38 shows a summary of the IF noise contributions in a

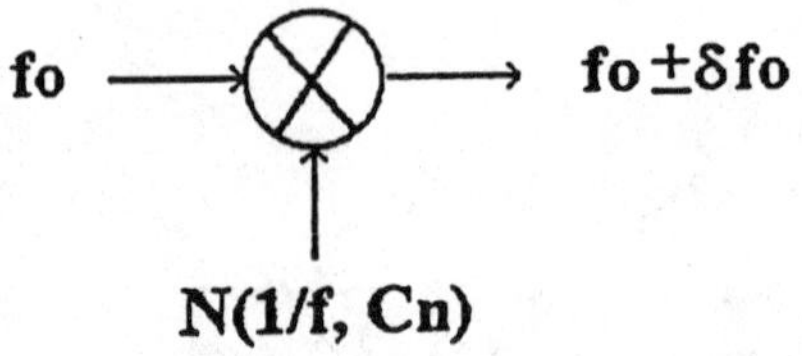

fo $\longrightarrow$ ⊗ $\longrightarrow$ fo $\pm\delta$fo

N(1/f, Cn)

Figure 7.36 The mixing arrangement within the oscillator whereby the carrier frequency (f_0), assumed noiseless, is modulated with various noise sources.

general nonlinear mixer circuit. Note the large number of contributing elements that make up the total noise at the output. The calculation of dN is performed by Equation 7.76, where the intermediate steps are given in [7.15]. (See also References [7.14] to [7.24].)

Calculation of the noise parameters proceeds as follows:

$$\left\langle |\delta \phi|^2 (f_d)\right\rangle = \frac{\left\langle |V_l(f_d)|^2\right\rangle + \left\langle |V_u(f_d)|^2\right\rangle - 2 \, \mathrm{Re}\left\{\left\langle V_l^*(f_d) \, V_u^*(f_d)\right\rangle \exp(2 \ j \ \phi_0)\right\}}{|V_0|^2}$$

(7.74)

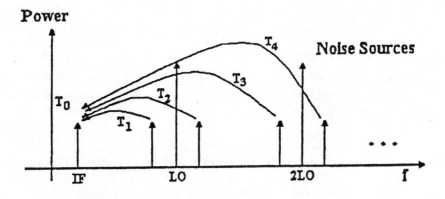

Figure 7.37 Summary of noise sources mixed to the IF. The noise at each sideband frequency contributes to the noise at the IF through frequency conversion.

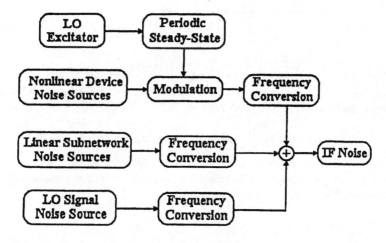

Figure 7.38 Summary of IF noise contributions.

$$\left\langle |\delta\,A|^2\,(f_d)\right\rangle = 2\frac{\left\langle |V_i(f_d)|^2\right\rangle + \left\langle |V_u(f_d)|^2\right\rangle + 2\,\mathrm{Re}\left\{\left\langle V_i^*(f_d)\,V_u^*(f_d)\right\rangle \exp(2\;j\;\phi_0)\right\}}{|V_0|^2}$$

(7.75)

$$d\,N(\omega_{IF}) = R_{IF}\sum_p T_{0p}\,C_L(\omega_{IF}+p\,\omega_0)\,T_{0p}^{\;*}$$

$$+ R_{IF}\sum_{p,\,q} T_{0p}\left[\sum_s H_{p-s}\,C_{dc}\,(\omega_{IF}+s\,\omega_0)\,H_{s-q}\right]T_{0q}^{\;*}$$

$$+ R_{IF}\sum_{p,\,q} Y_p^{\,s}\left[\begin{matrix}\left\langle |V_u(\omega_{IF})|^2\right\rangle & \left\langle V_u(\omega_{IF})\,V_i^*(\omega_{IF})\right\rangle \\ \left\langle V_u^*(\omega_{IF})\,V_i(\omega_{IF})\right\rangle & \left\langle |V_i(\omega_{IF})|^2\right\rangle\end{matrix}\right]Y_q^{\,S*}$$

(7.76)

Where:
T_{0X} = the sideband-to-IF conversion matrices
H_x = the spectral modulation components of the device
p, q, r, and s = sideband spectral indices
R_{IF} = the IF load
Y = a conversion admittance matrix between the LO noisy source and the IF
load at the IF frequency

ω_{IF} = a small frequency deviation in the neighborhood of the baseband frequency

The first term represents the noise contribution of the linear network, the second term is the noise contribution from the modulated nonlinear devices, and the third term is the noise contribution of the noisy LO.

7.5.11 Calculation of Total Carrier Phase Noise Spectrum

Assuming that the noise signals are small compared to the oscillator signal, the noise analysis is carried out as a perturbative analysis. The manner in which we determine the contributions is as follows:

- Far-carrier noise analysis:

$$J_{BB} \, \delta \, X_B = -N_B \, (\omega) \tag{7.77}$$

- Near-carrier noise analysis:

$$J_{HH} \, \delta \, X_H = -N_H \, (\omega) \tag{7.78}$$

The near-carrier noise is the noise-inducted jitter of the oscillatory steady state. The $\delta\omega$, one of the entries of δX_H, can be solved as

$$\delta \, \omega(\omega) = T_F \, [N_u(\omega) + N_L(\omega)] \tag{7.79}$$

and the resulting phase noise is:

$$\left\langle |\delta\phi_k \, (\omega)|^2 \right\rangle = \left\langle k^2/\omega^2 \right\rangle \Big[T_F \left\langle N_U \, (\omega) \, N_U \, (\omega)^+ \right\rangle T_F{}^+$$
$$+ T_F \left\langle N_L(\omega) \, N_L{}^+(\omega) \right\rangle T_F{}^+ + 2 \, \mathrm{Re} \Big\{ T_F \left\langle N_U \, (\omega) \, N_L{}^+(\omega) \right\rangle T_F{}^+ \Big\} \Big] \tag{7.80}$$

In the linear case, the assumed simplification did not consider the AM-to-PM conversion and the change of the loaded Q as a function of the oscillator condition.

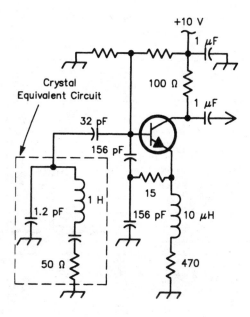

Figure 7.39 Simplified circuit of a 10-MHz crystal oscillator.

The approach shown here is the exact solution for the calculation of the oscillator phase noise.

7.5.12 Practical Examples

Next, we will present verification examples of the nonlinear mathematical approach of calculating phase noise, given previously.

A. Example 1: A 10-MHz crystal oscillator

Figure 7.39 shows the abbreviated circuit of a 10-MHz crystal oscillator. It uses a high-precision, high-Q crystal. Oscillators of this type are made by several companies and are intended for use as both frequency and low-phase-noise standards. In this particular case, the crystal oscillator being considered is part of the HP 3048 phase-noise measurement system. Figure 7.40(a) shows the measured phase noise of this frequency standard by HP, and Figure 7.40(b) shows the predicted phase noise using the mathematical approach outlined previously.

B. Example 2: A 1-GHz ceramic resonator VCO

A number of companies have introduced resonators built from ceramic materials with an ε ranging from 20 to 80. The advantage of using this type of resonator is

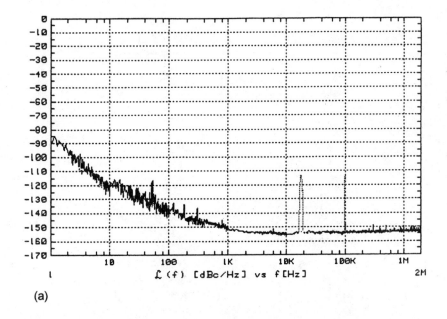

(a)

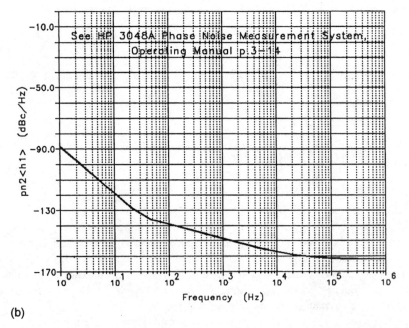

(b)

Figure 7.40 Performance of the oscillator shown in Figure 7.39: (*a*) phase noise as measured by HP and (*b*) simulated phase noise performance.

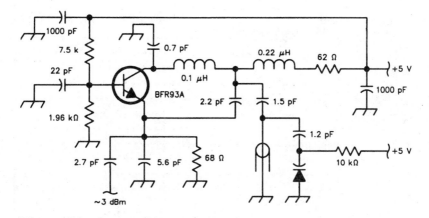

Figure 7.41 A typical test circuit for use with a ceramic resonator. These resonators are available in the 500 MHz to 2 GHz range.

that they are a high-Q element that can be tuned by adding a varactor diode. Figure 7.41 shows a typical test circuit for use in a ceramic resonator. These resonators are available in the range of 500 MHz to 2 GHz. (For higher frequencies, dielectric resonators are recommended.) Figure 7.42 shows the measured phase noise of the oscillator. The noise pedestal above 100 kHz away from the carrier is the result of the reference oscillator, a model HP 8662 generator.

Figure 7.43 shows the predicted phase noise of the 1-GHz ceramic resonator VCO without a tuning diode, and Figure 7.44 shows the predicted phase noise of the VCO with a tuning diode attached. Note the good agreement between the measured and predicted phase noises.

C. Example 3: A low-phase-noise FET oscillator

A number of authors recommend the use of a clipping diode to prevent the gate-source junction of an FET from becoming conductive, thereby lowering the phase noise. Claims also have been made that the diode was necessary to obtain long-term stability. In reality, these are misconceptions. The popular VCO circuit shown in Figure 7.45 was analyzed with and without the diode.

Figure 7.46 shows the measured phase noise of an oscillator of this type, and Figures 7.47 and 7.48 show the simulated phase noise of the oscillator, with and without a clipping diode, respectively. Note the degradation of the phase noise if the diode is used. (These two plots do not have the same vertical scale.)

D. Example 4: Recommended circuits for higher frequency applications

In this section we examine a group of VCOs that are ideal for low-phase-noise oscillators. Figure 7.49 shows a circuit using a 3-dB power divider at the output. The loop filter for the synthesizer application also is shown.

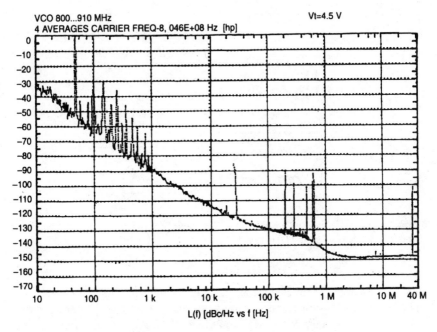

Figure 7.42 Measured phase noise of the oscillator shown in Figure 7.41.

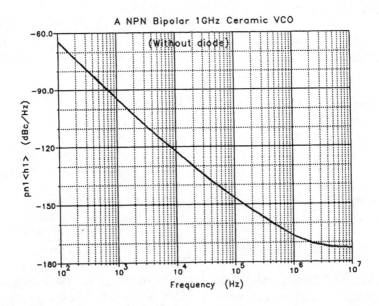

Figure 7.43 Predicted phase noise of the 1-GHz ceramic resonator VCO without the tuning diode attached.

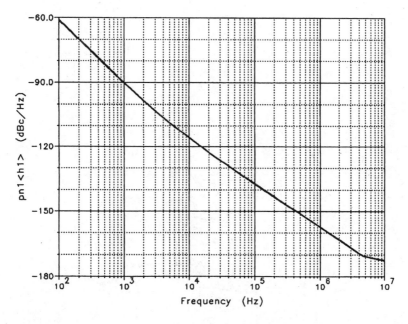

Figure 7.44 Predicted phase noise of the 1-GHz ceramic resonator VCO with the tuning diode attached.

Figure 7.50 shows a high-power, low-phase-noise VCO system recommended for the frequency range from 400 to 700 MHz. Note that the tuning element again uses several diodes in parallel.

Figure 7.51 shows a recommended VCO circuit covering the frequency range from 700 MHz to 1 GHz. The rule of thumb is that FETs do not have enough gain for high-Q operation in oscillators above 400 MHz, and bipolar transistors are a better choice. Only at frequencies above 4 or 5 GHz should GaAs FETs be considered because of their higher flicker noise contribution.

7.6 Noise and Performance Analysis of PLL Systems

To illustrate the effectiveness of CAD tools in PLL analysis, let us consider the design of a PLL synthesizer operating from 110 to 210 MHz. A reference frequency of 10 kHz is used, and the tuning diode has a capacitance range from 6 to 60 pF. For performance reasons we select a type 2 third-order loop. The phase calculations use Leeson's model [7.27] for oscillator noise and the following equation:

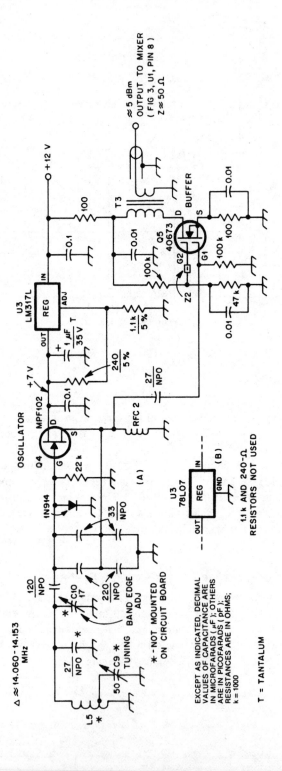

Figure 7.45 A 20-m VFO circuit reproduced from the 1993 *ARRL Handbook. (Courtesy of ARRL.)*

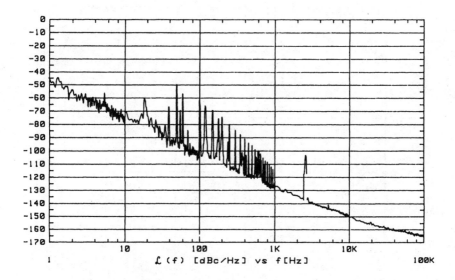

Figure 7.46 Measured phase noise of the type of oscillator shown in Figure 7.45.

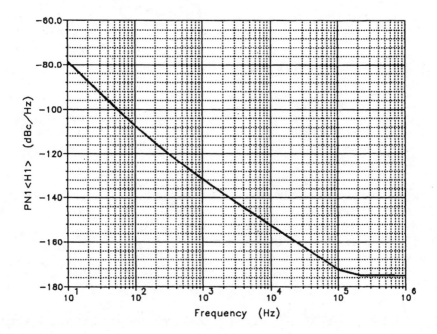

Figure 7.47 Simulated phase noise of the oscillator with a clipping diode installed.

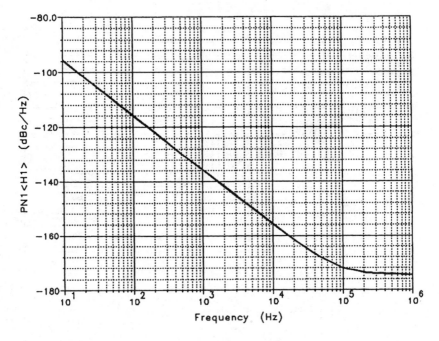

Figure 7.48 Simulated phase noise of the oscillator without the clipping diode installed.

$$L(f_m) = 10 \log \left\{ \left[1 + \frac{f_0^2}{(2 f_m \, q_{load})^2} \right] \left(1 + \frac{f_c}{f_m} \right) \frac{F \, k \, T}{2 \, P_{s \, av}} + \frac{2 \, K \, T \, R \, K_O^2}{f_m^2} \right\} \qquad (7.81)$$

where
$L(f_m)$ = ratio of sideband power in 1-Hz bandwidth at f_m to total power in dB
f_m = frequency offset
f_0 = center frequency
f_c = flicker frequency of the semiconductor
Q_{load} = loaded Q of the tuned circuit
F = noise factor
$kT = 4.1 \times 10^{21}$ at 300 K (room temperature)
$P_{s \, av}$ = average power at oscillator input
R = equivalent noise resistance of tuning diode
K = oscillator voltage gain

The lock-up time of a PLL may be defined in many ways. In the digital loop, we prefer to define it by separating the frequency lock, or *pull-in*, and the phase

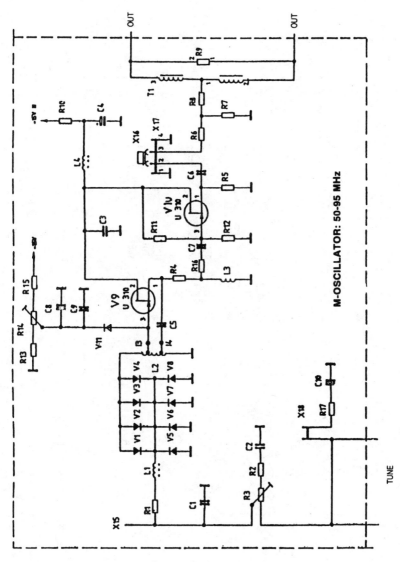

Figure 7.49 Schematic diagram of a low-phase-noise oscillator.

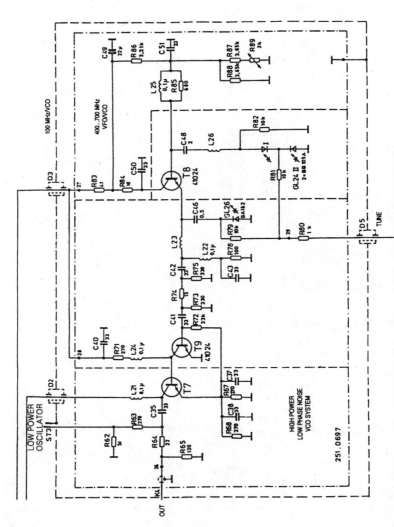

Figure 7.50 A high-power low-phase-noise VCO circuit recommended for the 400 to 700 MHz frequency range.

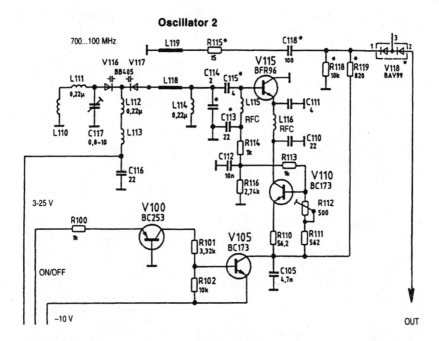

Figure 7.51 A recommended low-phase-noise VCO circuit covering the 700 MHz to 1 GHz frequency range.

lock and adding the two separate numbers. To determine the pull-in time, a statistical approach may be used, defining a new gain constant $K^2 = V_B/2\pi f$, where V_B is the supply voltage and f the frequency offset. The phase-lock time is determined from the Laplace transform of the transfer function (see [7.4, pp. 32-36]).

7.6.1 Design Process

A set of programs written around the preceding equations has been used in the design example[1]. Table 7.5 is the printout of input data and information on the lock-up time and reference frequency suppression. Based on the frequency range and the tuning diode parameters, a wide-band VCO is required. The computer program interactively determines the component values shown in Table 7.6. Depending upon the frequency range, the PLL design kit has four different recommended oscillator circuits, including narrow-band and wide-band VCOs

---

1 Available from Communications Consulting Corporation, 52 Hillcrest Drive, Upper SaddleRiver, NJ 07458.

TABLE 7.5 Input Data and Outputs on Lock-up Time and Reference Suppression (PLL Design Kit)

```
INPUT DATA:

REFERENCE FREQUENCY IN Hz =1000
NATURAL LOOP FREQUENCY IN Hz =50
PHASE DETECTOR GAIN IN V/rad =.9
VCO GAIN CONSTANT IN Hz/V =1.00E+07
DIVIDER RATIO =160000
VCO FREQUENCY IN Hz =          1.6E+8
PHASE MARGIN IN deg =45


THE LOCK-UP TIME CONTANT IS:      1.63E-02 sec
REFERENCE SUPPRESSION IS:             44.4 dB
ASSUMING PHASE DETECTOR OUTPUT PULSE AMPLITUDE = 6 V
PULSE WIDTH = 1uS THEN THE SPURIOUS SUPPRESSION
IS  64.05  dB
BROADER PULSES WILL WORSEN SPURIOUS
```

TABLE 7.6 VCO Design Parameters (PLL Design Kit)

```
CALCULATION OF VCO TUNING RANGE:
Fmin= 110  MHz    Fmax= 210  MHz
CENTER RANGE IS  160  MHz      TUNING RATIO =  1.909
Cmin (at Vmax) OF TUNING DIODE= 6  pF    Cmax (at Vmin) OF TUNING DIODE=  60   pF
FET CHOSEN:
CISS = 2  pF TRANSISTOR IS OPERATED AT Id= 11.5  mA ;Vc= 13  V
Gm= 17.5  mS
CUT-OFF FREQUENCY OF FET = 1.4  GHz ;
THEORETICAL OUTPUT POWER BASED OF FOURIER ANALYSIS IS  49.8 mW OR 17 dBm
BOARD STRAY CAPACITANCE = 1.2 Pf
Cmin OF DIODE COMBINATION= 5.44  pF; Cmax OF DIODE COMBINATION= 29.6  pF
COUPLING CAPACITOR Cs= 58.5  pF       ; REQUIRED INDUCTANCE IS  .0628  uH
FEEDBACK CAPACITOR OF 1 pF CHOSEN
PARALLEL TRIMMING CAPACITANCE CT = 2.5  pF
```

with lumped constants and half-wave and quarter-wave line VCOs for UHF. The selected circuit configuration is shown in Figure 7.52.

Having chosen the circuit and component values for the VCO, the program then calculates the SSB phase noise, following the interactive input of the additional parameters required (Table 7.7). The program provides a table of VCO SSB noise level as a function of frequency (Table 7.8) and a plot of the phase noise of the VCO, as shown in Figure 7.53. The close-in phase noise of the free-running oscillator is inherently poor. However, it is improved when the oscillator has been imbedded in the type 2 third-order loop. The resultant composite phase noise is

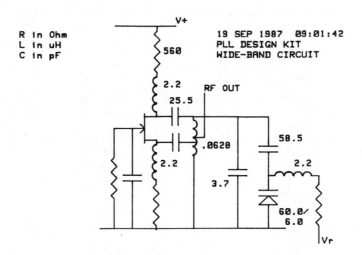

R in Ohm
L in uH
C in pF

19 SEP 1987 09:01:42
PLL DESIGN KIT
WIDE-BAND CIRCUIT

V+

560

2.2
25.5

RF OUT

.0628

2.2

58.5

2.2

3.7

60.0/
6.0

Vr

Figure 7.52 Schematic diagram of the selected VCO configuration (PLL design kit).

plotted in Figure 7.54. The close-in noise now has improved, and the free-running VCO and composite loop graphs meet at approximately 400 Hz.

The program examines the stability, using the Bode plot, as shown in Figure 7.55. The phase lock-up time appears in Table 7.9. It is good practice to define the lock-up time as the point where the phase error is less than 1°. Based on the 50-Hz loop frequency, this value is 32 ms. For the total lockup time, we must add the frequency pull-in time of 16.3 ms, resulting in a total lock-up time of approximately 50 ms. Finally, the program package provides the circuit of the active integrator for the loop, as shown in Figure 7.56.

Detailed information on oscillator design practices for a variety of systems can be found in References [7.28] to [7.39].

TABLE 7.7 SSB Phase Noise Calculation (PLL Design Kit)

```
SSB PHASE NOISE CALCULATION


LO-POWER = 0 dBM, LO NF= 10  dB
The rms noise voltage per sqr(1 Hz) bandwidth =1.30E-08 V
Rn= 10000  Ohm
F= 10  dB
VCO GAIN (Hz/V)=1.00E+07
SSB NOISE AT FREQUENCY OFFSET IN Hz= 10000
CENTER FREQUENCY = 160  MHz
LOADED RESONATOR Q = 120
FLICKER FREQUENCY IS  150 Hz

The ssb phase noise in 10000 Hz offset is -106.78 dBc/Hz
```

TABLE 7.8 SSB Phase Noise as a Function of Frequency Offset (PLL Design Kit)

	SSB NOISE TABLE	
FREQUENCY (Hz)		PHASE NOISE (dBc/Hz)
1.00E+00		−25.97
3.16E+00		−36.68
1.00E+01		−46.77
3.16E+01		−56.78
1.00E+02		−66.78
3.16E+02		−76.78
1.00E+03		−86.78
3.16E+03		−96.78
1.00E+04		−106.78
3.16E+04		−116.78
1.00E+05		−126.78
3.16E+05		−136.78
1.00E+06		−146.78
3.16E+06		−156.75
1.00E+07		−164.00

7.7 Fractional Division Loop

A conventional single-loop synthesizer uses a frequency division ratio N, an integer value between one and several hundred thousand. The step size is equal to the frequency reference. In other words, $f_{out} = Nf_{ref}$. The settling time is determined by the loop filter bandwidth, and the loop filter bandwidth is a fraction of

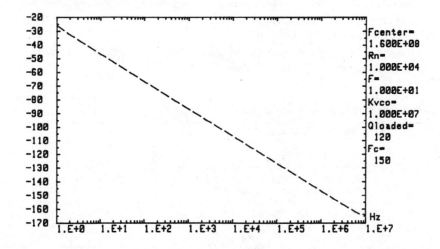

Figure 7.53 Plot of oscillator SSB phase noise (PLL design kit).

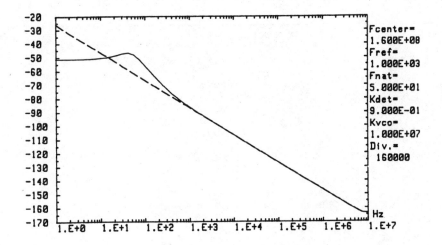

Figure 7.54 Plot of overall loop phase noise (PLL design kit).

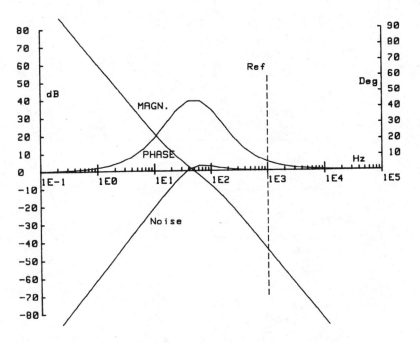

Figure 7.55 Bode plot of PLL (PLL design kit).

TABLE 7.9 Phase Deviation as a Function of Lock-up Time (PLL Design Kit)

```
            LOCK-IN FUNCTION :

       TIME/s              PHASE DET.DEV./deg
       0              3.60E+02
        .0016            2.58E+02
        .0032            1.16E+02
        .0048            2.94E+00
        .0064           -7.36E+01
        .008            -1.12E+02
        .0096           -1.20E+02
        .0112           -1.07E+02
        .0128           -8.40E+01
        .0144           -5.95E+01
        .016            -3.79E+01
        .0176           -2.12E+01
        .0192           -9.55E+00
        .0208           -2.37E+00
        .0224            1.45E+00
        .024             2.99E+00
        .0256            3.17E+00
        .0272            2.67E+00
        .0288            1.93E+00
        .0304            1.22E+00
        .032             6.45E-01
        .0336            2.45E-01
        .0352            3.35E-03
        .0368           -1.17E-01
        .0384           -1.57E-01
        .04             -1.49E-01
        .0416           -1.19E-01
        .0432           -8.34E-02
        .0448           -5.13E-02
        .0464           -2.65E-02
```

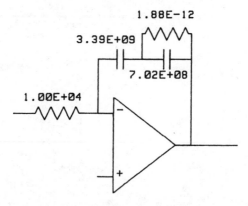

Figure 7.56 Active integrator design (PLL design kit).

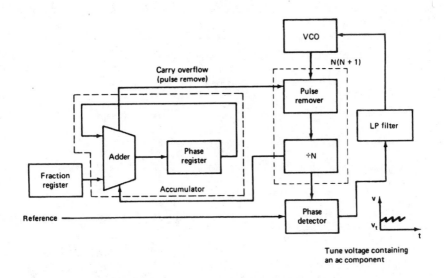

Figure 7.57 Block diagram of a fractional-N loop using a 10/11 two-modulus counter. (*Courtesy of Hewlett-Packard.*)

the reference. If it were possible to increment N in fractional steps, e. g., by setting $N = 11.5$ or 11.000001, the frequency resolution of f_{out} could be made arbitrarily small. The digital counter cannot provide a fractional count, but a method is available to modify the overflow externally to make such a fractional count possible.

If we assume that the programmable divider chain uses a dual-modulus counter, such as 10/11, it is possible to divide the output every M cycles by 11 and divide by 10 for the rest of the time. The effective division ratio is then $N + 1/M$, and the average output frequency is given by $f_{out} = (N + 1/M)f_{ref}$. Thus, the output frequency can be varied in fractional increments of the output reference frequency by varying M. The fractional part of the division is implemented by using a phase accumulator (Figure 7.57). For example, consider the problem of generating 72.455 MHz using a fractional k loop with 100 kHz reference frequency. The integral part of the division is $N (= 724)$. The fractional part is $1/M (= 0.55)$, yielding $M = 1.82$. M is not an integer; the VCO output frequency is to be divided by 725 ($N + 1$) every 1.82 cycles, or 55 times every 100 cycles. This can be implemented by adding the number 0.55 to the contents of an accumulator every cycle. Each time the accumulator overflows (the content equals or exceeds 1), the divider divides by 725 rather than by 724. Only the fractional value is retained by the accumulator after overflow. Arbitrarily fine resolution can be obtained by increasing the length of the phase

TABLE 7.10 Accumulator Contents and Count Size for One Complete Cycle of a Fractional Overflow Counter

Count number	Overflow at start of count	Count	Accumulator contents	
			Start	End
1	Yes	725	0	55
2	No	724	55	10
3	Yes	725	10	65
4	No	724	65	20
5	Yes	725	20	75
6	No	724	75	30
7	Yes	725	30	85
8	No	724	85	40
9	Yes	725	40	95
10	No	724	95	50
11	Yes	725	50	5
12	Yes	725	5	60
13	No	724	60	15
14	Yes	725	15	70
15	No	724	70	25
16	Yes	725	25	80
17	No	724	80	35
18	Yes	725	25	90
19	No	724	90	45
20	Yes	725	45	0

Repeat from count 1.

accumulator. For example, with a 100-kHz reference, a resolution of 1 Hz can be obtained by using a five-digit BCD accumulator.

The fractional technique, accomplished by *pulse swallowing* or deletion, as in the dual-modulus counter, has the unpleasant side effect of causing modulation sidebands to appear. In our example, 55 kHz is generated with the fine loop portion of the fractional overflow. This is accomplished by dividing by 724 at some times and by 725 at others (whenever the accumulator overflows). Generally the accumulator overflows after every second count, but after 10 counts the accumulator is at 0.5, so that it also overflows on the 11th count, leaving a residue of 0.05. By the end of the 19th count the accumulator has a residue of 0.45. At the 20th count it overflows with a residue of zero, and the cycle is ready to restart. The sequence of this complete cycle is listed in Table 7.10.

The total count from the VCO in this cycle includes 11 overflows. The total count is $(724 \times 9) + (725 \times 11) = 14491$. This occurs in 20 reference frequency cycles. The average frequency of the VCO is $(14491/20) \times 0.1 = 72.455$ MHz, as required. However, the phase detector produces regular negative pulse inputs at a 100-kHz rate, while the VCO produces its pulses with an irregularity in time determined by whether the count is 724 or 725. The result of these irregularities is that although the output of the phase detector has a zero dc component, it also has a periodic component at an average frequency of 100 kHz, modulated in phase, with a fundamental modulating frequency of 5 kHz. Fourier analysis would reveal a number of harmonics of 5 kHz, peaking in the vicinity of 55 kHz, resulting in many spurious outputs from the phase detector with frequencies equal to 100 ± 5 L, where L is integral. The components in the vicinity of 45 kHz (100 to 55) will tend to be strongest, but components can be as low as 5 kHz.

The condition becomes aggravated when the desired frequency is near a harmonic of the reference. For example, consider the case of 72.499 MHz. In this example, a complete cycle of the VCO counts requires 100 reference counts. Thus, the fundamental has been reduced to 1 kHz. The spurious components in this case will tend to be distributed with relatively uniform amplitude about the 100-kHz carrier, because the modulation appears as an impulse at a 1-kHz rate. Thus this technique can produce spurious signals at frequencies as low as the minimum step. While not large, the spurious outputs from the VCO are larger than normally tolerated.

We have seen that the natural frequency of the loop has to be much smaller than the reference in order to assure loop stability. But the loop filter cannot be used to eliminate these low-frequency spurs without reducing the closed-loop bandwidth so far that the lock-up time and the reduction of noise sidebands of the VCO are adversely affected. When the waveform generated by the phase detector at low frequencies is considered, it is seen that the frequency of the oscillator is higher than the value Nf_{ref}, so that the phase from the divided VCO gradually increases relative to the reference phase. However, when a pulse is deleted, the phase of the VCO drops behind the reference. Over a sufficient period, the increases and decreases average to produce the average phase of zero. The information needed to compensate for these phase changes in the output of the phase detector is available in the accumulator and may be used to generate a canceling waveform (for the particular type of phase detector in use), which may be subtracted from the phase detector output before it is applied to the loop filter to eliminate PM of the VCO, which causes the unwanted spurs. Figure 7.58 shows a block diagram of a fractional-N synthesizer with this feature added.

In practice, sideband cancellation using such deterministic cancellation schemes is limited to about 40 dB. Higher values may be obtained but are not usually stable under temperature and voltage drifts. If such a loop is used, it is advisable to divide the output of the fractional loop by at least 10 to obtain an additional 20 dB of suppression, for a total of at least 60 dB. By use of the cancellation technique, the spurs can be made acceptable for many applications. The lock-up time is determined only by the loop filter and not by the smallest step size, as in the conventional loop. Lock-up times in the vicinity of 1 ms are achievable, together with a resolution better than 1 Hz.

As the frequency range is increased, the system becomes increasingly complex, and the speed requirements for the digital circuitry increase. Advancements in

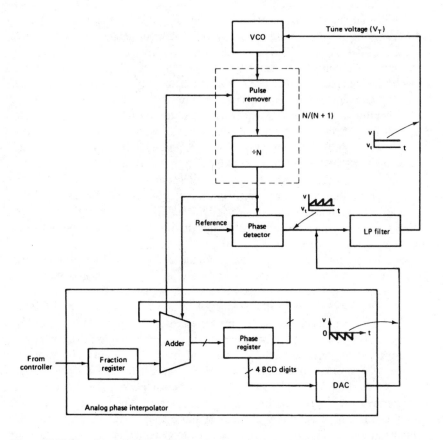

Figure 7.58 Fractional-N PLL with circuitry to cancel the PM of the VCO. (*Courtesy of Hewlett-Packard.*)

digital signal processing have made common use of fractional division synthesizers practical. Such devices can be found, for example, in cellular telephones.

7.8 Multiloop Synthesizers

To avoid the limitations of the single-loop synthesizer, synthesizers are often designed to employ more than one loop. Figure 7.59 shows a block diagram of a multiloop synthesizer. The first LO, operating from 81.4 to 111.4 MHz, is a two-loop synthesizer using a frequency translation stage. It comprises a 70- to 80-MHz loop, a divider, two frequency translators, and an output loop at the final LO frequency. Two single-loop synthesizers are also used later in the receiver, but our discussion will be confined to the multiloop unit.

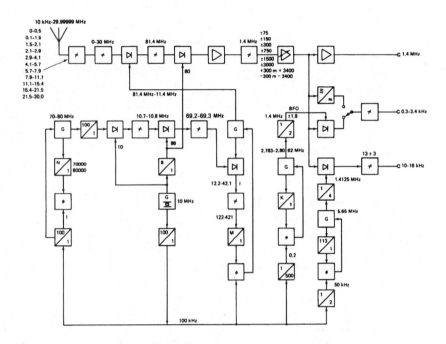

Figure 7.59 Block diagram of a multiloop synthesizer. (*Courtesy of Rohde and Schwarz.*)

A 10-MHz crystal oscillator is used as the standard to which all of the internal oscillator frequencies are locked. A divide-by-100 circuit reduces this to a 100-kHz reference used in both loops of the synthesizer. The 100-kHz reference is further divided by 100 to provide the 1-kHz reference for the 70- to 80-MHz loop. The output of this loop is then further divided by 100 to provide 10-Hz steps between 0.7 and 0.8 MHz. This division improves the noise sidebands and spurious signal suppression of the loop by 40 dB.

The 0.7- to 0.8-MHz band is converted to 10.7 to 10.8 MHz by mixing it with the 10-MHz reference. A crystal filter is used to provide the necessary suppression of the two inputs to the mixer. The resultant signal is translated to 69.2 to 69.3 MHz by further mixing with a signal of 80 MHz, the eighth harmonic of the 10-MHz frequency standard. The 80-MHz signal can be generated either by a frequency multiplier and crystal filter or by using an 80-MHz crystal oscillator under phase-lock control of the standard. In the former case, the noise sideband performance of the standard is degraded by 18 dB over the standard. Another possibility is the use of an 80-MHz crystal oscillator standard, followed by a divide-by-8 circuit to produce the 10-MHz internal reference. However, it is not possible to build crystal oscillators at 80 MHz with as good long- and short-term frequency stability as at

10 MHz. Hence, the phase-locked crystal oscillator approach was used to achieve high stability.

The 69.2- to 69.3-MHz output frequency from the mixer, after filtering, is mixed with the final VCO output frequency to produce a signal of 12.2 to 42.1 MHz, which after division by M is used for comparison in the final PLL with the 100-kHz reference. The M value is used to select the 0.1-MHz steps, while the value of N shifts the 70- to 80-MHz oscillator to provide 10-Hz resolution over the 0.1-MHz band resulting from its division by 100. This synthesizer provides the first oscillator frequency for a receiver with 81.4-MHz first IF, for a band of input frequencies up to 30 MHz, with 10-Hz step resolution.

This multiloop synthesizer illustrates the most important principles found in communications synthesizers. A different auxiliary loop could be used to provide further resolution. For example, by replacing the 10.7- to 10.8-MHz loop by a digital direct frequency synthesizer, quasi-infinite resolution could be obtained by a microprocessor-controlled low-frequency synthesizer. Such a synthesizer has a very fast lock-up time, comparable to the speed of the 100-kHz loop. In the design indicated, the switching speed is determined by the 1-kHz fine resolution loop. For a loop bandwidth of 50 Hz for this loop, we will obtain a settling time in the vicinity of 40 ms. The longest time is needed when the resolution synthesizer is required to jump from 80 to 70 MHz. During this frequency jump, the loop will go out of both phase and frequency lock and will need complete reacquisition. This results in an audible click, because the time to acquire frequency lock is substantially more than the time to acquire only phase lock for smaller frequency jumps. Thus, each time a 100-kHz segment is passed through, there is such a click. The same occurs when the output frequency loop jumps over a large frequency segment. The VCO, operating from 81.4 to 111.4 MHz, is coarse tuned by switching diodes, and some of the jumps are audible. The noise sideband performance of this communication synthesizer is determined inside the loop bandwidth by the reference noise multiplied to the output frequency, and outside the loop bandwidth by the VCO noise. The latter noise can be kept low by building a high-quality oscillator.

Another interesting multiloop synthesizer is shown in Figure 7.60. A 13- to 14-MHz singleloop synthesizer with 1-kHz steps is divided by 10 to provide 100-Hz steps and a 20-dB noise improvement. A synthesizer in this frequency range can be built with one oscillator in a single chip. The output loop, operating from 69.5 to 98 MHz, is generated in a synthesizer loop with 50-kHz reference and 66.6- to 66.7-MHz frequency offset generated by translation of the fine loop frequency. The reference frequencies for the two loops are generated from a 10.7 MHz *temperature-compensated crystal oscillator* (TCXO) standard. An additional 57.3- MHz TCXO is used to drive a second oscillator in the receiver as well as to offset the fine frequency loop of the synthesizer. An increase in frequency of this oscillator changes the output frequency of the first LO (synthesizer) in the opposite direction to the second LO, thereby canceling the drift error. This drift-canceling technique is sometimes referred to as the *Barlow-Wadley principle*.

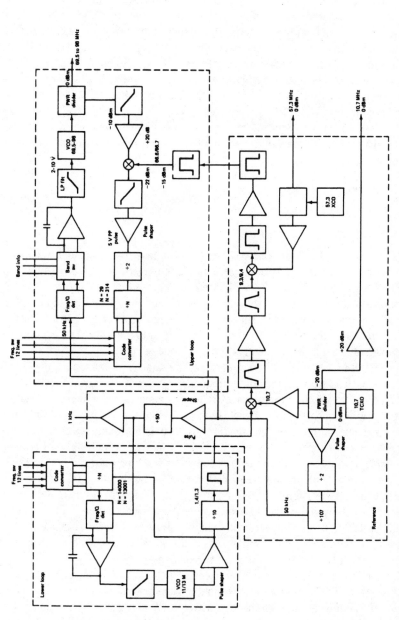

Figure 7.60 Block diagram of a multiloop frequency synthesizer incorporating a drift-canceling technique. (*Courtesy of Rohde and Schwarz.*)

7.9 Direct Digital Synthesis

Direct digital frequency synthesis (DDFS) consists of generating a digital representation of the desired signal and then using a D/A converter to convert the digital representation to an analog waveform. Recent advances in high-speed microelectronics, particularly the microprocessor, make DDFS practical at frequencies in the high-frequency band and below (as of this writing). Systems can be compact, use low power, and provide very fine frequency resolution with virtually instantaneous switching of frequencies. DDFS is finding increasing application, particularly in conjunction with PLL synthesizers.

DDFS uses a single-frequency source (clock) as a time reference. One method of digitally generating the values of a sine wave is to solve the digital recursion relation as follows:

$$Y_n = [2 \cos (2 \pi f t)] Y_{(n-1)} - Y_{(n-2)} \tag{7.82}$$

This is solved by $Y_n = \cos (2\pi fnt)$. There are at least two problems with this method, however. The noise can increase until a limit cycle (nonlinear oscillation) occurs. Also, the finite word length used to represent $2 \cos (2\pi ft)$ places a limitation on the frequency resolution. Another method of DDFS, direct table lookup, consists of storing the sinusoidal amplitude coefficients for successive phase increments in memory. The continuing miniaturization in size and cost of ROM make this the most frequently used technique.

One method of direct table lookup outputs the same N points for each cycle of the sine wave, and changes the output frequency by adjusting the rate at which the points are computed. It is relatively difficult to obtain fine frequency resolution with this approach, so a modified table look-up method is generally used. It is this method that we describe here. The function $\cos (2\pi ft)$ is approximated by outputting the function $\cos (2\pi fnT)$ for $n = 1, 2, 3, \ldots$, where T is the interval between conversions of digital words in the D/A converter and n represents the successive sample numbers. The sampling frequency, or rate, of the system is $1/T$. The lowest output frequency waveform contains N distinct points in its waveform, as illustrated in Figure 7.61. A waveform of twice the frequency can be generated, using the same sampling rate, but outputting every other data point. A waveform k times as fast is obtained by outputting every kth point at the same rate $1/T$. The frequency resolution, then, is the same as the lowest frequency f_L.

The maximum output frequency is selected so that it is an integral multiple of f_L, that is, $f_U = kF_L$. If P points are used in the waveform of the highest frequency, then $N (=kP)$ points are used in the lowest frequency waveform. The number N is limited by the available memory size. The minimum value that P can assume is usually taken to be four. With this small value of P, the output contains many harmonics of the desired frequency. These can be removed by the use of low-pass

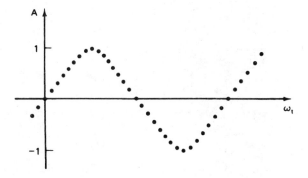

Figure 7.61 Synthesized waveform generated by direct digital synthesis.

filtering in the D/A output. For $P = 4$ the period of the highest frequency is $4T$, resulting in $f_U = 4f_L$. Thus, the highest attainable frequency is determined by the fastest sampling rate possible.

In the design of this type of DDFS, the following guidelines apply:

- The desired frequency resolution determines the lowest output frequency f_L.

- The number of D/A conversions used to generate f_L is $N = 4k = 4f_U/f_L$ provided that four conversions are used to generate f_U ($P = 4$).

- The maximum output frequency f_U is limited by the maximum sampling rate of the DDFS, $f_U \leq 1/4T$. Conversely, $T \leq 1/4f_U$.

The architecture of the complete DDFS is shown in Figure 7.62. To generate nf_L, the integer n addresses the register, and each clock cycle kn is added to the content of the accumulator so that the content of the memory address register is increased by kn. Each knth point of the memory is addressed, and the content of this memory location is transferred to the D/A converter to produce the output sampled waveform.

To complete the DDFS, the memory size and the length (number of bits) of the memory word must be determined. The word length is determined by system noise requirements. The amplitude of the D/A output is that of an exact sinusoid corrupted with the deterministic noise due to truncation caused by the finite length of the digital words (quantization noise). If an $(n + 1)$-bit word length (including one sign bit) is used and the output of the A/D converter varies between ± 1, the mean noise from the quantization will be:

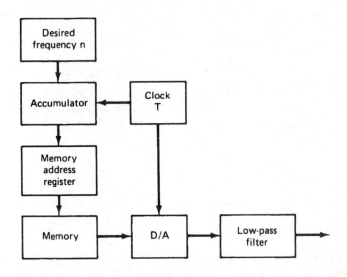

Figure 7.62 Block diagram of a direct digital frequency synthesizer. (*From [7.4]. Reprinted with permission.*)

$$\sigma^2 = \frac{1}{12}\left(\frac{1}{2}\right)^{2n} = \frac{1}{3}\left(\frac{1}{2}\right)^{2(n+1)} \tag{7.83}$$

The mean noise is averaged over all possible waveforms. For a worst-case waveform, the noise is a square wave with amplitude $\frac{1}{2}(\frac{1}{2})^n$ and $\sigma^2 = \frac{1}{4}(\frac{1}{2})^{2n}$. For each bit added to the word length, the spectral purity improves by 6 dB.

The main drawback of the DDFS is that it is limited to relatively low frequencies. The upper frequency is directly related to the maximum usable clock frequency. DDFS tends to be noisier than other methods, but adequate spectral purity can be obtained if sufficient low-pass filtering is used at the output. DDFS systems are easily constructed using readily available microprocessors. The combination of DDFS for fine frequency resolution plus other synthesis techniques to obtain higher-frequency output can provide high resolution with very rapid settling time after a frequency change. This is especially valuable for frequency-hopping spread-spectrum systems.

7.10 Monolithic PLL Systems

The rapid advancements in microprocessor and digital signal processing technologies have permitted large-scale integration of oscillator systems for receivers of all types. Reduction of a complex PLL system to a single device (plus a small number of supporting external components) offers a number of design benefits, not the least of which are reduced system complexity and lower overall cost.

An example of the trend in receiver design toward highly integrated systems through the use of dedicated chipsets can be found in the IS-54 *time division multiple access* (TDMA) chipset for cellular radios from Philips. TDMA, standardized as IS-54, is an evolving application for cellular radio. TDMA uses the same 30-kHz channel spacing as the current North American analog *frequency division multiple access* (FDMA) system, but multiplexes users over time; at different time intervals, multiple users may be present on the same frequency. The benefit of this technology is an increase from one to three users per channel.

An overall block diagram of the radio is given in Figure 7.63. The chipset combines all of the necessary RF and IF functions into four integrated devices. For the purposes of this chapter, we will focus on the dual frequency synthesizer chip (SA7025).

A detailed block diagram of the low-power synthesizer is shown in Figure 7.64. The IC is fabricated using the QUBiC (Philips) BiCMOS technology. The SA7025 features fractional-N with selectable modulo five or eight implemented in the main synthesizer. This allows the phase detector comparison frequency to be five or eight times the channel spacing. Therefore, if the channel spacing is 30 kHz, which is what is used for AMPS and IS-54, it becomes possible to use a comparison frequency of 150 kHz (modulus 5) or 240 kHz (modulus 8). The use of a higher comparison frequency moves spurs further away from the fundamental, thus allowing the use of a wider loop bandwidth filter. The result is more rapid switching times, as required by IS-54.

A triple-modulus high-frequency prescaler (divide by 64/65/72) is integrated on the chip, with a maximum input frequency of 1 GHz. The use of the prescaler lowers the main divider ratio, providing more flexibility in synthesizing channels, which, in turn, eliminates the possibility of blind channels in the system. Programming and channel selection are controlled by a high-speed serial interface.

7.10.1 Microwave PLL Synthesizer

The rapid growth of business and consumer applications in the low microwave frequencies is driving the production of a number of devices specifically targeted for operation in the neighborhood of 2 GHz. Applications such as wireless local area networks and the emerging personal communications services have generated a great deal of interest in microwave receiver circuits of all types. Although above the frequency range of receivers generally covered in this book, such circuits merit review.

For the purposes of illustration, we will examine a 2.2-GHz monolithic IC PLL frequency synthesizer designed for low power operation [7.40]. The chip features a selectable 64/65 or 128/129 dual-modulus prescaler, an internally regulated charge pump with tristate capability, and a microcontroller-compatible serial data

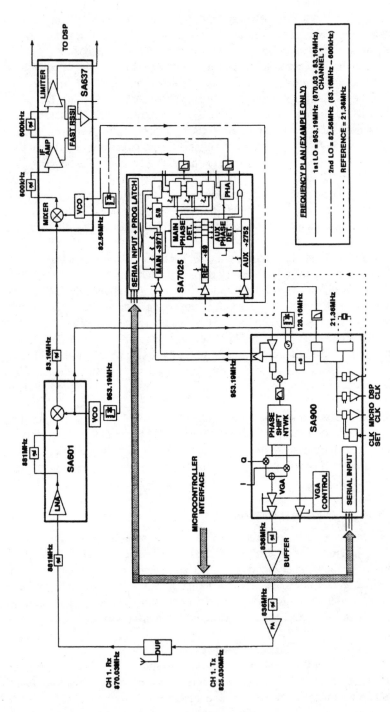

Figure 7.63 Overall block diagram of a TDMA IS-54 chip set. (*Courtesy of Philips.*)

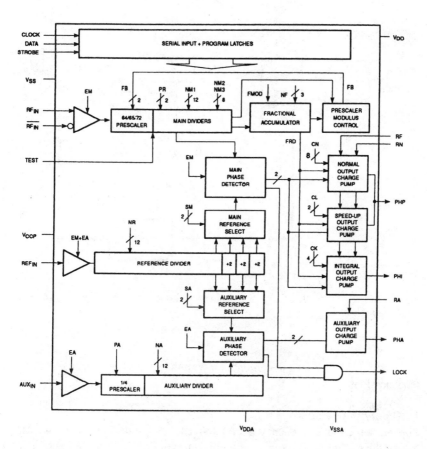

Figure 7.64 Block diagram of the SA7025 frequency synthesizer. (*Courtesy of Philips.*)

interface. The National Semiconductor device is shown in block diagram form in Figure 7.65. The major operational elements include:

- A 14-bit reference frequency divider
- An 18-bit dual-modulus high-frequency (*N*) divider
- Serial control register for loading the two counters
- A phase comparator-charge pump block

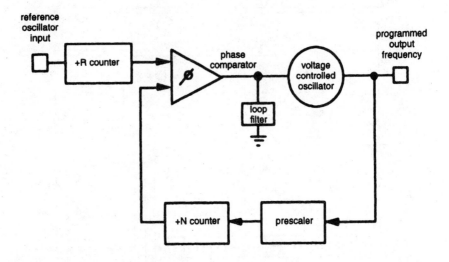

Figure 7.65 2.2-GHz PLL frequency synthesizer logic diagram. (*Courtesy of National Semiconductor.*)

An external VCO is used in conjunction with the synthesizer circuits to close the feedback loop.

A. Theory of operation

The reference signal frequency is first divided down to the desired tuning resolution of the system. This tuning resolution is usually some divisible fraction of the spacing between radio channels. The phase comparator drives the frequency of the external VCO in the direction which—when divided down by the N counter—equals the stepping resolution established by the R counter. The phase comparator accomplishes this by detecting the arrival of phase edges from the two counters and issuing a correction signal to the VCO that is proportional to the difference in their phases. When the phase and frequencies of the two counter outputs agree, the frequency of the VCO will be the selected N-modulus multiple of the tuning resolution.

The R counter is implemented entirely in CMOS and is composed of ripple toggle flip flops drawing less than 0.5mA at 30MHz. The dual-modulus N-counter is implemented using a selectable bipolar 64/65 or 128/129 prescaler followed by 18 CMOS counter bits [7.41]. The front-end of the prescaler, shown in Figure 7.66, consists of a 4/5 divider block followed by five toggle flip flops. The 2.2-GHz input buffer provides a sensitivity range of −15 to +6 dBm.

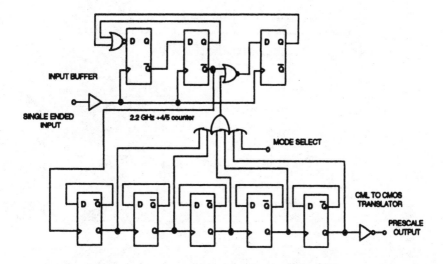

Figure 7.66 The prescaler system front-end. (*Courtesy of National Semiconductor.*)

The type 4 digital phase comparator-charge pump block incorporates two features that contribute to its low phase noise and spurious characteristics [7.42] and [7.43]. The first is a *deadband elimination* circuit, shown in Figure 7.67, which ensures that a charge pump dead zone cannot occur near zero phase offset [7.44]. This technique guarantees that the charge pump generators have in fact responded to the phase comparator before allowing the phase comparator reset logic to activate. Phase comparator deadband is one of the primary sources of jitter in phase-locked loops of this type.

The second charge pump feature minimizes reference frequency spurs. Similar current generating structures for the pump up and down circuits ensure that the magnitudes of the pump up and down current sources and the turn on and off times are matched. This matching minimizes the momentary pump up or down excursions on the charge pump line, which contribute to VCO FM at the reference rate. The deadband elimination circuitry inherently also ensures that reference frequency sideband spurs are minimized. The charge pump feedback signals allow the generators on only long enough to eliminate potential deadband yet not to contribute any excess active pump time. Excess pump time adds directly to the up or down excursions on the charge pump line by an amount equal to the absolute magnitudes of the current generators.

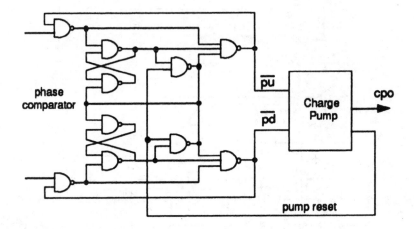

Figure 7.67 Phase comparator reset logic tree. (*Courtesy of National Semiconductor.*)

7.11 Oscillator Design

The frequency of a superheterodyne receiver is determined by its LOs. Even the DDFS is controlled by the oscillator in its frequency standard. Our discussion of the oscillator is drawn generally from [7.4]. There are also many texts on oscillators (References [7.11] to [7.15]), and extensive references for further reading are available in them, as well as from [7.4] and [7.5].

An electronic oscillator converts dc power to a periodic output signal (i. e., ac power). If the output waveform is approximately sinusoidal, the oscillator is referred to as a sinusoidal or *harmonic oscillator*. There are other oscillator types, often referred to as *relaxation oscillators*, whose outputs deviate substantially from sinusoidal. Sinusoidal oscillators are used for most radio applications because of their spectral purity and good noise sideband performance. Oscillator circuits are inherently nonlinear; however, linear analysis techniques are useful for the analysis and design of sinusoidal oscillators. Figure 7.68 is a generic block diagram of an oscillator. It is a feedback loop, comprising a frequency-dependent amplifier with forward loop gain $G(j2\pi f)$ and a frequency-dependent feedback network $H(2\pi f)$. The output voltage is given by

$$V_O = \frac{V_I\, G(j\, 2\, \pi f)}{1 + G(j\, 2\, \pi f)\, H(j\, 2\, \pi f)} \tag{7.84}$$

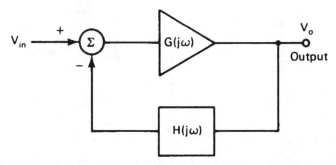

Figure 7.68 Block diagram of an oscillator showing forward and feedback loop components; $(\omega = 2\pi f)$.

To sustain oscillation, the output V_O must be nonzero even when the input signal V_I is zero. Because $G(j2\pi f)$ is finite in practical circuits, the denominator may be zero at some frequency f_0, leading to the well-known Nyquist criterion for oscillation, i. e., at f_0,

$$G(j\,2\,\pi\,f_0)\,H(j\,2\,\pi\,f_0) = -1 \tag{7.85}$$

The magnitude of the open-loop transfer function $G(j2\pi f)\,H(j2\pi f)$ equals unity, and its phase shift is 180°, that is, if in a negative feedback loop the open-loop gain has a total phase shift of 180° at some frequency f_0, the system will oscillate, provided that the open-loop gain is unity. If the gain is less than unity at f_0, the system will be stable, while if the gain is greater than unity, it will be unstable. In a positive-feedback loop, the same gain conditions hold for a phase shift of 0° (or 360°). This statement is not accurate for some complex systems, but it is correct for the simple transfer functions normally encountered in oscillators.

7.11.1 Simple Oscillator Analysis

The following analysis of the simple oscillator circuit shown in Figure 7.69 illustrates the design method. In the simplified linear equivalent circuit shown in Figure 7.70, the transistor parameter h_{rb} has been neglected, and $1/h_{ob}$ is assumed to be much greater than R_L and is also ignored. The transistor is connected in the common-base configuration, which results in no output phase inversion (positive feedback). The circuit analysis can be greatly simplified by the assumptions that the Q of the load impedance is high and that

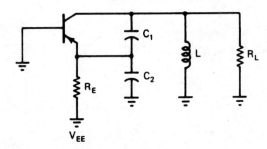

Figure 7.69 Oscillator with capacitive voltage divider feedback. (*From* [7.4]. *Reprinted with permission.*)

$$\frac{1}{2\pi f(C_1 + C_2)} \ll \frac{h_{ib} R_E}{h_{ib} + R_E} \tag{7.86}$$

In this case, the circuit reduces to that shown in Figure 7.71 with the following values:

$$V = \frac{V_0 C_1}{C_1 + C_2} \tag{7.87}$$

$$R_{eq} = \frac{h_{ib} R_E}{h_{ib} + R_E}\left(\frac{C_1 + C_2}{C_1}\right)^2 \tag{7.88}$$

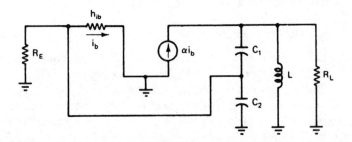

Figure 7.70 Linearlized and simplified equivalent of the circuit shown in Figure 7.69. (*From* [7.4]. *Reprinted with permission.*)

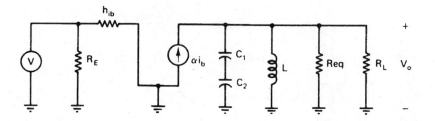

Figure 7.71 Further simplification of Figure 7.69, assuming high-impedance loads. (*From [7.4]. Reprinted with permission.*)

Then,

$$G(j\,2\,\pi f) = \frac{\alpha\,Z_L}{h_{ib}} \tag{7.89}$$

$$H(j\,2\,\pi f) = \frac{C_1}{C_1 + C_2} \tag{7.90}$$

and Z_L is the load impedance in the collector circuit, the parallel combination of $(C_1 + C_2)$, L, R_{eq}, and R_L. Because H has no frequency dependence, the phase of GH depends only on that of Z_L. With positive feedback, the phase of Z_L must be zero. This occurs only at the resonant frequency of this circuit as follows:

$$f_0 = \frac{1}{2\,\pi \left[\dfrac{L\,C_1\,C_2}{C_1 + C_2} \right]^{1/2}} \tag{7.91}$$

At this frequency,

$$Z_L = \frac{R_{eq}\,R_L}{R_{eq} + R_L} \tag{7.92}$$

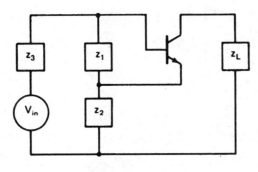

Figure 7.72 Generalized circuit for an oscillator using the amplifier model. (*From [7.4]. Reprinted with permission.*)

$$GH = \frac{\alpha R_{eq} R_L C_1}{h_{ib}(R_{eq} + R_L)(C_1 + C_2)} \tag{7.93}$$

The second condition for oscillation is that GH equal unity.

A direct analysis of the circuit equations is frequently simpler than the block diagram (Figure 7.68) interpretation, especially for single-stage amplifiers. Figure 7.72 shows a generalized circuit for such a single-stage amplifier. The small-signal equivalent circuit (with h_{re} neglected) is shown in Figure 7.73. The condition for oscillation is that the currents must exist even when V_I is zero. For this to occur, the determinant of the network equations must be zero. With h_{oe} negligible, this can be shown to require the following:

$$(Z_1 + Z_2 + Z_3) h_{ie} + Z_1 Z_2 \beta + Z_1(Z_2 + Z_3) = 0 \tag{7.94}$$

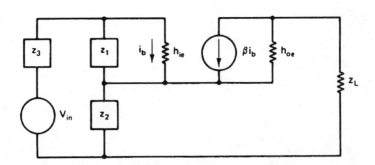

Figure 7.73 Small-signal equivalent circuit of Figure 7.72. (*From [7.4]. Reprinted with permission.*)

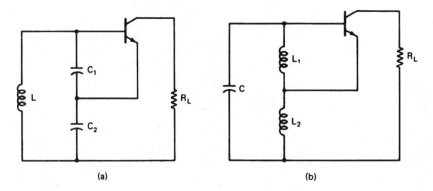

Figure 7.74 Oscillator circuits: (*a*) Colpitts oscillator and (*b*) Hartley oscillator. (*From [7.4].*
Reprinted with permission.)

Because the variables are complex, this results in two real equations that must
be satisfied. When h_{ie} is real (a valid approximation for operation below 50 MHz)
and Z_1, Z_2, and Z_3 are pure reactances, Equation (7.94) results in the following pair
of equations:

$$h_{ie}(Z_1 + Z_2 + Z_3) = 0 \qquad\qquad (7.95)$$

$$Z_1[(1+\beta)Z_2 + Z_3] = 0 \qquad\qquad (7.96)$$

Because β is real and positive, the reactances Z_2 and Z_3 must be of opposite sign.
In addition, because h_{ie} is nonzero, Z_1 and Z_2 must have the same sign. Two
possible oscillator connections (neglecting dc connections) satisfying these conditions are the Colpitts and Hartley circuits shown in Figure 7.74.

7.11.2 Negative Resistance

If we consider that the positive feedback in the oscillator is applied to compensate
for the losses in the tuned circuit, the amplifier and feedback circuit, in effect,
create a negative resistor. When Z_1 and Z_2 are capacitive, the impedance across the
capacitors can be estimated, using Figure 7.75:

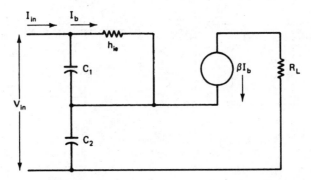

Figure 7.75 Small-signal circuit to find the equivalent impedance connected across Z_3 in Figure 7.72. (*From [7.4]. Reprinted with permission.*)

$$Z_I = \frac{V}{I} = \frac{(1 + \beta) X_{C1} X_{C2} + h_{ie} (X_{C1} + X_{C2})}{X_{C1} + h_{ie}} \qquad (7.97)$$

If $X_{C1} < h_{ie}$ then

$$Z_I \approx \frac{1 + \beta}{h_{ie} \, 4 \, \pi^2 \, f^2 \, C_1 \, C_2} = \frac{j(C_1 + C_2)}{2 \, \pi f C_1 \, C_2} \qquad (7.98)$$

The input impedance is a negative resistor in series with C_1 and C_2. If Z_3 is an inductance L with series resistor R_e, the condition for sustained oscillation is:

$$R_e = \frac{1 + \beta}{h_{ie} \, 4 \, \pi^2 \, f^2 \, C_1 \, C_2} = \frac{g_m}{4 \, \pi^2 \, f^2 \, C_1 \, C_2} \qquad (7.99)$$

The circuit corresponds to that in Figure 7.74(*a*) and the frequency is in accordance with Equation (7.91).

This interpretation of the oscillator provides additional design guidelines. First C_1 should be large so that $X_{C1} \ll h_{ie}$. Also, C_2 should be large so that $X_{C2} \ll 1/h_{oe}$. With both of these capacitors large, the transistor base-emitter and collector-emitter capacitances will have negligible effect on circuit performance. However, Equation (7.99) limits the maximum value of the capacitance, because $g_m/R \geq 4\pi^2 f^2 C_1 C_2$. This relationship is important because it shows that for oscillations to be maintained, the minimum permissible reactances of C_1 and C_2 are a function of the transistor mutual conductance g_m and the resistance of the inductor.

7.11.3 Amplitude Stabilization

The oscillator amplitude stabilizes as a result of nonlinear performance of the transistor. There are several mechanisms involved, which may act simultaneously. In most transistor amplifiers, the dc biases are substantially larger than the signal voltages. In an oscillator, however, we are dealing with a positive-feedback circuit. The energy generated by initial switch-on of the circuit is amplified and returned to the input in such a manner that the oscillation would increase indefinitely unless some limiting or other form of stabilization occurs. The following two mechanisms are responsible for limiting the amplitude of oscillation:

- Gain saturation and consequent reduction of the open-loop gain to just match the oscillation gain criterion. This saturation may occur in a single-stage amplifier or in a second stage introduced to ensure saturation.
- Bias generated by the rectifying mechanism of either the diode in the bipolar transistor or the grid conduction diode in a vacuum tube or in the junction of a JFET. In MOSFETs, an external diode is sometimes used for this biasing.

A third mechanism sometimes employed is an external AGC circuit.

The self-limiting process, using a diode-generated offset bias to move the operating point to a low-gain region, is generally quite noisy. As such, this process is not recommended for the low noise required in the early LOs of a receiver. In the two-stage emitter-coupled oscillator of Figure 7.76, amplitude stabilization occurs as a result of current limiting in the second stage. This circuit has the added advantage that its output terminals are isolated from the feedback path. The emitter signal of Q_2, having a rich harmonic content, is often used as the output. Harmonics of the frequency of oscillation can be extracted from the emitter by using an appropriately tuned circuit.

7.11.4 Phase Stability

The frequency or phase stability of an oscillator is usually considered in two parts. The first is *long-term stability*, in which the frequency changes over minutes, hours, days, weeks, or even years. This stability component is usually limited by circuit element changes with ambient conditions (such as input voltage, temperature, and humidity) and with aging. Second there is the *short-term stability* measured in seconds or tenths of seconds.

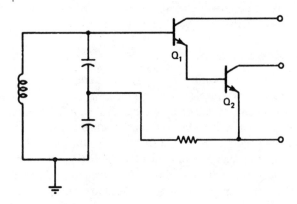

Figure 7.76 Two-stage emitter-coupled oscillator. (*From [7.4]. Reprinted with permission.*)

One form of short-term stability degradation is caused by phase changes within the system. In this case, the frequency of oscillation reacts to small changes in phase shift of the open-loop system. The system with the largest open-loop rate of change versus frequency $d\phi/df$ will have the best frequency stability. Figure 7.77 shows phase versus frequency plots of two open-loop systems used in oscillators. At the system crossover frequency, the phase is $-180°$. If some external influence causes a change in phase, such as a 10° phase lag, the frequency will shift until the total phase shift is again reduced to zero. In this case, the frequency will decrease to the point at which the open-loop phase shift is 170°. In Figure 7.77 the change in frequency associated with open-loop response GH_2 is greater than that for GH_1, whose phase slope changes more rapidly near the open-loop crossover frequency.

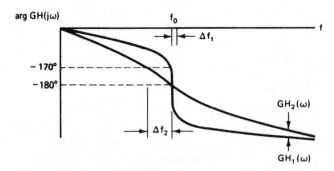

Figure 7.77 Phase versus frequency plot of two open-loop systems with resonators having different Q values. (*From [7.4]. Reprinted with permission.*)

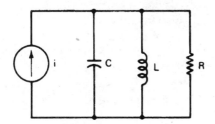

Figure 7.78 Parallel-tuned circuit for phase-shift analysis. (*From [7.4]. Reprinted with permission.*)

Consider the simple parallel tuned circuit shown in Figure 7.78. The circuit impedance is:

$$Z = \frac{R}{1 + j\, Q[(f/f_0) - (f_0/f)]}$$ (7.100)

where, as usual, $2\pi f_0 = [LC]^{-\frac{1}{2}}$ and $Q = R/2\pi fL$. From this, we may determine:

$$\frac{d\phi}{df} = \frac{f_0\, Q(f^2 + f_0^2)}{(f_0 f)^2 + Q^2(f^2 - f_0^2)^2}$$ (7.101)

At the resonant frequency, this reduces $[d\phi/df]_{f0} = 2Q/f_0$. The frequency stability factor is defined, relative to the fractional change in frequency df/f_0 as:

$$S_F = [F_0(d\phi/df)]_{f0} = 2\,Q$$ (7.102)

where S_F is a measure of the short-term stability of an oscillator. Equation (7.102) indicates that the higher the circuit Q the higher the stability factor. This is one reason for using high-Q circuits in oscillators. Another reason is the ability of the tuned circuit to filter out undesired harmonics and noise.

7.11.5 Low-Noise Oscillators

Oscillator circuits have circuit noise and amplifier noise like all other amplifier circuits. In the case of the oscillator, the noise voltage provides an input to the feedback loop and creates random PM about the average output frequency. The noise arises from thermal noise from the circuit, thermal and shot noise contributed by the device, and flicker noise in the device. If there is an electronic tuning device in the oscillator circuit, thermal noise from that circuit can also phase-modulate the oscillator to contribute to the phase noise. The measurement of oscillator noise commonly used is $L(f_m)$, the ratio of the SSB power of the phase noise in a 1-Hz bandwidth, f_m hertz away from the carrier frequency, to the total signal power of the oscillator. An expression for this noise was given in Equation (7.81), where the last term on the right-hand side represents the noise added by the tuning diode.

That expression is significant because the following guidelines to minimize phase noise can be derived from it.

1. Maximize the unloaded Q of the resonator.

2. Maximize the reactive energy by means of a high RF voltage across the resonator and use a low LC ratio. The limits are set by the breakdown voltage of the active devices, by the breakdown voltage and forward-bias condition of the tuning diodes, and by overheating or excess stress of the crystal resonators.

3. Avoid device saturation and try to achieve limiting or AGC without degradation of Q . Isolate the tuned circuit from the limiter or AGC devices. Use the *antiparallel* (back-to-back) tuning diode connection to help avoid forward bias.

4. Choose an active device with lowest NF. The NF of interest is that which is obtained under actual operating conditions.

5. Choose high-impedance devices, such as FETs, where the signal voltage can be made very high relative to the equivalent noise voltage. In the case of a limiter, the limited voltage should be as high as possible.

6. Choose an active device with low flicker noise. The effect of flicker noise can be reduced by RF feedback. For a bipolar transistor, an unbypassed emitter resistor of 10 to 30 Ω can improve flicker noise by as much as 40 dB. The proper bias point of the active device is important, and precautions should be taken to prevent modulation of the input and output dynamic capacitances of the device.

7. The output energy should be decoupled from the resonator rather than another portion of the active device so that the resonator limits the noise bandwidth. The output circuits should be isolated from the oscillator circuit and use as little of the oscillator power as possible.

The lower limit of the noise density is determined by inherent circuit thermal noise, not by the oscillator. This value is kT, which is −204 dBW/Hz or −174 dBm/Hz. However, this is a theoretical lower limit, and a noise floor of −170 dBm/Hz is rarely achieved from an oscillator.

7.11.6 Stability with Ambient Changes

Long-term instability is the change of the average center frequency over longer periods and is caused by changes in the components from aging or changing ambient conditions. Such drifts result from the following:

1. Change in the resonator characteristics as a result of aging or because of changes in temperature, humidity, or pressure. The changes in electrical characteristics of inductors or capacitors in *LC* resonators are much greater than the changes in quartz crystals.

2. Gain changes in the active device. Capacitive and resistive components of the input and output impedance of a transistor, for example, change with temperature and operating point, which may vary with changes in the supply voltage.

3. Mechanical changes in the resonator circuit, as a result of vibration or shock. Such mechanical effects are often referred to as *microphonics*. Many of the changes that occur in aging are believed to result from long-term easing of mechanical strains initially in the components.

4. Lack of circuit isolation, so that changes in other circuits react on the oscillator to produce frequency instability.

Aging effects can sometimes be reduced by initial burn-in periods, in which the oscillator may be subjected to higher than normal temperatures and, possibly, mechanical vibrations. Changes from humidity can be reduced by keeping stray capacitances through the air as low as possible, and by treating circuit boards to inhibit moisture absorption. Hermetic sealing of critical oscillator components or the entire oscillator may be used. A sealed compartment filled with dry gas under pressure might also be used in extreme cases. Care in the selection and mounting of components can minimize microphonics. Instability from supply voltage changes can be reduced by careful regulation of the supply voltage. Circuit design changes can be made to increase isolation from other circuits. Several possibilities exist for reducing instability from temperature changes.

Coils made from normal materials tend to increase their dimensions with increasing temperature, thus increasing inductance. Powdered iron and ferrite cores used in constructing some RF coils also usually increase inductance with temperature. Some of these materials cause much greater changes than others. It is possible through the selection of materials and by means of physical structures designed to move cores or portions of a coil to reduce the inductance change, or to cause it to be reversed. Such techniques are usually costly and difficult to duplicate, and should be considered only as a last resort.

Air capacitors used for tuning tend to increase dimensions similarly, increasing capacitance directly with temperature. Most solid dielectrics used for fixed capacitors have dielectric constants that increase with frequency. Over the temperature ranges where a receiver must operate, most inductance and capacitance changes are relatively linear. The variation is usually measured as the percent of change of the parameter per degree centigrade. For example, a capacitor whose capacitance changes from 100 to 101 pF when the temperature changes from 20 to 70°C has the following percentage change,

$$\alpha_c = \frac{\delta C}{C \delta T} = \text{i. e.,} \quad \frac{101 - 100}{100(70 - 20)} = \frac{0.01}{50} = 2 \times 10^{-4}/°C \qquad (7.103)$$

Here α_c is the temperature coefficient of the capacitance and is generally measured in parts per million per degree centigrade (ppm/°C). The temperature coefficients of inductance or resistance are similarily defined.

Air capacitors, like inductors, can be designed to have low temperature coefficients through the use of special materials (invar, for example), or by using materials with different temperature coefficients that cause in the adjacent plates a tendency to withdraw from each other slightly as they expand. This type of compensation is not recommended because of the need for costly manufacturing control of the materials and structure. Fortunately ceramic dielectrics are available with temperature coefficients from about +100 ppm/°C to lower than −1000 ppm/°C. Small capacitor sizes up to more than 100 pF may be obtained from many manufacturers. These capacitors may be included in resonator circuits to provide temperature compensation for the usual positive coefficients of inductance and capacitance of the other components. For the temperature ranges of receiver operation, negative coefficient compensating capacitors are nonlinear, becoming more so as the coefficient becomes more negative. This limits their compensation ability.

If we consider an oscillator using the parallel resonant circuit shown in Figure 7.78, the oscillator frequency is $f_0 = [2\pi(LC)^{1/2}]^{-1}$, and hence the temperature coefficient of frequency change is

$$\alpha_f \equiv \frac{d f_0}{f_0 \, d T} = \frac{-d L}{2 L \, d T} - \frac{d C}{2 C \, d T} = -\frac{1}{2}\alpha_L - \frac{1}{2}\alpha_C \qquad (7.104)$$

In order to compensate α_L, C may be broken into two (or more) parts, at least one of which has a negative temperature coefficient. If C is made up of two parallel capacitors, by differentiating the expression for the total capacitance, $C_t = C_1 + C_2$, with respect to T and dividing by the total capacitance, we arrive at:

$$\alpha_{Ct} = \frac{C_1 \, \alpha_{C1} + C_2 \, \alpha_{C2}}{C_1 + C_2} \qquad (7.105)$$

If C is made up of two capacitors in series, the expression must be derived from the expression $C_t = C_1 C_2 / (C_1 + C_2)$ and yields:

$$\alpha_{Ct} = \frac{C_2 \, \alpha_{C1} + C_1 \, \alpha_{C2}}{C_1 + C_2} \qquad (7.106)$$

If one of the capacitors is variable, it may be necessary to try to achieve optimum compensation over its tuning range. In this case, both series and parallel compensators might be included in the total capacitance. This provides sufficient degrees of freedom to achieve a desired overall temperature coefficient at two frequencies within the tuning range. At other frequencies, a small error can be expected. However, overall performance of the oscillator can usually be much improved over the entire tuning range. While one can use equations such as Equations (7.105) and (7.106) to guide compensation efforts, usually there is no detailed knowledge of the inductor and some of the capacitor coefficients; thus, compensation becomes an empirical process of cut and try until the optimum arrangement of compensators is achieved.

During periods of warm-up, or when the ambient temperature changes suddenly, steady-state temperatures of the various components may be attained at different rates, giving rise to transient frequency drift, even when the steady-state compensation is good. Therefore, the layout of the oscillator parts must be carefully planned to provide transient temperature changes. Use of a large thermal mass, near which the parts are mounted, can help this uniform heating and cooling. In sufficiently important applications, the entire oscillator can be mounted in a well-insulated oven whose temperature is maintained by a proportionally controlled thermostat. Usually these extreme measures are reserved for oscillators with crystal resonators that are used as an overall frequency standard. Frequency standards are discussed in more detail in a later section of this chapter.

7.12 Variable-Frequency Oscillators

The first LO in most receivers must be capable of being tuned over a frequency range, offset from the basic receiver RF tuning range. In some cases, this requires tuning over many octaves, while in others, a much more narrow frequency range may suffice. Prior to the advent of the varactor diode and good switching diodes, it was customary to tune an oscillator mechanically by using a variable capacitor with air dielectric or, in some cases, by moving a powdered iron core inside a coil to make a variable inductor. This resulted in tuning over a range of possibly as much as 1 to 1.5 octaves. For greater coverage, mechanical wafer switches were used to switch among different coils and/or fixed capacitors.

At higher frequencies, transmission lines may be used for tuning. Short-circuited lines less than a quarter-wave long present an inductive reactance at the open end and may be tuned by a variable capacitor shunted across this end. Open lines shorter than a half-wave may be tuned to resonance at one end by connecting a variable capacitor across the other.

Whether the oscillator is tuned mechanically or electrically, the capacitor is most often the source of small frequency changes, while the coil generally is switched for band changes. This can be achieved more or less easily in many different oscillator circuit configurations. Figure 7.79 shows a number of circuits used for VFOs. Different configurations are used in different applications, depending on the range of tuning, whether the oscillator is for a receiver or a transmitter, and whether the tuning elements are completely independent or have a common element (such as the rotor of a tuning capacitor).

Figure 7.80 is the schematic diagram of an oscillator in the Rohde and Schwarz signal generator SMDU. This oscillator is mechanically tuned by C_{12}, and may be frequency-modulated using varactors GL_2 and GL_3. The design exhibits extremely low noise sidebands. The measured noise at 1.0 kHz from carrier is -100 dB/Hz, and at 25 kHz from carrier it is -150 dB/Hz. The overall long-term stability, in general use, is in the vicinity of 100 Hz/h. The oscillator frequency range is 118 to 198 MHz.

7.12.1 Voltage-Controlled Oscillators

Newer receivers control the oscillator band and frequency by electrical rather than mechanical means. Tuning is accomplished by voltage-sensitive capacitors (varactor diodes), and band switching is accomplished by diodes with low forward conductance. For high-power tuning, inductors having saturable ferrite cores have been used; however, these do not appear in receivers. Oscillators that are tuned by varying the input voltage are referred to as VCOs.

The capacitance versus voltage curves of a varactor diode depend on the variation of the impurity density with the distance from the junction. When the distribution is constant, there is an *abrupt junction* and capacitance follows the law, $C = K/(V_d + V)^{1/2}$, where V_d is the contact potential of the diode and V is the applied voltage. Such a junction is well approximated by an alloyed junction diode. Other impurity distribution profiles give rise to other variations, and the previous equation is usually modified to

$$C = \frac{K}{(V_d + V)^n} \tag{7.107}$$

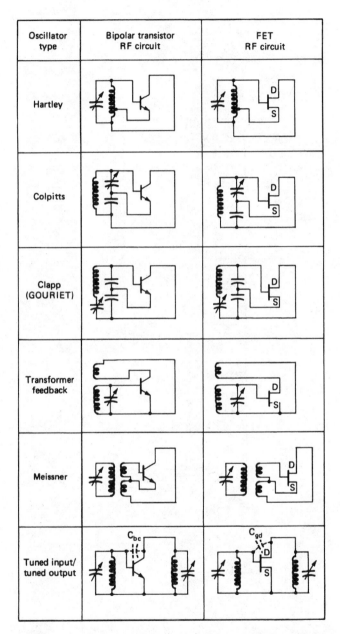

Figure 7.79 Schematic diagrams of RF connections for common oscillator circuits. For simplicity, dc and biasing circuits are not shown.

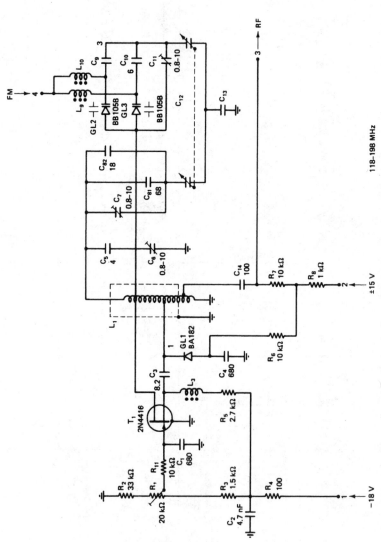

Figure 7.80 Schematic diagram of an 118- to 198-MHz oscillator from the Rohde and Schwarz SMDU signal generator. (*Courtesy of Rohde and Schwarz.*)

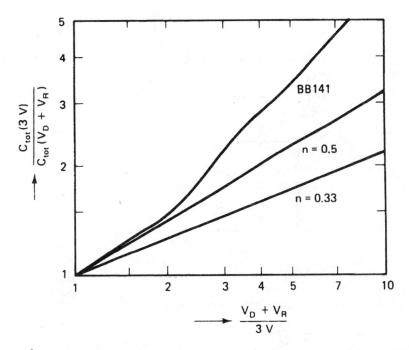

Figure 7.81 Capacitance-voltage characteristics for graded-junction ($n = 0.33$), abrupt-junction ($n = 0.5$), and hyperabrupt-junction (BB141) varactor diodes. (*Courtesy of ITT Semiconductors, Freiburg, West Germany.*)

where n depends on the diffusion profile and $C_0 = K/V_n^d$. A so-called graded junction, having a linear decrease in impurity density with distance from the junction, has a value of $n = \frac{1}{3}$. This is approximated in a diffused junction.

In all the cases, these are theoretical equations, and limitations on the control of the impurity pattern can result in a curve that does not have such a simple expression. In such a case, the coefficient n is thought of as varying with voltage. If the impurity density increases away from the junction, a value of n higher than 0.5 can be obtained. Such junctions are called *hyperabrupt*. A typical value for n for a hyperabrupt junction is 0.75. Such capacitors are used primarily to achieve a larger tuning range for a given voltage change. Figure 7.81 shows the capacitance-voltage variation for the abrupt and graded junctions, as well as for a particular hyperabrupt junction diode.

Varactor diodes are available from a number of manufacturers, such as Motorola, Siemens, and Philips. Maximum values range from a few to several hundred picofarads, and useful capacitance ratios range from about 5 to 15.

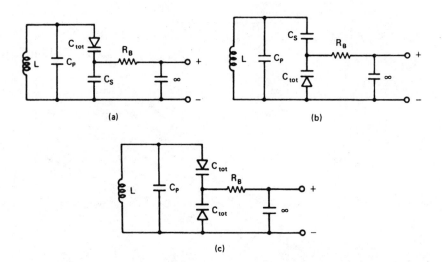

Figure 7.82 Typical circuits for use of varactor tuning diodes: (*a*) single diode in circuit low side, (*b*) single diode in circuit high side, and (*c*) two diodes in series back-to-back arrangement. (*From [7.4]. Reprinted with permission.*)

Figure 7.82 shows three typical circuits that are used with varactor tuning diodes. In all the cases, the voltage is applied through a large resistor R_e. The resistance is shunted across the lower diode and may be converted to a shunt load resistor across the inductance to estimate Q. The diode also has losses that may result in lowering the circuit Q at high capacitance, when the frequency is sufficiently high. This must be considered in the circuit design.

The frequency is not always the value determined by applying the dc tuning voltage to the variable V in Equation (7.107). If the RF voltage is sufficient to drive the diode into conduction on peaks, an average current will flow in the circuits of Figure 7.82, which will increase the bias voltage. The current is impulsive, giving rise to various harmonics in the circuit. Even in the absence of "conduction," Equation (7.107) deals with the small-signal capacitance only. When the RF voltage varies over a relatively large range, the capacitance changes. In this case we must change Equation (7.107) to:

$$\frac{dQ}{dV} = \frac{K}{(V + V_d)^n} \tag{7.108}$$

Here Q is the charge on the capacitor. When this relation is substituted in the circuit differential equation, it produces a nonlinear differential equation, dependent on the parameter n. Thus, the varactor may generate direct current and harmonics of the fundamental frequency. Unless the diodes are driven into conduction at some point in the cycle, the direct current must remain zero.

The current of the circuit shown in Figure 7.82(c) can be shown to eliminate the even harmonics and permits a substantially larger RF voltage without conduction than either circuit in Figure 7.82(a) or (b). When $n = 0.5$, only the second harmonic is generated by the capacitor, and this can be eliminated by the back-to-back connection of the diode pair. Integrating Equation (7.108), we have:

$$Q + Q_A = \frac{K}{1-n}(V + V_d)^{1-n} \tag{7.109}$$

$$Q + Q_A = \frac{C_v}{1-n}(V + V_d) \tag{7.110}$$

where C_v is the value of Equation (7.107) for applied voltage V, and Q_A is a constant of integration. If we let $V = V_1 + v$ and $Q = Q_1 + q$, where the lower case letters represent the varying RF and the uppercase letters indicate the values of bias when RF is absent, we have:

$$q + Q_1 + Q_A = \frac{K}{1-n}[v + (V_1 + V_d)]^{1-n} \tag{7.111}$$

and

$$1 + \frac{v}{V'} = \left(1 + \frac{q}{Q'}\right)^{V(1-n)} \tag{7.112}$$

where $V' = V_1 + V_d$ and $Q' = Q_1 + Q_A$. For the back-to-back connection of identical diodes, $K_{11} = K_{12} = K_1$, $V_1' = V_2' = V'$, $Q_1' = Q_2' = Q'$, $q = q_1 = -q_2$, and $v = v_1 - v_2$. Here the new subscripts 1 and 2 refer to the top and bottom diodes, respectively, and v and q are the RF voltage across and the charge transferred through the pair in series. With this notation, we have:

$$\frac{v}{V'} \equiv \frac{v_1 - v_2}{V'} = \left(1 + \frac{q}{Q'}\right)^{1/(1-n)} - \left(1 - \frac{q}{Q'}\right)^{1/(1-n)} \tag{7.113}$$

For all n, this eliminates the even powers of q, hence even harmonics. This can be shown by expanding Equation (7.113) in series and performing term-by-term combination of the equal powers of q. In the particular case $n = 1/2$, $v/V' = 4q/Q'$, and the circuit becomes linear.

The equations hold as long as the absolute value of v_1/V' is less than unity, so that there is no conduction. At the point of conduction, the total value of v/V' may be calculated by noticing that when $v_1/V' = -1$, $q/Q' = -1$, then $q_2/Q' = 1$, $v_2/V = 3$, and $v/V = 4$. The single-diode circuits conduct at $v/V' = -1$, so the peak RF voltage should not exceed this. The back-to-back configuration can provide a four-fold increase in RF voltage handling over the single diode. For all values of n, the back-to-back configuration allows an increase in the peak-to-peak voltage without conduction. For some hyperabrupt values of n, such that $1/(1 - n)$ is an integer, many of the higher-order odd harmonics are eliminated, although only $n = 1/2$ provides elimination of the third harmonic. For example, $n = 2/3$ results in $1/(1 - n) = 3$. The fifth harmonic and higher odd harmonics are eliminated, and the peak-to-peak RF without conduction is increased eightfold; for $n = 3/4$ the harmonics 7 and above are eliminated, and the RF peak is increased by 16 times. It must be noted in these cases that the RF peak at the fundamental may not increase so much, because the RF voltage includes the harmonic voltages.

Because the equations are only approximate, not all harmonics are eliminated, and the RF voltage at conduction for the back-to-back circuit may be different than predicted. For example, abrupt junction diodes tend to have n of about 0.46 to 0.48, rather than exactly 0.5. Hyperabrupt junctions tend to have substantial changes in n with voltage. The diode illustrated in Figure 7.81 shows a variation from about 0.6 at low bias to about 0.9 at higher voltages, with wiggles from 0.67 to 1.1 in the midrange. The value of V_d for varactor diodes tends to be in the vicinity of 0.7 V.

7.12.2 Diode Switches

Because they have a low resistance when biased in one direction and a very high resistance when biased in the other, semiconductor diodes may be used to switch RF circuits. Sufficiently large bias voltages may be applied to keep the diode on when it is carrying varying RF currents or off when it is subjected to RF voltages. It is important that in the forward-biased condition, the diode add as little resistance as possible to the circuit and that it be capable of handling the maximum RF plus bias current. When it is reverse-biased, not only should the resistance be very high to isolate the circuit, but the breakdown voltage must be higher than the combined bias and RF peak voltage. Almost any diodes can perform switching, but at high frequencies, PIN diodes are especially useful.

Diodes for switch applications are available from various vendors, for example, ITT, Microwave Associates, Motorola, Siemens, and Unitrode. The diode switches BA243, BA244, BA238 from ITT, and the Motorola MPN 3401 series were especially developed for RF switching. Diodes can also be employed to advantage in audio switching. The advantage of electronically tuning HF, VHF, and UHF circuits using varactor diodes is only fully realized when band selection also takes place electronically. Diode switches are preferable to mechanical switches because of their higher reliability and virtually unlimited life. Mechanically operated switching contacts are subject to wear and contamination, which do not affect the diodes. Diode switches can be controlled by application of direct voltages and currents, which obviates the need for mechanical links between the front panel control and the tuned circuits to be switched. This allows the RF circuits to be located optimally with regard to electrical performance and thermal influence. It also frees the front panel design from restrictions on control location, permits remote control to be simplified, and enables automatic frequency adaptation through computer control. Electronic switching allows much more rapid band switching than mechanical switching, thus reducing the time required for frequency change. This is useful in frequency-hopping spread-spectrum systems. Figure 7.83 shows a number of common connections for diode band switching. The most complete isolation is provided when two diodes (or more) are used per circuit to provide a high resistance in series with the switched-out component plus a low shunt resistance in parallel with it.

Figure 7.84 is an example of an oscillator circuit using complete electronic tuning, with series switching diodes permitting additional tuned elements to be switched in parallel to provide coarse tuning and the varactor diode to provide fine tuning. In this oscillator circuit, the frequency may be set within a few hundred kilohertz by the switching diodes, which essentially add capacitance in parallel with the main tuning circuit. Figure 7.85 shows a circuit where the tuning inductors rather than the capacitors are switched in and out, thereby reducing the degree of gain variation of the oscillator.

7.13 Crystal-Controlled Oscillators

Short-term frequency stability is a function of the resonator Q, and long-term stability is a function of the drift of the resonant frequency. Piezoelectric quartz crystals have resonances that are much more stable than the LC circuits previously discussed, and also have very high Q. Consequently for very stable oscillators at a fixed or only slightly variable frequency, quartz crystal resonators are generally used. Other piezoelectric crystal materials have been used in filters and to control oscillators. However, their temperature stability is considerably poorer than that of quartz, so for oscillator use, quartz is generally preferred. A piezoelectric material is one which develops a voltage when it is under a mechanical strain or is placed under strain by an applied voltage. A physical piece of such material, depending

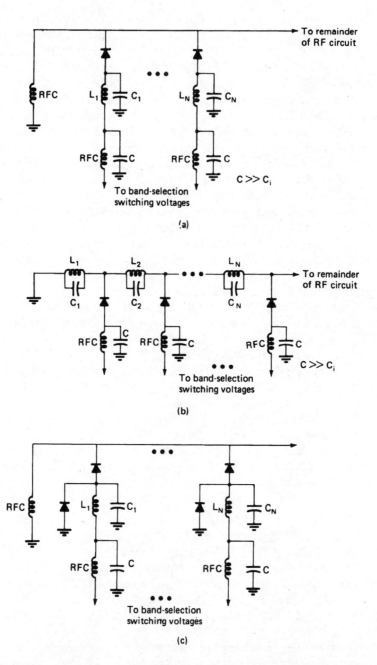

Figure 7.83 Typical circuits for diode band switching: (*a*) series-diode arrangement, (*b*) shunt-diode arrangement, and (*c*) use of both series and shunt diodes.

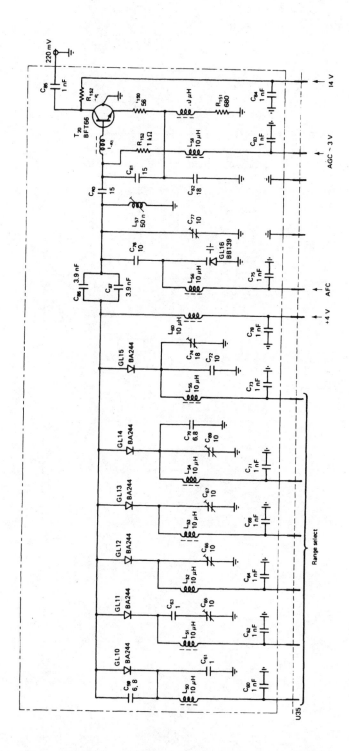

Figure 7.84 Schematic diagram of a coarse-tuned VCO from the Rohde and Schwarz EK070 receiver. (*Courtesy of Rohde and Schwarz.*)

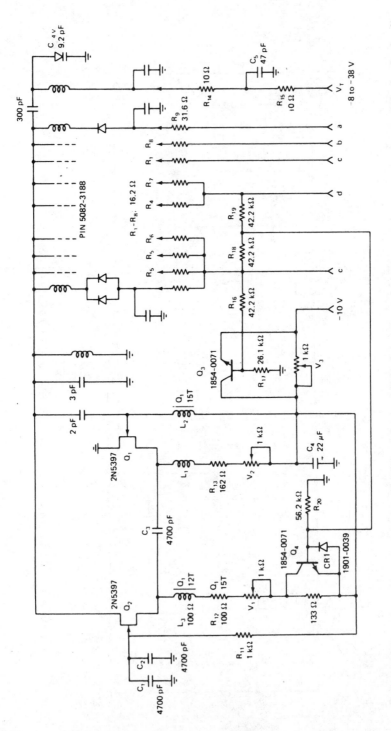

Figure 7.85 Schematic diagram of a 260- to 520-MHz VCO from the Hewlett-Packard HP8662A signal generator. (*Courtesy of Hewlett-Packard.*)

TABLE 7.11 Designations for Quartz Crystal Cuts (*After [7.33]*)

Vibrator designation	Usual reference	Mode of vibration	Frequency range
A	AT cut	Thickness shear	0.5 to 250 MHz
B	BT	Thickness shear	1 to 30 MHz
C	CT	Face shear	300 to 1000 kHz
D	DT	Face or width shear	200 to 750 kHz
E	+5°X	Extensional	60 to 300 kHz
F	−18°X	Extensional	60 to 300 kHz
G	GT	Extensional	100 to 500 kHz
H	+5°X	Length-width flexure	10 to 100 kHz
J	+5°X (2 plates)	Duplex length-thickness flexure	1 to 10 kHz
M	MT	Extensional	60 to 300 kHz
N	NT	Length-width flexure	10 to 100 kHz
K	X-Y bar	Length-width or length-thickness flexure	2 to 20 kHz

upon its shape, can have a number of mechanical resonances. By appropriate shaping and location of the electrodes, one or other resonant mode of vibration can be favored, so that the resonance may be excited by the external voltage. When the material is crystalline, such as quartz, the properties of the resonator are not only affected by the shape and placement of the electrodes, but also by the relationship of the resonator cut to the principal crystal axes. Table 7.11 lists designations of various quartz crystal cuts.

The temperature performance of a quartz crystal resonator varies with the angles at which it is cut. Figure 7.86 shows how different cuts react to temperature variations. Most crystals used in oscillators above 1 MHz use AT cuts for maximum stability; the CT and GT cuts are used at much lower frequencies. Other cuts, such as X and Y, will be found in some applications but are not much used nowadays. A relatively new cut, referred to as *stress-compensated* (SC), has been found to provide better temperature stability than the AT cut. The SC cut also has an easily excited spurious resonance near the frequency for which it is designed. Care must be taken in the design of the external circuits to avoid exciting the spurious mode. Most crystals have many possible resonant modes. In the AT cut, those modes near the desired resonance are usually hard to excite. Overtones may be excited in crystals to provide oscillations at higher frequencies than could otherwise be provided. The overtones are not exactly at harmonic frequencies of the lowest mode, so such crystals must be fabricated to provide the desired overtone frequency.

The crystal has, at its frequency of oscillation, an equivalent electrical circuit as shown in Figure 7.87. The series resonant circuit represents the effect of the crystal vibrator, and the shunt capacitance is the result of the coupling plates and of capacitance to surrounding metallic objects, such as a metal case. The resonant

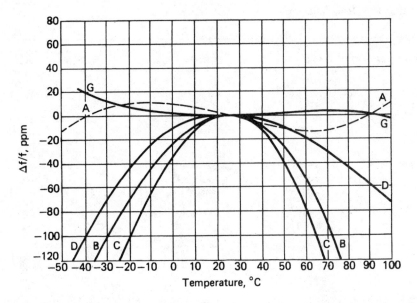

Figure 7.86 Frequency-temperate characteristics of various quartz vibrators. Letters on the curves are explained in Table 7.11. (*After [7.16].*)

circuit represents the particular vibrating mode that is excited. If more than one mode can be excited, a more complex circuit would be required to represent the crystal. Table 7.12 lists values for the equivalent circuit for various AT-cut fundamental crystals.

The most common type of circuit for use with a fundamental AT crystal is an *aperiodic oscillator*, which has no selective circuits other than the crystal. Such circuits, often referred to as parallel resonant oscillators, use the familiar Pierce and Clapp configurations (Figure 7.88). The crystal operates at a frequency where it exhibits a high-Q inductive reactance. When one terminal of the crystal is connected to the collector, the collector resistor must have high resistance to avoid loading

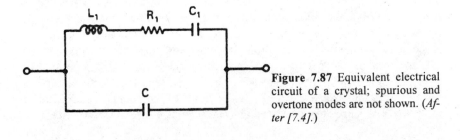

Figure 7.87 Equivalent electrical circuit of a crystal; spurious and overtone modes are not shown. (*After [7.4].*)

TABLE 7.12 Equivalent Data for AT-Cut Fundamental Crystal Resonators

Shape of crystal	Frequency range, MHz			Typical equivalent data*			
	HC-6/U	HC-25/U	HC-35/HC-45	C_0	C_1	Q	R_1
Biconvex	0.75–1.5	—	—	3–7 pF	8 fF	>100,000	100–500 Ω
Planoconvex	1.5–3	2.7–5.2	—	4–7 pF	10 fF	>100,000	< 200 Ω
Planoparallel with bevel	2–7	4.5–10.5	10–13	5–7 pF	20 fF[10 fF]	>50,000	10–100 Ω
Plane	7–20 (30)	10.5–20 (30)	13–20 (30)				

* Value in brackets is for HC-35/HC-45.
From [7.4]. Reprinted with permission.

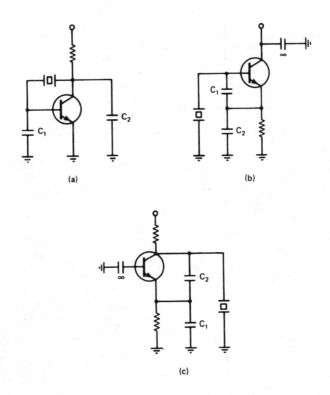

Figure 7.88 Common parallel-resonant circuits for use in fundamental crystal oscillators: (a) Pierce circuit; (b) Clapp circuit, collector grounded; and (c) Clapp circuit, base grounded. (*From [7.4]. Reprinted with permission.*)

the crystal, but low resistance to avoid excessive voltage drop. One way to solve this problem is to replace the resistor by an RF choke; another good way is by using a Darlington stage, as shown in Figure 7.89. Because of the high input impedance, capacitors C_1 and C_2 can have fairly large values. This reduces the effect of the transistor stage on the crystal. The pulling capacitor, typically about 30 pF, is set in series with the crystal and serves to adjust the oscillator exactly to frequency. Typical values for C_1 and C_2 for such an oscillator are shown in Figure 7.89. A disadvantage of the aperiodic oscillator is the occasional tendency to oscillate at the third or higher overtone of the crystal, or at some other spurious resonance. In difficult cases, this can be overcome by replacing capacitor C_2 with a resonant circuit detuned to present a capacitive reactance at the nominal frequency.

Generally speaking, for the simple crystal oscillator without an AGC or limiting stage, the positive feedback should not be greater than that needed for starting and

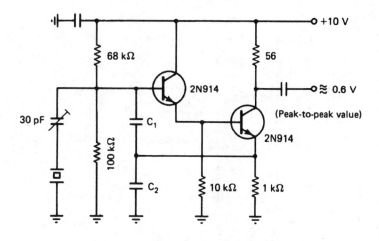

Frequency (MHz)	C_1 (pF)	C_2 (pF)
3–6	560	470
6–15	560	220
15–30	220	100

Figure 7.89 Oscillator with Darlington stage, suitable for fundamental crystals. (*Courtesy of Kristall-Verarbeitung Neckarbischofsheim GmbH, West Germany.*)

maintaining stable oscillation. In the case of the circuit of Figure 7.88(*b*) or 7.89, the values of C_1 and C_2 can be derived from the following equations:

$$\frac{C_1}{C_2} = \left(\frac{r_{be}}{R_a}\right)^{1/2} \tag{7.114}$$

$$C_1 \, C_2 = \frac{g_m'}{4\,\pi^2\,f_0^2\,R_1'} \tag{7.115}$$

where r_{be} is the RF impedance between base and emitter (of the Darlington), R_a is the ac output impedance (measured at the common emitter), g_m' is the transconductance ($= 1/R_{in}$ for an emitter follower), and R_1' is the resonant resistance of the crystal, transformed by the load capacitance.

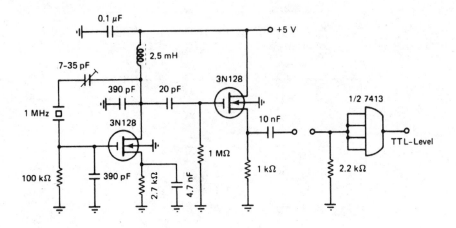

Figure 7.90 Crystal oscillator circuit using MOSFETs. (*Courtesy of Kristall-Verarbeitung Neckarbischofsheim GmbH, West Germany.*)

Figure 7.90 gives an example of a Pierce oscillator for 1 MHz, using MOSFET devices [7.17]. A TTL output level is available if the output of the crystal oscillator is to drive a Schmitt trigger (e.g., 7413). Such an oscillator is a suitable clock for frequency counters.

7.13.1 Overtone Crystal Oscillators

If the crystal disk is excited at an overtone in the thickness-shear mode, it will oscillate with several subareas in antiphase. Only odd harmonics can be excited. The fundamental frequency of an AT cut is inversely proportional to the thickness of the disk as follows:

$$f_0 \,(MHz) = \frac{1675}{thickness \,(\mu m)} \tag{7.116}$$

For example, a 20-MHz crystal will have a thickness of about 84 μm. If this crystal were excited at the third overtone (that is, 60 MHz), the effective electrical subdisk thickness would be about one-third of the disk thickness, or 28 μm.

The overtone frequencies are, however, not at exact multiples of the fundamental mode frequency. This anharmony decreases with higher-order overtones. It is important that crystals be ground to the frequency of the overtone at which they will be used. It is rather easy to operate crystal oscillators in overtone modes even up to frequencies on the order of 300 MHz, although the usual upper limit is at 200 MHz/ninth overtone. It is possible to operate at the eleventh or thirteenth overtone. The highest possible fundamental frequency (20 to 30 MHz) should be used so that the overtone modes are spaced as far as possible from one another. Typical data for AT overtone crystals are listed in Table 7.13. The motional capacitance C_1 reduces as the square of the overtone. The attainable Q value also falls with increasing frequency. For this reason, the values of R_1 increase, being typically in the range of 20 to 200 Ω. As the frequency increases, the static capacitance C_0 becomes an ever-increasing bypass for the crystal. For this reason, compensation for the static capacitance should be made by using a parallel inductance, approximately resonant at the series frequency. A rule of thumb is that C_0 compensation should be provided when X_{c0} 5R_1, generally when oscillation is to be in excess of 100 MHz. A compensating coil having a low Q (R_p 10R_1) is suitable, and the condition of compensation resonance need not be exactly maintained.

Aperiodic circuits do not operate reliably with overtone crystals. A resonant circuit should always be provided to prevent oscillation at the fundamental frequency. In the case of the Pierce circuit, the collector capacitor can be replaced by a circuit tuned below resonance so that it is capacitive at the overtone but inductive at lower overtones and the fundamental (and has relatively low impedance at the fundamental). Overtone crystals are usually aligned in series resonance; consequently the Colpitts derived circuits will not necessarily oscillate at the correct frequency, unless they are made with crystals ground to a specific customer specification. To pull the crystal frequency lower, an inductance may be connected in series with the crystal. However, it is possible for parasitic oscillations to arise across the inductance and the static capacitance C_0 of the crystal. These may prove difficult to suppress. Hence, it is better to use a series resonant circuit for an overtone crystal, as shown in Figure 7.91. The values of C_1 and C_2 are selected to provide sufficient loop feedback. The open-circuit gain is reduced by the divider C_1/C_2 and by the voltage division across the crystal impedance and the input impedance at the emitter. When selecting a suitable transistor, a rule of thumb is that the transit frequency should be at least ten times that of the oscillator. In addition, transistors should be used that have a high dc gain h_{fe} at a low base resistance $r_{bb'}$.

There is sometimes confusion between the designations of series- and parallel-resonance oscillators, and the series and parallel resonances of the crystal itself. In the case of series-resonance crystal oscillators, the circuit will resonate together with its pulling elements at a low-impedance resonance. Another example of such an arrangement is the Butler oscillator shown in Figure 7.92. This type of oscillator remains a series-resonant oscillator even when the crystal is pulled with the aid of a series capacitor, or even when (at higher frequencies) the phase angle of the transistor deviates from 0 or 180°. A series-pulling capacitor C_L reduces the inherent crystal series resonance to:

TABLE 7.13 Equivalent Data for AT-Cut Overtone Crystal Resonators

Overtone	Frequency range, MHz			Typical equivalent data*			
	HC-6/U	HC-25/U	HC-35/HC-45	C_0	C_1	Q	R_1
3	18–60 (80)	20–60 (90)	27–60 (90)	5–7 pF [2–4 pF]	2 fF [1 fF]	$> \dfrac{4 \times 10^6}{f \,(\text{MHz})}$	20 Ω [40 Ω]
5	40–115 (130)	40–115 (150)	50–125		0.6–0.8 fF [0.4 fF]	$> \dfrac{5 \times 10^6}{f \,(\text{MHz})}$	40 Ω [80 Ω]
7	70–150	70–150	70–175		0.3–0.4 fF [0.2 fF]		100 Ω [150 Ω]
9	150–200	150–200	150–200		0.2–0.3 fF [0.1 fF]		150 Ω [200 Ω]

* Values in brackets are for HC-35/HC-45.
From [7.4]. *Reprinted with permission.*

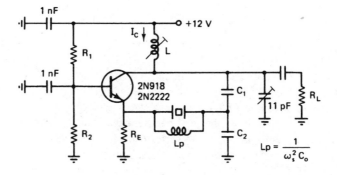

	75 MHz	120 MHz	150 MHz	200 MHz
C_1 [pF]	8	8	5	3
C_2 [pF]	100	50	25	20
I_c [mA]	25	25	5	5
R_E [Ω]	510	390	1.1 kΩ	1.1 kΩ
R_L [Ω]	470	300	600	600
Lp [μH]	0.25	0.10	0.08	0.05

Figure 7.91 Overtone crystal oscillator for operation up to 200 MHz. (*Courtesy of Kristall-Verarbeitung Neckarbischofsheim GmbH, West Germany.*)

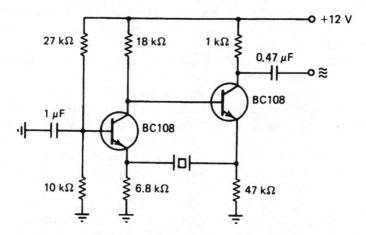

Figure 7.92 Butler oscillator for 50 to 500 kHz. (*Courtesy of Kristall-Verarbeitung Neckarbischofsheim GmbH, West Germany.*)

$$F_{CL} = f_s \left(1 + \frac{C_1}{C_0 + C_L} \right)^{1/2} \approx f_s \left(1 + \frac{C_1}{2(C_0 + C_L)} \right) \qquad (7.117)$$

In the case of a parallel-resonant oscillator, the circuit operates at a high-impedance resonance together with its adjacent (pulling) elements. In the case of the oscillator shown in Figure 7.89, series-connected C_1 and C_2 are in parallel with the crystal. For the general case, the total parallel capacitor C_L reduces the parallel resonance by the same factor as the series capacitance reduces the series resonance. In both cases, the crystal behaves like a high-Q inductance.

7.13.2 Dissipation in Crystal Oscillators

The power dissipation of crystals exhibits the following values in various oscillator circuits:

- TTL oscillator 1 to 5 mW
- Transistor oscillator 10 μW to 1 mW (100 μW typical)
- CMOS oscillator 1 to 100 μW

Because the crystal frequency and resonant impedance have some load dependence, a nominal load should be specified for low-tolerance crystals.

Crystal drive levels between 2 mW and 1 μW are recommended. Higher drive levels will cause deterioration of stability, Q, and aging characteristics. In the case of LF crystals and very small AT crystals (HC-35/U and HC-45/U, for example), 2 mW will be too much. Because the reactive power equals Q times the effective power, a reactive power of 200 W will be present periodically in the crystal reactances at a drive level of 2 mW with a Q of 100,000. Too low a drive can cause difficulties in starting oscillation, since a certain minimum amount of energy is required. This minimum value varies as a result of unavoidable variations in the crystal-electrode transition (at the submicroscopic level) and other damping influences. This can cause problems in certain CMOS and other very low power oscillators. The transistor parameters are valid only for low signal magnitudes, and the linear analysis is valid only as long as the transistor operates class A. In the case of a self-limiting oscillator, the transistor operates in the highly nonlinear saturation range. In this case it is virtually impossible to calculate the expected crystal drive level.

To determine the actual drive in a crystal measuring setup, either the RF current to the crystal or the RF voltage across the crystal is measured with the aid of a thermistor, oscilloscope, or RF voltmeter. If the equivalent data parameters of the crystal (C_0, C_1, R_1) are known, it is possible for the phase angle to be calculated from the oscillator frequency. From this, the actual crystal power can be determined. The value is much lower than an estimate made without consideration of the phase angle.

7.13.3 Voltage-Controlled Crystal Oscillator

A series capacitance can cause a change in frequency for a series-resonance type of crystal oscillator, and an analogous change occurs in a parallel-resonance oscillator when the capacitance in parallel with the crystal is changed. As a result, varactor diodes are used in crystal oscillator circuits to adjust the frequency, to produce either a *crystal-controlled VCO* (VCXO) or angle modulation of a crystal-controlled oscillator. Use of Equation (7.117), in conjunction with the capacitive effect of the diode, enables us to determine the oscillator modulation sensitivity or VCO gain for these applications. Varactors can also be used in conjunction with temperature-sensing components to provide temperature compensation for the frequency drift of crystal oscillators, known as TCXOs.

In all of these circuits, it is necessary to match the varactor type and its padding to attain the desired sensitivity and linearity of frequency change versus voltage. Because of the added circuits, the inherent stability of the crystal oscillator—both short- and long-term—is somewhat degraded. However, VCXOs are far more stable than *LC* VCOs, and are often used where a fixed-frequency oscillator locked to a standard is required. In the case of frequency modulators, it is essential to design for linear modulation versus voltage. The center frequency can be stabilized to a standard by a feedback loop with cutoff frequency well below the modulation frequencies. If the modulated oscillator is operated at a sufficiently low frequency, its drift may be only a small part of the overall transmitter drift, even if the center frequency is not locked to the standard.

7.13.4 Frequency Stability of Crystal Oscillators

The long-term stability of crystal oscillator circuits is affected by the aging of the external components, especially with regard to the Q of resonant circuits and the damping effect of the transistors on the Q of the crystal. It is, of course, also dependent on the aging of the crystal, which differs according to crystal type and drive level. The aging will result in typical frequency changes of 1 to 3×10^{-6} during the first year. Aging decreases logarithmically as a function of time, so it is possible to reduce this value by previous aging of the crystal—if possible, by the manufacturer—at a temperature between 85 and 125°C.

The drive level of the crystal should be as low as possible (1 to 20 μW) if the oscillator is to have good long-term stability. Because of their superior temperature characteristics, AT-cut crystals are preferred. When very stable oscillators are required, relatively low frequency overtone AT-cut crystals should be used because of their high Q and L_1/C_1 ratio. In this case, crystals operating at their third or fifth overtone are used, at a frequency of 5 to 10 MHz.

The short-term stability of crystal oscillators has increased in importance in recent years because of the widespread use of synthesizers in receivers for HF, VHF, and UHF applications. Oscillator chains for microwave frequencies are also locked to crystal oscillators, or derived from them by multiplication. There are a number of general guidelines that should be followed when designing crystal oscillators for short-term stability. In contrast to the design for long-term stability, the drive level of the crystal should be relatively high (100 to 500 μW) for this application. The effect of the oscillator circuit in reducing the crystal Q should be minimized. In the case of single-stage self-limiting oscillators, for example, the effective Q will be only 15 to 20 percent of the crystal Q, so this type of oscillator should be avoided. Usually series-resonance oscillators reduce Q less than parallel-resonance oscillators.

The noise in bipolar transistors is mainly dependent on the base-emitter path. The noise of PNP transistors is lower than that of the complementary NPN types. MOSFETs have a very high noise level, with $1/f$ noise dominating at low frequencies and drain-source thermal noise at high frequencies. JFETs have lower noise levels than either bipolar transistors or MOSFETs. Therefore a power FET with high current, such as type CP643 or P8000, is recommended for low-noise crystal oscillators [7.18]. If bipolar transistors are to be used, the types with highest possible dc gain h_{fe} , but with very low base resistance r_{bb}', should be selected (typical of transistors designed for VHF applications) and operated at low collector current.

As noted previously, single-stage crystal oscillators should be avoided. Moreover, an amplitude control loop is unfavorable because it is likely to generate additional phase noise. The best means of improving short-term stability is the use of RF negative feedback. A well-proven circuit was introduced by Driscoll [7.36], using a third-overtone 5-MHz crystal, and similar circuits possessing very good short-term stability up to 100 MHz have been developed since. The basic circuit is shown in Figure 7.93(a) . A cascaded circuit with low internal feedback is used as an amplifier. The first transistor is provided feedback in the emitter circuit by the crystal with a compensating shunt inductor. The first transistor T_1 operates stably in class A. Transistor T_2 is isolated from the crystal and operates at a quiescent current of only 0.8 mA. This stage is the first to limit and determines the oscillator amplitude. The higher the series-resonant resistor (for a given Q) of the crystal, the better the short-term stability, as this condition increases the negative feedback of T_1. Figure 7.93(b) shows the results of phase noise measurements [7.37].

Amplitude limiting can also be achieved by using biased Schottky diodes connected in opposition at the output of T_2. This is possible because of the low $1/f$ noise of these diodes. A low-noise oscillator designed according to this principle, at 96 MHz, is shown in Figure 7.94. Type P8000 power FETs are used in a stable class A cascode configuration. A relatively high value of C_1 keeps the oscillation feedback low. Diodes D are Schottky type, such as HP2800. For initial alignment, the limiting diodes are disconnected and the crystal is short-circuited. The self-excited frequency is then adjusted to 96 MHz with the trimmer capacitor. After

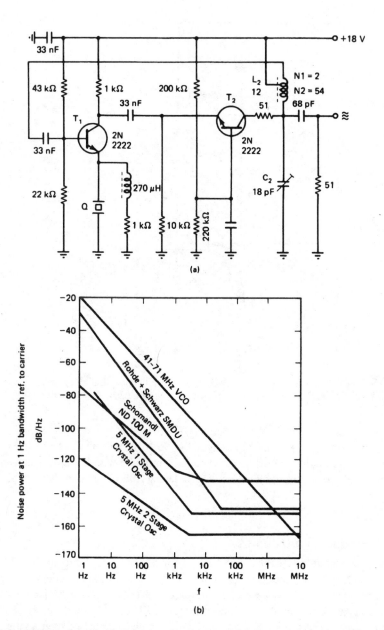

Figure 7.93 (*a*) Schematic diagram of a 5-MHz third-overtone crystal oscillator with exceptional short-term stability. (*b*) Sideband noise of various signal generators compared to the oscillator shown in *a*, according to [7.37]. (*Courtesy of Kristall-Verarbeitung Neckarbischofsheim GmbH, West Germany.*)

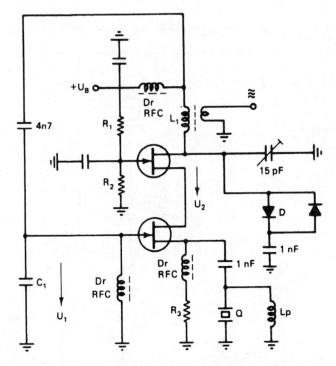

Figure 7.94 Schematic diagram of a VHF crystal oscillator with high short-term stability. (*Courtesy of Kristall-Verarbeitung Neckarbischofsheim GmbH, West Germany.*)

connecting the crystal, coil L_p is aligned to 96 MHz with the aid of a dip meter while the oscillator is switched off. The RF amplitude with the diodes connected is about half the self-limiting value. More details on this circuit have been published [7.38].

7.14 Frequency Standards

Frequency standards are the heart of frequency synthesizers and local clock timing for most modern receivers. The accuracy of the required standard depends on the application. In most cases, sufficient accuracy for radio receiver frequency determination can be obtained from quartz crystal oscillators. However, in some applications greater precision may be required than can be maintained over a long time by even the best available quartz standards. In this case, atomic frequency standards can be used. In these standards, an atomic resonance is used as the primary standard and a stable crystal oscillator is locked to the atomic standard. There are two atomic standards available at present, the *cesium atomic beam* type and the *rubidium gas cell* type. The cesium atomic beam type uses a true atomic resonance and is a primary standard. This type of unit is used by standards organizations, and is generally quite costly. The rubidium gas cell also employs an atomic resonance, but the frequency depends to some extent on the gas mixture and the gas pressure in the cell. For precise results, it must be calibrated from time to time against a cesium standard. However, its long-term stability is about two orders of magnitude better than that of the best quartz crystal oscillator standards. It is less costly and more easily portable than the cesium atomic standard.

Depending on performance and cost requirements, we find three classes of crystal oscillator used as standards:

- Oscillators of the type already discussed, with no further embellishments

- TCXOs, which permit operation over wide temperature ranges with a minimum change in frequency, when the extra weight and power of ovens cannot be afforded

- Oven-mounted crystal oscillators with controls to maintain a fixed internal temperature despite wide changes in the ambient temperature of the equipment

Depending upon the need, ovens can vary from singly insulated containers with simple thermostatic control to doubly insulated structures with proportional temperature control.

TCXOs are used for applications where the limitations of oven-controlled oscillators cannot be tolerated. Ovens require power, which must be supplied to the equipment. In backpack equipment, for example, this would either increase the weight of the batteries carried or reduce their operating life. For this application, the added weight and size of an oven would also be disadvantageous. To maintain its stability, an oven-mounted standard must be kept continuously at the controlled temperature. The power cannot be turned off without introducing strains that may cause subsequent frequency offsets. If the power is turned off, there is a relatively long warm-up period before the standard reaches its final frequency. In many applications, the chance of such delays is not acceptable. For such applications, TCXOs can often be used.

A TCXO requires as good an oscillator design as possible, with a crystal having low temperature drift (AT cut). Compensation has been attempted using temperature-sensitive capacitors and temperature-sensitive mechanical forces on the crystal. However, the approach preferred at present is the use of a varactor, the voltage across which is changed by a network of temperature-sensitive resistors (thermistors). Figure 7.95 is the schematic diagram of such a network containing one variable-capacitance diode and two thermistors [7.38]. Figure 7.96 illustrates the improvement that is available by compensation. More complex circuits may be

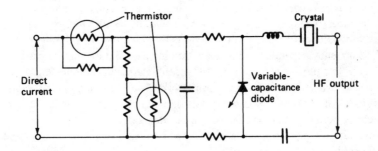

Figure 7.95 Temperature-compensation circuit containing one variable-capacitance diode and two thermistors. (*After [7.16].*)

used to achieve improved compensation. For high-precision compensation, each oscillator must be separately measured and compensated. This is practical with computer control of the process. It has been found possible to reduce the frequency deviation to 1 ppm over a temperature range of –30 to +60°C using the technique shown in the figure. With more complex designs, a change of 1 in 10^7 from –30 to +50°C has been achieved.

TCXOs are available from several vendors. Table 7.14 shows the typical specification for a TCXO with 1-ppm compensation over the temperature range

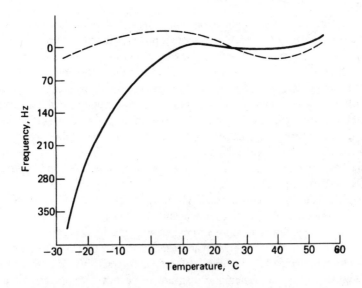

Figure 7.96 Effects of compensation on the frequency of a 25-MHz crystal unit by using an inductor-capacitor-thermistor network. (*After [7.16].*)

TABLE 7.14 Specifications for a Typical Temperature-Compensated Crystal Oscillator (After [7.4])

Temperature stability	-20 to $+70°C$
Center frequency	10 MHz
Output level	1 V rms into 1000 Ω
Supply voltage	15 V $\pm$ 5%
Current	5 to 15 mA
Aging	5×10^{-7} per year, 3×10^{-9} per day average
Short-term stability	1×10^{-9} per second under constant environment
Stability versus supply	2×10^{-8} per 1% change in supply voltage
Frequency adjustment	Range sufficient to compensate for 5 to 10 years of crystal aging; setable to $<1 \times 10^{-7}$
Electronic tuning	Permits remote frequency adjustment or locking onto an external frequency source

shown. Because of the compensating network, phase noise is poorer in TCXOs than in other standards. Also there can be wider variations in frequency over the medium term when there are sudden temperature changes. The crystal and the compensation network can have transient temperature differences that cause a transient reduction in compensation.

For high-performance synthesizers without the limitations that lead to the use of TCXOs, we should use a high-performance crystal oscillator in a proportional oven. Table 7.15 gives the specifications for such a standard. Characteristics of interest for accurate timing have been compared for the various standard types in Table 7.16.

TABLE 7.15 Specifications for a High-Power Oscillator Used in High-Performance Standard Rohde and Schwarz XSF (Courtesy of Rohde and Schwarz.)

Frequency	5 MHz
Crystal	5 MHz fifth overtone
Frequency error	1×10^{-10} per day after 30 days of continuous operation; 5×10^{-10} per day after 5 days of operation

Short-term stability	Averaging time, sec.	Stability
1.5×10^{-10}	10^{-3}	
1.5×10^{-11}	10^{-2}	
$5 \times 10_{-12}$	10^{-1}	
5×10^{-12}	10^{0}	
5×10^{-12}	10^{1}	
1×10^{-11}	10^{2}	

TABLE 7.16 Relative comparison of Time-Frequency Standards

Standard type	Comparative cost	Approximate accuracy/ stability	Potential delay between two clocks	
			10 min	2 h
Uncompensated quartz crystal oscillator	1 (ref)	$\pm 1 \times 10^{-5}$	± 12 ms	± 144 ms
Temperature-compensated quartz crystal oscillator	5–10	$\pm 2 \times 10^{-6}$	± 2.4 ms	± 28.8 ms
Simple oven-controlled quartz crystal oscillator	5–10	$\pm 1 \times 10^{-7}$	$\pm 120\ \mu s$	± 1.44 ms
High-grade oven-controlled quartz crystal standard	100–200	$\pm 1 \times 10^{-8}$	$\pm 12\ \mu s$	$\pm 144\ \mu s$
Rubidium standard	500–800	$\pm 3 \times 10^{-11}$	± 36 ns	± 432 ns
Cesium standard	900–1200	$\pm 1 \times 10^{-12}$	± 7 ns	± 86 ns

7.15 References

7.1. V. F. Kroupa, *Frequency Synthesis: Theory, Design and Applications*, Wiley, New York, 1973.

7.2. J. Gorski-Popiel, *Frequency Synthesis, Techniques and Applications*, IEEE Press, New York, 1975.

7.3. V. Manassewitsch, *Frequency Synthesizers, Theory and Design*, 2d ed., Wiley, New York, 1980.

7.4. U. L. Rohde, *Digital PLL Frequency Synthesizers*, Prentice-Hall, Englewood Cliffs, NJ, 1983.

7.5. J. Smith, *Modern Communication Circuits*, McGraw-Hill, New York, 1986.

7.6. A. B. Przedpelski, "Analyze, Don't Estimate Phase-Lock-Loop Performance of Type 2 Third Order Systems," *Electron. Design*, no. 10, p. 120, May 10, 1978.

7.7. A. B. Przedpelski, "Optimize Phase-Lock-Loop to Meet Your Needs—or Determine Why You Can't," *Electron. Design*, no. 19, p. 134, September 13, 1978.

7.8. A. B. Przedpelski, "Suppress Phase Locked Loop Sidebands without Introducing Instability," *Electron. Design*, no. 19, p. 142, Sept. 13, 1979.

7.9. U. L. Rohde, "Modern Design of Frequency Synthesizers," *Ham Radio*, p. 10, July 1976.

7.10. D. B. Leeson, "A Simple Model of Feedback of Oscillator Noise Spectrum," *Proceedings of the IEEE*, 1966, p 329-330.

7.11. D. A. Howe, "Frequency Domain Stability Measurements: A Tutorial Introduction," NBS Technical Note 679, March 1976.

7.12. M. C. Fischer, "Frequency Stability Measurement Procedures," Eighth Annual Precise Time and Time Interval Applications and Planning Meeting, December 1976.

7.13. U. L. Rohde, C. R. Chang, and J. Gerber, "Parameter Extraction for Large Signal Noise Models and Simulation of Noise in Large Signal Circuits Like Mixers and Oscillators," *Proceedings of the 23rd European Microwave Conference*, Madrid, Spain, Sept. 6-9,1993.

7.14. C. R. Chang, "Mixer Noise Analysis Using the Enhanced Microwave Harmonica," *Compact Software Transmission Line News*, June 1992, Vol. 6. No. 2, pp. 4-9.
7.15. T. Antognetti and G. Massobrio, *Semi-Conductor Device Modeling with SPICE*, McGraw-Hill, New York, NY, p. 91, 1988.
7.16. R. J. Hawkins, "Limitations of Nielsen's and Related Noise Equations Applied to Microwave Bipolar Transistors, and a New Expression for the Frequency and Current Dependent Noise Figure," *Solid State Electronics*, 20, pp. 191-196, 1977.
7.17. T.-H. Hus and C. P. Snapp, "Low Noise Microwave Bipolar Transistor Sub-Half-Micrometer Emitter Width, *IEEE Transactions Electron Devices*, Vol. ED-25, pp. 723-730, June 1978.
7.18. R. A. Pucel and U. L. Rohde, "An Accurate Expression for the Noise Resistance R_n of a Bipolar Transistor for Use with the Hawkins Noise Model," *IEEE Microwave and Guided Wave Letters*, Vol. 3. No. 2, February 1993, pp. 35-37.
7.19. G. Vendelin, A. M. Pavio, and U. L. Rohde, *Microwave Circuit Design: Using Linear and Nonlinear Techniques*, John Wiley and Sons, New York (1990).
7.20. R. A. Pucel, W. Struble, R. Hallgren, and U. L. Rohde, "A General Noise De-embedding Procedure for Packaged Two-Port Linear Active Devices," *IEEE Transactions on Microwave Theory and Techniques*, Vol. 40, No. 11, Nov 1993, pp. 2013-2024.
7.21. U. L. Rohde, "Improved Noise Modeling of GaAs FETS Parts 1 and ll: Using an Enhanced Equivalent Circuit Technique," *Microwave Journal*, November 1991, p. 87 and December 1991, p. 87.
7.22. V. Rizzoli, F. Mastri, and C. Cecchefti, "Computer-Aided Noise Analysis of MESFET and HEMT Mixers," *IEEE Transactions on Microwave Theory and Techniques*, Vol. MTT-37, pp. 1401-1410, Sept 1989.
7.23. V. Rizzoli and A. Lippadni, "Computer-Aided Noise Analysis of Linear Multiport Networks of Arbitrary Topology," *IEEE Transactions on Microwave Theory and Techniques*, Vol. MTT-33, Dec 1985, pp. 1507-1512.
7.24. V. Rizzoli, F. Mastri, and D. Masotti, "General-Purpose Noise Analysis of Forced Nonlinear Microwave Circuits," *Military Microwave*, 1992.
7.25. V. Rizzoli, F. Mastri, and C. Cecchetti, "Computer-aided Noise Analysis of MESFET and HEMT Mixers," *IEEE Transactions on Microwave Theory Techniques*, Vol. MTT-37, pp. 1401-1410, Sept 1989.
7.26. D. Scherer, "Design Principles and Test Methods for Low Phase Noise RF and Microwave Sources," RF and Microwave Measurement Symposium and Exhibition, Hewlett-Packard.
7.27. D. B. Leeson, "A Simple Model of Feedback Oscillator Noise Spectrum," *Proc. IEEE*, vol. 54, p. 329, Feb. 1966.
7.28. W. A. Edson, *Vacuum Tube Oscillators*, Wiley, New York, 1953.
7.29. W. Herzog, *Oszillatoren mit Schwingkristallen*, Springer, Berlin, 1958.
7.30. D. Firth, *Quartz Crystal Oscillator Circuits*, Design Handbook, Magnavox Co., Ft. Wayne, Ind., 1965; available from NTIS as AD 460377.
7.31. M. R. Frerking, *Crystal Oscillator Design and Temperature Compensation*, Van Nostrand Reinhold, New York, 1978.
7.32. J. Markus, *Guidebook of Electronic Circuits*, McGraw-Hill, New York, 1960.
7.33. E. A. Gerber and R. A. Sykes, "State of the Art Quartz Crystal Units and Oscillators," *Proc IEEE*, vol. 54, p. 103, Feb. 1966.
7.34. R. Harrison, "Survey of Crystal Oscillators," *Ham Radio*, vol. 9, p. 10, Mar. 1976.
7.35. M. Martin, "A Modern Receive Converter for 2 m Receivers Having a Large Dynamic Range and Low Intermodulation Distortion," *VHF Commun.*, vol. 10, no. 4, p. 218, 1978.
7.36. M. M. Driscoll, "Two-Stage Self-Limiting Series Mode Type Quartz Crystal Oscillator Exhibiting Improved Short-Term Frequency Stability," *Proc. 26th Annual Frequency Control Symp.*, 1972, p. 43.

7.37. U. L. Rohde, "Effects of Noise in Receiving Systems," *Ham Radio*, vol. 10, p. 34, Nov. 1977.

7.38. B. Neubig, "Extrem Rauscharme 96-MHz-Quarzoszillator for UHF/SHF," presented at the 25th Weinheimer UKW-Tagung, Sept. 1980 (VHF Communications, 1981).

7.39. M. Sera, "Chip Set Addresses North American Digital Cellular Market, *RF Design*, Intertec Publishing, Overland Park, KS, March 1994.

7.40. T. Luich, et. al., "A 2.2 GHz PLL Frequency Synthesizer," *Proceedings of the Portable by Design Conference*, pp. 68-71, Santa Clara, February 1994.

7.41. F. Gardner, *Phaselock Techniques*, p. 78, Wiley, New York, 1966.

7.42. R. Best, *Phase-Locked Loops: Theory, Design and Applications*, pp. 7-14, McGraw-Hill, New York, 1984.

7.43. M. O'Leary, "Practical Approach Augurs PLL Noise in RF Synthesizers," *Microwaves & RF*, pp. 185-194, September 1987.

7.44. D. Byrd, G. Tietz, and C. Davis, "Guaranteed Zero Deadband Phase Comparator/Charge Pump with Minimized Phase Offset," U.S. patent #4,814,726 1989.

Additional Suggested Reading

"All About Phase Noise in Oscillators: Part I," *QEX*, Dec. 1993.

"All About Phase Noise in Oscillators: Part II," *QEX*, Jan. 1994.

"All About Phase Noise in Oscillators: Part III," *QEX*, Feb. 1994.

"Analysis and Optimization of Oscillators for Low Phase Noise and Low Power Consumption," *RF Design*, p. 70, Mar. 1995.

"Computer-Aided Design of Phase-Locked Loops," presented at the RF Technical Exposition, Anaheim, Calif., Jan 23, 1985.

"Crystal Oscillator Provides Low Noise," *Electronic Design*, Oct. 11, 1975.

"Designing and Optimizing Low Phase Noise Oscillators Using Harmonic Balance Simulators and Advanced Parameter Extraction," Session B3-2, 2nd IEEE Joint Chapter Workshop in conjunction with M94 CAE, Modeling and Measurement Verification, Wembley Conference Centre, London, UK, Oct. 24, 1994.

"Designing SAW and DRO Oscillators Using Nonlinear CAD Tools," 1995 IEEE International Frequency Control Symposium, San Francisco, May 30, 1995.

"Developing Non-Linear Oscillator Models Using Linear Design Tools," *Proceedings: RF Expo East*, Boston, Nov. 10,1986.

"Eight Ways to Better Radio Receiver Design," *Electronics*, Feb. 20, 1975.

"Evaluating Noise Sideband Performance in Oscillators," *Ham Radio* magazine, Oct. 1978.

"Low-Noise Frequency Synthesizers Fractional-N Phase-Locked Loops," SOUTHCON/81, Jan. 1981.

"Mathematical Analysis and Design of an Ultra-Low Noise 100 MHz Oscillator with Differential Limiter and Its Possibilities in Frequency Standards," *Proceedings of the 32nd Annual Symposium on Frequency Control*, Ft. Monmouth, NJ, May 1978.

"Modern Design of Frequency Synthesizers," *Ham Radio* magazine, July 1976.

"Noise Prediction in Oscillators," *Proceedings: RF Technology 87*, Anaheim, Calif., Feb 11, 1987.

"Stable Crystal Oscillators," *Ham Radio* magazine, June 1975.

"The Accurate Simulation of Oscillator and PLL Phase Noise in RF Sources," Wireless '94 Symposium, Santa Clara, Feb. 1994.

Chapter 8
Demodulation and Demodulators

8.1 Introduction

The function of the receiver is to recover the original information that was used to modulate the transmitter. This process is referred to as *demodulation*, and the circuits that perform the recovery are called *demodulators*. The term *detector* is also used, and the demodulators in single superheterodyne receivers are sometimes called *second detectors*. However, today the term detector is seldom used this way.

Because of thermal, atmospheric, and man-made interference as well as transmission and circuit distortions, the demodulated signal is a distorted version of the modulating signal and is corrupted by the addition of noise. In the case of analog demodulators, we wish to minimize the distortion and noise so that the output signal waveform is as close to the original waveform as possible. In the case of digital demodulation, the objective is to produce a digital output of the same type as that at the transmitter input, with as few errors as possible in the digital output levels, with the correct signaling rate, and without addition or deletion of any symbols. Consequently, the performance measures for analog- and digital-signal demodulators differ. Often the digital demodulator is located separately in a modem, which also incorporates the digital modulator for a companion transmitter. In this chapter, we treat demodulators for analog and digital signals separately.

8.2 Analog Demodulation

Modulation types for analog-modulated waves include: AM and its various derivatives (DSB, SSB, VSB, etc.), angle modulation (PM and FM), and various sampled pulse systems. Demodulation will be covered in that order, although only a few of the pulse systems are of interest below microwave frequencies.

8.2.1 AM

An AM signal comprises an RF sinusoid whose envelope varies at a relatively slow rate about an average (carrier) level. Any type of rectifier circuit will produce an output component at the modulation frequency. Figure 8.1 illustrates two of the simple diode rectifier circuits that may be used along with idealized waveforms.

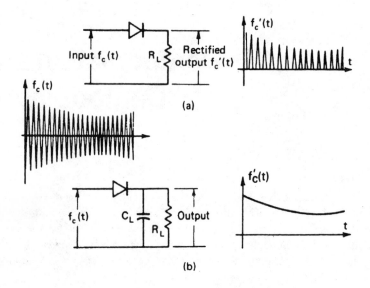

Figure 8.1 AM demodulators with idealized waveforms: (*a*) average demodulator and (*b*) envelope demodulator.

The average output of the rectifier of Figure 8.1*a* is proportional to the carrier plus the signal. The circuit has, however, a large output at the RF and its harmonics. A low-pass filter is therefore necessary to eliminate these components. If the selected filter incorporates a sufficiently large capacitor at its input, the effect is to produce a peak rectifier, with the idealized waveforms of Figure 8.1*b*. In this case the demodulated output is increased from an average of a half sine wave (0.637 peak) to the full peak, and the RF components are substantially reduced. The peak rectifier used in this way is often referred to as an *envelope detector* or *demodulator*. It is the circuit most frequently used for demodulating AM signals. The *balanced demodulator* offers some performance improvements over the simple envelope detector. Common balanced demodulator circuit implementations are shown in Figure 8.2.

Generally diodes are used as AM demodulators; however, the important requirement is for a nonlinear response, especially one with a sharp cutoff. Some tube demodulators have used the nonlinearity of the grid-cathode or plate-cathode characteristics. Bipolar transistor demodulators can use the nonlinearity of the base-emitter or base-collector characteristics. Analogous nonlinearities in FETs can also be used for demodulation.

Real devices do not have responses that are perfectly linear with a sharp cutoff. When the input voltage is small, the rectification takes place as a result of the square-law variation of the diode (or other device). Figure 8.3 shows the difference

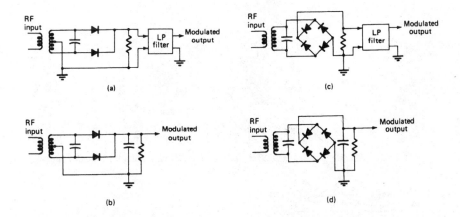

Figure 8.2 Balanced AM demodulator circuits: (*a*) average type, balanced grounded input, (*b*) envelope type, similar to *a*, (*c*) average type, using a diode bridge, and (*d*) envelope type, similar to *c*.

between average currents for devices that are essentially linear with sharp cutoff and for those where the cutoff is gradual, as in real devices. In the second case, the demodulated output has a somewhat distorted waveform.

As with the mixer, the principle of the square-law demodulator is relatively easy to analyze. Let us assume a diode connected as shown in Figure 8.1*a*, with a diode characteristic expressed as:

$$i_d = k_0 + k_1 V_d + k_2 v_d^2 \qquad (8.1)$$

Let $v_d = A [1 + ms(t)] \cos 2\pi f_c t$. This implies a load resistance sufficiently small that most of the voltage is developed across the diode. We may then write:

$$i_d = k_0 + k_1 A[1 + ms(t)] \cos 2\pi f_c t + k_2 A^2 [1 + ms(t)]^2 \cos^2 2\pi f_c t \qquad (8.2a)$$

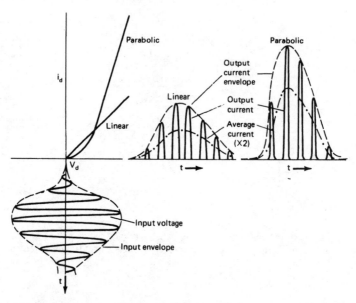

Figure 8.3 Demodulation with linear and parabolic demodulator output voltage characteristics.

$$i_d = k_0 + k_1 A[1 + ms(t)] \cos 2 \pi f_c t + \frac{k_2 A^2}{2}$$

$$+ \left\{ k_2 A^2 ms(t) + \frac{k_2 A^2}{2} m^2 s^2(t) \right\}$$

$$+ \frac{k_2 A^2}{2} [1 + ms(t)]^2 \cos(4 \pi f_c t) \tag{8.2b}$$

In Equation (8.2b) only the terms in braces on the right-hand side are modulation terms, the first being the desired signal and the second being the distortion. The other terms are direct current or RF. In order to keep distortion low, the peak modulation index m must be kept small.

As R_L becomes larger, the output current is reduced and the response becomes more linear. For small inputs, the resulting current can be expanded as a Taylor series in the input voltage. The second-order term still gives rise to demodulation (and distortion), and the higher-order terms generally introduce additional distortion. When the input becomes large, its negative swings cut off the current, but if

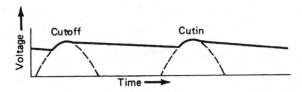

Figure 8.4 Representation of a small segment of the envelope.

the load resistor is sufficiently large, the linear response of Figure 8.3 is approximated. The output voltage still has large RF components that must be filtered. For that reason most diode demodulators use the circuit of Figure 8.1*b*.

When the carrier frequency is much higher than the highest modulation frequency, a small section of the output waveform will appear, as shown in Figure 8.4. Here the dashed line represents the voltage applied to the circuit input, and the heavy line represents the capacitor voltage. During each cycle, the capacitor is charged to the peak of the input voltage, and at some time after the peak of the cycle, the diode cuts off. The capacitor then discharges until the point in the next cycle when the input voltage reaches the capacitor voltage level. Then the diode begins again to conduct. The voltages of cutoff and cutin depend on the values of the load resistor and capacitor, and to a much lesser extent on the diode resistance during conduction. A factor that is neglected in the figure is the contact voltage of the diode, which sets the diode conduction point at a small negative voltage rather than zero.

The diode may be considered a switch with internal resistance. During the nonconduction interval, the switch is open, and the capacitor discharges at a rate determined by the capacitor and load resistor time constant. During the conduction interval, the switch is closed, and the circuit is driven by the input voltage modified by the voltage divider effect of the diode resistor and load resistor. The charging time constant is assumed to be small. The switch opens when the current through the diode would otherwise reverse. This occurs when the rate of change of the input voltage applied to the capacitor just equals the rate of change that would exist across the capacitor if the diode were opened. A straightforward analysis leads to a transcendental equation for the angle of conduction, which may be solved by the method of successive approximations. Because of the diode conductance and the load discharge, the output voltage reaches a peak somewhat below the peak input and discharges a small amount before the next cycle. The average voltage is close to the peak, and its ratio to the input peak voltage is referred to as the *rectifier efficiency*, which in most demodulators is 90 percent or above. The variations are close to a sawtooth voltage at the carrier frequency and represent residual RF and its harmonics, which are eliminated by the response of subsequent amplifiers.

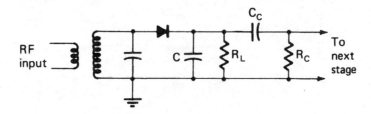

Figure 8.5 Schematic diagram of an envelope demodulator with a high-pass coupling circuit to the next stage.

Depending on the circuit requirements, these high frequencies at the demodulator output are typically 20 to 40 dB below the rectified carrier level.

The average output from the demodulator is proportional to the time-varying envelope of the input wave. In the previous figures it appears constant because the RF is assumed much higher than the maximum modulating frequency. If the time constant of the demodulator is increased, the residual RF output is decreased. However, if we increase it too much, the output will not be able to follow the higher modulating frequency changes in the envelope. It is necessary that $2\pi f R_L C$ be maintained low enough that the discharge can follow the rate of decrease in the input level at these frequencies. For example, a time constant that is 10 μs yields $2\pi f R_L C = 0.188$ at a modulating frequency of 3 kHz, resulting in an output reduction $[1 + (2\pi f R_L C)^2]^{-1/2}$ to 0.983, or 0. 15 dB. At 10 kHz these figures reduce to 0.847, or 1.45 dB. Thus, the extent to which the time constant can be increased depends on the specified requirements for response at higher modulating frequencies.

In addition to the distortion introduced by the response of the RC filter network, an additional nonlinear distortion occurs at high modulation indexes because of the impedance of the subsequent circuit. This circuit is usually coupled by a series capacitor and resistor combination (Figure 8.5). For speech reception, this high-pass arrangement is useful to eliminate residual low-frequency components (hum) as well as the direct current from the carrier. At low frequencies, the demodulator load impedance acts essentially as described previously, except for the extra capacitance of the coupling capacitor. However, at the modulating frequencies the impedance represents the shunt combination of the load resistor and the coupling resistor.

Figure 8.6 illustrates the result of this type of circuit. In this figure there are a series of diode current versus voltage curves displaced from the origin by various dc levels corresponding to the direct current demodulated at various peak carrier levels. The straight line *B* represents the load line ($I = E/R_L$). The intersections of the curves with this line represent the dc operating points for various input peak voltages V_I. At the modulating frequency, however, the load no longer is R_L, but the lower value resulting from the shunting by R_C. The straight line *C* has the slope

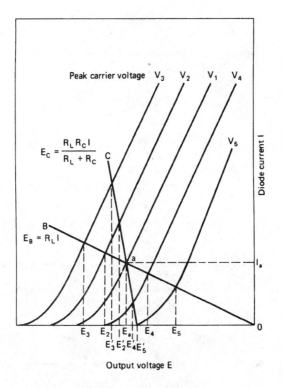

Figure 8.6 Voltage-current curves for an envelope demodulator with load.

of this line and is passed through the peak carrier voltage curve V_1. If we are receiving at this carrier level and the signal is modulated, the output current will follow curve C (which is determined by the lower impedance resulting from the shunt load). If the peak-to-peak modulation is from V_2 to V_4 or from V_3 to V_5, the output voltage is reduced to outputs E'_2, E'_4, E'_3, and E'_5. They will also be substantially distorted. If the input should swing below V_5, the diode will cut off and maintain a voltage of E'_5. The result is a clipping of the modulation waveform whenever the modulating voltage drops below the input voltage corresponding to V_5.

If the diode characteristics were linear, corresponding to a resistance $R_{d'}$, then for sinusoidal modulation, the modulation index at which distortion begins is:

$$m_d = \frac{(R_L + R_C)\, R_{d'} + R_L\, R_C}{(R_L + R_C)\,(R_L + R_{d'})} \tag{8.3}$$

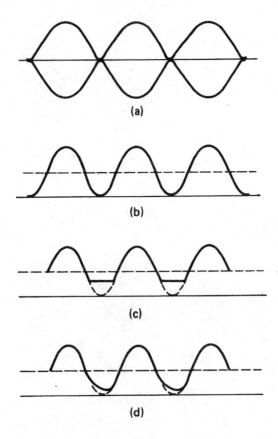

Figure 8.7 Nonlinear distortions in an envelope detector: (*a*) envelope of modulated wave; (*b*) diode output voltage, no distortion; (*c*) diode output voltage, negative peaks clipped; and (*d*) diode output voltage, diagonal clipping.

A greater modulation index would cause clipping of the low modulation levels. Actual diode characteristics are not linear, so that this relationship is not precise, nor is the shunt impedance Z_c always resistive. If the impedances are such that the efficiency is high over the range of amplitudes represented by the modulation envelope, distortion will be small if $|Z_c|/R_L \leq m$, where $|Z_c|$ is the effective magnitude of impedance at the modulating frequency. Equation (8.3) reduces to this form when $R_d' \ll R_L$ and $|Z_c| = R_C$. Although clipping is sharp when the impedance is resistive, if there is an angle associated with Z_c, the result is a diagonal clipping. Figure 8.7 illustrates these various distortions.

We should note that because the envelope demodulator takes power from the input circuit, it therefore presents a load to that circuit. The input impedance to the carrier is simply R_L divided by the demodulator efficiency; to the sideband frequency it is Z_c divided by the efficiency, where Z_c is the value at the baseband frequency corresponding to the particular sideband. The angle of the sideband impedance is the same as that of the baseband frequency for the USB and the negative of this angle for the LSB. The demodulator impedance is of importance only if the driving-source impedance is not substantially lower than the demodulator impedance. In this case, additional linear distortion of the signal is likely to occur.

A. AM interference

The envelope of a signal may be distorted by the filter circuits that provide receiver selectivity. Both in-phase and quadrature distortion may occur in the RF waveform, which can affect the envelope. To avoid quadrature distortion, we try to tune the carrier to the center frequency of the filter, and design our bandpass filters with complex zeros and poles distributed symmetrically about the filter center frequency. Then the demodulated signal is distorted linearly in amplitude and phase by the characteristics at the USB frequency offset from the carrier, and no quadrature component is introduced. Standard filter designs discussed earlier are based on seeking such characteristics.

However, with simple resonators, the bandpass response tends to be based on the variable $f/f_0 - f_0/f$, which is only approximately proportional to $f - f_0$. Moreover, coupling variations may introduce terms that vary as a positive or negative power of f. These effects cause dissymmetries in filters that can introduce quadrature distortion. When the signal bandwidth is much smaller than the center frequency, the distortion is negligible. When the bandwidth is a substantial fraction of the carrier, the distortion can become substantial. In this case, it is necessary to modify the filter design to achieve the proper zero and pole relationships. These effects can be analyzed using standard circuit analysis theory.

The input to the demodulator includes not only the desired AM signal, but also interference [8.1]. The envelope demodulator produces an output proportional to the overall envelope of the resulting sum of desired and undesired signals. The envelope of the sum of signals is affected not only by the levels of the individual envelopes, but also by the phases. While the phase of the desired AM signal remains constant, the phase of the interferers usually varies. Thermal noise of the receiver system has a gaussian distribution, with atmospheric noise being more or less impulsive, the impulses tending to bunch and man-made noise being often periodic with impulsive envelope. The rates of variation in amplitude and phase of the interferers are constrained by the filter bandwidth. Thermal noise becomes colored gaussian with Rayleigh envelope distribution and random phase. The rates of change of the phase and envelope are limited by the receiver bandwidth. Similarly, short impulses of atmospheric or man-made origin are broadened in the envelope

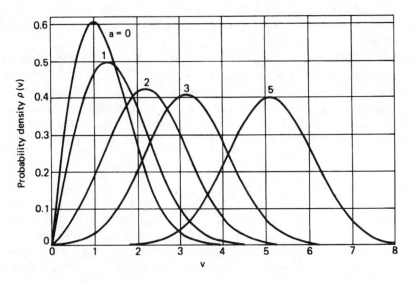

Figure 8.8 Probability density of an envelope sine wave with amplitude P and added noise I_n. (*From [8.2]. Reprinted with permission.*)

by the receiver filter, and the rate of angle modulation is constrained. Longer-duration impulsive noise usually controls the resultant envelope and phase so long as it is stronger than the signal. This causes interruptions in the desired demodulated waveform, leading to loss of information and, in the case of audio or visual signals, unpleasant outputs that can tire the user. These effects can be reduced more or less effectively by noise limiters or blankers that reduce the impulsive energy.

Demodulation is essentially a nonlinear process on the received signal. We would like the output to have a high S/N, little signal distortion, and the minimum subjective effect from whatever noise is present. In envelope detection of AM with added thermal noise, it is customary to begin with the examination of the combined envelope of the unmodulated carrier and the gaussian noise. The envelope distribution as a function of S/N has been calculated (References [8.2] and [8.3]). Figures 8.8 and 8.9 show the density and distribution of the combined envelope for various values of input S/N.

Using this distribution, it can be shown that the mean envelope is:

$$\overline{R} = \left(\frac{\pi \, \psi_0}{2}\right)^{1/2} \exp\left(\frac{-x}{2}\right)\left[(1 + x)\, I_0\left(\frac{x}{2}\right) + x\, I_1\left(\frac{x}{2}\right)\right] \tag{8.4}$$

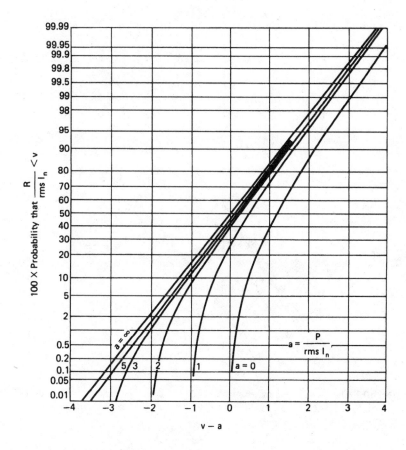

Figure 8.9 Distribution function corresponding to Figure 8.8. (*From [8.2]. Reprinted with permission.*)

where ψ_0 is the noise power, x is the input S/N, and $I_0(z)$ and $I_1(z)$ are Bessel functions of imaginary argument. The mean-square envelope is:

$$\overline{R^2} = 2\,\psi_0\,(1+x) \tag{8.5}$$

This analysis can be extended to modulated waves, demodulators that produce output shapes at powers of the envelope other than the first power, and for various narrow bandpass filter shapes preceding the demodulator [8.4]. Figure 8.10 shows the linear case, corresponding to the envelope demodulator, for sinusoidal modu-

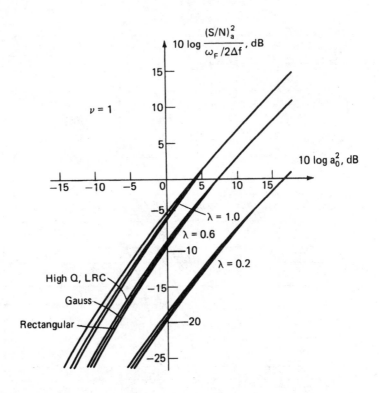

Figure 8.10 Output S/N versus input C/N for an envelope detector, with the carrier sinusoidally modulated at various indicies λ and for several bandpass filter shapes. (*From [8.4]. Courtesy of D. Middleton.*)

lation at various indexes λ_i and different bandpass filter shapes. The parameter $a_0{}^2$ is the C/N, ω_F is a factor needed to normalize the input noise power for equal white-noise densities, and Δf is the baseband cutoff frequency, assumed to be abrupt.

For large values of C/N, the S/N varies linearly, whereas for small values of C/N, S/N varies as a square law (slope of logarithmic curve = 2). For large values of C/N, when the normalizations are carried out, the output S/N (for 100 percent modulation) is 3 dB greater than C/N′, where N′ is the noise power in $2\Delta f$. For small values of C/N, the signal is suppressed; however, it is already useless for most communications purposes. When the signal is very small, only the noise envelope produces significant output. As the carrier increases, the noise output also increases, but at a slower rate. The noise output ultimately increases by 3.7 dB over the no-carrier state. This phenomenon of noise rise is obvious when tuning in a carrier on standard AM receivers. Figure 8.11 illustrates this phenomenon.

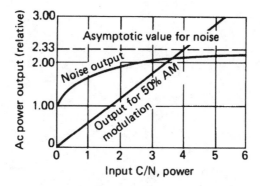

Figure 8.11 Modulation and noise output levels for an envelope demodulator with constant input noise level and increasing carrier level. (*After [8.5]. Reprinted with permission.*)

When the carrier is strong enough to produce a good S/N, the spectrum of the demodulated signal component is the spectrum of the original modulation, modified by the selective filtering that has occurred. If the filters are properly symmetrical and in tune with the signal, the effect is to modify the spectrum in accordance with the filter response at the equivalent sideband offset. When the filtering is off tune, the in-phase component is reduced, and a small quadrature component is added. The quadrature component has only a small effect on the amplitude so long as it remains small compared to the carrier. When the dissymmetry in the filtering is excessive, the quadrature component can produce substantial distortion. In specific cases, the effects can be evaluated using Fourier analysis.

The noise components for high C/N are colored gaussian noise, whose output frequency spectrum is 0.707 of the rss of the relative receiver amplitudes at the USB and LSB frequencies. For symmetrical filtering this is simply the USB filter response, reduced by 3 dB at each frequency. The noise is not difficult to calculate for detuning or dissymmetrical filters. In most cases, proper system design allows the symmetrical filter calculation to be made if the output noise spectrum is needed. When C/N is not high, the spectrum of the noise becomes much more complex. However, conditions for poor C/N should be rare in good system designs. Where possible, measured performance rather than calculations should be used.

B. Transmission distortion

The principal transmission distortion affecting AM demodulation is multipath. If the multipath components have short delays, compared with the reciprocal bandwidth, the RF phase difference among the sideband components does not vary greatly and all of the resultant sideband components tend to increase and decrease simultaneously, usually at a rate that is slow compared to the modulation. We may combat this type of fading by the AGC design. Then when the signal fades, the

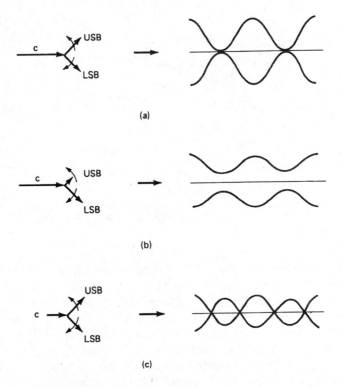

Figure 8.12 Effects of selective fading on demodulation of an AM signal by an envelope demodulator: (*a*) undistorted signal, (*b*) USB reduced, and (*c*) carrier reduced.

output tends to remain constant, although the background noise rises and falls in inverse relation to the fading depth.

When the multipath delay difference becomes substantial, different frequencies within the band experience different fading and phase shifting. This "selective" fading can cause sidebands to become unbalanced, resulting in changes in amplitude and phase of the corresponding components of the demodulated signal. The principal applications of AM are for voice and music transmission. The effect of slow variations of the output frequency response resulting from selective sideband fading, while detectable, is usually tolerable to the user. However, occasionally the carrier fades without significantly affecting the sidebands. This results in an increase in modulation percentage and can cause severe distortion at negative modulation peaks.

Figure 8.12 illustrates this effect. The undistorted envelope is shown in Figure 8.12*a*. When the fading causes reduction of one sideband, the result is envelope distortion, as shown in Figure 8.12*b*. When the fading reduces the carrier, the envelope becomes severely distorted, as in Figure 8.12*c*. This type of distortion is of particular importance for AM on HF, where the objective is long-range transmission and severe multipath is often present. In the broadcast band, this occurs at night on distant stations, beyond the commercial interest range of most broadcast stations.

A technique called *exalted* or *enhanced carrier* was devised to combat the effect of selective fading on AM signals. Because the carrier is separated from the useful voice sidebands by more than 100 Hz, it may be filtered from the remaining signal, clipped, and amplified. With a suitable phase shift to compensate for delays in the circuits, the amplified carrier is added to the original signal before it is applied to the demodulator. The increased resultant carrier reduces the modulation index so that input carrier fading cannot cause the negative modulation index to approach 100 percent. The narrow-band filter response carrier continues even during short periods of complete carrier fading.

C. Coherent demodulation

AM signals may also be demodulated by using a *coherent* or *synchronous demodulator* (at one time referred to as a *homodyne detector*). This type of demodulator circuit uses a mixer circuit, with an LO signal synchronized in frequency and phase to the carrier of the AM input (Figure 8.13). The synchronous oscillation may be generated by the technique described previously for exalted carrier demodulation, or an oscillator may be phase-locked to the carrier. These techniques are illustrated in Figure 8.14. Another alternative is the use of the Costas loop, which is described later in this chapter.

The coherent demodulator translates the carrier and sidebands to baseband. As long as the LO signal is locked to the carrier phase, there is no chance of overmodulation, and the baseband noise results only from the in-phase component of the noise input. Consequently, the noise increase and S/N reduction, which occur at low levels in the envelope demodulator, are absent in the coherent demodulator. The recovered carrier filtering is narrow-band, so that phase lock can be maintained to C/N levels below useful modulation output levels. This type of circuit, while better than an envelope demodulator, is not generally used for AM demodulation because it is far more complex.

The AM signal can also be demodulated as an SSB signal, using either sideband and ignoring the other sideband. Both sidebands may be independently demodulated, and the better one is selected in a switching diversity technique. Such demodulators are useful if the receiver is designed for SSB or ISB signals, but are not used normally since the principal reason for AM transmission is the simplicity of the receivers in broadcast applications.

8.2.2 DSB Demodulation

DSB suppressed-carrier (DSB-SC) modulation cannot be demodulated satisfactorily by an envelope demodulator because the envelope does not follow the modulation waveform, but is a full-wave rectified version it. Enhanced carrier tech-

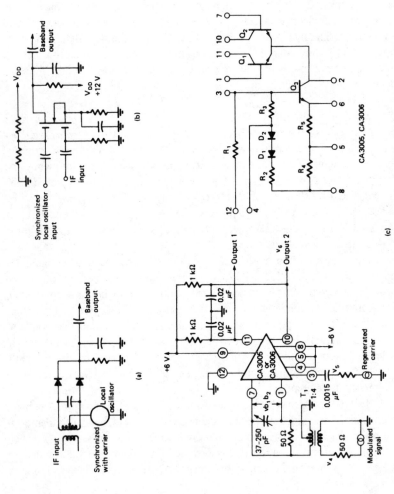

Figure 8.13 Coherent demodulator types: (*a*) diode, (*b*) dual-gate MOSFET, and (*c*) bipolar integrated circuit. (*Courtesy of RCA.*)

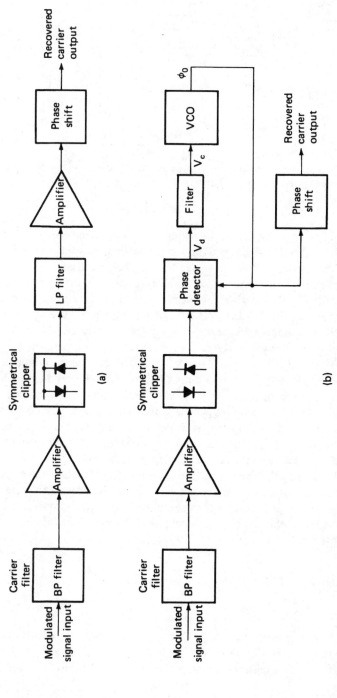

Figure 8.14 AM carrier recovery: (*a*) filter, clipper, and amplifier, and (*b*) filter, clipper, and PLL.

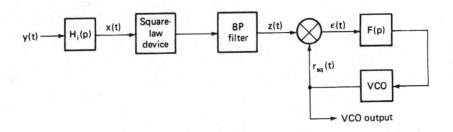

Figure 8.15 Carrier recovery for a coherent demodulator using frequency doubling. (*From* [8.6]. *Reprinted by permission of the authors.*)

niques could be used, but coherent demodulation is the commonly used demodulation method. The DSB-SC signal has a constant frequency within the envelope, but every axis crossing of the envelope causes a 180° phase change, so that there is no component of carrier frequency for locking the LO. This problem is solved by passing the signal through a nonlinear device to produce a component at the double frequency (Figure 8.15), where the phase change at envelope crossover is 360°. A filter at double frequency can separate this second harmonic. Subsequently division by two produces a signal synchronous in frequency with the missing carrier, but either in phase or 180° out of phase with the missing carrier. This phase ambiguity produces an output that is either the original modulating signal or its negative. In audio applications such a reversal is of no consequence.

A clever circuit that accomplishes the same recovery is the Costas loop [8.7], shown in Figure 8.16. The input signal is passed through two quadrature coherent demodulators (the LO signal to demodulator is one shifted 90° relative to the other). The output signals are multiplied and, after low-pass filtering, control the VCO that provides the LO signals. The two outputs give the original modulating signal $m(t)$, multiplied by the cosine and sine, respectively, of the LO phase difference θ from the received carrier. When the two are multiplied together, they produce a signal that is the square of the modulating signal times 1/2 sin (2θ), as follows:

$$\varepsilon = \frac{1}{2}[m(t)]^2 \sin(2\theta) \qquad (8.6)$$

The correct polarity connection of ε to the VCO will cause it to be driven to zero, corresponding to θ = 0 or 180°. Values of 2θ = 90 or 270° also produce zero output, but at those angles, the polarity of θ is opposite to that required for the

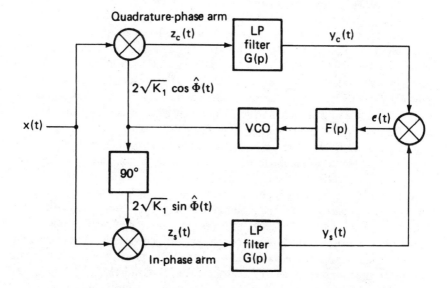

Figure 8.16 Block diagram of a Costas loop. (*From [8.6]. Reprinted by permission of the authors.*)

VCO to drive ε to zero. The equilibria are unstable. The term $[m(t)]^2$ is positive. So neither stable equilibrium is preferred. The output may be either a correct or an inverted replica of the modulating wave. While this circuit seems to differ from the frequency-doubling arrangement, it has been shown [8.8] that the two are equivalent in performance.

As long as the LO signal remains locked and free of excessive phase noise, the S/N from the DSB-SC demodulator is the same at high levels as that from an envelope demodulator. At all levels, it is the same as AM output from a coherent demodulator. The coherent demodulator responds only to the in-phase component of the signal and also of the noise. Thus S/N output is 3 dB better than the input C/N. The input power required for a particular output S/N is much lower than for AM because the carrier has been eliminated (or reduced). At 100 percent modulation index, only one-third of the total AM power is in the sidebands, and at lower indexes, there is much less. For the same S/N output in the presence of random noise, the input signal power required for DSB-SC is thus a minimum of 4.8 dB below the total power required for AM. The actual power saving depends on the statistics of the modulation signal. A large saving can be made in transmitter size and power by using DSB-SC in a system rather than AM, at the expense, however, of receiver complexity and cost. It is advantageous for commercial broadcasters to use AM, so as to have the maximum possible audience. The cost of coherent demodulation has been reduced substantially by the advent of integrated circuits.

Linear distortion for DSB-SC is essentially the same as for AM. Filter symmetrical amplitude distortion with frequency translates to the baseband, as does antisymmetrical phase distortion. Other filter distortion produces a quadrature signal, which reduces the amplitude of the recovered signal and thus affects output S/N. On the other hand, the coherent demodulator does not respond to the quadrature component, so that the net distortion is reduced. For the same output signal level, the lower power of the DSB-SC signal results in a lower level of IM products, and less chance of distortion by nonlinear amplifiers or mixers. The effect of clipping on negative modulation troughs by the envelope demodulator is absent.

The effects of selective fading in causing nonlinear demodulation distortion are absent, but the carrier recovery circuit must be designed to hold phase over a period that is long compared to the time required for a selective fade to pass through the carrier frequency. Otherwise, the phase of demodulation may vary sufficiently to cause interference from the quadrature components generated by the fade and the other mechanisms mentioned previously. Flat fading presents a problem to the DSB-SC signal because the lack of a carrier makes AGC more difficult. The power of the DSB-SC signal varies continually during modulation, and there can be frequent periods of low or no power. Under such conditions, the AGC must have a fast attack time to prevent overloading, but a relatively slow release so as not to cancel modulation or cause rapid rises of noise during low-modulation intervals. This makes it more difficult for the AGC circuit to distinguish between fades and modulation lows. The only remedy is to give the user some control over the AGC time constants, and a manual gain setting in case none of the AGC settings proves satisfactory.

8.2.3 SSB and ISB Demodulation

SSB and ISB transmissions can be demodulated by several techniques. The technique almost universally used for SSB is indicated in Figure 8.17. The received signal is filtered at IF to eliminate noise and interference in the region of the missing sideband, and then translated to baseband by the coherent demodulator. ISB is demodulated similarly, except that separate LSB demodulators are used.

The LO should ideally be at the frequency and phase of the missing carrier. However, SSB and ISB have two principal uses. The first is speech transmission, where phase errors are undetectable and frequency errors of up to about 50 Hz are nearly undetectable to most users. Errors of up to several hundred hertz can occur before difficulties are encountered with intelligibility and speaker identification, although the resulting speech sounds odd. Even a 50-Hz error is unsatisfactory if the modulation is a musical program. The second use of SSB (and ISB) is to transmit frequency-multiplexed digital data channels. In this case, the recovery of frequency and phase of the individual data channels is accomplished in the data demultiplexer and demodulator, which can correct errors of 50 to 75 Hz in the SSB demodulation.

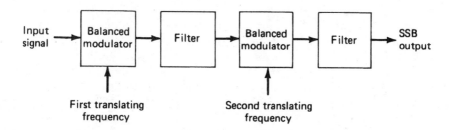

Figure 8.17 SSB demodulator using a sideband filter and coherent demodulation. (*From [8.9].*)

If SSB is to be used in a situation where frequency and phase accuracy are required in demodulation, a reduced carrier or other pilot tone related to the carrier may be sent with the transmission. If the modulating signal contains distinctive features that are sensitive to phase and frequency errors, it is possible that these may be used for correcting the demodulator's injection frequency. An example of this is given in Chapter 9.

The output S/N for SSB for a particular input thermal S/N is the same as for DSB-SC. In DSB-SC, the noise bandwidth is about twice that required for SSB, so the demodulated noise is twice as great. However, the two sidebands add coherently to produce the in-phase signal for demodulation, while the two noise sidebands add incoherently. This results in the 3-dB noise improvement mentioned previously. The SSB signal has the full power in one sideband, producing 3 dB less output than DSB-SC from the coherent demodulator. However, the bandwidth required is one-half that of DSB-SC, resulting in a 3-dB reduction in noise. The two effects offset each other, resulting in equal output S/N for the same input power, regardless of whether DSB-SC or SSB is used.

Distortion caused by the envelope demodulator is absent with SSB. However, effects that give rise to quadrature distortion are not negligible, since the demodulation is usually not coherent with the absent carrier. Because this distortion only modifies the phase and amplitude of the demodulated components, it has only a small effect in the usual speech applications of SSB. The nonlinear distortion effects of selective fading are absent because of the use of the product demodulator. The problems of finding suitable AGC time responses to fading are the same as for DSB-SC. In both cases, a reduced carrier can provide both control signals for AGC and AFC.

Two other techniques for demodulating SSB are referred to as the *phase method* [8.9] and the *third method* [8.10], respectively. The same techniques can be used for the generation of SSB. Because they are seldom used in receivers, we shall not discuss them here. For more information the reader is referred to the references.

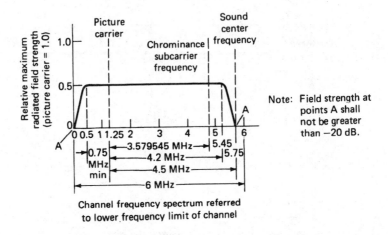

Figure 8.18 RF amplitude characteristics of a television picture transmission (not to scale); shaping to the left of the picture carrier is performed by a VSB filter. (*From [8.11]. Reprinted with permission.*)

8.2.4 VSB Demodulation

VSB is used where it is necessary to send low frequencies, but to use less spectrum than AM or DSB. The largest application is for television broadcasting (Figure 8.18). In this case, sufficient carrier is included so that the signal may be demodulated by an envelope demodulator. While there can be some waveform distortion, the filters are designed so that this is not harmful to the picture. The distortion resulting from poor receiver tuning and multipath can be considerable, so that most modern television receivers use an AFC circuit to adjust the carrier to the proper spot in the IF band.

It is possible to use a coherent demodulator to demodulate the VSB signal, locking the injection oscillator to the carrier in frequency and phase. This has been reported to improve the signal quality through improvement of S/N and also by reducing multipath and impulse noise effects on the sweep recovery circuits. If VSB with completely suppressed carrier were used, it would be necessary to recover the carrier accurately from some characteristic in the modulation so as to maintain the capability of demodulating low frequencies with correct phase. While VSB has been used in some digital data modems, they have been used primarily for wireline rather than radio applications.

8.2.5 Angle Demodulation

The passage of angle-modulated signals through linear networks can result in nonlinear distortion of the modulation and crosstalk between the angle and amplitude modulation. The standard angle-demodulation technique, referred to as *dis-*

crimination, uses this distortion intentionally to convert the angle modulation to AM for demodulation by envelope demodulators. The calculation of transmission distortion for narrow-band angle modulation through linear networks is much more difficult than for AM. While the basic linear network equations hold, the output angle is a more complex function than the envelope, even in simple cases.

In specific cases, the computations can be carried through and are best evaluated by computer. However, this does not provide a good intuitive understanding of the effects of transmission distortion. Before computers were readily available, a number of approximation techniques were devised to treat these problems. They include the following:

- Periodic modulation
- Approximation of the transmission characteristic by either a Taylor expansion or a Fourier expansion
- Quasistationary approximation
- Use of impulse or step modulation
- Series expansion of the modulated signal phase

These are reviewed in the following sections since they give some feel for the distortion effects and, in some cases, a quick way to evaluate what is to be expected.

A. Periodic modulation

If the input signal angle modulation $\phi(t)$ is assumed periodic, then $\exp[\,j\phi(t)]$ is also periodic and can be expanded as a Fourier series. Similarly, if $\phi(t)$ is a sum of periodic functions $\phi(t) + \phi(t) + ...$, all with different periods, then:

$$\exp[\,j\ \phi(t)] = \exp[\,j\ \phi_1(t)]\,\exp[\,j\ \phi_2(t)]\,...\qquad(8.7)$$

and the resultant is the product of a number of periodic series. The individual product terms may be expanded using trigonometric identities so that the end result is a multiple sum of sinusoidal terms whose amplitudes are the spectrum amplitudes and whose phases are the spectrum phases of the overall spectrum of the modulated signal. These are modified by the transmission response for each component frequency to produce the spectrum of the distorted output wave. From this spectrum, expressions for the amplitude, frequency, and phase of the output wave may be derived. If the input and network functions are sufficiently simple, output distortion can be approximated.

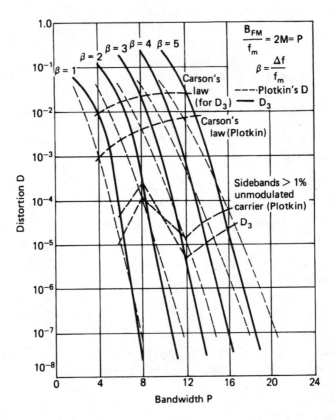

Figure 8.19 FM bandwidth and distortion for single sine wave modulation. (*After [8.16].* *Reprinted with permission of IEEE.*)

The method has been used to provide the spectra of a single sine wave modulation, of a square wave modulation, or of the sum of two sine waves (References [8.12] to [8.14]). Crosby [8.14] demonstrated that for the sum of two modulating frequencies, the output contained not only harmonics of each constituent sine wave, but also components with all orders of IM distortion. Medhurst [8.15] gave an expression for the distortion that results from truncating the RF spectrum (sharp cutoff filter with linear phase) of a multiple sine wave modulation and approximations for the resultant distortion from a perfect demodulator. An example of the use of such results to estimate the distortion of sharp filter cutoff in the case of single sinusoid modulation was plotted by Plotkin (with subsequent discussion by Medhurst and Bucher) and compared with the rule-of-thumb Carson's bandwidth. Figure 8.19 [8.16] indicates these results.

B. Taylor or Fourier expansion

An expansion of the transmission function in a Taylor series can sometimes provide good approximation in the pass band. The Taylor expansion of the transmission function $h(z)$ can be expressed as:

$$H(z) = H(0) + \left(\frac{dH}{dz}\right)_0 z + \left(\frac{d^2 H}{dz}\right)_0 z^2 + \ldots \tag{8.8}$$

where z is the frequency offset $f - f_c$. The output waveform is approximated by a replica of the original waveform, plus distortion terms that have the shape of the time derivatives of the original waveform multiplied by coefficients that, depending on the parameters, may converge rapidly enough to provide a useful approximation.

An alternative approximation is obtained if the transmission characteristic is assumed periodic about the center frequency. The period must be sufficiently long that no significant frequency components of the signal fall beyond the first period. In this case, $H(z)$ can be approximated as a complex Fourier series in frequency:

$$H(z) = \Sigma\, c_n \exp\left(\frac{j 2 \pi n z t}{M}\right) \tag{8.9}$$

where c_n are complex coefficients and M is the period of H. The output comprises a series of replicas of the input wave with amplitudes and phase shifts determined by c_n and delayed by amounts n/M.

These techniques have been applied mainly when the transmission characteristic over the band of the signal is close to ideal but with a small deviation from constant amplitude or linear phase that can be approximated by a power series or a sinusoidal shape. For example, the amplitude of $H(z)$ can be represented by $1 + \alpha \cos (\pi \tau z)$, while the phase remains linear. The response can be shown to have the form of a delayed version of the original wave, accompanied by two replicas, one preceding and one following the main response, by times $\pm \tau/2$, with amplitudes $\alpha/2$ times the amplitude of the main response. Similarly, if the phase departs from linearity by a small sine term, two small echoes are produced, but with opposite polarities and somewhat more complex amplitudes. In this case, the amplitude of the main response is somewhat reduced. This approach is called the *method of paired echoes*, for obvious reasons. Figure 8.20 illustrates this effect for the amplitude, when the input is a pulse and there are small variations in both transmission amplitude and phase, but with different τ for the delays.

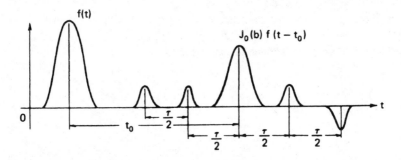

Figure 8.20 Paired echos of a pulsed signal with small sinusoidal distortions in amplitude and phase. (*From [8.5]. Reprinted with permission.*)

The use of small deviations approximated by a few terms of a power series has been used in estimating IM distortion of frequency-multiplexed signals in an FM situation. To achieve an acceptable level of IM, the transmission distortion must be very low for this technique to prove useful. Design curves for linear and quadratic amplitude and delay distortion are given by Sunde [8.17].

C. Quasistationary approximation

If an unmodulated sinusoid of frequency $f + f_c + \delta f$ is passed through a narrow-band transmission network, the amplitude and phase of the output are the network response. If δf is varied slowly, the output follows the input in frequency, with the envelope and phase modified by the amplitude and phase of the transmission function. Consequently, for sufficiently low rates of modulation, the FM response is the same as the stationary response of the network. This gives rise to the concept of a quasistationary response of the network. The main question is how slow the modulation rate must be for such an approximation to remain valid. Analytical approaches to yield the quasistationary approximation (References [8.18] to [8.20]) show that this approximation is satisfactory when:

$$\left| \frac{\dot{\phi}(t_1)}{2} \right|_{max} \times \left| \left[\frac{d^2 H(\Omega)}{d \Omega^2} \right]_{\dot{\phi}(t_1)} \right| \ll |H[\dot{\phi}(t_1)]| \qquad (8.10)$$

where $\phi(t)$ is the input phase modulation, with dots representing time derivatives, and $\Omega = 2 \pi z$. This approach is useful when the maximum modulation rate is small compared to the bandwidth of the network.

D. Impulse or step modulation

When the input frequency change is of an impulsive or step type, it can be approximated by a discontinuous impulse or step. Continuous pulse or step modulations are used in sampled-signal transmission and especially for digital signal transmission. Single pulses or steps in phase and frequency (a step function in phase is equivalent to an impulse function in frequency) can be generalized to a sum of steps occurring at different times and with different coefficient amplitudes and polarities.

A frequently worked example is that of a frequency step through a single resonant circuit (References [8.21] and [8.22]). Oscillograms of tests with this type of circuit and input are given in Figure 8.21 [8.23] for a wide variety of parameters. In this figure x_i equals $4\pi/Q$ times the initial frequency offset, and x_f has the same relationship to the final frequency offset. All but oscillogram f show the instantaneous output frequency. Figure 8.21f shows that an amplitude null accompanies the transitional phenomenon in oscillogram e. In oscillograms l, m, and n, the input is a square pulse, and x_1 and x_2 designate $4\pi/Q$ times the frequency extremes.

From Figure 8.21, note that for small deviations the response is very close to that of a step input of a low-pass RC circuit with the same time constant $2/Q$. As the deviation begins to increase, there is initially some increase in rise time and an overshoot. At higher deviation, the output becomes badly distorted. Under some conditions (oscillograms e and f), the envelope may vanish, giving rise to bursts of noise in the output. Oscillograms l through o illustrate that the responses of the filter to rises and falls in frequency are not the same, unless the wave is modulated symmetrically about the filter mean frequency. These observations have been confirmed for more complex filters than the single tuned circuit.

E. Series expansion

In the quasistationary approximation, the modulation angle $\theta(t)$ is expanded in a Taylor series, and the resulting product of exponentials is further expanded, keeping only a few terms. The process is useful where the frequency rate of change is small compared to the bandwidth of the circuit. For rapid changes in frequency, such as those found in digital data modulation, a different expansion leads to a more useful approximation (References [8.24] to [8.26]). When the total phase change during the rise or fall of the frequency is small, only a few terms are required to estimate the output amplitude or phase. Under this condition, the phase is expressed as $\Delta f_p S(t)$, where Δf_p is the maximum frequency deviation and the exponential is expanded as a series in Δf_p. This results in a convergent series of terms, with the powers of Δf_p being multiplied by rather complex expressions involving the network response to $S(t)$ and powers of it. For small phases, only a few terms need be retained. Figures 8.22 and 8.23 show some results from such approximations. In Figure 8.22, we see the output frequency from a single tuned circuit for various square pulses and steps in frequency; Figure 8.23 shows the response of an ideal gaussian filter to similar modulation.

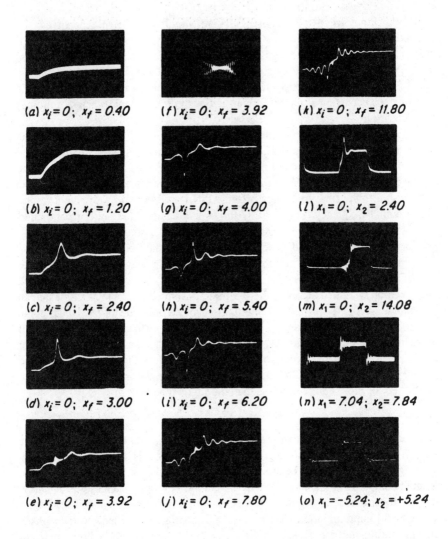

Figure 8.21 Oscillograms of the response of single-tuned circuits to frequency-step modulation for various frequency offsets and deviations. (*From [8.23]. Reprinted with permission.*)

From these results, we note that as long as the frequency deviation remains within the 3-dB bandwidth of the transmission characteristic, good approximations to the output frequency response are obtained using just two terms of the expansion. The first term is the response of the equivalent low-pass filter to the input frequency waveform. This explains why for low deviations, the output frequency for FM

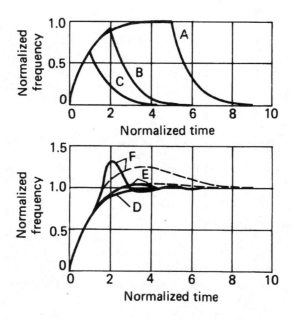

Figure 8.22 Output frequency response of a single-tuned circuit to an FM signal with rectangular FM. (*From [8.24].*)

A and *B*—Input modulation of 0.3 normalized time unit width and, respectively, zero and one-half bandwidth peak deviation.

C and *D*—0.831 unit width and, respectively, zero and one-half bandwidth peak deviation.

E and *F*—Step input and, respectively, zero and one-half bandwidth peak deviation.

resembles the output envelope for AM passed through the same filter. The difference in shape with the deviation index, even when the peak deviation reaches the 3-dB attenuation frequencies, is sufficiently small that the first term is a reasonable approximation for many applications. The second term is more complex, involving the difference of two components. This correction term is not difficult to evaluate, but for any circuits other than simple ones, the process is tedious. If computer evaluation is to be used, the complete equations might as well be evaluated, instead of the approximation.

The approximation for the envelope requires more terms for reasonable accuracy. However, it was noted in the few cases investigated that the output amplitude response could be well estimated by using the quasistationary approach, but with the instantaneous frequency of the output used, rather than that of the input. The

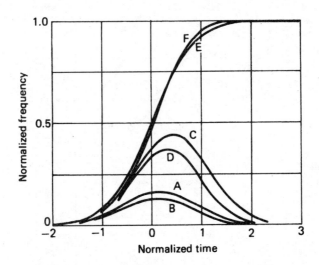

Figure 8.23 Approximate output frequency response of a gaussian filter to a signal with rectangular FM. (*From [8.24].*)

A and *B*—Input modulation of 0.3 normalized time unit width and, respectively, zero and one-half bandwidth peak deviation.

C and *D*—0.831 unit width and, respectively, zero and one-half bandwidth peak deviation.

E and *F*—Step input and, respectively, zero and one-half bandwidth peak deviation.

reason for this has not been analyzed, so such an approximation should be used with care.

8.2.6 PM Demodulators

Phase demodulation presents a difficulty because of the multiple values of phase that give rise to the same signal. If the PM varies more than ±180°, there is no way for a demodulator to eliminate the ambiguity. Phase demodulators, based on product demodulation with derived local reference, have a range of only ±90°. With digital circuits, the range can be extended almost to the ±180° limit. For PM with wider deviation, the recovery must be by integration of the output of an FM demodulator. For analog communications applications, FM demodulators are most suitable. Product phase demodulators are generally used for PSK digital data transmission with limited deviation per symbol.

Figure 8.24 shows the schematic diagram of a product demodulator used as a phase demodulator. The input signal $a_0 \cos [2\pi f_c t + \phi(t)]$ is applied through the balanced transformer to the two diode rectifiers. A local unmodulated reference at

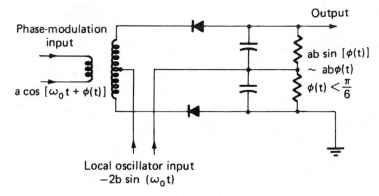

Figure 8.24 Schematic diagram of a product-type phase demodulator.

the correct frequency and with reference phase displaced 90°, $e_{10} \sin (2\pi f_c t)$, is applied at the center tap of the balanced mixer at such a level as to produce a product demodulator. The output is proportional to $\sin [(\phi(t)]$, the sine of the phase deviation. For small deviations, the sine approximates the phase but has substantial distortion as the deviation approaches 90°. The output is also proportional to the amplitude of the input signal. Therefore, the input must be limited in amplitude by earlier circuits to eliminate changes in phase from incidental envelope modulation during fading.

The sinusoidal distortion can be eliminated by replacing the sinusoids by square waves. Because the input signal must be limited in any event, the demodulator bandwidth may be made sufficiently broad to produce a square wave input. The same is true of the LO, which should be of such amplitude that it causes the diodes to act as switches. With the square wave inputs, the output varies linearly with the input phase. This same effect can be achieved using digital circuits, as shown in Figure 8.25. The flip-flop is keyed on by the positive transition of the reference square wave and keyed off by the positive transition of the signal square wave. The area under the output pulses is directly proportional to the phase difference between the circuits. A low-pass filter can serve as an integrator to eliminate components at the carrier frequency and above. The output area increases from near zero to a maximum output voltage level, as the phase difference between input and LO signals varies over 360°. For the reference phase to be at the center of this variation, its phase difference should be 180° rather than 90° as in Figure 8.24. The response then is linear over ±180°. A dc offset is required to bring the reference voltage to zero. The flip-flop must be chosen to have good square wave transitions at the particular IF.

The most recent demodulators use digital signal processing of samples from A/D converters. The least amount of processing occurs when samples of the in-phase

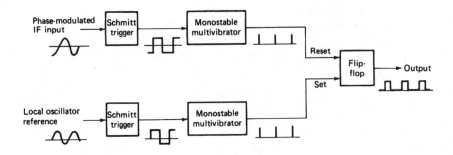

Figure 8.25 Block diagram of a digital phase demodulator.

and quadrature components of the modulation are used. Such samples may be obtained by using a reference LO at the IF, and applying the signal to two product demodulators with quadrature LO references prior to A/D conversion (Figure 8.26a). An alternative is sampling the IF signal with two sample trains offset by one-fourth of the IF period (Figure 8.26b). Another approach, when a sufficiently fast processor is available, is to sample the IF waveform above the Nyquist rate and perform both filtering and demodulation functions at the high rate.

The processing required for the output from Figure 8.26 is first to store the I and Q samples. The sign of each is then separated; the ratio of the smaller to the larger is taken and the identity of the larger is noted. The value of the ratio is used as an address to an arctan table for inputs between zero and unity. The resultant is an angle θ. If the I sample is the larger, this angle is stored; if the Q sample is the larger, it is subtracted from 90° before storage. For I and Q with opposite signs, 90° is added to the result. For a positive value of I, the resultant angle is stored as the output sample; for a negative I, its negative value is stored. Thus, the accumulated difference phase samples fall between zero and 180°. Low-pass digital filtering recovers the modulation signal.

The output is correct if the LO has the proper reference phase and frequency. More generally, the samples must be processed to establish the correction voltage samples for application to the LO through a digital filter and D/A converter. The processor then serves as part of the reference LO PLL to acquire and track the phase, within the usual ambiguity. Alternatively, instead of correcting the LO, an internal algorithm can be used to shift the phase of the incoming samples continually at the difference frequency and phase, using the trigonometric relationships for sum and difference angles. Thus, the derivation of the reference and correction can all be accomplished in the processor as long as the frequency offset does not require too great a phase correction per sample. The sampling rate and the LO frequency

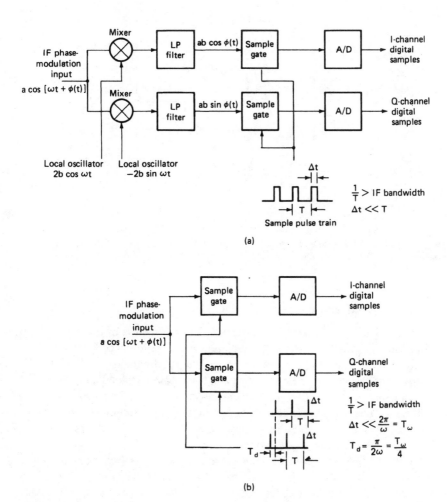

Figure 8.26 Block diagram of the derivation of I and Q demodulated samples for digital processing: (a) sampling I and Q signals from quadrature phase demodulators and (b) sampling of IF with offset sampling pulse trains.

accuracy must be chosen so that the phase change per sample is much less than 360°.

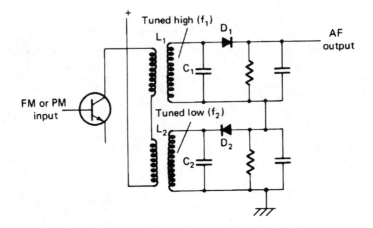

Figure 8.27 Schematic diagram of a Travis discriminator. (*From [8.27]. Reprinted with permission.*)

8.2.7 FM Demodulators

The most common technique for FM demodulation is the use of linear circuits to convert the frequency variations to envelope variations, followed by an envelope detector. Another technique used for linear integrated circuits is to convert the frequency variation to a phase variation and use a phase demodulator. Other FM demodulators employ PLLs and *frequency-locked loops*, i. e., *FM feedback* (FMFB) circuits, or counter circuits whose output is proportional to the rate of zero crossings of the wave. Frequency demodulators are often referred to as *discriminators* or *frequency detectors*.

While the inductor provides a linear amplitude versus frequency response, resonant circuits are used in discriminators to provide adequate sensitivity to small-percentage frequency changes. To eliminate the dc component, two circuits may be used, one tuned above and one below the carrier frequency. When the outputs are demodulated by envelope demodulators and subtracted, the dc component is eliminated and the voltage sensitivity is doubled compared to the use of a single circuit. The balanced circuit also eliminates all even-order distortion so that the first remaining distortion term is third-order. For minimum output distortion in the balanced circuit, the circuit Q and offsets should be chosen to eliminate the third-order term. This occurs when the product of Q and the fractional frequency offset for the circuits x equals ± 1.225. Figure 8.27 shows a schematic diagram for one implementation of this scheme, known as the *Travis discriminator*. In the design, we must be careful to ensure that the dc voltages of both circuits are identical and that the circuit parameters are such as to provide the same slope at the optimum offsets.

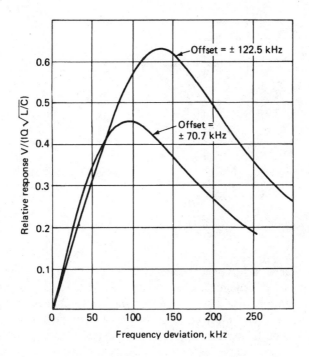

Figure 8.28 Example of responses of a Travis discriminator.

Figure 8.28 shows curves of output voltage versus frequency deviation for a particular example. In this case, a 30-MHz IF with 8-kHz peak deviation was required. Two conditions were assumed: the offset function x was chosen (1) for maximum sensitivity ($x = 0.707$) and (2) for minimum third-order distortion. The parameters of the circuit and drive were selected to produce a 1.69-V peak across each circuit at resonance, and offsets of ±70.7 and ±122.5 kHz, respectively, for the two conditions. The greater sensitivity in one case and the greater linearity in the other are obvious. In most applications, the more linear case would be most suitable. Because the Travis discriminator circuit depends on the different amplitude responses of the two circuits, it has sometimes been called an *amplitude discriminator*.

Another, more prevalent, circuit is the *Foster-Seeley discriminator* (Figure 8.29). In this circuit, the voltage across the primary is added to the voltage across each of the two halves of the tuned secondary. At resonance, the secondary voltage is in quadrature with the primary voltage, but as the frequency changes, so do the phase shifts. The voltages from the upper and lower halves of the secondary add to the primary voltage in opposition. As the frequency rises, the phase shift increases, and as the frequency falls, it decreases. The opposite phase additions cause the

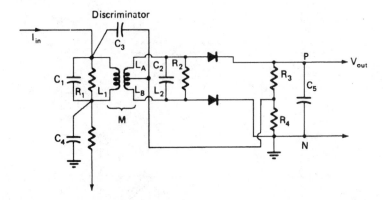

Figure 8.29 Foster-Seeley discriminator circuit with a tuned primary.

resultant amplitudes of the upper and lower voltages to differ, as shown in Figure 8.30, producing the discriminator effect. When the primary circuit is also tuned to the center frequency (which produces much higher demodulation sensitivity), the phase of the primary voltage also varies slightly, as does its amplitude. In this case, the proper selection of the coupling factor is needed to produce the optimum sensitivity and linearity of the discriminator. Because of the method of arriving at the amplitude difference in this demodulator, it is sometimes referred to as a *phase discriminator*.

The typical Foster-Seeley discriminator shown in Figure 8.29 might be driven by an FET, for example. The usual design has a secondary voltage twice the primary voltage [8.26]. Equal primary and secondary Qs are used, and Q is determined by $f_c/2f_l$, where f_l is the range of substantially linear operation. The transformer coupling coefficient Qk is set at 1.5 to give a good compromise between linearity and sensitivity, but is set at 2.0 if better linearity is required. From $V_2/V_1 = Qk$ $(L_2/L_1)^{1/2}$, we find that $L_2 = 1.77 L_1$ for $Qk = 1.5$, and $L_2 = L_1$ for $Qk = 2.0$. The discriminator sensitivity is:

$$S = A Q^2 L_1 \, \varepsilon \, g_m \, v_g \text{ (V/kHz)} \tag{8.11}$$

where ε is the diode detection efficiency, g_m is the transductance of the FET, v_g is the gate voltage, and A is a constant, which is 5.465 for the first case ($Qk = 1.5$, $L_2/L_1 = 1.77$) and 3.554 for the second case ($Qk = 2$, $L_2/L_1 = 1$). The more linear circuit has a sensitivity loss of 3.74 dB. Figure 8.31 shows curves of the relative response in the two cases. For actual responses, the curves must be multiplied by the sensitivity factor given previously. Only half of the curves are shown since the

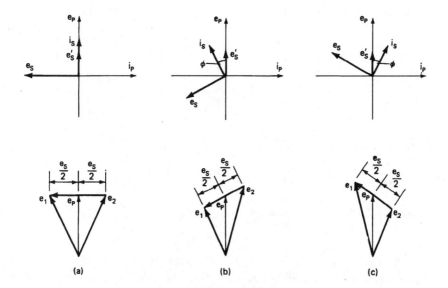

Figure 8.30 Phase relationships in a Foster-Seeley discriminator: (*a*) at resonance, (*b*) below resonance, (*c*) above resonance. (*From [8.28]. Reprinted with permission.*)

other half is antisymmetrical. By the choice of f_c and f_l for a specific application, Figure 8.31 can be scaled for the application.

The ratio detector [8.29] is a variant of the phase discriminator, which has an inherent degree of AM suppression. The circuit tolerates less-effective limiting in the prior circuits and thus reduces the cost of the receiver. Figure 8.32 shows the basic concept of the ratio detector. It resembles the Foster-Seeley circuit, except that the diodes are reversed. The combination R_1, R_2, and C_3 has a time constant that is long compared to the lowest modulation frequency (on the order of 0.1 s for audio modulation). The result is that during modulation, the voltage to the (grounded) center tap across the load resistor R_2 is $(E_1 + E_2)/2$, and across R_1 it is $-(E_1 + E_2)/2$. Following the circuit from ground through R_2 and C_2, we see that the voltage at the center tap of the capacitors is $(E_1 + E_2)/2 - E_2 = (E_1 - E_2)/2$, or half the value of the Foster-Seeley discriminator.

The long time constant associated with C_3 reduces the required current from the diodes when $E_1 + E_2$ drops and increases it when $E_1 + E_2$ rises. This changes the load on the RF circuit and causes higher drive when the output falls and lower drive when it rises. It tends to further stabilize the voltage $E_1 + E_2$ against incidental AM. The sum voltage can also be used to generate an AGC, so that the prior circuits need not limit. This can be advantageous when the minimum number of circuits is required and the selectivity is distributed.

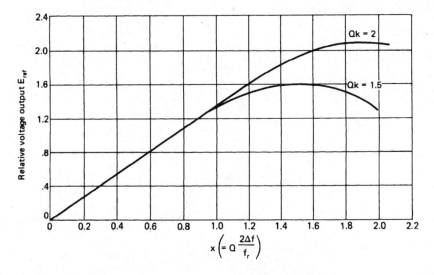

Figure 8.31 Generalized response curves for a Foster-Seeley discriminator with Qk of 1.5 and 2.0. (*From* [8.29]. *Courtesy of Amalgamated Wireless Ltd., Australia.*)

Figure 8.33 shows several implementations of the ratio detector. In practice, the primary is tuned; the coupling of the tuned secondary is similar to the earlier circuit and the untuned tertiary, when used, is tightly coupled to the primary to provide a lower voltage than appears across the primary. It may be replaced by a tap, isolating capacitor, and RF choke to get the same effect. A lower-impedance primary can achieve a comparable performance, except for the gain. The use of the tertiary allows the primary to be designed for optimum gain from the driving amplifier. Figure 8.34 shows typical input-output curves for the ratio detector, illustrating the

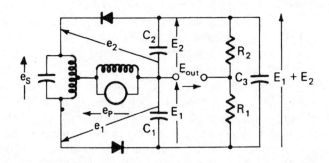

Figure 8.32 Schematic diagram of a basic ratio detector circuit.

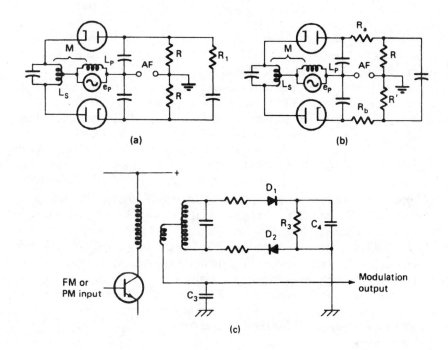

Figure 8.33 Methods for stabilizing ratio detector dc component: (*a* and *b* from [8.29] and *c* from [8.27]). (*Reprinted with permission.*)

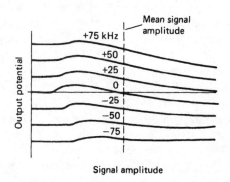

Figure 8.34 Input-output curves for a ratio detector. (*From [8.29]. Courtesy of Amalgamated Wireless Ltd., Australia.*)

small residual changes with signal amplitude. There is usually residual amplitude modulation response because of unbalances in the actual implementation. Figure 8.33*a* and *b* shows circuits that effectively shift the center point of the output to the correct value to eliminate the unbalance. A third approach (Figure 8.33*c*) uses resistors in series with the diodes. This also reduces rectification efficiency. Further details on design may be found in [8.28] and [8.29].

The phase comparison or phase coincidence detector shown in Figures 8.24 and 8.25 can also be used as an FM demodulator if the in-phase and quadrature signals from a phase discriminator are applied to the two inputs. This type of circuit has found particular use in integrated-circuit designs, such as the CA3089E (Figure 8.35) that, with a few external parts, provides all of the IF functions required for an FM receiver. This classic device will serve to illustrate the basic operation of this type of demodulator. The balanced product demodulator, which is used as the phase comparison detector, includes Q_{27}, Q_{28}, and Q_{22} to Q_{25} [8.30]. The limited in-phase input is fed to Q_{27} and Q_{28}, while Q_{29} to Q_{31} provide the limited RF output to develop the quadrature signal for application to Q_{22} and Q_{25}. A circuit for producing the quadrature signal is driven from pin 8 of the device, and the quadrature signal is applied between pins 9 and 10. Various circuits can be used to provide the quadrature signal. Figure 8.36 shows a single resonator, coupled through an inductance. Figure 8.37*a* shows a coupled pair used to reduce distortion and increase range. The resultant discriminator response is shown in Figure 8.37*b*. Such integrated circuits are available from a number of vendors.

FM demodulators can be made by replacing LC resonators with transmission line resonators at higher frequencies. Also, the improved stability desirable from a high-quality communications receiver can be achieved with the use of quartz crystal resonators in the discriminator circuit.

Either amplitude or phase discriminators can be made using crystal resonators if the frequency and the Q of the crystals are compatible. The phase comparison detector is most easily adapted to such an application. A particularly convenient arrangement is to use a monolithic crystal with two poles. Spacing between the two resonators on the substrate determines the coupling. The voltage across the two resonators differs in phase by 90° at the center frequency, as required for the phase coincidence detector. Because of the high Q of crystal resonators, the bandwidth of such discriminators is comparatively narrow. Figure 8.38 shows an application suggested by one vendor, using the CA3089E described previously.

When linearity is the primary consideration, an FM demodulator based on the *zero crossing counter* scheme may be used. This sort of design was described in the early days of FM [8.31]. Because the number of zero crossings per second is equal to the instantaneous frequency, the result of this circuit is to produce an output whose average voltage is proportional to the frequency. The circuit has a center voltage proportional to the unmodulated carrier, so it requires a low center frequency for reasonable sensitivity. This type of circuit can be balanced [8.32], as shown in Figure 8.39. Table 8.1 lists the characteristics of two experimental

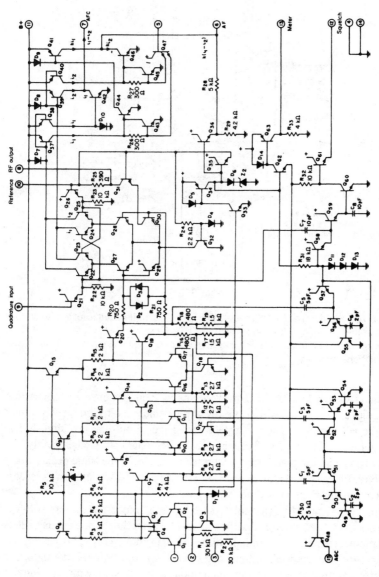

Figure 8.35 Functional schematic of a CA3089 multifunction integrated circuit. (*From [8.30].*)

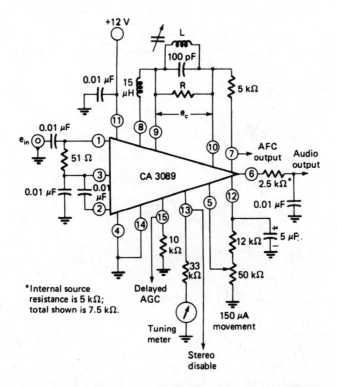

Figure 8.36 Test circuit for a CA3089 showing the use of a single-tuned resonator to provide quadrature voltage. L tunes with 100 pF at 10.7 MHz; $Q_{LOADED} \approx 14$; and $R \approx 3.9$ kΩ is chosen for $e_c \approx 150$ mV to provide proper operation of the squelch circuit. (*From [8.30]. Reprinted with permission.*)

demodulators of this kind and Figure 8.40 shows their response characteristics. Other circuits use the IF transitions to key a monostable multivibrator to produce the required equal pulses. The monostable can have an on-time just slightly less than the period of the highest frequency to be demodulated. Consequently, the sensitivity can be higher than with the simpler approach for the same peak voltage on the driver. Because of its lower sensitivity, this type of discriminator is seldom used for communications receivers.

As with filters and AM demodulators, the trend in modern receivers is to use digital processing hardware or algorithms for FM demodulation. Because filters can be implemented readily by digital FIR or IIR algorithms, Travis, Foster-Seeley, or phase comparison filter algorithms can be developed. Similarly, an algorithm for zero crossings is possible. Direct calculation of the frequency modulation can be derived from the I and Q samples by first differentiating I and Q, evaluating the quantity $(I'Q - Q'I)/(I^2 + Q^2)$, and then low-pass filtering the result. Differentiation may be achieved by a digital differentiator. Alternatively, the phase difference

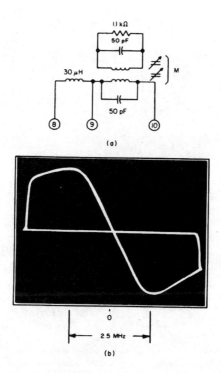

Figure 8.37 Double-tuned 10.7-MHz quadrature circuit for improved performance. Peak-to-peak separation = 2.5 MHz; distortion 0.05 percent; and output at 75-kHz deviation = 250 mV. (*From [8.30]. Reprinted with permission.*)

between samples can be used as an approximation to angular frequency. If the sampling rate is sufficiently high, $I_1Q_2 - Q_1I_2$ is the sine of the difference angle multiplied by the product of the amplitudes of the two sampled pairs. If the sampling rate is high compared to the rate of phase change, the sine can approximate the angle; and if the signal has previously been limited, these samples can be low-pass filtered to eliminate the effects of noise. Limiting is achieved by dividing I and Q by $(I^2 + Q^2)^{1/2}$.

8.2.8 Amplitude Limiters

Amplitude-limiting circuits are essential for angle demodulators using analog circuits. Although both tube and solid-state amplifiers tend to limit the signal when the input signal level becomes large, limiters that make use of this characteristic often limit the envelope dissymmetrically. For angle demodulation, symmetrical limiting is desirable. AGC circuits, which can keep the signal output constant over wide ranges of input signal, are unsuitable for limiting, since they

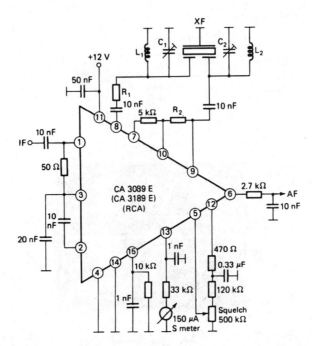

Figure 8.38 Narrow-band FM demodulator using a dual-crystal resonator as the discriminator. (*Courtesy of Kristall-Verabeitung Neckarbischofsheim GmbH, West Germany.*)

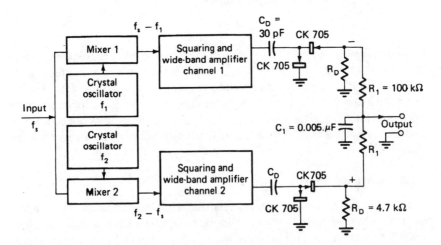

Figure 8.39 Block diagram of a double-counting discriminator. (*From [8.32]. Reprinted with permission.*)

TABLE 8.1 Experimental Demodulator Parameters (*After [8.32]*)

Characteristics	Circuit A	Circuit B
Center frequency	4.0 MHz	4.5 MHz
Crystal frequencies		
f_1	3.92 MHz	4.0 MHz
f_2	4.08 MHz	5.0 MHz
Peak separation	0.16 MHz	1.0 MHz

cannot be designed with a sufficiently rapid response time to eliminate the envelope variations encountered in angle modulation interference. One or more cascaded limiter stages are required for good FM demodulation.

Almost any amplifier circuit when sufficiently driven provides limiting capabilities. However, balanced limiting circuits produce better results than those that are not. In general, current cutoff is more effective than current saturation in producing sharp limiting thresholds. Nonetheless, overdriven amplifiers have been used in many FM systems to provide limiting. If the amplifier is operated with a low supply voltage and near cutoff, it becomes a more effective limiter. The standard transistor differential amplifier of Figure 8.41*a* is an excellent limiter when the bias of the emitter load transistor is adjusted to cause cutoff to occur at small base-emitter input levels.

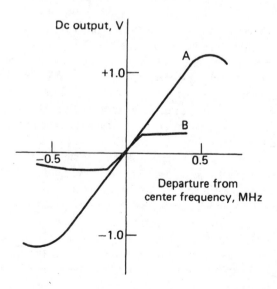

Figure 8.40 Characteristics of two experimental counting discriminators.

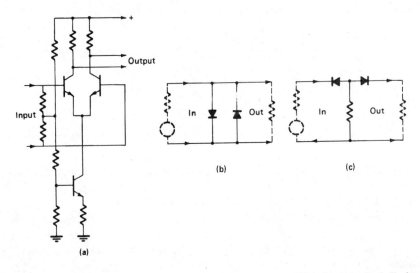

Figure 8.41 Typical limiter circuits: (*a*) balanced transistor amplifier, (*b*) shunt diode limiter, and (*c*) series diode limiter.

The classic shunt diode limiter circuit is shown in Figure 8.41*b*. The diodes may be biased to cut off if the resistance from contact potential current is too low for the driver source. It is important that the off resistance of the diodes be much higher than the driving and load impedances, and that the on resistance be much lower. Figure 8.41*c* shows the classic series diode limiter. In this case, the diodes are normally biased on, so that they permit a current flow between driver and load. As the RF input voltage rises, one diode is cut off and, as it falls, the other is cut off. The effectiveness of limiting is determined by the difference in the off and on resistances of the diode compared to the driving and load impedances. If biasing circuits are used to increase this ratio, care must be taken that the arrangement does not upset the balance, and that associated time constants do not cause bias changes with the input signal level.

Interference can occur from adjacent-channel signals, in-band signals, local thunderstorms, or electric machinery. Whatever the source, the rates of variation in the frequency and envelope are limited by the channel filters. When the interference has an envelope comparable to that of the desired signal, the resultant signal can have amplitude and phase modulation rates much faster than either component. A limiter eliminates the amplitude variations, but the resultant output bandwidth can be substantially increased. The limiter output bandwidth must be designed to be substantially wider than the channel bandwidth to avoid the elimination of high-frequency spectrum components of the limited wave, a process that would restore some amplitude variations.

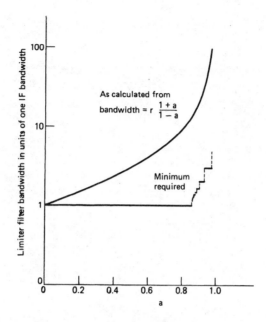

Figure 8.42 Plots of "sufficient" limiter bandwidth $(1 + a)/(1 - a)$ to capture at interference-to-signal ratio a, and of "necessary" bandwidth. (*From [8.33]. Reprinted with permission.*)

An analysis under idealized assumptions [8.33] gives rise to Figure 8.42. This figure indicates the required limiter bandwidth to preserve the stronger signal FM in the output. The lower curve is estimated from a detailed analysis of sidebands resulting from the limiting process. The upper curve is based upon earlier analysis of the envelope of the limiter output signal. In practice, some margin over the lower curve is required when real filters are used, but two to three IF bandwidths is normally adequate. If the limiter bandwidth cannot be as large as desired, the output envelope variation is still less than the input. Consequently, cascading limiters can further reduce the interference. If the envelope reduction by a single limiter is small, an excessive number of limiters may be required. When a problem of this sort occurs in design, an experimental trade-off between limiter bandwidth and number of cascaded circuits should be performed.

8.2.9 FM Demodulator Performance

The performance of FM demodulators can be understood by analyzing the effects of interference between the desired signal and a second interfering signal. Because both signals have passed through the narrow-band channel filter prior to demodulation, they are each narrow-band signals, with envelopes and phases that change slowly compared to the mean frequency of the filter (which should coincide with

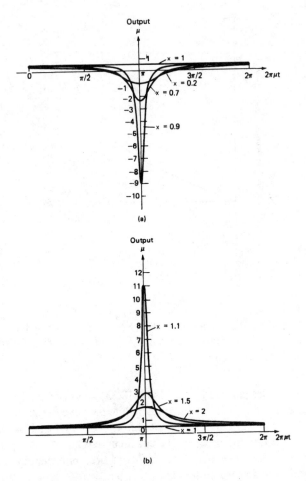

Figure 8.43 Frequency output for the sum of two unmodulated waves, one at carrier frequency f_c, the other separated by frequency μ: (a) $x < 1$ and (b) $x > 1$. *(After [8.34].)*

the carrier frequency of the desired signal). Thus, the rates of change of the individual envelope or phase are approximately equal to or smaller than the reciprocal of the IF bandwidth. Corrington [8.34] has given an extensive treatment of the result of adding two narrow-band signals. Figure 8.43 shows the instantaneous output frequency over a cycle of the difference frequency between the two waves. The value x is the ratio of the weaker to the stronger signal, and μ is the frequency difference between the waves. The same curves hold when the interfering signal is stronger, provided that $1/x$ is substituted for x. Figure 8.43a indicates how the frequency of the resultant varies when the interfering signal is weaker than the desired signal, and Figure 8.43*b* gives the frequency when the interfering signal is stronger.

Two things are of particular interest. First, the average frequency of the output is the same as the frequency of the stronger signal. This is the basis for the phenomenon of *capture* in FM receivers, whereby the stronger of two cochannel signals tends to suppress the output from the weaker one, except when the two are almost identical in signal strength. The output frequency variation is accompanied by an envelope variation. The demodulator output can differ from that indicated if the envelope variations are not completely suppressed. In particular, capture is much less sudden as the envelope ratio changes if the limiting circuits preceding the discriminator do not limit the signal adequately. In this case, the interferer will be objectionable further below envelope equality.

The second effect worth examining is the sharp frequency pulses that occur when the envelope ratio is close to unity. This is responsible for the impulse noise encountered when the envelope of the gaussian noise accompanying a wide-deviation FM signal is nearly equal to the signal envelope. The effect occurs near the threshold where the output S/N slope suddenly changes with the input S/N of the FM receiver. Also, for digital PM signals, this can be troublesome, since whenever the noise envelope is greater than the signal envelope at the time of the impulse, there is a phase discontinuity of 2π.

Corrington gives a number of characteristics of the envelope and frequency of the resulting wave that can prove useful in assessing interference effects. For both signals unmodulated we have:

$$\text{Average envelope} = \frac{1}{\pi} 2\, e_1\, (1 + x)\, E\left[\frac{2(\sqrt{x})}{1 + x}\right] \tag{8.12}$$

$$\text{Rms envelope} = e_1\, (1 + x^2)^{\frac{1}{2}} \tag{8.13}$$

$$\text{Average frequency} = 0 \text{ for } x < 1; = \mu \text{ for } x > 1 \tag{8.14}$$

$$\text{Rms frequency} = \frac{x\,\mu}{[2\,(1 + x^2)]^{\frac{1}{2}}} \text{ for } x < 1 \tag{8.15a}$$

$$\text{Rms frequency} = \mu \left[\frac{2\,x^2 - 1}{2\,(1 + x^2)}\right]^{\frac{1}{2}} \text{ for } x > 1 \tag{8.15b}$$

$$\text{Peak-to-peak frequency} = \frac{2\,x\,\mu}{x^2 - 1} \tag{8.16}$$

where e_1 is the level of the desired signal envelope and $E[z]$ is the complete elliptic integral of the second kind with modulus z. Expressions for the envelope and frequency, Equations (8.12) to (8.16), are given in [8.34] for unmodulated waves, for modulated desired signal and unmodulated interferer, and for both waves modulated.

Sensitivity is measured with thermal noise as the only interference. In the case of envelope demodulation of AM, it will be recalled that for high C/N, the output S/N was proportional to C/N. However, below a threshold region near 0-dB C/N, the output became proportional to the square of C/N. A similar but more pronounced threshold occurs for a frequency demodulator and is especially noticeable in systems designed for wide deviation ratios. The noise spectrum of an envelope demodulator above the threshold follows the equivalent low-pass characteristic of the composite receiver filters. For FM, the spectrum is much different. The theoretical performance of FM demodulators in the presence of gaussian noise has been analyzed extensively (References [8.35] to [8.37]).

Figure 8.44 shows the spectrum of the FM output noise when the input noise has been passed through a gaussian-shaped filter. While the curves apply in the case when the signal is unmodulated, they give useful approximations even when the wave is modulated. Maximum deviation would seldom exceed the 3-dB point on the filter, which corresponds to $f/\sigma = 1.18$. The modulating signal bandwidth will generally be one-half this value or less, so that only a fraction of the noise spectrum need appear in the output. At high-input C/N, these curves indicate that the output noise power density varies as the square of the frequency in the normal baseband. As C/N drops below 7 dB, the spectrum density at low frequencies begins to rise rapidly, so that the variation with frequency tends to flatten. The f^2 rise in output noise density is the reason for using a preemphasis network for high-quality FM broadcasting and multichannel multiplexing on FM, so as to maintain comparable output S/N over much of the baseband spectrum.

Total output S/N is obtained by integrating the noise density. Figure 8.45 shows the output noise as a function of the input noise-to-signal ratio for a gaussian IF filter, designed for a peak deviation $\Delta\omega$. The baseband is sharply filtered at frequency f_a. The threshold effect is plainly visible in these curves, occurring very close to 10-dB S/N, for the larger deviation indexes. When the input S/N drops below about 6 dB, there is a suppression of the modulation by the noise. This variation in output signal power is $[1 - \exp(-S/N)]^2$. The resulting output amplitude variation is plotted in Figure 8.46 and is used in estimating the output S/N at and below the threshold.

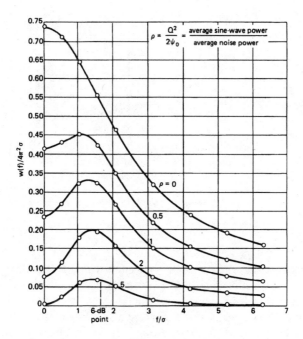

Figure 8.44 Power spectrum of output angular frequency for an unmodulated carrier plus noise. (*After [8.35].*)

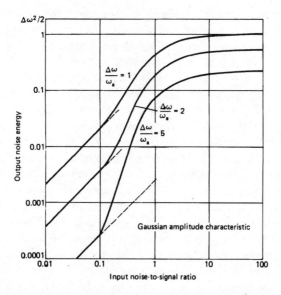

Figure 8.45 Output noise power of angular frequency for unmodulated carrier plus noise; a gaussian IF filter is used. (*From [8.36]. Reprinted with permission.*)

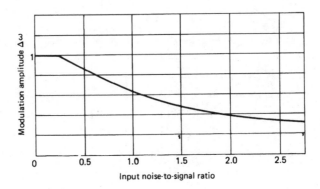

Figure 8.46 Suppression of FM by noise. (*From [8.36]. Reprinted with permission.*)

Rice [8.37] developed a simplified method of estimating FM performance based on the "click" theory, where a click is an impulse that increments the phase by ±2π. This approximation theory has proved a useful tool for estimating FM performance. Figure 8.47 shows an example of the estimate of (S/N)_{out} versus (C/N)_{in}, both when modulation is neglected and when it is not. Modulation increases the input S/N required for threshold by about 1 dB while improving the output S/N by about 0.5 dB above the threshold. This approximation technique has also proved valuable in estimating error rates for digital signals (References [8.38] and [8.39]).

Impulse noise is one of the most troublesome types of interference. It is characterized by high-level bursts of short duration, which may occur at random, in groups, or periodically, depending on the source. This interference has a bandwidth much wider than the receiver bandwidth, and appears to the receiver as impulse excitation of the receiver filters. The filter outputs are narrow-band signals with large amplitude, random phase, and a duration determined by the filter bandwidth. Whenever the envelope of the impulse is greater than the signal envelope, the resulting angle is determined primarily by the noise.

Very large noise input pulses determine the resultant envelope and phase; the desired signal produces small distortion components. In the early stages of the receiver, where the bandwidth is wide, the pulse duration is short. As the impulse passes through filtering devices, its envelope amplitude decreases approximately proportionately to the bandwidth reduction, and its width increases in inverse proportion to the bandwidth reduction. As long as the amplitude remains higher than the signal, the effect of filtering simply increases the duration of the interference. By limiting the signal amplitude before the final bandwidth is reached, we can eliminate more energy than by filtering. Hence, the amount of interference is reduced. Distributed limiting and filtering may thus provide better performance against impulse interference than lumped filtering preceding the limiters. Even if

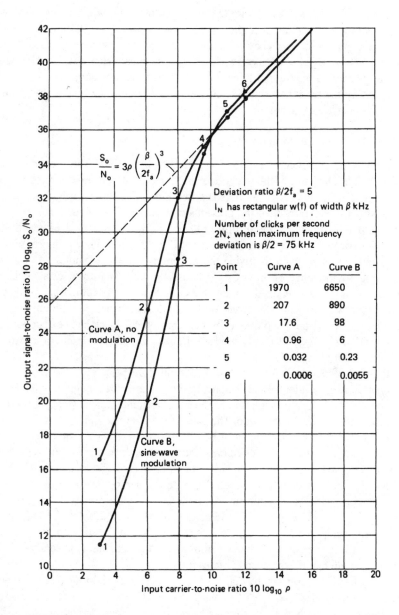

Figure 8.47 Output S/N versus input S/N for example using the click method of approximation. (*After [8.37]. Courtesy of S. O. Rice and M. Rosenblatt.*)

the selectivity is all achieved prior to the limiter, the pulse width will generally be considerably shorter than the highest modulation period by virtue of the bandwidth required to pass peak frequency deviation sidebands. The greater the peak deviation for which the design has been made, the shorter the interference pulse will be. Much of the spectrum of the output interference is well above the baseband, and so it is removed by the baseband filter. Thus, the larger the designed modulation index, the less the effect of impulse noise.

8.2.10 Threshold Extension

As the FM signal deviation ratio is increased, the output S/N is improved for high input S/N, but because of the wider bandwidth required, threshold also occurs at a higher level of input C/N, as we noted earlier. Figure 8.48 shows this characteristic somewhat more clearly than the previous curves. Because of the bending of the curves, it is difficult to define threshold precisely. However, in Figure 8.48 it appears to occur from an input S/N of about 12 dB for wide bandwidths to about 4 dB at minimum bandwidth (twice the maximum baseband width). As systems requiring a wide range of input C/N (troposcatter and satellite repeater) began to come into use, it was realized that the threshold in receivers could be improved by using negative-frequency feedback to reduce deviation before the nonlinear de-modulation operation. Such frequency-compressive feedback had been considered much earlier (References [8.41] and [8.42]), but from the standpoint of improving distortion at high input S/N.

Enloe (References [8.40] and [8.43]) described the principles of FM frequency feedback and gave rules for the design. Figure 8.49 is a block diagram of the FMFB demodulator. The analysis is based on the use of an unmodulated signal and a linear approximation to the FMFB loop. The basic idea is simple. The VCO is frequency-modulated by the filtered output of the discriminator to provide a negative feedback of modulation, so that the modulation at the output of the mixer is compressed, and the IF bandpass filter and frequency demodulator deal with a narrow-band FM whose threshold is thereby lowered. Because the demodulated noise is also fed back, the output S/N above threshold is not reduced by the process. The maximum useful compression just reduces the IF bandwidth to twice the baseband width, since this width is required to pass signals with even very small deviations. Care must be taken, as in all feedback amplifiers, to minimize delays and maintain adequate phase and amplitude margins so as to avoid oscillation.

Enloe found experimentally that the threshold of the closed-loop circuit occurred when the rms deviation of the VCO reached about 1/3 rad. This appeared to be true for widely varying loop parameters, so he selected the condition as the closed-loop threshold and expressed the threshold input S/N (based on the linearized model) as:

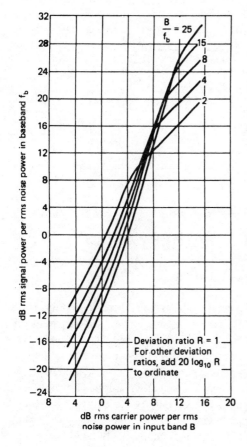

Figure 8.48 FM threshold curves. (*After [8.2] and [8.40].*)

$$\left(\frac{S}{N}\right)_{TF} = 4.8\left(\frac{F-1}{F}\right)^2 \qquad\qquad (8.17)$$

Enole provided guidance for designing an FMFB demodulator, using a two-pole IF filter, and recommended that:

- Full feedback be maintained over all baseband frequencies
- The closed-loop noise bandwidth be as small as possible
- Open-loop threshold and feedback (closed-loop) threshold coincide

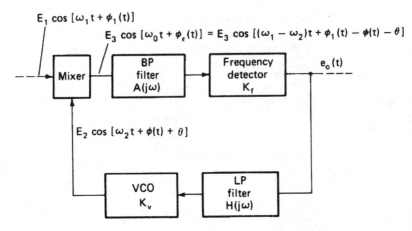

$E_1 \cos [\omega_1 t + \phi_1(t)]$

$E_3 \cos [\omega_0 t + \phi_\epsilon(t)] = E_3 \cos [(\omega_1 - \omega_2)t + \phi_1(t) - \phi(t) - \theta]$

$E_2 \cos [\omega_2 t + \phi(t) + \theta]$

Figure 8.49 Block diagram of a frequency feedback demodulator. (*From [8.40]. Reprinted with permission.*)

While the open-loop threshold can be obtained from the earlier curves, Figure 8.50 presents this information directly. Based on the use of the Carson bandwidth, Enloe gives for the open-loop noise bandwidth of a single-pole filter $B_n/2f_b = (\pi/2)$ $(1 + M/F)$. He indicates that this bandwidth is pessimistic in practice and substitutes the 3-dB bandwidth at peak-to-peak deviation, $B_n/2f_b = (\pi/2) (M/F)$. Using Figure 8.50 and the relationships shown, curves of open-loop threshold as a function of the feedback and modulation index may be developed. Figure 8.51 shows such a curve for the second bandwidth.

For the feedback loop, the response of the maximally flat two-pole network is modified by the addition of excess phase shift from parasitics and delays around the loop. For adjustment, zeros of the response are included at two frequencies af_b and bf_b. Factor a is chosen to provide minimum closed-loop noise bandwidth B_c for a feedback factor F and excess phase ϕ. Factor b is chosen to make the open- and closed-loop thresholds occur simultaneously. Figure 8.52 gives values of a and b for various values of F and ϕ. The resultant transfer function is:

$$H_0(s) = \frac{(s/b\,\omega_b + 1)\,(s/a\omega_b + 1)}{(s/\omega_b)^2 + \sqrt{2}\,s/\omega_b + 1} \tag{8.18}$$

A network to provide this response is shown in Figure 8.53. The minimum closed-loop bandwidth is determined by eliminating the term with b in the nu-

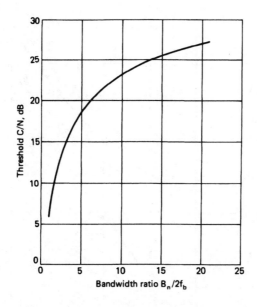

Figure 8.50 Plot of threshold C/N points referred to a bandwidth equal to twice the baseband versus the ratio of IF noise bandwidth to twice the baseband. (*From [8.43]. Reprinted with permission.*)

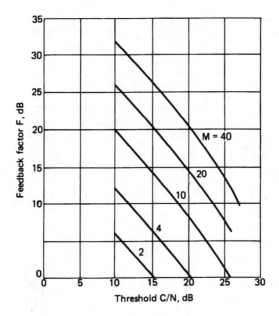

Figure 8.51 Plots of open-loop threshold versus modulation index M and feedback factor F, assuming $B_n/2f_b = (\mu/2)\,(M/F)$. (*From [8.43]. Reprinted with permission.*)

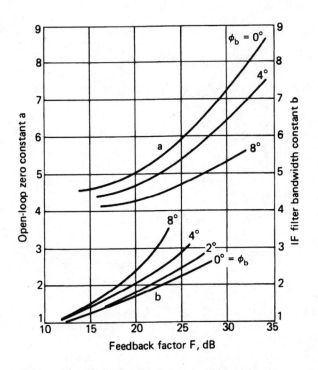

Figure 8.52 Plots of open-loop zero constant a and IF filter-bandwidth constant b versus feedback factor F. (*From [8.43]. Reprinted with permission.*)

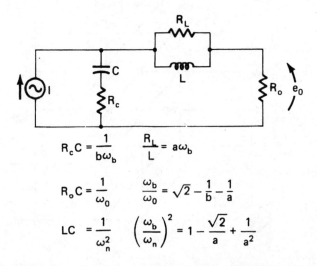

$$R_c C = \frac{1}{b\omega_b} \qquad \frac{R_L}{L} = a\omega_b$$

$$R_o C = \frac{1}{\omega_0} \qquad \frac{\omega_b}{\omega_0} = \sqrt{2} - \frac{1}{b} - \frac{1}{a}$$

$$LC = \frac{1}{\omega_n^2} \qquad \left(\frac{\omega_b}{\omega_n}\right)^2 = 1 - \frac{\sqrt{2}}{a} + \frac{1}{a^2}$$

Figure 8.53 Network having the prescribed transfer function. (*From [8.43]. Reprinted with permission.*)

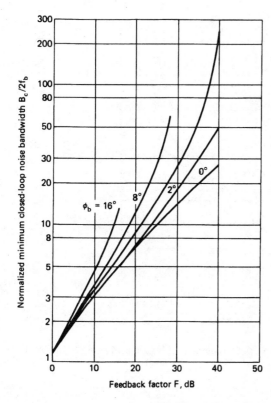

Figure 8.54 Plots of normalized minimum closed-loop bandwidth versus feedback factor F and excess phase ϕ_b. (*From [8.43]. Reprinted with permission.*)

merator of Equation (8.18) and setting factor a to the value shown in Figure 8.52. The resulting values of $B_c/2f_b$ for various values of F are shown in Figure 8.54.

When Figure 8.51 is overlaid on Figure 8.54, the value of F is at the intersection of the curves for the selected values of ϕ and M, as shown in Figure 8.55. The resultant output S/N at the threshold is determined by multiplying the input S/N by $3M^2$. This is plotted in Figure 8.56 for various values of F, along with similar results for a conventional discriminator. The difference indicates the expected reduction in input S/N for a required output S/N at the threshold using this design technique. The technique is reported to be good for $M/F \geq 1$ and $F \geq 5$.

Schilling and Billig [8.44] developed a different model of the FMFB demodulator using Rice's click model of FM noise. Frutiger [8.45] also developed a model of the FMFB demodulator, using an approximate formula for FM noise. An FMFB may be designed using any of these techniques as a starting point. Final selection of the components will be determined experimentally, since the prediction of the excess phase shift through the system can only be approximate and suitable phase equalization is required to permit stable feedback. This is included explicitly in Enloe's design technique for establishing the threshold. The others implicitly

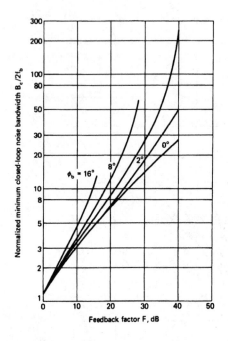

Figure 8.55 Superposition of Figure 8.51 replotted on Figure 8.54 using Equation 8.17. (*From [8.43]. Reprinted with permission.*)

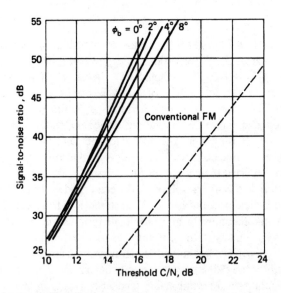

Figure 8.56 Baseband S/N versus input S/N at the threshold as determined from Figure 8.55. (*From [8.43]. Reprinted with permission.*)

include feedback stability, and Frutiger shows a phase-equalizing network in his block diagram, although he does not use the correction explicitly in his technique. Figure 8.57 shows measured performance under a variety of design parameters of an FMFB demodulator that was designed for multiple applications, primarily for an earth station receiver in space links [8.46].

The PLL, which is sometimes used as a frequency demodulator, also has threshold-extending properties (References [8.44] and [8.47]). This is understandable since, like the FMFB, the PLL uses a feedback FM oscillator that is mixed with the incoming FM signal to reduce the deviation in the output. The design differs in that the IF has been reduced to zero, and the demodulator is a phase rather than a frequency demodulator. Therefore, for the VCO to follow the incoming signal completely, it must be able to follow the instantaneous phase of the incoming signal plus noise without the phase error exceeding the detector's limits ($\pm 90°$ to $\pm 180°$, depending on the type of phase detector used).

Moderate nonlinearity in the phase detector can be compensated by the feedback; but should the error exceed the monotonic limit of the phase detector, the loop can temporarily lose lock. When a stable lock is regained, the VCO may have gained or lost a multiple of $360°$, resulting in spikes in the output signal. If this happens frequently, as when the signal and noise envelopes are nearly equal, the output noise increases more rapidly than the input noise, resulting in a typical demodulation threshold.

The usual linear model of the PLL is not adequate to predict threshold. With the usual product-type phase demodulator, Develet [8.47] developed a quasilinear model from which the performance of PLLs could be predicted to below threshold. The performance was predicted for a signal phase-modulated by gaussian noise with rms σ_m and using a PLL with an optimum filter. The performance is substantially better than has been observed in real PLLs. Develet also used his model to predict results using a second-order transfer function (Figure 8.58). Because the results are for PM, they should resemble those from gaussian noise modulated FM with preemphasis. The relationship of input and output noise at threshold is given by:

$$\left(\frac{S}{N}\right)_I = 4.08 \left[\left(\frac{S}{N}\right)_O \right]^{1/3} \tag{8.19}$$

This curve is shown in Figure 8.58, as are the curves obtained with the optimum filter and those that predict maximum possible output S/N as a function of input S/N based on Shannon's theorem [8.48], as follows:

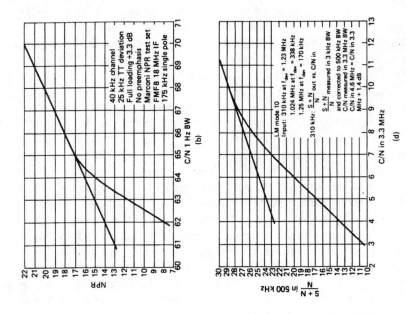

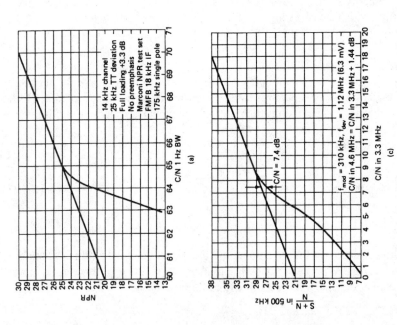

Figure 8.57 Measured performance for various design parameters of an FMFB demodulator designed for space applications. *(From [8.46]. Reprinted with permission.)*

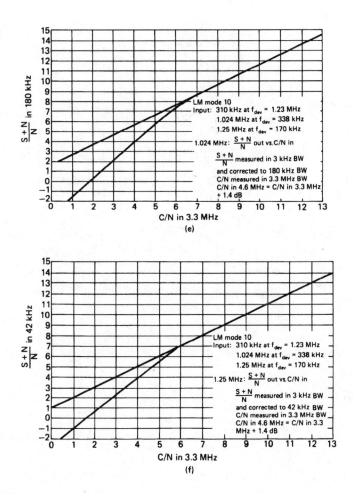

Figure 8.57 *continued*

$$\left(\frac{S}{N}\right)_I = \frac{1}{2} \ln\left[1 + \left(\frac{S}{N}\right)_O\right] \tag{8.20}$$

Develet indicated that the results have been checked by limited experimentation.

Schilling and Billig [8.44] also use Rice's click model to predict the performance of PLLs as frequency demodulators. Their results for output S/N with very high loop gain and modulation indices of 5 and 10 are given in Figure 8.59. The model agrees better with predictions than did the similar FMFB model. Based on their two sets of results, Schilling and Billig conclude that the FMFB demodulator and the

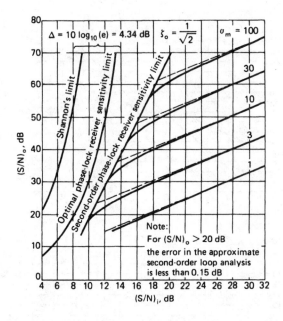

Figure 8.58 Predicted performance of a PLL demodulator with input signal phase modulated with white gaussian noise, and with second-order feedback loop transfer function. (*After [8.47].*)

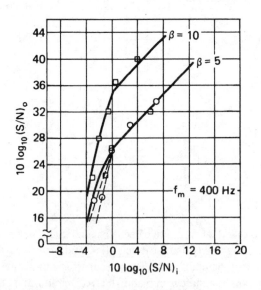

Figure 8.59 Comparison of performance of the PLL model with experimental measurements. (*After [8.44].*)

PLL demodulator perform similarly at peak deviation ratios of 5 and 10. However, for larger modulation indices, the FMFB threshold is expected to improve more rapidly than the PLL threshold.

8.3 Pulse Demodulation

As indicated in Chapter 1, most pulse modulation techniques are appropriate for multichannel transmission, which is not a subject for this book. Such techniques represent a secondary modulation, the primary modulation usually being AM or FM, which would be demodulated in the normal manner and passed to the demultiplexer for channel separation and pulse demodulation. Pulse duration modulation was proposed [8.48] for use in receiver circuits for aeronautical and marine applications using satellite repeaters. In such uses, radiated power is limited, and it is desirable to receive at S/N output levels below those normally tolerated in high-quality services. Figure 8.60 indicates the predicted performance for several alternative modulation techniques, using a 20-kHz IF bandwidth. The curve labeled PDM uses 6-kHz sampling of a 2.5-kHz speech baseband. The samples are converted to pulse duration modulation of the rear edges, and the front edge transitions are suppressed so that the result is a binary signal with each successive transition having the appropriate duration from the missing front edge reference. The suppression of the reference transitions reduces the required trans-mission bandwidth. The resulting binary wave is transmitted using binary PSK with coherent demodulation.

Figure 8.61 is a block diagram of a demodulator for this signal. It uses a Costas loop to recover the phase transitions. A VCO is locked at the sampling frequency by the average transition rate of the recovered signal. The clock is divided to provide the missing transitions and to regenerate the PDM from which the original signal is recovered. Figure 8.60 indicates that at test-tone-to-noise output ratios below about 18 dB, this technique has superior performance to other modulation schemes considered, including FM, FM with PLL as threshold extender, PCM-PSK, and DM-PSK. PCM and DM have an advantage over PDM for applications where secrecy or privacy is desired because they are regularly sampled binary signals and can be encrypted readily. They are true digital data signals.

8.4 Digital Data Demodulation

In many cases, digital data demodulation is not performed in the receiver, but rather in the demodulator section of a separate modem. The receiver provides an IF or baseband output to the modem for processing. When a baseband output is provided, it must be offset so that the lowest frequency in the translated data spectrum is sufficiently above zero frequency. The demodulator processes the signal to recover timing and to determine the transmitted symbols, making use of any constraints on the transmitted waveforms and any error correction or detection

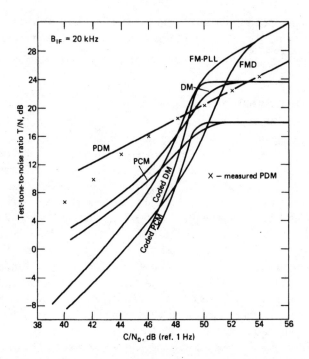

Figure 8.60 Performance comparison of various modulation-demodulation techniques for narrow-band transmission. (*From [8.48]. Reprinted with permission from* COMSAT Technical Review.)

codes that were added for the radio transmission. In the case of error detection, a separate output line may be provided to indicate the errors detected, or a special symbol, not of the transmitted set, may be output instead of the symbols in which the error was detected.

Some radio receivers include the digital data demodulator. Examples are links using digital speech encoding for encryption or privacy, or police radio where simple alphanumerical data are sent as a shorthand code or for license numbers. With the increasing use of digital processing in receivers, we expect to see more cases of this type in the future. Therefore, some simple digital data demodulators are treated in the following sections.

8.4.1 ASK

ASK is not very suitable for radio transmission circuits. It requires higher peak power than angle modulation and is more sensitive to fading. For simple on-off keying, an envelope demodulator may be used, followed by a threshold device. When the demodulated signal is above the threshold, the received signal is consid-

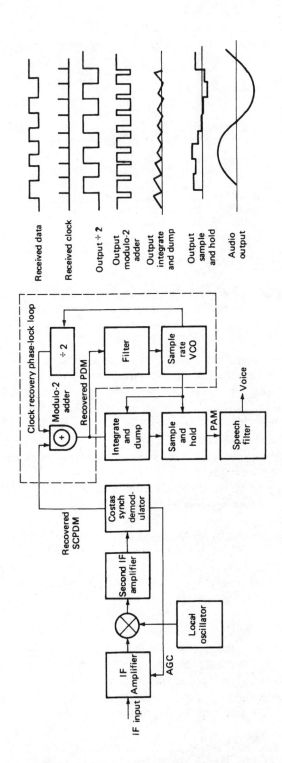

Received data

Received clock

Output ÷ 2

Output modulo-2 adder

Output integrate and dump

Output sample and hold

Audio output

Clock recovery phase-lock loop

÷ 2

Modulo-2 adder

Filter

Sample rate VCO

Recovered PDM

Integrate and dump

Sample and hold

PAM

Speech filter

→ Voice

Recovered SCPDM

Costas synch demod-ulator

Second IF amplifier

Local oscillator

IF Amplifier

AGC

↑ IF input

Figure 8.61 Block diagram of a suppressed-clock PDM demodulator with waveforms. (U.S. patent 3,667,046, issued to R. L. Schoolcraft and assigned to Magnavox Company, covers a voice transmission and receiving system employing pulse duration modulation with suppressed clock.) (*After [8.49].*)

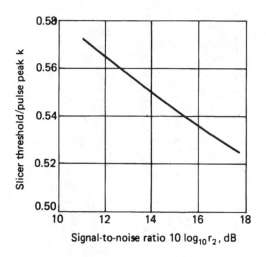

Figure 8.62 Threshold required for the minimum error rate for envelope demodulation of OOK with additive gaussian noise. (*From [8.50]. Reprinted with permission.*)

ered on; when it is below the threshold, it is off. Because the noise distribution of the envelope varies with the amount of signal present, the optimum value for the threshold is not one-half of the peak, but somewhat higher, depending on S/N. Figures 8.62 and 8.63 indicate the optimum threshold and the resultant error rate at optimum and 0.5 threshold levels. In the figures, M represents the on condition and S the off condition. The indicated signal power is the peak envelope power of the signal. The average power is 3 dB lower for square-wave transmission with equal frequency of marks and spaces.

In practice, the transmitted waveform will be shaped to minimize adjacent channel interference, and the receiver will be provided with similar IF filter shaping to optimize the S/N at the demodulator. *Nonreturn-to-zero* (NRZ) transmission provides the minimum bandwidth, so the composite of the transmitter and receiver filter should allow close to full rise in one bit period, i. e., the bandwidths, if equal, should be about $\sqrt{2}/T$. If the transmitter uses a sharp cutoff filter with broader-nose bandwidth to eliminate adjacent channel interference, rather than gradual cutoff, the receiver IF filter bandwidth can be made closer to $1/T$ so as to provide an improvement of S/N.

If m-ary ASK were used with an envelope demodulator, multiple thresholds would be required. To provide equal probability of adjacent threshold crossing requires not only unequal threshold separations, but differing amplitude changes from level to level at the transmitter. It is preferable to use coherent demodulation of the ASK signal with a product demodulator (which could also be used with OOK). Coherent demodulation requires that the phase of the received signal be

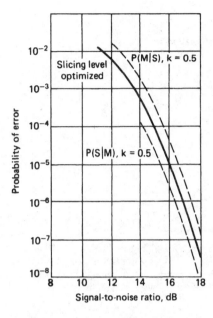

Figure 8.63 Probability of error for envelope demodulation of OOK with additive gaussian noise at the optimum threshold and at 0.50 threshold. (*From [8.50]. Reprinted with permission.*)

recovered. For an ASK signal, a hard limiter and filter will produce the carrier output, to which a PLL may be locked to further reduce noise on the carrier phase. Because of the small likelihood of using such a system, m-ary ASK demodulation will not be discussed further. However, many complex modulation systems use m-ary ASK with suppressed carrier, especially with modulation of a pair of quadrature carriers.

8.4.2 FSK

FSK customarily has been demodulated using either a limiter-discriminator frequency demodulator or narrow-band filters tuned to the shifted frequencies with comparison of output levels. A PLL demodulator may be used, and in special cases (such as the product of frequency shift and symbol period integral), coherent demodulation is applicable. Assuming a knowledge of initial phase, Kotel'nikov [8.51] showed that the optimum frequency separation for coherent demodulation of binary FSK is $0.715/T$, where T is the symbol period. The optimum performance predicted is shown in Figure 8.64. This procedure requires resetting of the starting phase at the beginning of each symbol, depending upon the demodulated value of the last symbol. With the availability of low-cost digital processing, such a procedure is possible. However, if orthogonal frequencies are chosen (separation n/T, with n integral), the starting phase for each symbol is identical. For binary FSK, the optimum separation provides a gain of about 0.8 dB over orthogonal separation, as indicated in Figure 8.64.

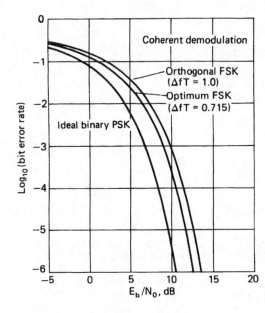

Figure 8.64 Predicted performance of optimum coherently-demodulated binary FSK with additive gaussian noise.

Orthogonal frequency separation is generally used for m-ary FSK signaling. Coherent demodulation compares each received signal to all of the reference frequencies, which are separated in frequency by $1/T$. The largest output is selected as the demodulated symbol. In such a receiver, the generated local references must be synchronized in phase to the incoming signal states, and the symbol timing must be correctly recovered. This permits matched filter detection. Figure 8.65 is a block diagram of an m-ary coherent FSK demodulator for orthogonal signals and Figure 8.66 shows the expected performance in white gaussian noise.

Noncoherent detection of m-ary FSK is also possible, and for large signal sets introduces only a small loss in performance. The signals may be separated by a bank of bandpass filters with bandwidth approximately $1/T$, spaced in frequency by a similar amount. The outputs are envelope-modulated, and the largest output is selected as the transmitted signal. Figure 8.67 indicates the expected performance in gaussian noise. Envelope demodulation may be accomplished by combining the outputs of two quadrature demodulators at each frequency rss. Addition of the absolute values of the two outputs may be substituted for rss processing, with a slight loss in performance and a simplification in processing.

A limiter discriminator preceded by a filter is often used for demodulating a binary FSK signal. This type of demodulator can achieve performance comparable to the optimum predicted by Kotel'nikov (References [8.51] to [8.59]. Figure 8.68 shows a comparison of experimental and simulation test results for various deviations and predemodulation filter bandwidths. The peak-to-peak deviation ratio of $0.7/T$ provides the lowest error rates. For all of the deviations examined, a prede-

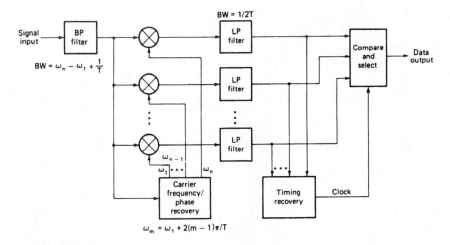

Figure 8.65 Block diagram of a coherent demodulator for *m*-ary orthogonal FSK signals.

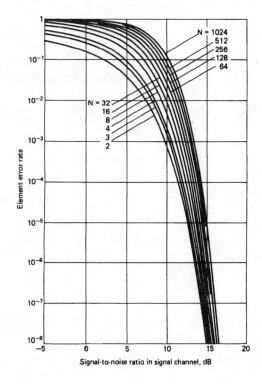

Figure 8.66 Predicted performance of a coherent demodulator for *m*-ary orthogonal signals with additive gaussian noise. (*After [8.52].*)

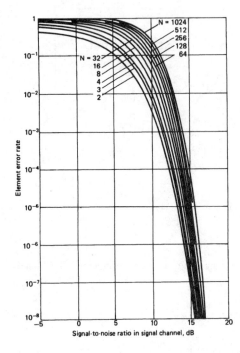

Figure 8.67 Predicted performance of a noncoherent demodulator for *m*-ary orthogonal signals with additive gaussian noise. (*After [8.52]*.)

modulation filter bandwidth of $1/T$ provides the best results. For these tests, the symbols had rectangular transitions. If transmitter premodulation shaping is used, we can expect slight differences in the results. Limiter-discriminator demodulation can be used for *m*-ary FSK, but the performance deteriorates because of the threshold effect in the wide bandwidth required.

Signal transitions can be distorted by multipath in the transmission medium. One technique to reduce errors from this source is to use a symbol interval that is much longer than the anticipated maximum multipath delay, and to gate out a segment equal to this maximum delay at each transition time. FSK is generally used for applications where bad multipath is expected or where simple demodulation is desired. While the performance in gaussian noise is somewhat poorer than that of PSK, FSK is somewhat more rugged when subjected to multipath. All modulations are comparably affected by impulse noise, which predominates in the HF portion of the spectrum and occurs at VHF and lower UHF.

8.4.3 PSK and Combined Modulations

Quadrature coherent balanced demodulators are generally used for PSK in its many variations. Such demodulators are, in fact, used with almost all complex signal constellations. It is, however, possible to demodulate PSK by using a frequency demodulator (limiter discriminator or PLL) and either integrating the

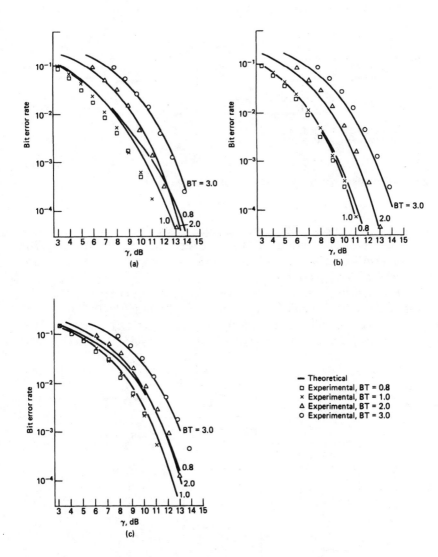

Figure 8.68 Comparison of simulation and experimental error rates for binary FSK: (a) h = 0.5, (b) h = 0.7, and (c) h = 1.0. (*After [8.53].*)

output before decision or changing the decision rules to those that apply to the derivative of the phase. One relatively simple PSK demodulation technique for binary PSK uses a delay line of one-symbol duration and a product demodulator (Figure 8.69). If the transmitted symbols have been encoded differentially, the output of this demodulator represents the distorted version of the input wave train. The signal may be optimally demodulated by matched filtering and sampling at

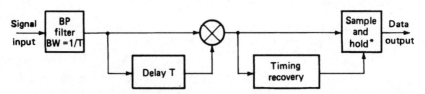

*May be replaced by integrate, sample, hold, and dump if
bandpass filter bandwidth is widened.

Figure 8.69 Block diagram of differential demodulation of PSK.

the resulting wave peaks. While the technique appears simple, caution must be
taken to produce a precise delay, so that there is no phase error in the delayed IF
wave. Otherwise an effective loss in the output signal amplitude will occur, equal
to the cosine of that phase error.

Figure 8.70 shows a generic block diagram for quadrature demodulation of PSK
signals and most higher-order *m*-ary signal constellations. The carrier recovery
circuit is varied, depending on the signal type. Some signals have carrier or other
steady frequency components that may be filtered to provide the reference, either

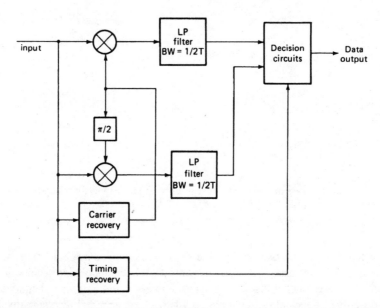

Figure 8.70 Generic block diagram of quadrature demodulation of signal constellations.

directly or by use of PLLs. In most cases, however, some form of nonlinear processing is required to get a reference to maintain carrier phase. For *binary PSK* (BPSK), for example, a squaring circuit (or, equivalently, a Costas loop) may be used. For *quaternary PSK* (QPSK), a fourth-order circuit or higher-order Costas loop may be used. Another technique sometimes used is *decision-aided feedback*, where the demodulated output is used to control remodulation of the incoming signal stream, eliminating the modulation to provide a reference carrier. An example of this is shown in Figure 8.71.

Whenever the signal constellation is balanced, carrier recovery is ambiguous. The carrier recovery circuits are thus suitable for tracking, but if unambiguous recovery is necessary for demodulation, the transmitted wave must be coded to facilitate it. This can be achieved by a preamble that generates the carrier, by occasional insertion of a specific sequence in the data stream at predetermined times, or by using data coding that produces correct results only when the correct phase position is attained. Differential coding of the data is another alternative when the signal constellation is symmetrical. The change between successive signals is then used to demodulate the data, obviating the need for absolute phase recovery. For differential PSK, demodulation using the ambiguous recovered carrier can be used or the demodulator can use a carrier of correct frequency with random phase. In the first case, each symbol is separately demodulated, using the recovered carrier, which is (relatively) noise free, and the data are recovered from the successive symbols. In the second case, the difference in the signal space is determined directly from the successive symbols. This process is slightly noisier and leads to somewhat poorer performance.

In most cases, symbol-timing-recovery is required as well as carrier recovery for accurate data demodulation. With binary signals, it is possible to regenerate the modulation data stream by using only a clipper set midway between the two states of the waveform, and for some applications this is adequate. However, the noise, error in clipper setting, and intersymbol interference can all cause undesirable jitter in the data transmissions. Symbol timing may be recovered more accurately with a PLL, often using a harmonic of the symbol rate for locking to the symbol transitions. The use of a recovered harmonic of the timing frequency facilitates the development of accurately delayed pulse trains at the symbol rate to produce sampling trains at either side of the transition interval for loop control. A sampling train can also be produced at the optimum delay from the transitions for best performance. An alternative to early-late sampling is to differentiate the signal waveform. Transitions provide positive or negative pulses that can be balance-rectified and used to lock the symbol-timing oscillator.

The decision circuits indicated in Figure 8.71 use the filtered in-phase and quadrature signal levels at the symbol sampling time to establish the output symbol, which may then be output in the required format. Each symbol is defined by a point in the I-Q plane. When the received signal falls within a region (usually a rectangle in the I-Q plane) surrounding this point, a decision is made that the particular

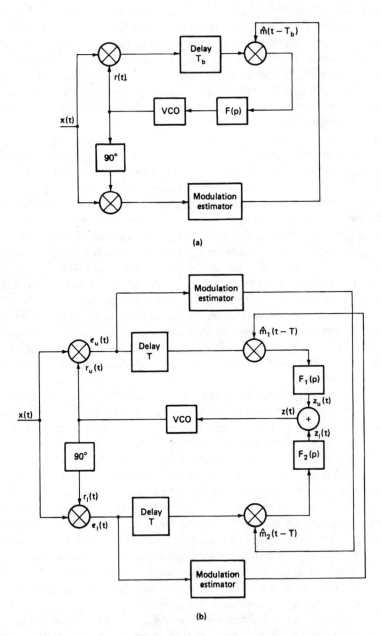

Figure 8.71 Block diagram of (*a*) binary and (*b*) quaternary PSK demodulators using decision-aided carrier recovery. (*From [8.6]. Reprinted with permission.*)

symbol was sent. In the case of differential decoding of coherently demodulated symbols, successive symbol decisions are retained so that the final data decision may be made based on the present and preceding demodulated symbols. In the case of differential demodulation, successive I-Q pair values are stored, and the data symbol is determined by equations that relate to the relative positions between the two. The particular algorithms will depend on how the difference relations are chosen. When large symbol sets are used, differential coding is unlikely to be used, since the use of large sets implies relatively stable media, low noise, and high data rates. In this case, accurate carrier recovery is desirable.

To permit comparison of the performance of different modulation systems, we will review error probability performances for additive gaussian noise with perfect receivers. This gives an indication of the relative performance of different systems. However, practical receivers may show losses of a few tenths to several decibels, depending on the accuracy of phase recovery, timing recovery, and residual intersymbol interference. The common comparison is symbol-error probability (or equivalent bit-error probability) versus bit-energy to noise power density ratio E_b/N_0. The data are given for systems without error correction coding, which can improve performance.

PSK is one of the simpler m-ary modulation formats, sending one phase from a selection of many, separated by $2\pi/m$, where m is the number of symbols. Binary and quaternary PM are antipodal signaling for a single carrier and orthogonal carriers, respectively, and each shows the optimum performance (without coding) when demodulated coherently. Coherent demodulation with differential coding is somewhat poorer for low E_b/N_0, but shows little difference at high E_b/N_0. Differential demodulation of differentially coded signals is still poorer. Figure 8.72 shows ideal bit-error probability for these three modulation schemes for 2-, 4-, 8-, and 16-ary PSK. Here it is assumed that the higher-order signals are coded so that adjacent symbols differ by only 1 bit (Gray coding).

Figure 8.73 gives an indication of the effect of timing and phase recovery on the BER in one channel of an offset-keyed QPSK modulation system. These data were obtained using the simulation of a receiver with a rate of 90 Mb/s in each of the quaternary channels (symbol duration 11.1 ns). The eye pattern of the output signal showed that a delay of 90.7 ns produced the largest opening (from an arbitrary reference: a transmission delay of approximately 53 ns). The aggregate of filters in baseband, RF, and IF in the system included a total of 80 poles. In curve 1, no limiting was used in the system, so that the 0.5 to 1-dB loss is solely the result of the intersymbol interference. In the other curves, hard-limiting was assumed in the system, and the effect of quadrature channel crosstalk introduced another 3.0-dB loss (curve 3). About 0.3 dB was recovered by a slight delay in sampling time (curve 2). Curves 2, 4, and 5 show the effects of progressively greater phase errors in the recovered carrier, the sampling time being adjusted in each case for the lowest average error probability.

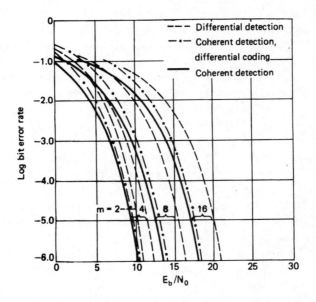

Figure 8.72 Ideal bit BER versus E_b/N_0 for m-ary PSK modulation-demodulation techniques.

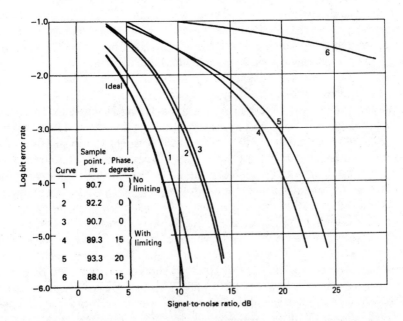

Figure 8.73 Effects of sampling and phase errors in BER.

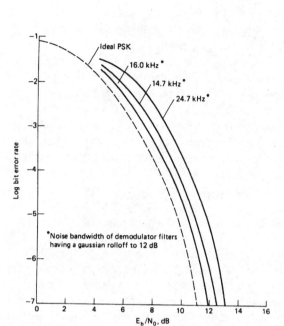

Figure 8.74 Performance of an MSK modem. (*From [8.55]. Reprinted with permission.*)

FSK with a peak-to-peak deviation index of 1/2 has been called MSK and *fast FSK* (FFSK). The main difference between the two techniques is the premodulation coding. In MSK, the input is coded so that the in-phase and quadrature channels carry two independent antipodal binary bit streams, while in FFSK, the input bit stream frequency modulates the signal. The resulting relationship between the in-phase and the quadrature bit streams eliminates the ambiguity in signal recovery at the expense of a slight loss in performance. The performance of MSK with the ambiguity (any of four phases) removed is ideally the same as ideal BPSK or QPSK. FFSK removes the ambiguity, but has a performance equivalent to the coherently demodulated DPSK. By premodulation coding and postdemodulation decoding, an MSK system can be converted to FFSK and vice versa. De Buda [8.54] proposed a scheme of clock and carrier recovery for FFSK, as well as methods of generating it stably. Figure 8.74 shows performance measurements made on an experimental modem built following this scheme.

Recently many phase-continuous digital modulation schemes have been proposed to try to improve on MSK by reducing either the bandwidth occupancy or the error rate, or both. TFM [1.40] and *gaussian MSK* (GMSK) [1.41] seek to reduce the bandwidth occupancy while minimizing the increase in error rate. In both cases, premodulation shaping is used prior to FM. The intersymbol interference is designed so that the BER performance reduction for additive gaussian noise, using normal quadrature demodulation techniques, is less than 1 dB, while adjacent channel interference is much reduced.

TFM and GFSK use only the data within a single symbol interval for decision. A number of other systems have been proposed that use information over the complete interval of spreading of the shaped input. (Shaping may be accomplished with either analog or digital filters.) In these cases, maximum likelihood decoders are used over the full interval of spreading of the individual symbol (References [8.56] and [8.57]). By proper choice of the signal shaping (or coding), it is possible to obtain performance improvement and reduced band occupancy. The price is a considerable increase in complexity of the demodulation (decoding) process, which generally uses a Viterbi decoder over the constraint length of possibly as much as six or seven symbol intervals. The expected performance and bandwidth requirements for continuous phase modulation have been summarized for both full response signaling (symbol occupying a single interval) [1.42] and partial response signaling (symbol spread over several intervals) [1.43]. The tradeoffs in the first case were indicated in Table 1.1. For partial response signaling, some tradeoffs are indicated in Figure 8.75. In this figure, the relative gain over MSK with additive gaussian noise is plotted against bandwidth at the –60-dB (from carrier) level. In all but the broken curves, the deviation is varied to produce the points on the curves.

The partial response systems can be expected to be of use primarily where the transmission either is relatively stable or can be equalized readily. Implementation is likely to be only by digital sampling and processing. Figure 8.76 shows a transmitter diagram for the generation of TFM. The receiver is similar to other quadrature demodulators, except that some care is required in the design of the low-pass filters following the product demodulators. Designs for Viterbi decoders are available in convolutional data decoder designs. These can be adapted for cases where the combined modulation and coding of partial-response continuous PM systems prove useful.

8.5 Digital Signal Processing

Digital signal processing (DSP) techniques are being applied to the implementation of various stages of communications receivers for a number of reasons, including:

- Improved cost-performance considerations
- Future product enhancement capabilities
- Greatly reduced alignment and testing requirements

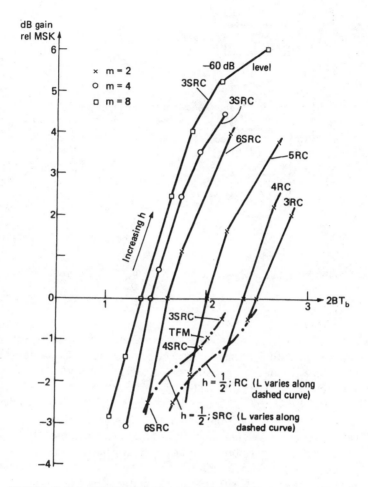

Figure 8.75 Bandwidth-power tradeoffs among partial-response continuous PM systems. SRC—raised cosine spectrum; RC—raised cosine time response; and 2, 3, ... 6—number of intervals over which symbol is spread. (*From [1.43]. Reprinted with permission of IEEE.*)

Receiver demodulators, decoders, and modems can be implemented incorporating various degrees of imbedded DSP. The most important parameters are signal bandwidth and S/N, which define, respectively, the required sampling rate and the effective number of bits required for the conversion. Additional design considerations include the stability of the sampling clock, quadrature channel matching, aperture uncertainty, and the cutoff frequency of the quantizer networks. Subject to satisfying these considerations, it is usually desirable to locate the A/D interface and processing circuits as early as possible in the receiver chain.

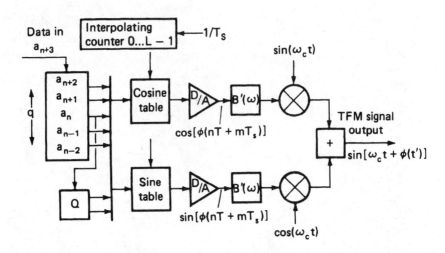

Figure 8.76 Block diagram of a TFM transmitter. (*From [1.41]. Reprinted with permission of IEEE.*)

DSP devices differ from microprocessors in a number of ways. First, microprocessors are typically built for a range of general-purpose functions and normally run large blocks of software. Second, microprocessors are usually not called upon to perform real-time computation. Usually, they are at liberty to shuffle the workloads and select an action branch, i. e., complete a printing job before responding to a new input command. The DSP, on the other hand, is dedicated to a single task or small group of related tasks. In a sophisticated receiver, one or more DSPs may be employed as attached processors, assisting a general-purpose host microprocessor, which manages the front panel controls.

One convenient way to classify DSP devices and applications is by their dynamic range. In this context, the dynamic range is the spread of numbers that must be processed in the course of an application. It takes a certain range of values, for example, to describe a particular signal and that range often becomes even wider as calculations are performed on the input data. The DSP must have the capability to handle such data without overflow.

The processor capacity is a function of its data width, i. e., the number of bits it manipulates, and the type of arithmetic that it performs (fixed or *floating point*). Floating point processing manipulates numbers in a form similar to scientific notation, which enables the device to accommodate an enormous breadth of data. Fixed arithmetic processing, as the name implies, restricts the processing capability of the device to a predefined value.

DSP manufacturers have developed families of devices, each intended for specific applications. Sixteen-bit fixed-point DSPs, such as the common Motorola 56100 family, are intended for voice-grade systems. High-fidelity stereo sound

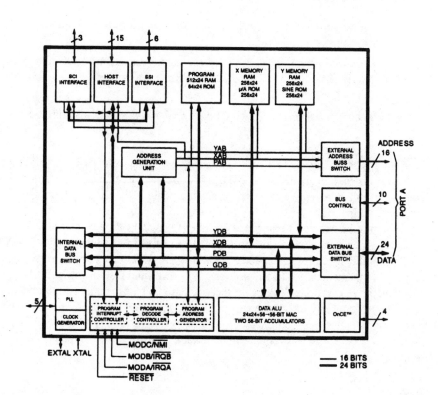

Figure 8.77 Simplified block diagram of the DSP56002. (*After [8.60].*)

applications, on the other hand, could incorporate a 24-bit fixed point DSP such as the 56001. As a point of reference, as of this writing general-purpose 24-bit DSP devices operating at 80 MHz are in production (Motorola 56300 series). Devices utilizing 32-bit architecture with floating-point capabilities are also available, but are primarily intended for graphics and scientific simulation applications (Motorola 96002).

A general block diagram of the 56002 is given in Figure 8.77. Although an examination of the operation of DSP systems is beyond the scope of this book, it is important to touch on a few key areas of operation. The 56002 is a general-purpose DSP composed of a 24-bit signal processing core, program and data memories, various peripherals, and support circuitry [8.60]. The core is fed by a program RAM, two independent data RAMs, and two data ROMs with sine and A-law and μ-law tables. The device contains several communications systems, including the following:

- *Serial communication interface* (SCI) for full-duplex asynchronous communications.

- *Synchronous serial interface* (SSI) to communicate with codecs and synchronous serial devices. Up to 32 time-multiplexed channels can be supported.

- Parallel *host interface* (HI) with *direct memory access* (DMA) support.

The 56002 is capable of up to 33 *million instructions per second* (MIPS), yielding a 30.3-ns instruction cycle at 66 MHz. The architecture includes four 24-bit internal data buses and three 16-bit internal address buses for efficient information transfer on-chip.

Conventional receiver architectures designed to leverage the advantages of DSP typically rely on separate functional blocks for the IF gain, demodulation, baseband filtering, *I* and *Q* sampling, clock generation, and local oscillator subsystems. Recent advancements in VLSI technologies in general, and DSP in particular, have permitted the integration of several of these functional blocks into a single device. Such designs typically offer excellent performance because of the elimination of the traditional interfaces required by discrete designs. The high level of integration also decreases the total parts count of the system, increasing the overall reliability of the receiver.

References

8.1. M. Schwartz, *Information, Transmission and Noise*, 2d ed., McGraw-Hill, New York, 1970.
8.2. S. O. Rice, "Mathematical Analysis of Random Noise," *Bell System Tech. J.*, vol. 23, p. 282, July 1944; vol. 24, p. 96, Jan. 1945.
8.3. M. Nakagami, "Study on the Resultant Amplitude of Many Vibrations whose Phases and Amplitudes Are Random," *Nippon Elec. Comm. Eng.*, vol. 22, p. 69, Oct. 1940.
8.4. D. Middleton, *An Introduction to Statistical Communication Theory*, McGraw-Hill, New York, 1960.
8.5. P. F. Panter, *Modulation, Noise and Spectral Analysis*, McGraw-Hill, New York, 1965.
8.6. W. C. Lindsey and M. K. Simon, *Communication Systems Engineering*, Prentice-Hall, Englewood Cliffs, N.J., 1973.
8.7. J. P. Costas, "Synchronous Communications," *Proc. IRE*, vol. 44, p. 1713, Dec. 1956.
8.8. W. C. Lindsey, *Synchronization Systems in Communications and Control*, Prentice-Hall, Englewood Cliffs, N.J., 1972.
8.9. D. E. Norgaard, "The Phase-Shift Method of Single-Sideband Reception," *Proc. IRE*, vol. 44, p. 1735, Dec. 1956.
8.10. D. K. Weaver, Jr., "A Third Method of Generation and Detection of Single-Sideband Signals," *Proc. IRE*, vol. 44, p. 1703, Dec. 1956.
8.11. *Reference Data for Engineers: Radio, Electronic, Computer and Communication*, 7th ed., Howard Sams, Indianapolis, Ind., 1985.

8.12. J. R. Carson, "Notes on the Theory of Modulation," *Proc. IRE*, vol. 10, p. 57, Feb. 1922.

8.13. B. Van der Pol, "Frequency Modulation," *Proc. IRE*, vol. 8, p. 1194, July 1930.

8.14. M. G. Crosby, "Carrier and Side-Frequency Relations with Multi-Tone Frequency or Phase Modulation," *RCA Rev.*, vol. 3, p. 103, July 1938.

8.15. R. G. Medhurst, "Bandwidth of Frequency Division Multiplex Systems Using Frequency Modulation," *Proc. IRE*, vol. 44, p. 189, Feb. 1956.

8.16. T. T. N. Bucher and S. C. Plotkin, "Discussion on FM Bandwidth as a Function of Distortion and Modulation Index," *IEEE Trans.* (Correspondence), vol. COM-17, p. 329, Apr. 1969.

8.17. E. D. Sunde, *Communication Systems Engineering Theory*, Wiley, New York, 1969.

8.18. J. R. Carson, "Variable Frequency Electric Circuit Theory," *Bell Sys. Tech. J.*, vol. 16, p. 513, Oct. 1937.

8.19. B. Van der Pol, "The Fundamental Principles of Frequency Modulation," *J. IEE*, vol. 93, pt. 3, p. 153, May 1946.

8.20. F. L. H. M. Stumpers, "Distortion of Frequency-Modulated Signals in Electrical Networks," *Commun. News*, Phillips, vol. 9, p. 82, Apr. 1948.

8.21. R. E. McCoy, "The FM Transient Response of Band-Pass Filters," *Proc. IRE*, vol. 42, p. 574, Mar. 1954.

8.22. I. Gumowski, "Transient Response in FM," *Proc. IRE*, vol. 42, p. 919, May 1954.

8.23. E. J. Baghdady (ed.), *Lectures on Communication Theory*, McGraw-Hill, New York, 1961.

8.24. T. T. N. Bucher, "Network Response to Transient Frequency Modulation Inputs," *AIEE Trans.*, vol. 78, pt. 1, p. 1017, Jan. 1960.

8.25. A. Ditl, "Verzerrungen in frequenzmodulierten Systemen (Distortion in Frequency-Modulated Systems)," *Hochfrequenztech. und Elektroakustik*, vol. 65, p. 157, 1957.

8.26. E. Bedrosian and S. O. Rice, "Distortion and Crosstalk of Linearly Filtered Angle-Modulated Signals," *Proc. IEEE*, vol. 56, p. 2, Jan. 1968.

8.27. S. W. Amos, "F. M. Detectors," *Wireless World*, vol. 87, no. 1540, p. 77, Jan. 1981.

8.28. S. W. Seeley and J. Avins, "The Ratio Detector," *RCA Rev.*, vol. 8, p. 201, June 1947.

8.29. F. Langford-Smith (ed.), *Radiotron Designer's Handbook*, 4th ed., Amalgamated Wireless Valve Company Pty. Ltd., Australia, 1952.

8.30. J. Avins, "Advances in FM Receiver Design," *IEEE Trans.*, vol. BTR-17, p. 164, Aug. 1971.

8.31. S. W. Seeley, C. N. Kimball, and A. A. Barco, "Generation and Detection of Frequency-Modulated Waves," *RCA Rev.*, vol. 6, p. 269, Jan. 1942.

8.32. J. J. Hupert, A. Przedpelski, and K. Ringer, "Double Counter FM and AFC Discriminator," *Electronics*, vol. 25, p. 124, Dec. 1952.

8.33. E. J. Baghdady, "Frequency-Modulation Interference Rejection with Narrow-Band Limiters," *Proc. IRE*, vol. 43, p. 51, Jan. 1955.

8.34. M. S. Corrington, "Frequency Modulation Distortion, Caused by Common and Adjacent Channel Interference," *RCA Rev.*, vol. 7, p. 522, Dec. 1946.

8.35. S. O. Rice, "Statistical Properties of a Sine Wave Plus Random Noise," *Bell Sys. Tech. J.*, vol. 27, p. 109, Jan. 1948.

8.36. F. L. H. M. Stumpers, "Theory of Frequency Modulation Noise," *Proc. IRE*, vol. 36, p. 1081, Sept. 1948.

8.37. S. O. Rice, "Noise in FM Receivers," in *Time Series Analysis*, M. Rosenblatt (ed.), chapt. 25, p. 395, Wiley, New York, 1963.

8.38. J. Klapper, "Demodulator Threshold Performance and Error Rates in Angle-Modulated Digital Signals," *RCA Rev.*, vol. 27, p. 226, June 1966.

8.39. J. E. Mazo and J. Salz, "Theory of Error Rates for Digital FM," *Bell Sys. Tech. J.*, vol. 45, p. 1511, Nov. 1966.

8.40. L. H. Enloe, "Decreasing the Threshold in FM by Frequency Feedback," *Proc. IRE*, vol. 50, p. 18, Jan. 1962.

8.41. J. G. Chaffee, "The Application of Negative Feedback to Frequency Modulation Systems," *Bell Sys. Tech. J.*, vol. 18, p. 403, July 1939.

8.42. J. R. Carson, "Frequency Modulation—The Theory of the Feedback Receiving Circuit," *Bell Sys. Tech. J.*, vol. 18, p. 395, July 1939.

8.43. L. H. Enloe, "The Synthesis of Frequency Feedback Demodulators," *Proc. NEC*, vol. 18, p. 477, 1962.

8.44. D. L. Schilling and J. Billig, "On the Threshold Extension Capabilities of the PLL and FDMFB," *Proc. IEEE*, vol. 52, p. 621, May 1964.

8.45. P. Frutiger, "Noise in FM Receivers with Negative Frequency Feedback," *Proc. IEEE*, vol. 54, p. 1506, Nov. 1966.

8.46. M. M. Gerber, "A Universal Threshold Extending Frequency-Modulated Feedback Demodulator," *IEEE Trans.*, vol. COM-18, p. 276, Aug. 1970.

8.47. J. A. Develet, Jr., "A Threshold Criterion for Phase-Lock Demodulation," *Proc. IRE*, vol. 51, p. 349, Feb. 1963.

8.48. S. J. Campanella and J. A. Sciulli, "A Comparison of Voice Communication Techniques for Aeronautical and Maritime Applications," *COMSAT Tech. Rev.*, vol. 2, p. 173, Spring 1972.

8.49. Magnavox Brochure on MX-230 Voice Modem R-2018A, Sept. 1970.

8.50. W. R. Bennett and J. R. Davey, *Data Transmission*, McGraw-Hill, New York, 1965.

8.51. V. A. Kotel'nikov, *The Theory of Optimum Noise Immunity*, transl. by R. A. Silverman, Dover, New York 1968, republication.

8.52. H. Akima, "The Error Rates in Multiple FSK Systems and the Signal-to-Noise Characteristics of FM and PCM-FS Systems," Note 167, National Bureau of Standards, Mar. 1963.

8.53. T. T. Tjhung and H. Wittke, "Carrier Transmission of Binary Data in a Restricted Band, *IEEE Trans.*, vol. COM-18, p. 295, Aug. 1970.

8.54. R. de Buda, "Coherent Demodulation of Frequency-Shift Keying with Low Deviation Ratio," *IEEE Trans.*, vol. COM-20, p. 429, June 1972.

8.55. E. J. Sass and J. R. Hannum, "Minimum-Shift Keying for Digitized Voice Communications," *RCA Eng.*, vol. 19, Dec. 1973/Jan. 1974.

8.56. J. B. Anderson and D. P. Taylor, "A Bandwidth Efficient Class of Signal Space Codes," *IEEE Trans.*, vol. IT-24, p. 703, Nov. 1978.

8.57. D. Muilwijk, "Correlative Phase Shift Keying—A Class of Constant Envelope Modulation Techniques," *IEEE Trans.*, vol. COM-29, p. 226, Mar. 1981.

8.58. C. E. Shannon, "Communication in the Presence of Noise," *Proc. IRE*, vol. 37, p. 10, Jan.1949.

8.59. A. A. Meyerhoff and W. M. Mazer, "Optimum Binary FM Reception Using Discriminator Detection and IF Shaping," *RCA Rev.*, vol. 22, p. 698, Dec. 1961.

8.60. Motorola Semiconductor Product Information, "Product Brief: 24-bit Digital Signal Processor," Motorola, 1994.

Additional Suggested Reading

"Eight Ways to Better Radio Receiver Design," *Electronics*, Feb. 20, 1975.

Chapter 9
Other Receiver Circuits

9.1 Introduction

There are a number of functions that are performed in specialized receivers that, dependent upon their particular use, are not present in all receivers. We have already addressed AGC, which is used in almost all AM receivers and some FM receivers. In this chapter, we address a number of other circuits. The first three—noise limiting and blanking, squelch, and AFC—have been used in many receivers. Diversity reception systems have been used extensively to counteract multipath fading. Adaptive processing and quality monitoring have begun to appear in many modern receivers, especially those that require frequency changes to optimize performance because of propagation or interference conditions.

9.2 Noise Limiting and Blanking

While the standard measurement of receiver sensitivity is performed with additive gaussian noise, pulse interference is often the limiting noise for communications receivers at frequencies through the lower portions of the UHF band. This unpleasant interference may be generated by many different sources. To provide a better understanding of this problem, we shall first review the typical sources and types of interfering pulses. We shall then discuss various types of noise-reducing schemes that have been used, and finally we will provide conceptual information regarding one particular high-performance solution. There have been numerous solutions proposed since the early days of radio [9.1], but much of the information in this section is based on more recent papers by Martin (References [9.2] to [9.6]), and is reproduced here with his permission and that of the publishers.

Noise impulses are generated by a variety of different sources. Their characteristics in the time domain (oscillograms) and in the frequency domain (spectrum analyzer) are illustrated in Figure 9.1 to indicate the effects that are encountered in the noise-limiting or blanking process. The sources illustrated include switching clicks (Figure 9.1*a*), commutator sparking (Figure 9.1*b*), ignition interference (Figure 9.1*c*), corona discharge (Figure 9.1*d*), lightning discharge (Figure 9.1*e*), precipitation static from raindrops or sandstorms (Figure 9.1*f*), and radar pulses (Figure 9.1*g*). While the details of the waveforms differ, they are all characterized by high peaks and wide bandwidths. Some types may be aperiodic, while many of them are periodic, at a wide variety of rates.

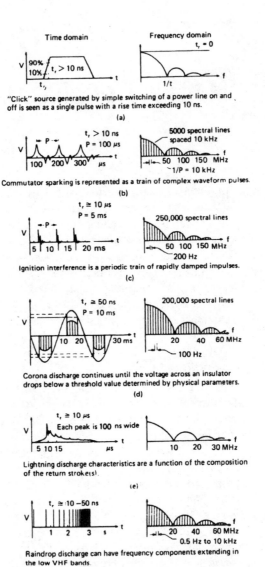

Figure 9.1 Various types of impulse interference displayed in the time and frequency domains. (*After [9.6].*)

A narrowband system with resonant circuit amplifiers will pass only those frequency components that fall within its passband range. An individual resonant circuit will be excited to oscillation at its resonant frequency by an impulse slope. The transient duration is dependent on the bandwidth, as determined by the circuit Q. In the case of multistage amplifiers or multipole filters, the output pulse delay is dependent on the circuit group delay t_g. This increases linearly with the number of resonators and is generally measured from the input impulse time t_0 to the time at which the output signal has risen to 50 percent of its peak amplitude. An approximate formula [9.7] gives $t_g = 0.35N/\delta f$, where N is the number of resonant circuits, δf is the bandwidth in hertz, and the rise time $t_r \approx 1/\delta f$. If stages of differing bandwidth are connected in series, the rise time will be determined mainly by the narrowest filter. The group delay results from the sum of the individual delay times.

If RF impulses of very short rise times are fed to an amplifier with much longer rise time, three different types of response can occur [9.7].

1. If the input pulse has a duration t_p longer than the transient time t_{rv}, the output signal will achieve the full amplitude ($U_a = VU_{in}$) and will maintain it for the duration of the drive time less the transient time, $t_p - t_{rv}$.

2. If $t_p = t_{rv}$, the output signal will achieve full amplitude during time t_{rv}, but immediately after it will commence to decay to zero in a period t_{rv}.

3. If the pulse is shorter than the transient time, the response will be essentially that of the transient, but with an amplitude that only reaches a portion of VU_{in}, i. e., the shorter the pulse, the smaller the amplitude produced by the amplifier. In other words, a substantial portion of the spectrum of the input pulse is not within the bandwidth of the amplifier and, therefore, does not contribute to the output amplitude.

In many systems, an AGC circuit is provided to ensure that a large range of input signal voltages are brought to the same level at the demodulator output. The AGC control voltage is generated subsequent to the narrowband IF filter. Typical communications receiver bandwidths range from about 0.1 to 50 kHz, corresponding to transient times of about 0.04 to 20 ms. The early receiver stages have much broader bandwidths. This means that steep input pulses can drive early amplifier stages into saturation before a reduction in gain is caused by the AGC (whose response time is generally longer than the narrowband filter transient response). This is especially true for receivers whose selectivity is determined in the final IF, often after substantial amplification of the input signal. Thus, the audible interference amplitudes may be several times stronger than the required signal level after the AGC becomes effective. The long AGC time constant increases the duration of this condition. It is only when the rise time of the input pulse is longer than the

AGC response time (e. g., during telegraphy with "soft" keying), that the output amplitude will not overshoot, as shown in Figure 9.2.

An effective method of limiting the maximum demodulator drive, and thus reducing the peak of the output pulse, is to clip the signal just in front of the demodulator, with symmetrically limiting diodes, to assure that the IF driver amplifier is not able to provide more than the limited output level, e. g., 1.5 V, peak to peak. It is necessary in this case that the AGC detector diode be delayed no more than a fraction of this level (0.4 V is a reasonable value) in order to ensure that the clipping process does not interfere with AGC voltage generation. Use of a separate AGC amplifier not affected by the limiter will also ensure this condition and can provide an amplified AGC system to produce a flatter AGC curve.

In the literature, three different methods have been tried to suppress interfering pulses. We designate these as *balancing*, *limiting*, and *blanking* (or *silencing*) [9.8]. Balancers attempt to reproduce the pulse shape without the signal in a separate channel and then perform a subtraction from the channel containing both the signal and the pulse. Limiters attempt to prevent the pulse level from becoming excessive. Blankers attempt to detect the onset of a pulse, and reduce to zero the gain of the signal amplifier chain at an early stage, for the duration of the pulse.

9.2.1 Balancers

Balancer systems are designed to obtain two signals in which the signal and noise components bear a relatively different ratio to one another. The two signals are then connected in opposition so as to eliminate the noise while a signal voltage remains. The main problems with this type of impulse noise suppression are obtaining suitable channels and exactly balancing out the noise impulse, which is generally many times stronger than the desired signal. Attempts have been made to use different frequency channels with identical bandwidths to get the same impulse shape but with the signal in only one channel. The difficulties of matching channels and finding interference-free channels make this approach unsatisfactory in most cases. Other approaches have attempted to slice the center from the pulse (to eliminate the signal) or to use a high-pass filter to pass only the higher frequency components of the pulse for cancellation. While some degree of success can be achieved with such circuits, they generally require very careful balancing and hence are not useful when a variety of impulses and circuit instabilities are encountered.

This type of circuit can be useful where the impulse source is a local one, which is physically unchanging. In this case, a separate channel (other than the normal antenna) can be used to pick up the pulse source with negligible signal component, and the gain of the pulse channel can be balanced carefully using stable circuits (and a feedback gain-control channel if necessary). It has also been found that modern adaptive antenna systems with sufficiently short response times can substantially reduce impulse noise coming from directions other than the signal

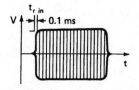

Input voltage first stage, hard keying

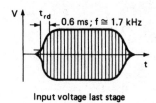

Input voltage last stage

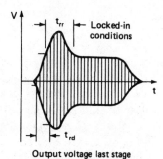

Output voltage last stage

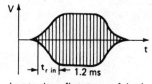

Input voltage first stage, soft keying

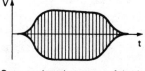

Output voltage last stage, soft keying

Figure 9.2 Impulse drive of RF amplifier with AGC. (*After [9.6].*)

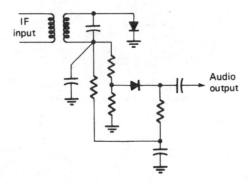

Figure 9.3 Schematic diagram of an automatic series noise limiter circuit.

direction. This is especially useful for those bands where narrowband signaling is the general rule (LF and below).

9.2.2 Noise Limiters

Because most of the noise energy is in the relatively large peak, limiters have been used, especially in AM sets, to clip audio signal peaks that exceed a preset level. It has been mentioned that an IF limiter to control maximum demodulator drive is effective in situations with overloading AGC circuits. Figure 9.3 shows a series limiter circuit at the output of an envelope demodulator, which has proven effective in reducing the audio noise caused by impulse interference. This type of circuit makes listening to the signal less tiring, but does not improve intelligibility of the received signal. The limiting level may be set to a selected percentage of modulation by adjusting the tap position of the two resistors feeding the anode of the limiter diode. If it is set below 100 percent, the limiting level also limits peaks of modulation. Since these occur seldom, such a setting is acceptable.

Because the impulse amplitude is higher and the duration shorter in the early stages of a receiver, limiting in such stages reduces the noise energy with less effect upon signal than limiting in later stages. Some FM receivers use IF stages that are designed to limit individually, while gradually reducing bandwidth by cascading resonant circuits. Such a design eliminates strong short impulses early in the receiver, before they have had a chance to be broadened by the later circuits. Such receivers perform better under impulse noise conditions than those that introduce a multipole filter early in the amplifier chain. We must remember, however, that wideband limiting can reduce performance in the presence of strong adjacent-channel signals.

The principles discussed are also applicable to data receivers. As long as the impulse interference is stronger than the signal, the signal modulation contributes little to the output. Generally the data symbol duration is longer than the duration of the input impulses. If the impulse can be reduced or eliminated before the establishment of final selectivity, only a small portion of the signal interval is distorted, and a correct decision is much more likely. Consequently, limiters at wide-bandwidth locations in the amplifying chain can result in a considerable reduction of the error rate in a data channel. Again, the possibility of interference from adjacent or other nearby channel signals must be considered.

9.2.3 Impulse Noise Blankers

Impulse noise blankers operate based on the principle of opposite modulation. In effect, a stage in the signal path is modulated so that the signal path is blanked by an AM process for the duration of the interference. It is also possible to use an FM method in which the signal path is shifted to a different frequency range. This latter procedure [9.9] uses the attenuation overlap of IF filters in a double super-heterodyne receiver. The second oscillator is swept several kilohertz from nominal frequency for the duration of the interference so that the gain is reduced to the value of the ultimate selectivity in accordance with the slope of the filter curves. This method is especially advantageous since the switching spikes, which often accompany off-on modulation, should not be noticeable. However, when using an FM modulator having high speed (wide bandwidth), components can appear within the second IF bandwidth from the modulation. The most stringent limitation of this method is the requirement for two identical narrow-band filters at different frequencies along with an intermediate mixer. Thus, the concept is limited to a double conversion superheterodyne receiver with a variable first oscillator.

When using an AM method, two types of processing are possible:

1. The interference signal is tapped off in parallel at the input of the systems and increased to the trigger level of a blanker by an interference channel amplifier having a pass bandwidth that is far different from the signal path. A summary of such techniques is given in [9.10]. This method is effective only against very wide band interference, since noticeable interference energy components must fall into the pass-band range to cause triggering. This method will not be effective in the case of narrowband interference, such as radar pulses, which are within or directly adjacent to the frequency range to be received.

2. The interference signal is tapped off from the required signal channel directly following the mixer (References [9.2] and [9.3]) and fed to a fixed-frequency second IF amplifier, where it is amplified up to the triggering level. Because there is a danger of crosstalk from the interference channel to the signal amplifier channel, it is advisable to use a frequency conversion in the interference channel. Thus, interference amplification occurs at a different frequency than the signal IF. Attention must be paid when using this method that there are no switching spikes generated during the blanking process that can be fed back to the interference

channel tap-off point. Otherwise there would be a danger of pulse feedback. The return attenuation must, therefore, exceed the gain in the interference channel between the tapping point and the blanker.

The blanker must be placed ahead of the narrowest IF filter in the signal path. It must be able to blank before the larger components of the transient have passed this filter. Therefore, we must assure a small group delay in the interference channel by using a sufficiently broad bandwidth and a minimum of resonant circuits. It is desirable to insert a delay between the tap-off point and the signal path blanker so that there is sufficient time for processing the interference signal. If this is done, it is not necessary to make the interference channel excessively wide, while still assuring the suppression of the residual peak.

Figure 9.4 is the block diagram of a superheterodyne receiver with this type of impulse noise blanker. Figure 9.5 illustrates its operation in the presence of a strong interfering radar pulse. An essential part of the blanker is the use of a gate circuit that can operate linearly over a wide dynamic range. Figure 9.6 shows such a gate, using multiple diodes. The circuit is driven by the monostable flip-flop, which is triggered by the noise channel. Figure 9.7 is a schematic of a noise blanker circuit designed following these principles. When the noise channel is wider than the signal channel, this type of noise reducer can also have problems from interfering signals in the adjacent channels.

9.3 Squelch Circuits

Sensitive receivers produce considerable noise voltage output when there is no signal present. This condition can occur when tuning between channels or when the station being monitored has intermittent transmissions. If the signal is being monitored for audio output, such noise can be annoying and, if repeated frequently, fatiguing. To reduce this problem, circuits are often provided to reduce the output when a signal is not present. Such circuits have been referred to as *squelch, muting,* and *quiet automatic volume control* (QAVC) systems. The circuits used differ, depending on the received signal characteristics.

Squelch circuits for AM receivers generally operate from the AGC voltage. When a weak signal or no signal is present, the voltage on the AGC line is at its minimum and receiver gain is maximum. When a usable signal is present, the AGC voltage rises to reduce the receiver gain. The voltage variation tends to rise approximately logarithmically with increasing signal levels. By using a threshold at a preset signal level, it is possible to gate off the audio output whenever the signal level drops below this point. Such a system can be used to mute the receiver during the tuning process. The threshold may also be set for the level of a particular signal with intermittent transmissions, so that noise or weaker interfering signals will not be heard when the desired signal is off. When the transmission medium causes

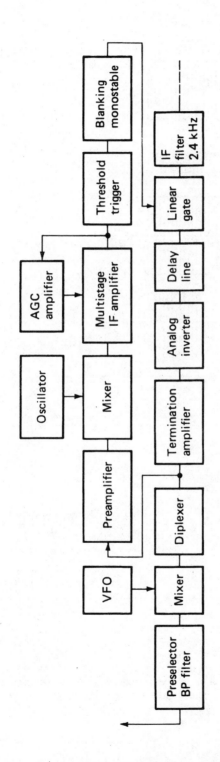

Figure 9.4 Block diagram of a superheterodyne receiver with noise blanker. (*After [9.6].*)

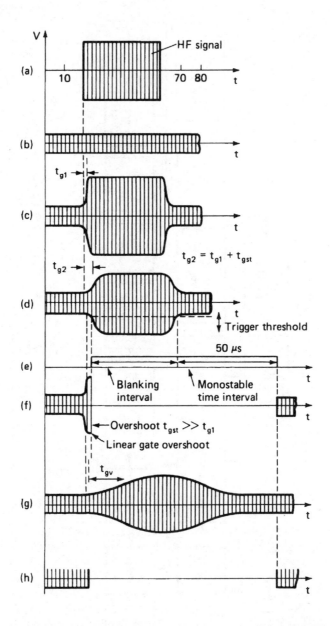

Figure 9.5 Waveforms illustrating the operation of a noise blanker: (*a*) interfering radar noise pulse of 40 μs-duration, (*b*) desired signal, (*c*) interference and signal after diplexer, (*d*) noise-channel output signal, (*e*) blanking monostable output, (*f*) linear gate output, (*g*) delayed version of linear gate input signal, and (*h*) delayed version of linear gate output signal at main channel. (*After [9.6].*)

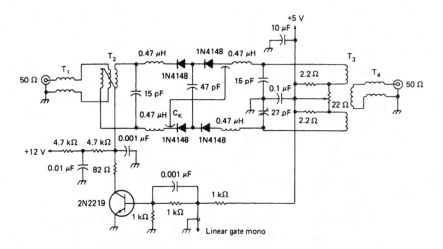

Figure 9.6 Schematic diagram of a blanker gate with high dynamic range. (*After [9.6].*)

signal fading, as is common at HF, squelch circuits are somewhat less effective for this use, since the threshold must be set low enough to avoid squelching the desired signal during its fades. This provides a smaller margin to protect against noise or weaker interfering signals.

Figure 9.8 shows the block diagram of an AM squelch system, and Figure 9.9 shows a typical schematic diagram, where a diode gate is used for reducing the output signal. Many types of switching have been used for this purpose, including biasing of the demodulator diode and biasing one element of a multielement amplifying device. The latter approach was frequently used with receivers that employed multigrid vacuum tube amplifiers. However, it can be applied to multigate FET amplifiers or balanced amplifier integrated circuits with current supplied by a transistor connected to the common-base circuit of the amplifying transistor pair. Figure 9.10 illustrates these alternative gating techniques.

Many FM receivers do not use AGC circuits but depend on circuit limiting to maintain the output level from the demodulator. In this case, squelch may be controlled by the variations in voltage or current that occur in the limiter circuits. Such changes occur when single-ended amplifiers are used for limiting, but in balanced limiter arrangements, these changes may not be so readily available. Furthermore, the wide range of threshold control provided by AGC systems is generally not available from limiters. This tends to make FM squelch systems, which are dependent on the signal level, more susceptible to aging and power supply instabilities than the AGC operated systems. Consequently, two other types of control have evolved for FM use; noise-operated and tone-operated. (The latter could be used for AM also.)

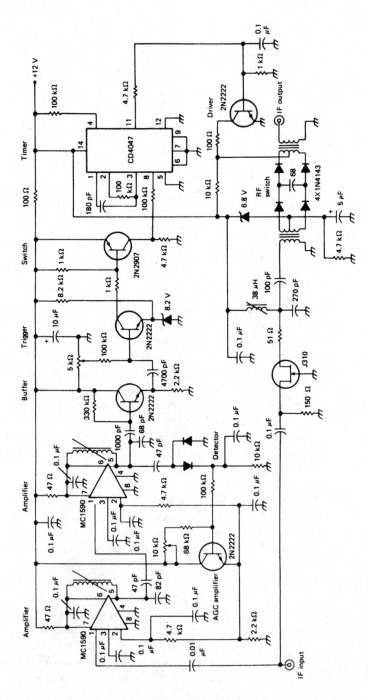

Figure 9.7 Schematic diagram of a noise blanker circuit.

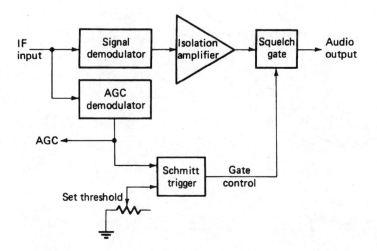

Figure 9.8 Block diagram of an AM squelch circuit.

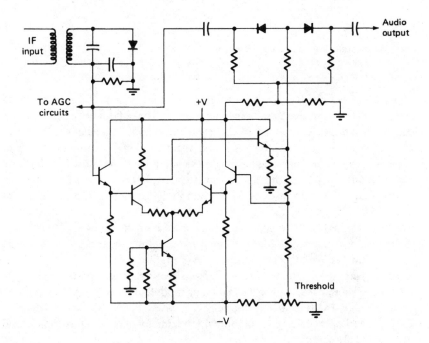

Figure 9.9 Schematic diagram of an AM squelch circuit.

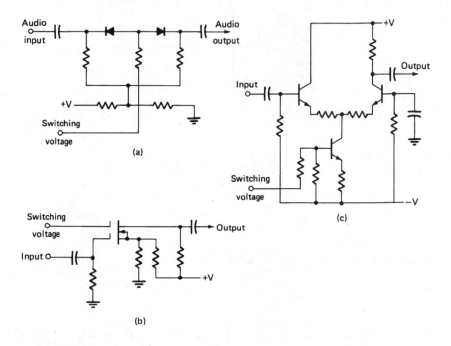

Figure 9.10 Common gate circuits for squelch applications.

Figure 9.11 is a block diagram for a noise-operated squelch. This system makes use of the fact that the character of the output noise from a frequency demodulator changes when there is no signal present. At the low output frequencies, when noise alone is present in the FM demodulator, there is a high noise level output, comparable to that at other frequencies in the audio band. As the strength of the (unmodulated) signal rises, the noise at low frequencies decreases, while the noise at higher frequencies decreases much less rapidly. This can be seen by referring to Figure 8.44, which shows the variation in FM output spectrum as the S/N varies from 0 to 7 dB. When there is no signal, the maximum output noise density is at zero frequency. The density drops off slowly as the output frequency increases. The peak deviation is not likely to exceed the 3-dB point of the IF filter (about $1.18 f/\sigma$ in the figure), and the peak modulating frequency is likely to be substantially less. Assuming $0.5 f/\sigma$ to be the maximum modulating frequency, at 7-dB S/N, this has a density 15 dB lower than the density at 0-dB S/N. At $0.05 f/\sigma$ the reduction is 22.7 dB. At $1.5 f/\sigma$ (three times or more the maximum modulating frequency) the reduction is only about 9 dB.

If in Figure 9.11, the squelch low-pass filter cuts off at $0.025 f/\sigma$ (150 Hz if we set $0.5 f/\sigma$ to 3 kHz), it will be uninfluenced by modulation components. If the gain

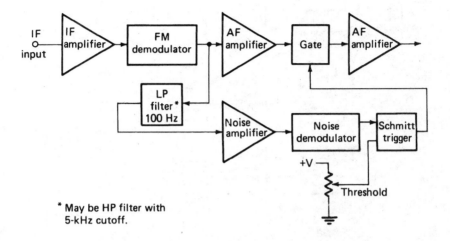

Figure 9.11 Block diagram of a noise-operated squelch circuit for an FM receiver.

of the squelch amplifier is set so that the squelch rectifier produces 5 V when S/N equals zero, then a 7-dB signal level will cause this output to drop to about 0.03 V. A threshold may readily be set to cause the squelch gate to open at any S/N level between −3 and 7 dB. Because the control voltage level is dependent on the gain of the RF, IF, and squelch amplifiers, variations in squelch threshold may occur as a result of gain variation with tuning or gain instabilities. If a second filter channel (Figure 9.12) tuned above the baseband is used, the two voltages can be compared to key the gate on. While both are subject to gain variations, their ratio is not, resulting in better threshold stability. A similar scheme has been used for SSB voice, where noise density is uniform, but modulation energy is greater below 1 kHz.

With the difference approach, the range of threshold control is, however, limited. Hence a weak interfering signal, which would produce negligible interference to the desired signal, may still operate the squelch gate. To overcome the problem of undesired weak interferers operating the squelch, the tone-operated squelch was devised. In this case, the transmitted signal has a tone below normal modulation frequencies added to the modulation, with a relatively small deviation. At the receiver, a narrowband filter is tuned to the tone, and its output is amplified and rectified to operate a trigger for the squelch gate. This scheme is quite effective as long as interferers do not adopt a similar system using the same frequency. In such a case, multitone coding could be employed, or digital modulation of the tone with a predetermined code could be used to assure the desired performance. For ordinary receivers, these more elaborate schemes have not been widely used.

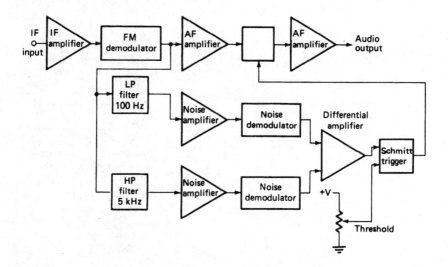

Figure 9.12 Block diagram of an improved noise-operated squelch circuit for an FM receiver.

However, in some systems with multiuser network operation on one frequency, a coding scheme known as *selective call* (Sel Call) has been implemented so that users need receive only those messages directed toward them. In this type of signaling scheme, the caller sends a multitone or digital code at the beginning of the message to indicate the identity of the called party or parties. Only if the receiver code matches the transmitted code is the output gate enabled to transmit the message to the receiver user. This type of system may be used with both analog and digital modulation, and is independent of the modulation type used for transmission (AM, FM, or PM). Such a scheme is more elaborate than a normal squelch system but performs an important function in multiuser networks.

9.4 AFC

AFC has been used for many years in some receivers to correct for tuning errors and frequency instabilities. This was of special importance when free-running LOs were used. Inaccuracies in the basic tuning of the receiver, and of drifts in some transmitters, could cause the desired signal to fall on the skirts of the IF selectivity curve, resulting in severe distortion and an increased chance of adjacent channel interference. Provision of a circuit to adjust the tuning so that the received signal falls at (or very near) the center of the IF filter enables the receiver to achieve low demodulation distortion, while maintaining a relatively narrow IF bandwidth for interference rejection.

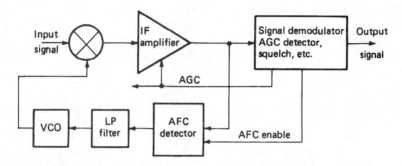

Figure 9.13 Typical AFC block diagram.

The need for such circuits has been almost eliminated by the advent of synthesized LOs under the control of very accurate and stable quartz crystal standards. However, some instances still arise where an AFC may be of value. In some tactical applications, it is not possible to afford the power necessary for temperature control of the crystal standard, so that at sufficiently high frequencies the relative drift between the receiver and the transmitter may be more than is tolerable in the particular application. Unstable oscillators in some older and less accurate transmitters have not yet been replaced. Current receivers may be required to receive signals from such transmitters for either communications or surveillance. In automobile FM broadcast receivers, the cost of synthesis is considerably higher than the cost of AFC, so that progress to high stability has been slower than for communications receivers. Finally, some modulations, notably SSB, require much better frequency accuracy for distortionless demodulation than AM or FM. As such modulations are extended to higher carrier frequencies, AFC can be more economical than still more accurate frequency control. Consequently, AFC circuits are still found in receivers, though much less frequently than in the past.

The basic elements of the AFC loop are a frequency (or phase) detector and a VCO. A typical block diagram is shown in Figure 9.13. If the received signal carrier is above or below the nominal frequency, the resulting correction voltage is used to reduce the difference. The typical problems of loop design occur. If a PLL is used, the design is comparable to that required by a synthesizer (see Chapter 7). The low-pass filter, however, should be capable of eliminating any FM or PM of the received signal. If a frequency-locked loop is used, then the error detector will be of the frequency discriminator type discussed in Chapter 8. Because it is the center frequency, not the modulation, that is of importance, the separation of the peaks of the discriminator curve will be much closer than is normal for FM. As in any other feedback system, attention is needed in design to the problems of response time and stability. As in the case of FM demodulators, the various processes can be achieved more accurately by digital processing than by analog circuits. If a

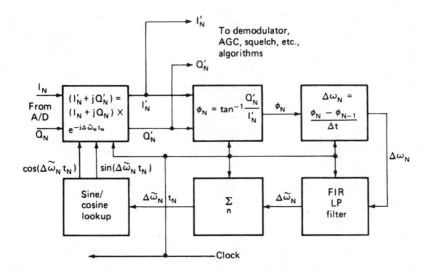

Figure 9.14 Block diagram of AFC algorithms for digital processor implementation.

sufficiently capable processor is available, the entire process, including frequency error detection and frequency correction, can be carried out digitally. Figure 9.14 shows the basic algorithms needed for such processing.

In designing AFC circuits, it is necessary to control carefully the pull-in and hold-in ranges of the circuit. For this application, it is desirable to have a limited pull-in range. Otherwise a strong interfering signal in a nearby channel may gain control of the loop and tune to the wrong signal. This is a common experience in low-quality FM auto receivers with AFC. The hold-in range should be as great as the expected maximum error between the receiver frequency and that of the desired signal. Generally the hold-in range exceeds the pull-in range, often by several times, so pull-in is likely to prove the larger design problem. Another problem that can occur for some modulation types, especially those having subcarriers, is lock on a sideband rather than the carrier when the frequency error is sufficiently large. Circuits have been devised for specific cases of sideband lock to distinguish between the sideband and the carrier. Because of the decreasing use of AFC circuits, we shall not go further into design principles and problems here. However, in the following paragraphs, we describe a digital AFC scheme for accurately tuning an SSB signal [9.11].

Figure 9.15 illustrates the block diagram of a digital HF radio. The portion implemented digitally is shown within dashed lines. Figure 9.16 illustrates, again in block format, how the SSB detection is performed. Figure 9.17 shows the signal spectra at various stages of the processing. So that we may develop a software solution to automatic SSB tuning, we use the method shown in Figure 9.18. The voice bandwidth signal input is transformed to a voice power spectrum, which is analyzed to detect harmonic relations and complex correlations. We process the resulting correlations to determine the frequency offset relative to the (suppressed) carrier frequency. A digital command is then generated to retune the frequency synthesizer to correct the offset. The resulting center frequency can then be stored in the processor along with the demodulated data.

Alternatively, the same process can be used to regenerate the suppressed carrier. If we refer to Figure 9.19, we see the results from a succession of processing steps, starting from the bottom. The lowest line appears to be pure noise. After sufficient digital processing it is possible to discover the actual suppressed carrier, and even 60-Hz hum sidebands. Such detail is of value, not only for tuning of the signal, but also in recognizing differences among specific pieces of transmitting equipment.

9.5 Diversity Reception

The term *diversity reception* refers to a receiving process that uses more than one transmission of the same information to obtain a better result than can be achieved in a single transmission. The first use of diversity radio reception probably occurred the first time that an operator received a garbled message and asked for a repeat. This form of diversity, which relies on transmissions repeated after a delay, is known as *time diversity*, since the second channel may use the same transmission medium during a later time interval. Simple *requests for retransmission* (RQs) sufficed until commercial long-distance transmission began to use HF radio for machine telegraphy or AM radiotelephony.

Multipath fading of HF channels resulted in the early use of *space diversity*, in which multiple antennas sufficiently separated could provide the diverse channels (References [9.12] and [9.13]). Later it was determined that diversity for narrowband telegraph channels could be provided by sending the same information over two separate channels with sufficient frequency separation. This is known as *frequency diversity*. Another form of diversity useful at HF when there is not sufficient space available (space diversity requires separation of antennas by a distance of many wavelengths to be effective), is *polarization diversity*. This technique uses two antennas that respond to vertical and horizontal polarization, respectively, of an incoming wave, and is possible because the two components of the wave tend not to fade simultaneously. Circularly polarized antennas can also provide clockwise and counterclockwise polarizations for polarization diversity. At higher frequencies, where circular polarization of antennas is more common, fades between different polarizations tend to be highly correlated, so that polarization diversity is ineffective.

A higher-power transmitter or longer-time transmission is required for time and frequency diversity than for space or polarization diversity to provide comparable performance. This is because in the former cases, the energy available from the

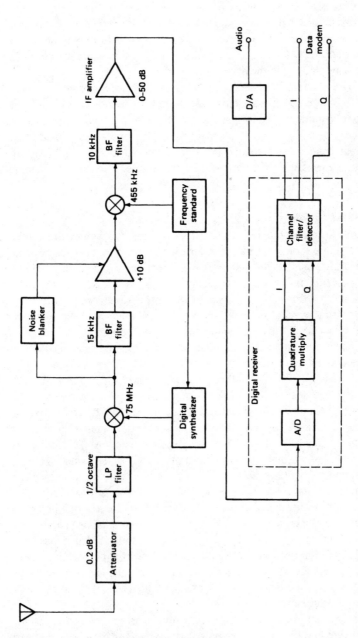

Figure 9.15 Block diagram of a typical HF radio. Digitally implemented portions are contained within the dashed lines.

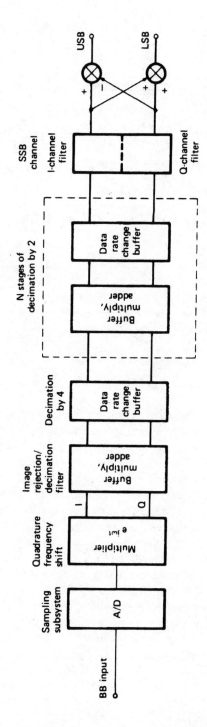

Figure 9.16 Block diagram of digital-processing functions in SSB demodulators.

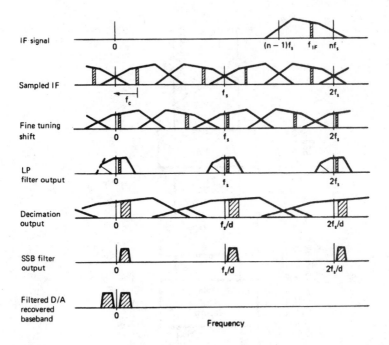

Figure 9.17 Signal spectra at various stages of processing.

transmitter must be divided among the multiple channels, whereas in the latter cases the diverse channels can be derived from a single transmission. In all cases, the effectiveness of diversity reception in improving transmission performance is dependent upon the independence of fading among the diverse signals and the method by which the receivers use these signals. Complete independence is assumed in most analyses. However, in practice there is often some degree of correlation. Analyses have showed that the correlation can be substantial (greater than 0.5) without causing major reduction in the diversity advantage.

Diversity techniques may be divided into switching and combining approaches. The former attempts to select for output the channel that has the higher overall level. The S/N would be a better measure, but it is much more difficult to determine. Because the system design should provide for a signal that on average has substantially higher amplitude than the noise, selection of the channel with larger amplitude provides a substantial improvement much of the time. The simplest system of this sort uses antenna switching, as shown in Figure 9.20. Whenever the signal drops below a predetermined level, the system switches to another antenna and if necessary and possible, to still others until a channel above the threshold is encountered.

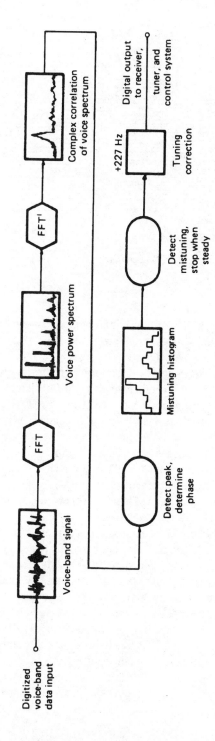

Figure 9.18 A method for automatic SSB tuning.

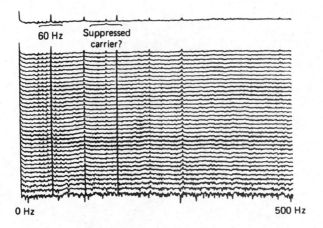

Figure 9.19 Exponentially averaged SSB spectra ($\alpha = 0.9$), mistuned upward by approximately 150 Hz.

The simple switching system, while providing adequate signal most of the time, occasionally may lose signal for a short time while finding a suitable antenna. For more than two antennas, the signal selected may not necessarily be the best available from the several antennas. The system can be improved by having a second receiver that is continually scanning the several channels and recording the levels, so that switching may be accomplished to the strongest channel on a regular basis without waiting for the current channel to fade below a satisfactory level. This control system and the second receiver introduce more complexity than the simple system, so that one of the other techniques may prove equal or better.

When multiple receivers are available, they may all be controlled to provide equal gain. The strongest-level output is then selected. A technique similar to the one given in [9.12] may be used for AM signals. This is illustrated in Figure 9.21 for two receivers, but it can be further extended. Switching is accomplished by using diode demodulators feeding a common load circuit. The strongest signal develops a demodulated voltage level across the load that prevents conduction in the diodes driven by weaker-amplitude signals. Under the condition that two or more of the receivers produce amplitudes that are nearly the same, all of these stronger signals contribute to the generation of the demodulated voltage, which prevents contribution by the weaker signals. For AM signals, the carrier level provides a dc component across the load, which can be used to develop a common AGC voltage for all the receivers. This maintains their gains equal, even though some fading occurs in the strongest signal, requiring a gain change to provide a constant output. For other modulation types, this simple combination technique is not generally possible, since the diode envelope detector is not an appropriate

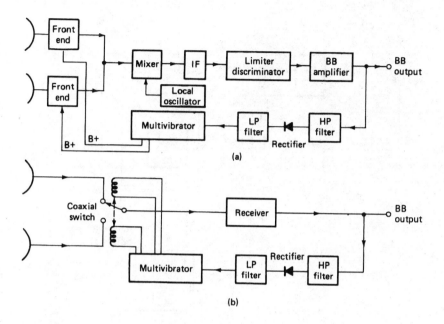

Figure 9.20 Block diagram of an antenna-switching diversity system. (*After [9.14].*)

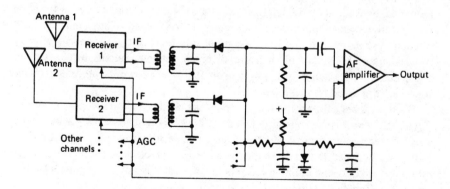

Figure 9.21 Switching diversity circuit for AM signals.

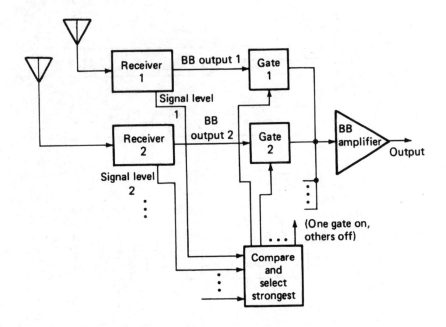

Figure 9.22 Switching diversity circuit for FM channels.

demodulator for such signals. In such cases, the amplitude-sensing, switching, demodulation, and AGC channels can be separated as shown in Figure 9.22.

Diversity system analysis is generally based on the assumption that multipath fading has an essentially Rayleigh distribution. The justification of this assumption is doubtful in many applications, since it depends on the law of large numbers, and often only two or three multipath components are present. A good summary discussion of the theory of different diversity schemes, along with a careful consideration of assumptions, is given in [9.15]. The results presented in the following are based on that reference.

What we have called switching diversity is called *selection diversity* in [9.15]. Figure 9.23 shows the probability that the output S/N will be below various levels when different numbers of diverse channels with identical noise and Rayleigh fading are switched so that the output with best S/N is selected. This curve is based on the fact that for the output to drop below a level S/N, all of the channels must do so. The Rayleigh distribution for the single channel is:

$$P_1 (x < S_1/N_1) = 1 - \exp(-S_1/N_1) \tag{9.1}$$

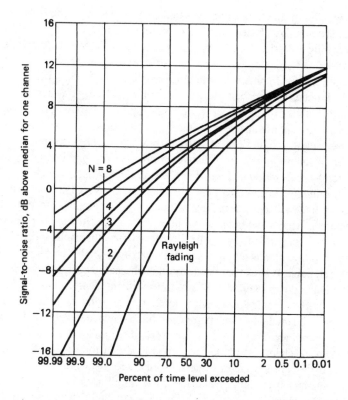

Figure 9.23 Selection diversity output distributions for n independent diversity channels with equal S/N.

For all n channels to be less than S/N, when the S/N distributions are equal and independent:

$$P_n (x < S/N) = [1 - \exp(-S_1/N_1)]^n \tag{9.2}$$

In combining techniques, the signals from all of the channels are multiplied by a weighting function and then added. The principle is based on the assumption that the signal amplitudes are all in phase so that they add arithmetically, while the noise amplitudes are all independent and add in an rms fashion. The two combining techniques in use are simple addition and maximal-ratio combining. In the first case, all weighting functions have the same constant value. In the latter case, they follow a particular law, based on channel measurements.

In switching diversity, the switches may be located either prior to demodulation or after it without affecting the output statistics, as long as the sensing circuits can function adequately. In practice, switching prior to demodulation, where the phases of the RF signals may vary, can cause undesirable switching transients. In the case of combining diversity, however, it is obvious that in combining before demodulation, the random phase differences among the channels will defeat the purpose of adding the weighted signal amplitudes, whereas combining after modulation does not suffer from this problem. The two techniques of combining are generally referred to as *predetection* and *postdetection combining*, respectively.

If the modulation is nonlinear, the advantage of the overall improvement in S/N from combining will be reduced by postdetection combining since the relative relationships of signal and noise amplitudes can be modified in the demodulation process. In FM, for example, when the S/N drops below threshold, the demodulated S/N rises rapidly. Predetection combining can reduce the probability that the composite signal fails to exceed the threshold, whereas in postdetection combining the individual signals may drop below threshold frequently, thus degrading the composite output signal. Predetection combining requires that the phases of the several signals be made the same before combining, hence increasing the complexity of the combiner. Figure 9.24 is a block diagram of an equal-gain (equal weighting in all channels) predetection combiner. In the case of the equal-gain combiner, if the noise voltages in each channel are assumed to have equal independent gaussian distributions, and the noise amplitudes independent Rayleigh distribution variables, the distribution for the combined S/N is given in Brennan [9.15, Equation (41)]. The resultant distributions for several orders of diversity are given in Figure 9.25.

A block diagram of *maximal-ratio combining*, introduced to radio communications by Kahn [9.17], is shown in Figure 9.26. This differs from the equal-gain combiner in that the weighting functions instead of being the same and constant in all channels, are now weighted by the factor x_i/σ^2, where x_i are the values of the ith signal amplitude and σ is the rms value of the channel gaussian noise, again assumed equal in all the channels. In this case, the resultant combined distribution function is easy to integrate in terms of tabulated functions (incomplete gamma function) and can be expressed as a sum of simple functions:

$$P\left(x < S/N\right) = 1 - \left\{\sum_{k=0}^{n-1}\left[\frac{(S/N)^k}{k!}\right]\right\}e^{-S/N}$$

$$(9.3)$$

$$P\left(x < S/N\right) = \left\{\sum_{k=n}^{\infty}\left[\frac{(S/N)^k}{k!}\right]\right\}e^{-S/N}$$

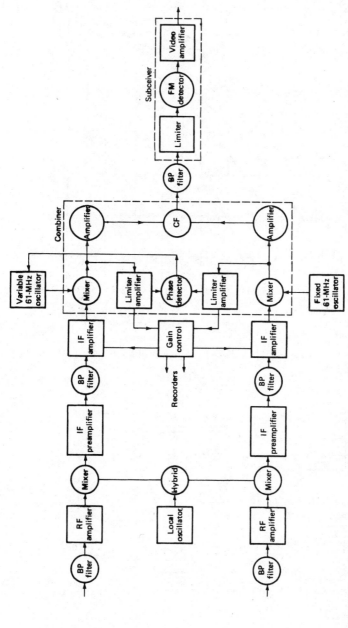

Figure 9.24 Block diagram of an equal-gain predetection combiner. *(From [9.16].)*

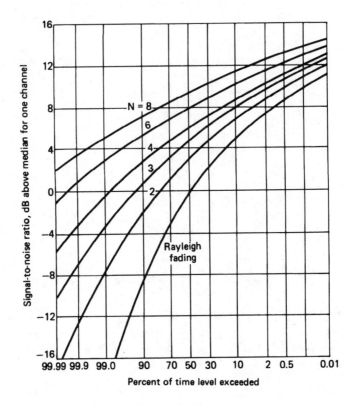

Figure 9.25 Equal-gain combiner diversity output distributions for *n* diversity channels having equal S/N.

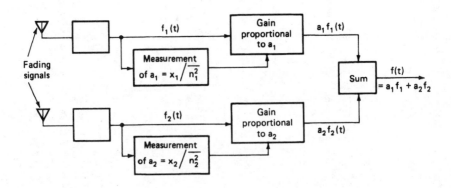

Figure 9.26 Block diagram of a maximal-ratio diversity combiner. (*From [9.15]*.)

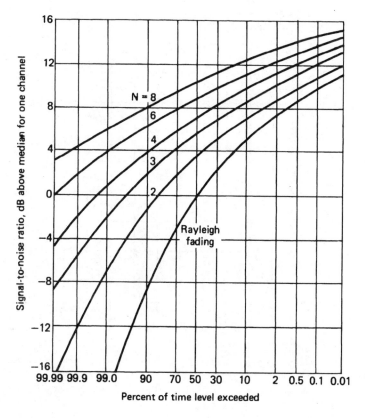

Figure 9.27 Maximal-ratio combiner distributions for n diversity channels having equal S/N.

Figure 9.27 shows the resulting distribution functions for various numbers of diversity channels.

In Figures. 9.23, 9.25, and 9.27, we note that the largest increment of diversity improvement occurs between no diversity and dual diversity, and the improvement increment gradually decreases as n grows larger. This is most obvious for switched diversity, where at 0.01 availability the improvement from $n = 1$ to 2 is about 8 dB, from $n = 2$ to 3, it is 3.7 dB, and from $n = 3$ to 4, it is 1.7 dB. This same trend is indicated in the improvement of average S/N level with diversity order, as shown in Figure 9.28. All of the foregoing theoretical estimates have assumed independent Rayleigh fading in the channels. The effects of other possible fading distributions have not been explored extensively, since experimental information on distributions of fading is limited and in many cases the data fits a Rayleigh distribution

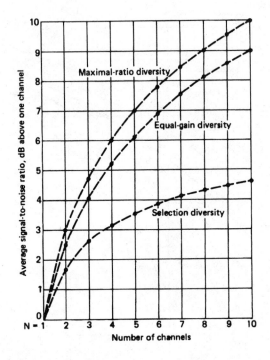

Figure 9.28 Diversity improvement in average S/N for different diversity techniques. (*From [9.15].*)

reasonably well over a limited range. Some calculations have been made on the effect of correlation between Rayleigh fading channels (Figure 9.29).

All of the techniques described here have been found effective in various multipath situations, although in some cases it is difficult to distinguish whether the implementation produces switching or combining diversity. Diversity is only of value for the relatively rapid fading caused by multipath. This kind of fading is encountered extensively at HF and for tropospheric and ionospheric scatter propagation at higher frequencies. Similarly, mobile vehicles passing through a static multipath field at VHF and UHF encounter such fading as a result of their motion. However, the long-term fading, which occurs as a result of diurnal, seasonal, or sunspot cycle variations, affects all direct radio channels between two points essentially equally, and hence cannot be improved by diversity reception.

9.6 Adaptive Receiver Processing

The term adaptive processing is applied to techniques intended to modify the receiver characteristics with changing signal environment so as to achieve im-

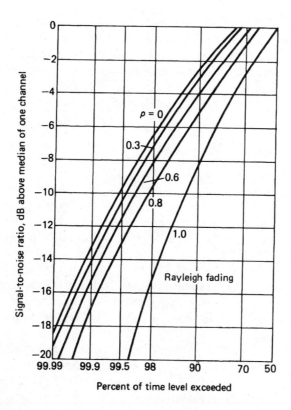

Figure 9.29 Dual switching diversity output distribution for correlated Rayleigh fading. (*From [9.15]*.)

proved performance. The use of diversity, discussed in the last section, may be considered a simple form of adaptation that operates on several samples of the signal and interference to produce a reduced outage fraction or reduced error rate under certain conditions of fading. HF frequency management, changing the frequency of a radio link in response to the diurnal, seasonal, and sunspot cycles, is another simple, manual form of adaptation that has been used for many years.

In recent years, two additional forms of adaptivity have appeared—*adaptive antenna processing* (sometimes referred to as *adaptive null steering*) and *adaptive equalization*. With the availability of low-cost low-power microprocessors, the implementation of automatic adaptive frequency management has become attractive, as well as other adaptive processes. Adaptive antenna processing arose out of the antijam requirements of radar systems, but is equally applicable to communications use, when several antenna elements are available. In adaptive antenna processing, each of the inputs are modified in amplitude and delay (or phase) prior to their combination to achieve a desired signal improvement. The adaptive control process by which this is accomplished and the techniques for input modification must be designed for the specific application. Adaptive equalization came about because of developments intended to counteract intersymbol interference generated in data transmission over the telephone network by channel distortion. The concepts are now being used to combat multipath and other distortion in radio communications.

9.6.1 Adaptive Antenna Processing

Much of the original work on adaptive antenna processing was related to military applications (References [9.18] to [9.22]). The techniques described in [9.19] to [9.22] are based on the fact that objectionable interference generally does not come from the same direction as the signal, so that separated antenna elements receive the two sets of radiation with different delays. By appropriate amplitude and/or delay variations in the two channels, a single interferer may be nulled, while the signal remains available. Because it is the differences in delays that are primarily responsible for the signal differences, at least when the antenna elements are identical, broadband processing requires that the amplitude and delay be adjusted in each channel. However, for narrowband signals, carrier phase difference may be substituted for delay.

To illustrate how this occurs, a simple two-element example from Widrow et al. [9.18] will be used before the more complex general expressions are given. Figure 9.30 shows the configuration. The antennas are assumed to be simple omnidirectional elements, with the signal arriving at angle θ and the interferer at angle Φ. The signal amplitude in each antenna element is the same, but the phase differs because of the arrival angle ϕ. Analogously, the interference amplitudes are the same, and phases differ because of angle θ. Each received channel is split into I and Q components, which are separately multiplied by different values or weights. This results in an effective amplitude and phase change in each channel. If we use the usual complex notation for the narrowband signals, we have for the signal and interferer components at the output:

$$S_{out} = W_1 s \exp[j(d \sin \phi)/2] + W_2 s \exp[-j(d \sin \phi)/2] \tag{9.4}$$

$$N_{out} = W_1 n \exp[j(d \sin \theta)/2] + W_2 n \exp[-j(d \sin \theta)/2] \tag{9.5}$$

where s and n are the amplitudes of the two waves, and the phases are referred to the phase at the center of the array. We would like to eliminate the interferer output N_{out} while keeping the signal output S_{out} equal to si, for example. This requires four real equations to be satisfied:

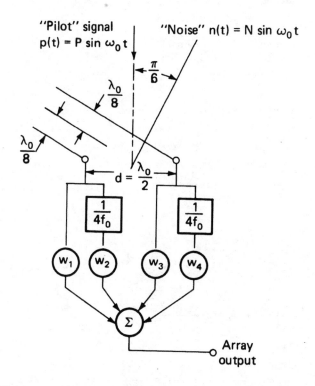

Figure 9.30 Simple array configuration. (*From [9.18]. Reprinted with permission of IEEE.*)

$$w_{1i} \cos\left(\frac{d \sin\theta}{2}\right) - w_{1q} \sin\left(\frac{d \sin\theta}{2}\right)$$

$$+ w_{2i} \cos\left(\frac{d \sin\theta}{2}\right) + w_{2q} \sin\left(\frac{d \sin\theta}{2}\right) = 0 \qquad (9.6)$$

$$w_{1q} \cos\left(\frac{d \sin\theta}{2}\right) + w_{1i} \sin\left(\frac{d \sin\theta}{2}\right)$$

$$+ w_{2q} \cos\left(\frac{d \sin\theta}{2}\right) - w_{2i} \sin\left(\frac{d \sin\theta}{2}\right) = 0 \qquad (9.7)$$

$$w_{1i} \cos\left(\frac{d \sin\phi}{2}\right) - w_{1q} \sin\left(\frac{d \sin\phi}{2}\right)$$

$$+ w_{2i} \cos\left(\frac{d \sin\phi}{2}\right) + w_{2q} \sin\left(\frac{d \sin\phi}{2}\right) = 1 \qquad (9.8)$$

$$w_{1q} \cos\left(\frac{d \sin\phi}{2}\right) - w_{1i} \sin\left(\frac{d \sin\phi}{2}\right)$$

$$+ w_{2q} \cos\left(\frac{d \sin\phi}{2}\right) + w_{2i} \sin\left(\frac{d \sin\phi}{2}\right) = 0 \qquad (9.9)$$

These are four linear equations with constant coefficients, and may be solved in the usual way to produce:

$$W_1 \equiv w_{1i} + j\, w_{1q}$$

$$W_1 = -\frac{\sin[(d \sin\theta)/2] + j \cos[(d \sin\theta)/2]}{2 \sin[d (\sin\phi - \sin\theta)/2]} \qquad (9.10)$$

$$W_1 = -\frac{j \exp[-j (d \sin\theta)/2]}{2 \sin[d (\sin\phi - \sin\theta)/2]}$$

$$W_2 \equiv w_{2i} + j\, w_{2q}$$

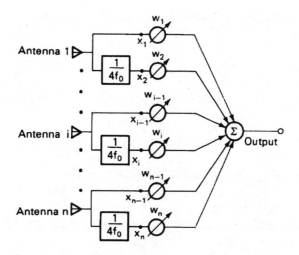

Figure 9.31 Generalized form of an adaptive array. (*From [9.18]. Reprinted with permission of IEEE.*)

$$W_2 = -\frac{\sin\left[(d\sin\theta)/2\right] - j\cos\left[(d\sin\theta)/2\right]}{2\ \sin\left[d(\sin\phi - \sin\theta)/2\right]} \qquad (9.11)$$

$$W_2 = \frac{j\exp\left[j\,(d\sin\theta)/2\right]}{2\ \sin\left[d(\sin\phi - \sin\theta)/2\right]}$$

The final forms on the right also arise from a direct solution of Equations (9.4) and (9.5) by letting $W_1 = W\exp\left[-j\,(d\sin\theta)/2\right]$ and $W_2 = -W\exp\left[j\,(d\sin\theta)/2\right]$, which satisfy Equation (9.5) with $N_{out} = 0$, and then solving for W in Equation (9.4). The problem of automatically controlling the coefficients to find these values remains and will be discussed after the forms of equation and solution have been indicated for the more general network of Figure 9.31.

In Figure 9.31 there are now n array elements providing separate inputs. The bandwidth is assumed narrow, so there are also n complex weights that must be adjusted to produce an optimum setting in some sense. In principle, using the approach given previously, n inputs should allow $n - 1$ interfering signals to be nulled, except at some angles where this may not be possible, while still retaining adequate signal strength. More generally, however, there may be more or fewer interferers, so that some criterion other than nulling may be desirable. To solve the problem, which becomes a solution of n-linear equations subject to some constraints, matrix algebra is convenient. The equations analogous to Equations (9.4) and (9.5) are:

$$S_{out} = \mathbf{W}_T \, \mathbf{S}_{in} \qquad (9.12)$$

$$N_{out} = \mathbf{W}_T \, \mathbf{N}_{in} \qquad (9.13)$$

The boldface characters are used to represent matrices, in this case one-dimensional matrices (vectors); the subscript T represents the matrix transpose and the superscript -1 represents the inverse. The inner product of two vectors may be written $\mathbf{A}_T\mathbf{B}$ or $\mathbf{B}_T\mathbf{A}$, since both forms produce the same scalar value. The vectors $\mathbf{W}$, $\mathbf{N}_{in}$, and $\mathbf{S}_{in}$ are defined as:

$$\mathbf{W} = \begin{Vmatrix} W_1 \\ W_2 \\ W_3 \\ . \\ . \\ . \\ W_n \end{Vmatrix} \quad \mathbf{N}_{in} = \begin{Vmatrix} \sum N_{1\,in} \\ \sum N_{2\,in} \\ \sum N_{3\,in} \\ . \\ . \\ \sum N_{n\,in} \end{Vmatrix} \quad \mathbf{S}_{in} = \begin{Vmatrix} S_{1\,in} \\ S_{2\,in} \\ S_{3\,in} \\ . \\ . \\ S_{n\,in} \end{Vmatrix}$$

In order to select W_i, it is necessary to place some further constraints on the system. In the simple example, it was assumed that N_{out} should be zero. This is not generally possible, especially when the random noise is added to the interferers or if the interferers happen to be more in number than the number of W_i. The two principal constraints that have been used in the past are the *minimum S/N* (MSN), also called the *minimum signal-to-interference ratio* (MSIR), and the *least-mean-square* (LMS) error. The former requires that it be possible to identify the signal from the interferers, the latter requires a local reference that has a higher correlation with the signal than with any of the interfering sources. Both criteria lead to similar solutions.

In the MSN case, we have:

$$P_m = S_{out} \cdot S_{out} = (\mathbf{W}_T \, \mathbf{S}_{in}) \cdot (\mathbf{S}_{in\,T} \, \mathbf{W}) \qquad (9.14)$$

$$P_n = E\{N_{out} \cdot N_{out}\} = E\{(\mathbf{W}_T \, \mathbf{N}_{in}) \cdot (\mathbf{N}_{in\,T} \, \mathbf{W})\} \qquad (9.15)$$

$$P_n = W_T \cdot E\{\mathbf{N}_{in} \cdot \mathbf{N}_{in\ T}\}\ \mathbf{W} \equiv \mathbf{W}_T \cdot \mathbf{MW}$$

where $E\{\ ...\}$ is the expectation of a random variable and $*$ indicates the complex conjugate. It has been shown [9.18] that to maximize the S/N:

$$\mathbf{MW} = \mu\mathbf{S} \tag{9.16}$$

$$\mathbf{M} = E\{\mathbf{N}_{in} \cdot \mathbf{N}_{in\ T}\} \tag{9.17}$$

where μ is an arbitrary constant. The weights $\mathbf{W}$ can be solved for by multiplying Equation (9.16) by $\mathbf{M}^{-1}$.

The drawback in this equation is that, in practice, the statistics of the interference are not necessarily known, so that the value of $\mathbf{M}$ cannot be determined a priori. Moreover, the interferers will change from time to time, so that no fixed values of $\mathbf{W}$ could be used. It is therefore necessary to estimate the various interferer expectations from previously received signal samples. A compromise must be reached between the number of samples required to approximate the expectations and the time in which the values of the statistics might change. The functional circuit suitable for implementing this MSN algorithm is shown in Figure 9.32.

Identical circuits are used for weighting each element. The circuit may be implemented using either digital or analog circuits, and applying the weights directly at RF or in the IF stages of the receiver. While it might seem that a more complex receiver is required to implement at IF, since n frequency converters and IF amplifiers are required, it must be realized that as a practical matter, the levels input to the control unit from both individual and sum channels must have sufficient level to overcome circuit noise and thresholds and provide adequate output power to perform the weighting operation. Alternatively, in a digital implementation they must have sufficient level to drive the A/D converters. Figure 9.33 gives diagrams indicating in somewhat more detail the analog and digital implementation of the control circuits.

In Figure 9.32, a fixed reference voltage is inserted in the control loop. In the absence of a reference matrix, such as this represents, the MSN algorithm attempts to null signals coming from all directions, including the desired signal. Suitable reference levels can prevent nulling in a preselected direction. This is important if there are only a few interferers and the signal has a comparable energy. Where the signal is comparatively weak, this is less important. This might be true, for example, in spread-spectrum cases where the spread has caused the general noise level and other users to provide a background noise level greater than the signal level (prior

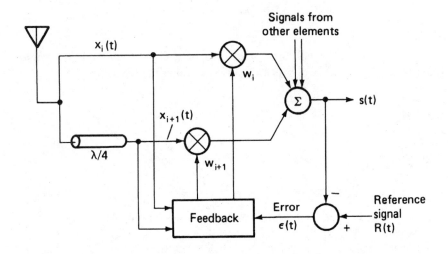

Figure 9.32 Functional block diagram of a control circuit for an adaptive array. (*From [9.23]. Reprinted with permission of IEEE.*)

to despreading to achieve the receiver's processing gain). Without the reference (or in other directions when there is a reference), the adaptation operates to invert the stronger signals' powers about the background noise level, the strongest first [9.20]. The speed with which the algorithm converges in the presence of strong interferers is also a problem with which the designer must be concerned.

The LMS algorithm differs from the MSN in that it assumes a capability to distinguish between signal and noise. To achieve this, Widrow et al. [9.18] have compared the summed signal with a local reference signal, which in its simplest form is a replica of the received signal. Figure 9.34 is a block diagram of this circuit, including diagrams of the control circuits. It will be noted that the form of these circuits is similar to those for the MSN algorithm. The solution for the LMS algorithm is given in [9.18] as:

$$\emptyset \, W = \emptyset_{xd} \tag{9.18}$$

where $\emptyset$ resembles **M**, except that it includes the received signal terms as well as the noise. $\emptyset_{xd}$ is the expectancy of the product of the input vector with reference signal d. When $d = 0$, the control algorithm reduces to that of the MSN. In [9.18], this case is shown with an added pilot signal, generated in a direction where it is

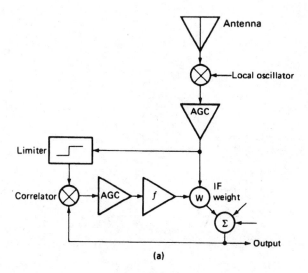

(a)

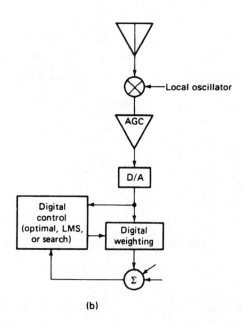

(b)

Figure 9.33 Block diagrams of analog and digital implementations of control loops for an MSN algorithm: (*a*) analog correlation control (analog IF weights), and (*b*) all-digital control and weighting. (*From [9.24]. Reprinted with permission of IEEE.*)

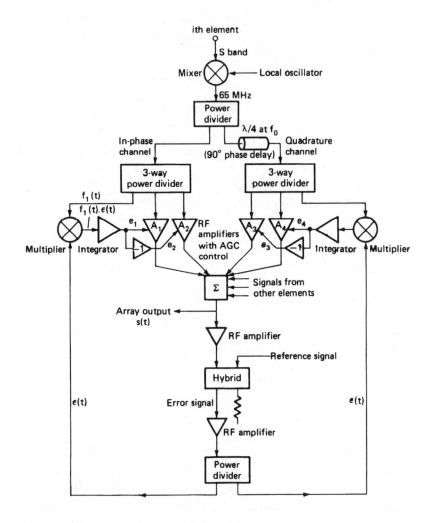

Figure 9.34 Block diagram of an LMS algorithm. (*From [9.24]. Reprinted with permission of IEEE.*)

desired not to reduce the output. This is equivalent to the MSN with an offset vector. Because of uncertainties in timing at a receiver, in many cases it is not possible to produce a good reference signal unless local demodulation is being accomplished. Demodulation may not be possible in the face of interference. Therefore, the reference is not available at the start of processing. In this case, a two-mode system may be used with an MSN while the interference is high, switching to an LMS when the reduction in the stronger interferers allows recovery of the signal. This depends strongly on the transient behavior of the adaptive algorithm.

9.6.2 Adaptive Equalization

Adaptive equalization of radio communications channels, while developed to improve the performance of a channel with a single input, has much in common with adaptive antenna processing. In adaptive equalization, the interferers are symbols sent at earlier or later times. As a result of multipath and other channel distortions, tails of prior symbols or precursors of symbols yet to come cause intersymbol interference that reduces or eliminates tolerance to noise that would exist in a channel free of these problems. One of the earliest adaptive techniques used to combat radio multipath was the RAKE system [9.25]. Subsequently, the use of an inverse ionosphere for combating multipath was suggested [9.26], and various other techniques were proposed for automatic equalization (References [9.27] and [9.28]). Meanwhile the problems of adapting the switched telephone system to transmit higher data rates in the face of unpredictable distortion had also led to the development of adaptive linear transversal filter equalizers [9.29]. The rapid development of digital processing technology in recent years has led to ever-growing efforts in the area. Several survey treatments (References [9.30] to [9.32]) have appeared, providing extensive reference and background material for further study. Here we attempt to present the general current approaches.

For channels with small distortion from phase or amplitude dispersion, or small amplitude multipath, the linear transversal equalizer is useful (Figure 9.35). As with adaptive antenna processing, a reference may be used in the control system. It is common to provide a reference transmission prior to data transmission to allow initial setting of the weighting vector and subsequently update the values using the differences between the output and the expected output of the equalizer (Figure 9.36). This type of equalizer proved useful in reducing error rates in telephone channels and thus allowed higher data rates with acceptable performance. After the line is connected, a training signal, whose characteristic is known at the receiver, is sent. This allows for the initial setting of the weighting coefficients, using an LMS-type algorithm. After training, the signal outputs are compared with the stored expected signal values, and an MSN or other algorithm is used to correct for slow changes. The same sort of algorithm can be useful to correct dispersion in propagation on a single-path channel or a multipath channel of the sort that has a strong main path and a number of weak multipaths.

This type of equalizer is not effective when there is substantial multipath, as in some HF channels (Figure 9.37). In such a case, a nonlinear decision feedback configuration added to the linear circuit is necessary. The block diagram for such a circuit is illustrated in Figure 9.38. In this case, the already decided output symbols are passed to a transversal filter, the outputs of which are weighted and summed at the input to the decision process along with the outputs from the weighted and the

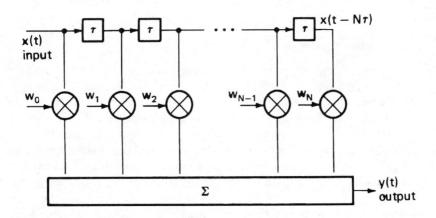

Figure 9.35 Block diagram of a linear transversal equalizer. (*From [9.31]. Reprinted with permission of IEEE.*)

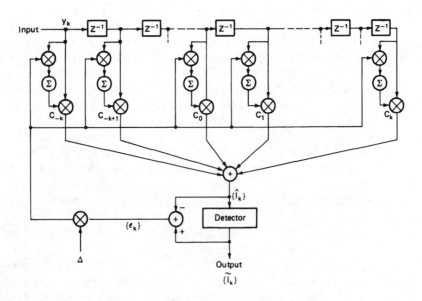

Figure 9.36 Block diagram of control functions for a linear transversal equalizer. (*From [9.30]. Courtesy of J. G. Proakis and Academic Press.*)

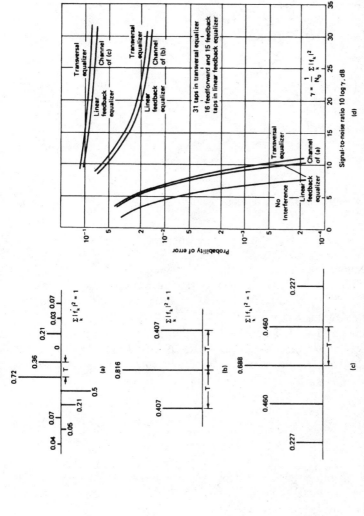

Figure 9.37 Effect of different degrees of channel distortion. *(From [9.30]. Courtesy of J. G. Proakis and Academic Press.)*

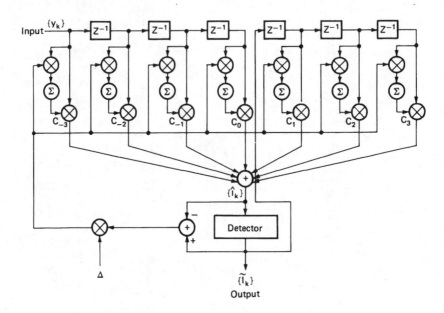

Figure 9.38 Nonlinear adaptive equalizer configuration. (*From [9.30]. Courtesy of J. G. Proakis and Academic Press.*)

summed outputs from the linear segment from the symbols yet to be decided. This nonlinear process represents a difference from the adaptive antenna structure, but results in much improved performance when channels with significant levels of much delayed multipath are present.

Referring now back to Figure 9.34, the similarity of the adaptive equalizer and the adaptive antenna processor is easy to see if we consider the n outputs of the transversal filter, the equivalent of the n separate antenna signals. If we adopt the notations of Widrow [9.21], the sum output S has the form:

$$S = \mathbf{W}_T \mathbf{X} \tag{9.19}$$

where the X_i are the signal variations at the various taps of the equalizer. If the stored reference signal is D, then when it is known that a training preamble is being sent, the error becomes:

$$\varepsilon = S - D \tag{9.20}$$

The LMS criterion applied to this situation gives rise to the same expressions and diagrams as shown previously. If an analog delay line is used, ε is a time function whose square must be integrated over the period of the delay line to obtain the equivalent of the expectation for the successive digital samples, in the case of a digital delay line.

Another algorithm that has been used [9.29] replaces the LMS criterion with a peak distortion criterion. The peak criterion is defined as the sum of the absolute values of the error at sample times other than that at which the maximum value of S occurs. This gives rise to a zero forcing algorithm, where the outputs at all sample times other than the maximum are forced to zero. The maximum value may be set to unity. This is a satisfactory criterion when the peak distortion is less than the maximum sample and when the noise is small, as in telephone channels or high-grade radio channels. However, it is not likely to be of use where distortion and noise are high. As with adaptive antenna processing, various modifications of the control algorithms have been tried to improve the convergence period so that more of the available transmission period can be used for information transmission.

The use of linear feedback equalization (i. e., the placement of the tap corresponding to the decision sample time within the equalizer, rather than at the right-hand side) proved of little value in good or bad channels [9.30]. Linear equalization, instead, is most useful in the feedforward mode. These observations led to the development of the nonlinear feedback equalizer (Figure 9.38). The fact that the feedback delay line is used for the already decided symbols leads to the setting of coefficients in that section on a different basis. With the LMS criterion, it can be shown that the circuit will completely eliminate the interference from the already decided symbols as long as the decision is correct. Because the time averaging or summing is generally slow enough that the weighting coefficients do not change significantly during a period of many output symbols, the effect of occasional errors is of little importance. The feedback equalization technique has been used to produce a high-speed digital transmission through HF multipath channels.

The adaptive equalizers discussed here use recursive methods of setting the weighting coefficients, so that every decision is affected by the n samples of the waveform in the delay line and a number of previous n or more sets of prior samples. This leads to the possibility of algorithms that accumulate kn^2 samples and make each decision on the basis of the entire statistics, using maximum-likelihood techniques [9.30]. As each new sample enters the system, the oldest sample is removed and the process is repeated. Properly designed, such a program should produce the best possible equalization for the number of points selected. Some experimental work of this type has been done using a *maximum-likelihood sequence estimation* (MLSE) or a Viterbi algorithm to reduce the computations. Figure 9.39 [9.33] indicates some comparisons of different equalization techniques in a simulated time-invariant channel.

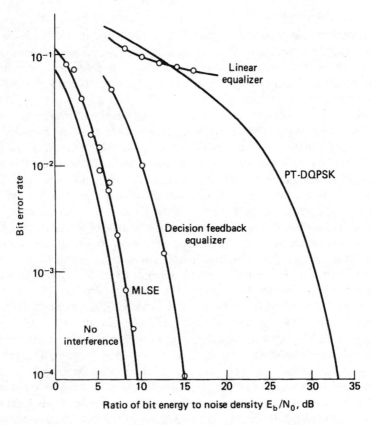

Figure 9.39 Simulation results of binary signaling through a time-invariant multipath channel. (*From [9.33]. Courtesy of Naval Ocean Systems Center.*)

In the previous discussions, the delay is the inverse of the symbol frequency. This is the maximum delay that can be used, and is used in most adaptive equalizer designs. In order for the equalizer to perform satisfactorily, symbol timing must be recovered very accurately. If the delay is made a fraction of a symbol, more timing error can be tolerated, but the processing load is increased. For binary symbols this is probably of minor significance, but if the symbols are higher-order, it may be necessary to use a *fractionally spaced* equalizer. A delay of one-half or one-third may prove more practical than improving symbol timing recovery.

9.6.3 Time-Gated Equalizer

The time-gated equalizer is a somewhat different approach to adaptive equalization than those described previously [9.11]. Its purpose is to achieve and maintain equalization in an HF band, even in the presence of frequency hopping at rates up to several hundred per second.

The technique is based on the premise that at HF, the major equalization problem arises from a small number of paths, individually having low distortion in a bandwidth of up to 10 kHz. Path delay separation is assumed much larger than the chip period (keying rates above about 1000 bits/s). This is certainly not true for every HF path, but is close enough for many applications. The assumptions dictate the need for feedback equalizer processing. To avoid the problem of realizing an array of coefficients, as in the usual transversal filter, the further assumption is made that the path delays change slowly compared to the keying rate, so that a measured delay may be useful for a comparatively long time.

Hulst [9.26] estimated the maximum rate of change for the F layer at about 7.5 m/s. Maximum delay occurs for vertical paths, so the maximum change for an N-hop wave for this value is $15N$ m/s, where N is the number of reflections. Usually the F-layer returns occur on long-range paths. The actual rate of oblique path change is much less than for a vertical path. For example, the maximum single-hop F range is about 4000 km. This reduces the rate of change to 2.6 m/s for a single hop. Multiple hops have increasingly greater rates of change, approaching $15\ N$ m/s for very large values of N. At 1200 Hz, a chip occupies 833 μs, or about 2.5×10^5 m; 9600 Hz, 3.1×10^4 m. The spacing change, for a single vertical reflection, in 10 min at maximum rate is 30 percent of the 9600-Hz period. Processor sampling periods usually do not exceed 4 times the chip rate. Thus, the delay of a particular multipath component is likely to remain at a particular sampling time for 8 or 9 min, in agreement with the assumption.

The basic concept for the multipath processor, based on these assumptions, is shown in Figure 9.40. The input wave is buffered, and when a signal is detected on a selected frequency, it is processed by the tap locator. Initially this uses correlation with the expected synchronization packet to locate the taps at the particular frequency, using the techniques indicated in Figure 9.41. Then the taps are located for other frequencies in the preamble. A limited number of measurements can be used to predict reasonably well the tap locations at the other expected hopping frequencies, provided that the preamble hops widely in the band.

To avoid the need for frequent synchronization packets, subsequent tracking and late entry make use of partial autocorrelation of the data packets. This is done as shown in Figure 9.41, except that only a single received frame is retained. The effects of this correlation are shown in Figure 9.42, based on simulator plots. The signal labeled "transmitted packet" represents the packet received with the shortest delay. While peaks appear in the autocorrelation signals at the multipath delay intervals, the data side lobes sometimes obscure them. To improve the estimates, averaging over multiple packets can be used. Figure 9.43 shows how averaging suppresses the data side lobes and causes the multipaths to be identified clearly.

In common implementations, a maximum of two multipaths (the larger pair if more than two exist) are used. Techniques include eliminating special synchroni-

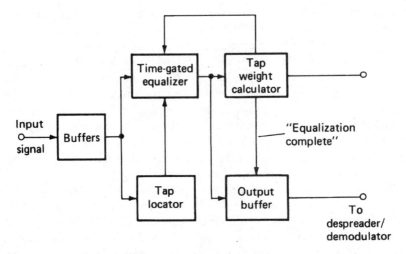

Figure 9.40 Block diagram of a multipath equalization processor. (*From [9.11]. Courtesy of RCA and* Ham Radio Magazine.)

zation preambles at the beginning of transmission, but including a small fraction of such bits in each packet. This puts the initial synchronization and late entry on an equal footing, and makes the averaging process somewhat more efficient in reducing the unwanted side lobes, since short cross correlation has replaced the partial autocorrelation process.

Referring to Figure 9.40, the tap location decisions are fed to the equalizer, shown in more detail in Figure 9.44. The tap locators are used to determine the

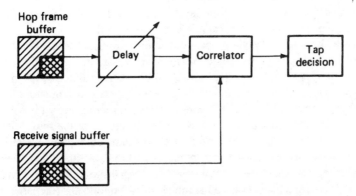

Figure 9.41 Correlation process. (*From* [9.11]. *Courtesy of RCA and* Ham Radio Magazine.)

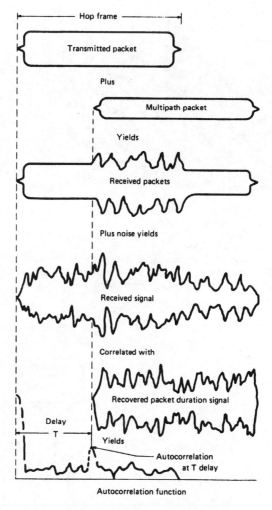

Figure 9.42 Time relationships in multipath and correlation process, using simulated waveforms. (*From [9.11]. Courtesy of RCA and* Ham Radio Magazine.)

delays for the feedback. Once they have been determined (for each hop frequency), the delays need only be updated if they change by more than one-half of the sample interval (substantial fractions of an hour). Before final processing of the packet, the tap weights must be determined. Because each frequency is revisited at an average rate of N_T/F_H, where N_T is the number of frequencies assigned for hopping and F_H is the number of hops per second, it could require one to several seconds to revisit each frequency. (Unless N_T is in the hundreds, the protection from jamming is not likely to be adequate.) At the maximum rate of path change, the phase of the RF

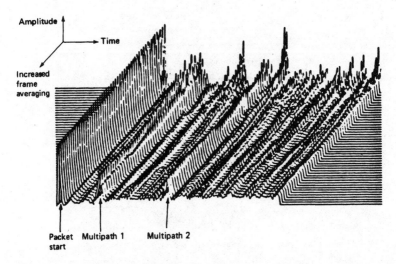

Figure 9.43 Reduction of autocorrelation side lobes by averaging. Correlation results averaged over 50 frames. Two multipaths: 1—delayed by 0.88 ms; 2—delayed by 1.76 ms. (*From [9.11]. Courtesy of RCA and* Ham Radio Magazine.)

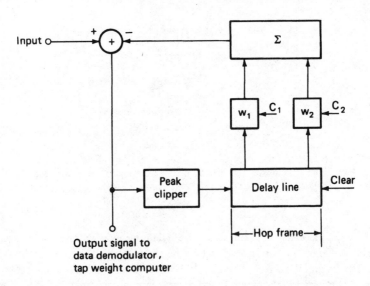

Figure 9.44 Time-gated feedback equalizer. (*From [9.11]. Courtesy of RCA and* Ham Radio Magazine.)

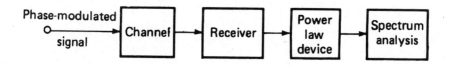

Figure 9.45 Distortion measurement approach. (*From [9.11]. Courtesy of RCA and* Ham Radio Magazine.)

signal can change at 5 to 50°/s in the HF band, depending on the frequencies that are in use during the communication (and based on the *F*-layer estimate). Consequently, the coefficients must be updated frequently. Updating of weights has been implemented for each received packet.

The estimation of the coefficients is based on the concept indicated in Figure 9.45. The received (stored) samples of the packet are passed through a power-law nonlinear process and subjected to spectrum analysis by an FFT. The power-law device must be capable of producing CW spectrum peaks. For example, with an MSK modulation, a square-law device produces two peaks separated by the chip frequency. The ratio of total power in the other components to that in the peaks is used as a quality measure Q. First, the larger multipath complex weight undergoes a series of steps in phase and in amplitude (starting at prior values) until the minimum Q is obtained. Then the second multipath is similarly treated. The final values of weights so obtained are used in the equalizer to process the packet. It has been found that with a moderate number of steps in amplitude and phase, sufficient improvement is obtained to provide good performance under otherwise impossible multipath conditions.

The processing required for these steps is far less than that required for the usual equalizers. Moreover, the process can operate in a frequency-hopping mode at rates of hundreds of hops per second. Figure 9.46 shows the results of one medium simulation with a single multipath, delayed by 0.5 ms from the main signal and of equal amplitude.

9.6.4 Link-Quality Analysis

A special aid to frequency management is link-quality analysis and automatic channel selection. This process uses capabilities readily obtainable in a digital frequency-hopping transceiver in conjunction with typical microprocessor control operations and speeds. The process is especially useful for HF frequency-hopping systems and can be of value for the selection of acceptable channels in other frequency bands as well.

In an HF system, the region of the band used must be changed, sometimes several times a day because of major changes in the ionospheric transmission. In a

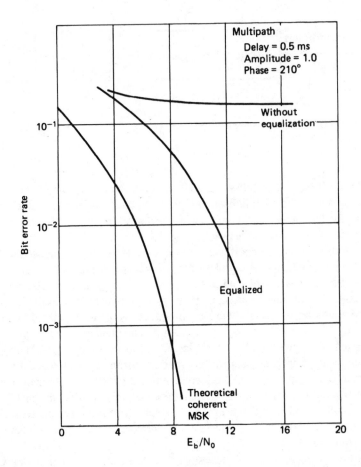

Figure 9.46 Results of multipath equalization simulation in a nonfading two-path channel.

frequency-hopping system, the users have a common dictionary of frequencies over which they may hop, and performance predictions may indicate which ones should be tried for the expected ionospheric conditions. However, there is no way of knowing in advance about the interference on the specific channels. It is important for reliable communications that the specific channels selected be as free of unintentional interference as possible, and that transmission among the users in the communications situation be good. This can be achieved by using regular spectrum scans of the assigned frequencies and by measuring the received quality of transmissions in the different channels.

Ionospheric sounding equipments are available using various techniques, which at a particular location can scan the condition of the ionosphere by sending signals and receiving the vertical returns. The sounding signals may be short high-power pulses, transmitted on successive carrier frequencies throughout the band. A low-power continuous slow scan sounding of the band has also been used, with a chirp-type receiver to detect returns. This causes less interference with others using the band than the pulsed-sounding signals. There are also receivers and processors that can measure the energy in channels throughout the band. If cooperative arrangements are made, receivers at a distance can use the transmitters of the sounder to measure the transmission over oblique paths. This type of equipment is useful for scientific studies of the ionosphere and for selecting frequencies for use between point-to-point fixed-frequency earth stations. However, such equipment provides much more information than is needed in a rapidly changing tactical situation, and its size and complexity often reflect this.

Where there is a network of frequency-hopping radios that must intercommunicate, cooperative sounding may be undertaken among them on a scheduled basis, or under the direction of a network controller. The frequency-hopping transmitter provides a sounder that can hop over all frequencies the network may use. A quality monitor is required in each receiver to estimate the utility of the channel. Each receiver assesses the S/N on all the channels for all the paths, for each propagating channel, and determines a measure of channel utility. For each link, the information is reported to other stations in the network, until at the end of the process the relative performance of all the links, in both directions and of all the stations, is known throughout the network. By establishing a selection rule, automatic selection (or rejection) of channels is made at each station. Every network member then knows the frequency group for use during the ensuing interval and until the next sounding is made. Sounding can occur during continuous information transmission by using a small fraction of the hopping time for sounding. For example, 10 percent sounding in a 150-hop/s system allows 1000 channels to be checked in slightly more than a minute.

Quality can be assessed by making measurements in the channel during the correlation interval. The output samples prior to correlation measure interference as well as side lobes. The amplitude of the successful correlation is affected by the signal-to-interference ratio. Such a spot S/N measurement, however, gives only limited data on a longer-term performance. For greater confidence, several checks should be made during sounding.

9.6.5 Automatic Link Establishment

Automatic link establishment (ALE) is a process by which automated, digital signal transmission techniques are used to improve communications system reliability and versatility (References [9.34] to [9.36]). Standards and practices relating to ALE have their roots in Federal (U. S.) Standard 1045 (FS-1045), which provides the foundation for a family of HF radio systems featuring—among other things—the ability to automatically adapt to ever-changing HF propagation conditions. Such systems, thus, are termed *adaptive HF radio links*. ALE technology enables radio stations to automatically initiate and establish bilateral connectivity. In the process of establishing links, an LQA is performed that allows the ALE radio to select the best available frequency.

FS-1045 provides the protocols and functions for a three-way handshake between two or more stations. The linking processing includes the following elements:

- The emission of a call. The emission waveform contains address information that selectively alerts a station in the system.

- A response signal, which is emitted if the station is operational, either scanning a number of frequencies or monitoring a specific frequency.

- An acknowledgment signal indicating that the proper code transactions have taken place.

The calling station initiates the call by transmitting a series of 24-bit words containing a "To" preamble, and concluding with a word containing a "This is" preamble and its own address. The called radio(s), which typically is scanning a number of channels, stops on the channel on which it hears the call and decodes the ALE words to determine if the call is intended for that unit. The called radio then answers with a short response beginning with two words containing the "To" preamble and the address of the calling station, and concludes by transmitting two words with a "This Is" preamble and its own address. When the origial calling station receives this response, it is assured of bilateral connectivity and sends an acknowledgment to the called station, thus completing the three-way handshake. This process is usually accomplished in 10 to 20 seconds, depending on the number of channels scanned by the radios.

In order to make the ALE system work, a set of basic operating rules was devised [9.37]. Those specifications, listed in order of precedence, are:

- Each ALE receiver is independent of all other receivers in the system
- The radio always listens for ALE signals
- It always responds to a call unless deliberately inhibited
- It always operates in the scanning mode if not otherwise in use
- It never interferes with active ALE channels unless operating in a priority mode, or otherwise forced to interrupt
- It always exhanges LQA data with other stations when requested, and always measures the signal quality of other stations
- The system respondes in preset/derived/directed time slots

- The radio always seeks and maintains track of connectivity with others in the system (unless inhibited)
- Linking ALE stations employ the highest mutual level of capability
- Users should minimize time on-channel
- As capable, radios should minimize the amount of transmitter output power so as to reduce spectrum congestion

As shown, the FS-1045 standard specifies the required protocols, timing, and technical definitions. It leaves the details of system implementation, however, to the individual equipment manufacturers. When fully implemented, the ALE system provides substantial improvement in radio communications efficiency and interoperability within and among various applications and groups of end-users [9.38]. The protocol also has the capability to exchange short digital text messages.

References

9.1. J. R. Carson, "Reduction of Atmospheric Disturbances," *Proc. IRE*, vol. 16, p. 966, July 1928.

9.2. M. Martin, "Die Störaustastung," *cq-DL*, p. 658, Nov. 1973.

9.3. M. Martin, "Modernes Störaustaster mit hoher Intermodulationsfestigkeit," *cq-DL*, p. 300, July 1978.

9.4. M. Martin, "Modernes Eingangsteil für 2-*m*-Empfänger mit grossem Dynamikbereich," *UK W-Berichte*, vol 18, no. 2, p. 116, 1978.

9.5. M. Martin, "Empfängereingangsteil mit grossem Dynamikbereich," *cq-DL*, p. 326, June 1975.

9.6. M. Martin, "Grossignalfester St"raustaster für Kurzwellen- und UKW-Empfänger mit grossem Dynamikbereiech," *UKW-Berichte*, vol. 19, no. 2, p. 74, 1979.

9.7. R. Feldtkeller, *Rundfunksiebschaltungen*, Hirzel Verlag, Leipzig, 1945.

9.8. T. T. N. Bucher, "A Survey of Limiting Systems for the Reduction of Noise in Communication Receivers," Int. Rep., RCA-Victor Div., RCA, June 1, 1944.

9.9. R. T. Hart, "Blank Noise Effectively with FM," *Electron. Design*, p. 130, Sept. 1, 1978.

9.10. J. S. Smith, "Impulse Noise Reduction in Narrow-Band FM Receivers: A Survey of Design Approaches and Compromises," *IRE Trans.*, vol. VC-11, p. 22, Aug. 1962.

9.11. U. L. Rohde, "Digital HF Radio: A Sampling of Techniques," presented at the 3d Int. Conf. on HF Communication Systems and Techniques, London, England, Feb. 26-28, 1985; also, *Ham Radio*, Apr. 1985.

9.12. H. H. Beverage and H. O. Peterson, "Diversity Receiving System of RCA Communications, Inc., for Radiotelegraphy," *Proc. IRE*, vol. 19, p. 531, Apr. 1931.

9.13. H. O. Peterson, H. H. Beverage, and J. B. Moore, "Diversity Telephone Receiving System of RCA Communications, Inc.," *Proc. IRE*, vol. 19, p. 562, Apr. 1931.

9.14. C. L. Mack, "Diversity Reception in UHF Long-Range Communications," *Proc. IRE*, vol. 43, Oct. 1955.

9.15. D. G. Brennan, "Linear Diversity Combining Techniques," *Proc. IRE*, vol. 47, p. 1075, June 1959.

9.16. F. J. Altman and W. Sichak, "A Simplified Diversity Communication System," IRE Trans. vol. CS-4, Mar. 1956.

9.17. L. R. Kahn, "Ratio Squarer," *Proc. IRE*, vol. 42, p. 1704, Nov. 1954.

9.18. B. Widrow, P. E. Mantey, L. J. Griffiths, and B. B. Goode, "Adaptive Antenna Systems," *Proc. IEEE*, vol. 55, p. 2143, Dec. 1967.

9.19. S. P. Applebaum, "Adaptive Arrays," *IEEE Trans.*, vol. AP-24, p. 585, Sept. 1976.

9.20. R. T. Compton, Jr., R. J. Huff, W. G. Swarner, and A. A. Ksienski, "Adaptive Arrays for Communication Systems: An Overview of Research at the Ohio State University," *IEEE Trans.*, vol. AP-24, p. 599, Sept. 1976.

9.21. B. Widrow and J. M. McCool, "A Comparison of Adaptive Algorithms Based on the Methods of Steepest Descent and Random Search," *IEEE Trans.*, vol. AP-24, p. 615, Sept. 1976.

9.22. W. D. White, "Cascade Preprocessors for Adaptive Antennas," *IEEE Trans.*, vol. AP-24, p. 670, Sept. 1976.

9.23. R. L. Riegler and R. T. Compton, Jr., "An Adaptive Array for Interference Rejection," *Proc. IEEE*, vol. 61, June 1973.

9.24. G. C. Rossweiler, F. Wallace, and C. Ottenhoff, "Analog versus Digital Null-Steering Controllers," *ICC '77 Conf. Rec.*, 1977.

9.25. R. Price and P. E. Green, Jr., "A Communication Technique for Multipath Channels," *Proc. IRE*, vol. 46, p. 555, Mar. 1958.

9.26. G. D. Hulst, "Inverse Ionosphere," *IRE Trans.*, vol. CS-8, p. 3, Mar. 1960.

9.27. E. D. Gibson, "Automatic Equalization Using Time-Domain Equalizers," *Proc. IEEE*, vol. 53, p. 1140, Aug. 1965.

9.28. M. J. Di Toro, "Communications in Time-Frequency Spread Media," *Proc. IEEE*, vol. 56, Oct. 1968.

9.29. R. W. Lucky, "Automatic Equalization for Digital Communication," *Bell Sys. Tech. J.*, vol. 44, p. 547, Apr. 1965.

9.30. J. G. Proakis, "Advances in Equalization for Intersymbol Interference," in *Advances in Communication Systems, Theory & Application*, A. V. Balakrishnan and A. J. Viterbi (eds.), Academic Press, New York, 1975.

9.31. P. Monsen, "Fading Channel Communications," *IEEE Commun. Mag.*, vol. 18, p. 16, Jan. 1980.

9.32. S. Qureshi, "Adaptive Equalization," *IEEE Commun. Mag.*, vol. 20, p. 9, Mar. 1982.

9.33. L. E. Hoff and A. R. King, "Skywave Communication Techniques," Tech. Rep. 709, Naval Ocean Systems Center, San Diego, Calif, Mar. 30,1981.

9.34. Institute for Telecommunication Sciences, National Telecommunications and Information Administration, U.S. Department of Commerce, *Federal Standard 1045, Telecommunications: HF Radio Automatic Link Establishment*, General Services Administration, Office of Information Resources Management, Jan. 24, 1990.

9.35. S. Horzepa, "ALE: A Cure for What Ails HF Communications," *QST*, p. 107, Nov. 1992.

9.36. R. T. Adair and D. Bodson, "A Family of Federal Standards for HF ALE Radios," *QST*, p. 73, May 1992.

9.37. R. T. Adair, D. F. Peach, and D. Bodson, "The Growing Family of Federal Standards for HF Radio Automatic Link Establishment (ALE), Part 1," *QEX*, p. 3, July 1993.

9.38. R. T. Adair, D. F. Peach, and D. Bodson, "The Growing Family of Federal Standards for HF Radio Automatic Link Establishment (ALE), Part 2," *QEX*, p. 7, Dec. 1993.

Additional Suggested Reading

"Eight Ways to Better Radio Receiver Design," *Electronics*, Feb. 20, 1975.
"Silence the Russian 'Woodpecker' with a Noise Blanker Having 80 dB Dynamic Range," *Electronic Design*, Mar. 1980.

Chapter 10
Receiver Design Trends

10.1 Introduction

In earlier chapters, we provided a guide to the design of communications receivers in accordance with the present state of the art. Page constraints have made it necessary to limit coverage in some areas, and at best, a book represents the situation at the time the manuscript was prepared. In this chapter, we touch on related areas of receiver design that we have not been able to cover in depth previously. In particular, we discuss three areas:

- Expanded digital implementation of receiver functions
- Spread-spectrum receivers
- The use of system simulation in design

10.2 Digital Implementation of Receiver Functions

The development of advanced integrated circuits has made digital logic functions relatively cheap and reliable, allowing complete microprocessors to be built on a single chip. This has led to the replacement of analog circuits in communications receivers with digital-processing circuits. Techniques for performing filtering, frequency changing, demodulation, error correction, and many other functions have been developed, and have been touched upon in earlier chapters.

Digital processing has a number of inherent advantages over analog processing. Greater accuracy and stability frees digital circuits from the drifts caused by temperature, humidity, pressure, and supply voltage changes. The possibility of long-time storage of signal samples makes repeated processing of the same data for more accurate detection and demodulation feasible. The economy of small-size but large-scale digital implementation makes practical optimum detection and decoding techniques that were once only theorists' dreams.

While there are many advantages to digital processing, there are limitations that will restrict implementation of all-digital radios in many applications for some years to come. Figure 10.1 shows the block diagram of a typical modern radio. Digital processing is used in the circuits following the first IF in the receiver, the

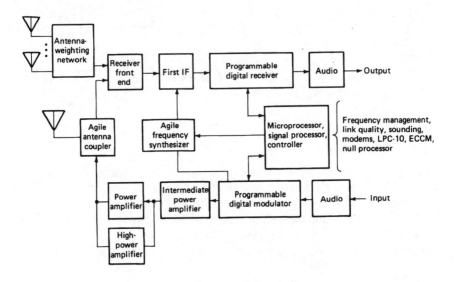

Figure 10.1 Block diagram of a radio set with digital signal processing and control.

exciter circuits in the transmitter, and the synthesizer. The particular product on which the diagram is based is an HF voice set, but it is equally applicable to other frequency bands and other modulations. All control, tuning, bandwidth, gain, modulation type, power level, antenna weighting functions, and the like are effected through the microprocessor. User decisions on these matters may be entered through a keypad, either locally or remotely.

In the transmitter, the low power level of digital circuits requires the use of analog power amplifiers with adequate filters to reduce undesired noise and harmonics. The digital modulator produces a sequence of numbers representing uniformly separated samples of the waveform levels. A D/A converter and filter produce the input for the analog power amplifiers. The level of quantization used in the digital numbers and converter must be such as to keep the transmitter noise level low in adjacent and nearby channels. Often it is convenient to use combined analog and digital techniques in the modulator, rather than analog conversion at the output.

In the receiver, an A/D conversion is made before digital processing. The quantization must be adequate to add minimally to the S/N. Yet there must be sufficient levels to handle the dynamic range. The sampling rate must also be high enough for the widest usable signal bandwidth to be processed. It is these stringent requirements that establish the limitation on digital receiver processing. The sample rate must be greater than twice the bandwidth to be handled, and the sampling width must be small compared to the period of the highest frequency to be handled

(References [3.13] and [3.14]). Because real filters do not completely eliminate frequencies immediately above cutoff, sampling rates tend to be 2.25 to 2.5 times the top frequency of interest. The amount of filter attenuation at one-half the sampling rate and above should be sufficient such that the higher frequency interference folded back into the band at that point is tolerable. The narrow sampling pulse waveform of period $1/f_m$ retains the filter output spectrum at baseband, but the output is augmented by the same spectrum translated to the various harmonics of f_m. This results in the need to filter the spectrum adequately to $f_m/2$. However, it also permits sampling of a bandpass spectrum at a rate of twice the pass bandwidth or more, as long as the sampling pulse is narrow enough to produce a substantial harmonic close to the pass band and as long as the spectra translated by adjacent sampling harmonics do not overlap it. The sampling pulse train can operate as both sampler and frequency translator.

The width of the sampling pulse may not occupy more than a small fraction of the period of the highest frequency in the pass band. This localizes the point of sampling of the waveform, and it leaves time for the sample to be retained while the remainder of the circuit digitizes it. If the outputs are digitized to m levels, the output of the A/D converter may be delivered on m buses at a rate of f_m or on one bus at a rate of mf_m, in both cases with an accompanying clock signal for sampling. The same narrow sampling width must be used whether the sampler operates above twice the highest frequency in the signal or much lower, at more than twice the highest bandwidth of the modulation band of the signal.

The active input circuits of the A/D converter contribute some amount of noise. For receivers in which digitization occurs relatively late in the system chain, this noise is of little concern. However, as the A/D converter moves toward the antenna, its NF becomes significant. In addition to the inherent NF of A/D converter circuits, quantizing noise is introduced by the digitizing process.

Linear A/D converters are used for signal processing to minimize the generation of IM products. (Some A/D converters for PCM voice use a logarithmic input-output relationship, but do not encounter further digital processing before reconversion by the inverse A/D converter.) A linear quantizer divides the input signal into a series of steps, as shown in Figure 10.2, which are coded to provide the digital output. The output voltage represents a series of steps, each q V high. An input voltage that falls between $(2k + 1)Q/2$ and $(2k + 3)Q/2$ is represented by an output kQ. The output voltage waveform is equivalent to the input waveform plus a random error waveform $e(t)$, which varies between $\pm Q/2$ and is distributed uniformly between these values. The rms voltage of $e(t)$, $Q/\sqrt{12}$ is the quantizing noise, and its spectrum is uniform.

When the level of the input signal waveform is too large, the quantizer acts as hard clipper at the levels $\pm NQ$. This can produce IM products and partial suppression of the weaker signals and noise. The AGC must be designed to avoid such clipping. The linear conversion is not always completely accurate; the input-output curve may show deviations that can produce IM products. The resulting quantizing

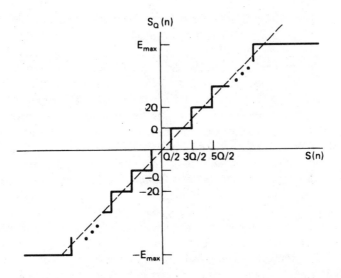

Figure 10.2 Linear quantizer input-output curve. (*From [3.14]. Reprinted with permission of Prentice-Hall.*)

noise may also increase slightly. The output level of the quantizer is coded into an m-bit binary number for further processing, $m = [\log_2(2N+1)]$, where the brackets indicate the next highest integer. The rms quantizing noise voltage as a fraction of the peak voltage NQ, for various values of m, is given in Table 10.1.

To achieve an all-digital receiver, the number of bits of quantization required must be sufficient to handle the dynamic range from somewhat below the input noise level to the maximum input power expected from the antenna network. To the extent that lossy circuits are used in coupling and filtering the input to the A/D converter, the dynamic range may need to be somewhat reduced. For example, assume an HF receiver with input bandpass of 2 to 30 MHz, requiring 11-dB NF

TABLE 10.1 Quantization Noise as a Function of Peak Voltage

	Quantizing noise relative to peak voltage	
m	Fraction	dB
3	0.096	-20.3
6	0.0093	-40.6
8	0.0023	-52.9
10	5.65×10^{-4}	-65.0
15	1.76×10^{-5}	-95.1
20	5.51×10^{-7}	-125.2

TABLE 10.2 Typical Operating Values of a Digital Communications Receiver

Sampling rate	≈ 70 MHz
Sampling width	≈ 1 ns
Input noise density	-204 dBW/Hz
Available for A/D NF and quantizing noise	-195 dBW/Hz
Minimum signal	
CW	0.05 μV (100-Hz bandwidth)
SSB	0.25 μV (3-kHz bandwidth)
30% AM	1.18 μV (6-kHz bandwidth)
Maximum interferers	10.0 V

and 50-Ω input impedance, and having 2-dB loss ahead of the A/D converter. We might select the values given in Table 10.2.

The quantizing noise plus the NF component of the A/D converter must not equal more than -195 dBW/Hz. If we divide this equally, -198 dBW/Hz to each, the NF must not be more than 6 dB. The total quantizing noise (distributed uniformly over 35 MHz) becomes -122.55 dBW, and across 50 Ω. This corresponds to 5.27 μV. The quantizing step then must be 18.24 μV. For a 10-V rms sinusoidal interference, the maximum voltage is 14.14 V, and the peak-to-peak voltage is 28.28 V, requiring a 21-bit A/D converter. If we allow some clipping on the peak, we might be able to use a 20-bit A/D converter. (Peak nonclipped voltage is now 4.8 V rather than 14 V.)

In our example system, the signals we wish to receive are well below the quantizing step size. This is no problem as long as the total signal, noise, and interference at the point of quantization is of the same order or greater than the step. In the example, the total noise density is -195 dBW/Hz, leading to a total noise in 35 MHz of 7.4 μV rms. This results in about 25.7 μV quasipeak (this noise is considered gaussian and has no absolute peak), or 51.5μV peak-to-peak, which is adequate to meet the criterion. If the many other signals in the HF band, even in the absence of very strong ones, are considered, there is little doubt that the small signals will not be lost because of the quantization.

A further disadvantage of quantizing at such a high rate is that initial processing will have to proceed at that rate and with that precision, leading to significant microprocessor demands. Once the band has been further constricted, the sampling rate can be decimated, but the initial filtering problem is still substantial. It is usually far easier and cheaper to use a few stages of frequency changing, analog filtering, and amplification before entering the digital processing environment.

When digital filter design was discussed in Chapter 3, the accuracy of filter designs with finite arithmetic was not addressed. Filters may be designed for either fixed-point or floating-point arithmetic, but in either case, the numbers can only be defined to the accuracy permitted by the number of digits in the (usually binary) arithmetic. Fixed-point additions are accurate except for the possibility of overflow, when the sum becomes too large for the number of digits with the selected *radix point* (decimal point in common base 10 arithmetic). Fixed-point multiplication can suffer from the need to reduce the number of places to the right of the radix point as well as from possible overflow. The number of places can be reduced by truncating or rounding. The former is easier to do, but the latter is more accurate. Floating-point arithmetic has no overflow errors, but adding is more difficult than for fixed-point arithmetic, and addition as well as multiplication is subject to the need for rounding or truncating.

There are several types of errors that must be addressed:

1. The original quantizing errors from the finite arithmetic are essentially like an input noise source.

2. Uncorrelated roundoff noise occurs when successive samples of the input are essentially independent. In this case, every point in the filter at which roundoff occurs acts as an independent noise source of the same general type as the quantizing noise. Overflow must be avoided in filters by scaling adequately to handle the dynamic range (or using floating-point arithmetic).

3. Inaccuracies in filter response result from quantization of filter coefficients.

4. Correlated roundoff noise results from nonindependent input waveforms. The effects of these errors differ, depending on the type of filter used—FIR, IIR, or DFT.

Uncorrelated roundoff errors cause noise that is similar to quantizing noise. Various attempts at analysis have been made with various levels of success. However, simulation is probably a more useful and practical tool for a specific evaluation. At lower frequencies, where there is likely to be correlation in the roundoff, simulation is quite essential. Because roundoff errors produce increased noise in the filter output, their minimization is desirable. Some guides for filter design rearrangement, location, and sizing of scaling are given in [3.14], for example.

Coefficient quantization is another source of error in filters. The coefficients occur in multiplications and thus affect scaling and, through it, rounding off. Coefficient errors can also affect the filter performance, even in the absence of other quantization. Filter coefficients are carefully chosen to provide some specific response, usually filtering in the frequency domain. Small changes in these coefficients can, for example, reduce the loss in the stop band, which generally depends on accurate balancing of coefficient values. In recursive filters, errors in coefficients can make the filter unstable by shifting a pole outside of the unit circle, so that the output grows until it is limited by the overflow condition at some point in the filter.

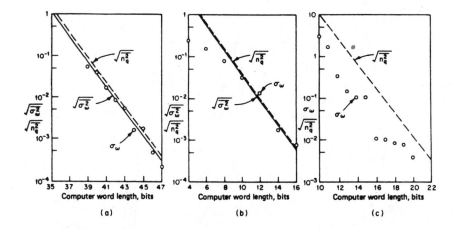

Figure 10.3 Theoretical and measured error variances for several recursive implementations of a twenty second-order bandstop filter: (*a*) direct programming errors, (*b*) parallel programming errors, and (*c*) cascade programming errors. (*From [10.1]. Adapted from [3.14] by permission of Prentice-Hall.*)

In narrowband filters, the poles have small negative real parts. To avoid instability in IIR structures, whose coefficients define these poles, it is necessary to reexamine the poles resulting from the quantized coefficients to assure that the quantization has not changed the negative to a positive value or zero. Shifts in pole locations, even without instability, can change the filter response, causing undesirable transient response, intersymbol interference, and lower reject band attenuation. A study of the effects of this roundoff on the rms error in frequency response and the rms noise resulted in theoretical prediction formulas. A twenty-second-order band-stop elliptical filter was designed, using direct configuration, parallel single- and two-pole sections, and cascaded single- and two-pole sections. Only in the first two cases could the theory be evaluated; in all the cases, simulations were made and the error was measured. The results are shown in Figure 10.3. From this it is noted that the parallel form has the smallest error and noise, while the direct form has the largest.

One would like to answer the question, "How many bits of quantization are required to assure that maximum deviations from the desired response do not exceed some preassigned value, which may vary from point to point?" In the work noted previously, coefficient rounding was accomplished by a fixed rule. This is not necessary; each coefficient could be rounded either up or down. A measure of filter performance deviation from a desired response as well as an optimization procedure for coefficient rounding based on minimizing the maximum value of this measure was defined in Avenhaus and Schüssler [10.2]. Figure 10.4 shows the

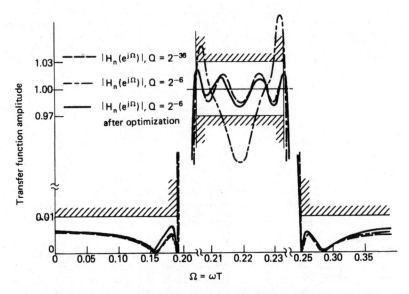

Figure 10.4 Frequency response effects of optimization of filter coefficient rounding. (*From [10.2]. Reprinted with permission from* Archiv für Elektronik und Übertragungstechnik *and the authors*.)

results of one test of this approach. The dashed curve was obtained with 36-bit implementation of an eighth-order elliptic filter (with standard rounding). The minimum acceptable filter, using standard rounding (up for 0.5 or more and down for less than 0.5 of the last retained digit) had 11-bit quantization. When the coefficients were rounded to 8 bits ($Q = 2^{-6}$ in the figure) using standard rounding, the dot-dash curve resulted, which is obviously unacceptable. By use of the optimization technique for the coefficients, the solid-line result was obtained. Thus, 3 bits were saved by rounding optimally. It is interesting to note that the variations from reducing the number of bits occurred primarily in the pass band.

Many receiver functions can be performed by digital processing. The most obvious candidates include the following:

- All control functions, such as tuning, gain control (both automatic and manual), bandwidth, squelch (off, on, or threshold), demodulator-type selection, diversity selection, antenna-array null direction (manual or automatic), error detection and correction (off, on, or type selection), time constant selection (AGC, squelch, attack, and release), and address recognition or change. All may be implemented digitally and conveniently using a microprocessor for local or remote control. Increasing the amounts of logic or memory on a single chip and decreasing the cost per function has made many new functions practical.

- The encoding of source signals to remove redundancy (such as vocoders and complex delta modulators), which may be designed into the receiver or the terminal instruments to increase information throughput. Complex error-control coding techniques may be implemented—coding with higher-order alphabets, maximum-likelihood decoding approximations, such as the Viterbi algorithm, and so on.

- Optimum diversity techniques, such as adaptive equalizers, which can be used to combat channel distortion. Adaptive antenna array processing can favor the desired signal direction over others from which interfering signals arrive.

In short, digital processing has opened up greater capabilities for receivers at a reasonable cost. Designs are continually improving and new concepts are being developed to make use of the increasing capabilities of digital integrated circuits. However, complexity has its price; more digital hardware implies larger and larger programs, with their attendant development costs and lengthy periods of debugging. We must beware, even in the new digital age, of overdesign. The good designer uses the latest state of the art to its fullest, but only to achieve the actual needs of a product at the lowest overall design and implementation cost.

10.2.1 Digital Receiver Design Techniques

Traditional radios, both receivers and transceivers, use analog parts that are bulky, expensive, subject to manufacturing tolerances, and require alignment in production. The number-crunching power of digital processing allows the replacement of many of these analog functions and the movement of the digital portions of the system closer to the antenna. Efforts to implement DSP into communications equipment, specifically HF and VHF equipment, date back to the late 1970s. At that time, the number-crunching capabilities were insufficient to accommodate the requirements that the analog parts could provide. In those days, microprocessors were too power hungry, expensive, and mostly found only in state-of-the-art designs for validation of the principles of the approach.

Today, a number of manufacturers supply sufficiently powerful DSP processors. The two best known are the Texas Instrument TSM 3232 series and the Motorola 56002 series. While there are many similarities between these two devices, there are also subtle differences. The key difference is the availability of floating point arithmetic versus integer calculations, which determines the price and power consumption of the device. Many of the commercial/military and radio amateur transceivers rely mostly on the Motorola DSP chip, which is available at various computation speeds.

Figure 10.5 shows a block diagram of a modern HF radio incorporating a moderate amount of DSP technology. For various reasons, such as the number-crunching power of the digital logic used, the last IF is chosen to be 25 kHz or less

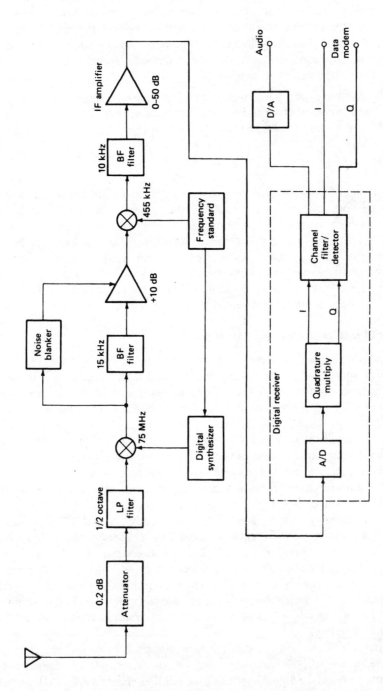

Figure 10.5 Block diagram of a typical modern HF radio. The digitally-implemented portions are contained within the dashed lines.

and the number of intermediate IF analog stages is minimized. In general, modern radios are not only used for point-to-point communication, but also for radio monitoring. Based on the complexity of the signals, the actual radio can become quite complex. Figure 10.6 shows a complete surveillance system, which can easily be expanded in frequency range. It does not require much imagination to envision the software requirements to collect all the data for such a system.

The transition for existing radios such as the Rohde & Schwarz EK 890 to DSP-based radios like the EK 895 is done by replacing the conventional analog IF section and the audio portion with the appropriate DSP. Figure 10.7 shows the block diagram of the EK 895, in which the performance increased in the DSP area is more dramatic than the improvements in the hardware of the front-end. The weakest links are the input filter switching, the input mixer and its roofing filter (being too wide for cost reasons; narrow filters cost 3 to 5 times more), and the synthesizer.

The jury is still out on whether DDS-based synthesizers are the best approach or whether the fractional division synthesizer is more appropriate. In the case of multimode transceivers, it is possible to apply fast modulation to the fractional division N synthesizer, but care must be taken not to violate some of the patents around this area. Figure 10.8 shows a synthesizer that was recently published by Rohde [10.3]. Its phase noise requirements are quite sufficient compared to the needs of shortwave dynamic requirements.

The use of MOS switches in mixers has also improved the dynamic range. The schematic shown in Figure 10.9 is a quad circuit with a measured 3rd order intercept point of 40 dBm and better with a 2nd order intercept point of 77 to 80 dBm and insertion loss of about 6.5 dB. Its upper frequency limit of about 100 MHz is determined by the actual physical construction and the capacitance of the MOS transistors.

The roofing filter, which is driven by the mixer, is also in the critical path for communications applications with FM capabilities. This filter is generally found to be too wide and deteriorates the performance. A ±5 kHz filter is still rather expensive for some applications, but commercial radios using *narrowband frequency modulation* (NBFM) applications take advantage of it.

The final technical challenge is the switching mechanism of the input filters. For low loss, relays are the best choice, however, for more affordable applications a technique using JFETs as heavily-biased switches with essentially no DC supply voltage may be used. Figure 10.10 shows the simple gating arrangement. The only complication that arises for the designer involves the generation of a positive and negative voltage to bias the gate. Several companies manufacture small inexpensive DC-to-DC converters and, therefore, single power supply radios can easily be updated. Having shown intermod performance that is better than the mixer, the diodes are transparent in the design.

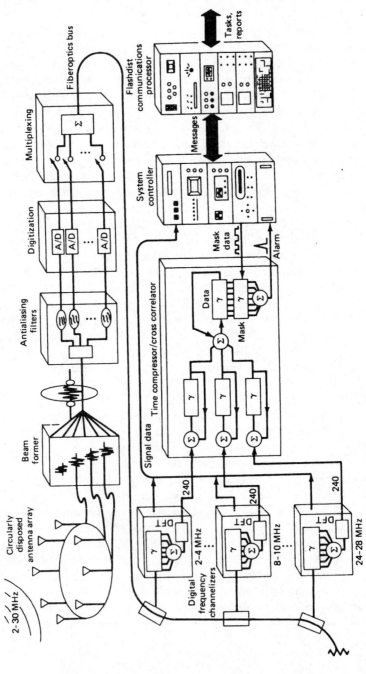

Figure 10.6 Block diagram of a complete surveillance system.

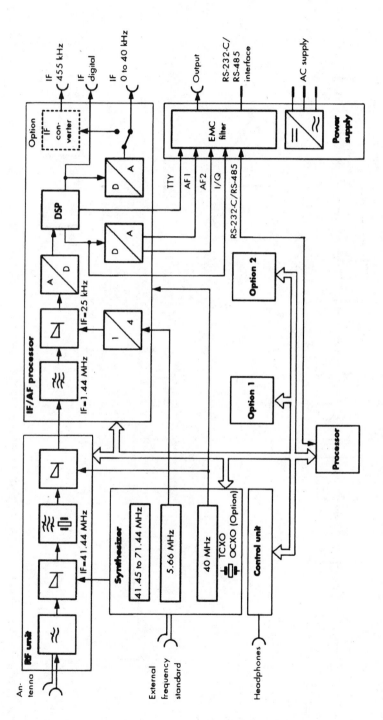

Figure 10.7 Block diagram of the EK-895/EK-896 VLF-HF receiver.

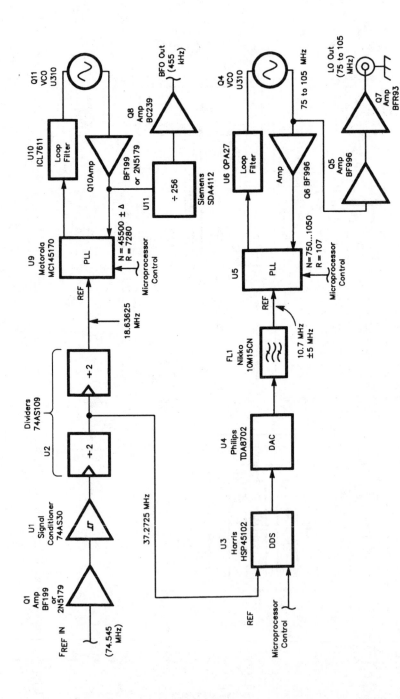

Figure 10.8 A high-performance hybrid synthesizer for shortwave receivers.

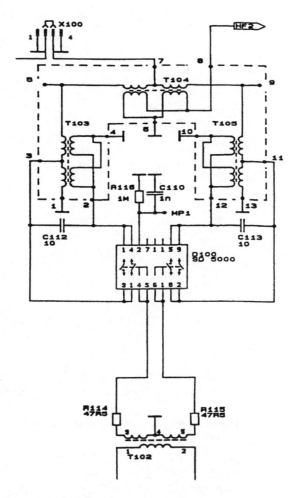

Figure 10.9 A high intercept point mixer using FET switches.

10.2.2 DSP Considerations

The first truly affordable all-DSP transceiver was the TS 870 made by Kenwood. Figure 10.11 shows a block diagram of the radio. It uses a combination of well-established techniques and moves the input frequency from a first IF of 73.5 MHz to 8.8 MHz, 455 kHz, and finally down to about 11 kHz. In the case of high-end transceivers such as the Rohde & Schwarz XK2100, the second IF jumps down 25 kHz directly. This can be established by using an image reject mixer. This scheme avoids the close-in intermodulation distortion (IMD) problems. Figure 10.12 shows a detailed block diagram of a digital SSB receiver discussed in

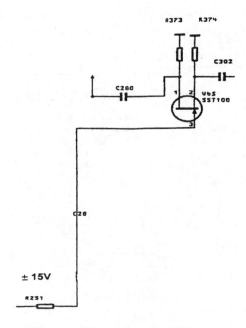

Figure 10.10 Improved filter switching system using JFETs.

[10.4]. After review, it becomes apparent that the digital portion of the receiver starts at the A/D converter and the overall performance is highly dependent upon this device.

Referring to the Rohde & Schwarz XK2100, the block diagram would show the following properties:

- The gain from the antenna to the input of the A/D converter is 48 dB with an AGC range (analog) of about 34 dB. The A/D converter has an imped-ance input point of 15 kΩ, so part of the gain is due to the impedance transformation from 50 Ω to 15 kΩ.

- The A/D conversion part of the DSP system is a 16-bit device with a 96 dB S/N from 5V peak-to-peak input maximum voltage.

- The A/D converter operates at a 3.2-MHz reference sampling frequency externally and then internally reduces this to 100-kHz sampling.

These types of A/D converters are also referred to as *sigma delta* ($\Sigma \Delta$) convert-ers, which consists of a system wherein high accuracy is obtained at a relatively slow sampling rate. This is accomplished by grossly over-sampling with a 4-bit

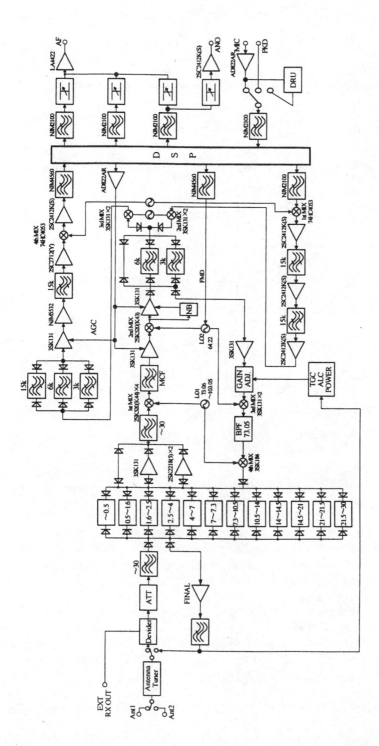

Figure 10.11 Block diagram of the TS–870 receiver. (*All information regarding the TS–870 supplied by Yukio Kawana of Kenwood Corp.*)

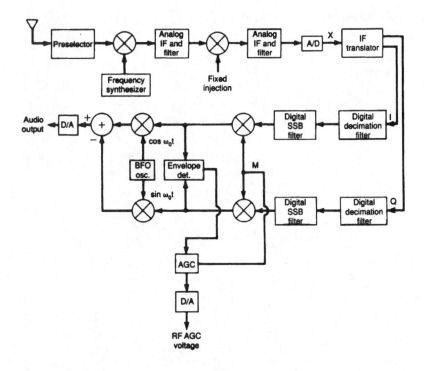

Figure 10.12 Block diagram of a digital SSB receiver. (*From [10.4]. Used with permission.*)

A/D converter or higher in a feedback loop. At the same time, the requirements of the analog anti-aliasing filter are considerably relaxed. The associated low-pass filter needs to be only half of the sampling frequency and, therefore, is 50 kHz. By using a mixing process that occurs by under-sampling, we can translate the channels to a baseband with a narrower decimating filter. This decimating process is dependent on the actual bandwidth. As the bandwidth becomes more narrow, the sampling moves from 25 kHz to 12.5 kHz to 6.25 kHz and ultimately down to 1.78 kHz. The sampling frequency is reduced to maintain the same amount of number crunching and, therefore, the same delay. While maintaining 16-bit arithmetic, the filter types chosen still have up to 20 mS delay to calculate. Such a system will be difficult for ARQ operation. The type of filters calculated by the DSP system are called *finite impulse response* (FIR) filters. Figures 10.13 and 10.14 show the characteristic appearance. Rather than going into the noise floor, these filters have a stop-band ripple, which is typical for these types of digital filters. A more dramatic presentation is shown in Figure 10.15. The use of FIR filters allows for the design

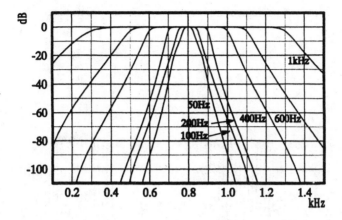

Figure 10.13 CW filter width plot.

of arbitrary selectivity curves and orders as high as 100. The main advantage these DSP filters offer is that the time delay is absolutely flat and constant with varying frequencies. It is practically impossible to build such filters in analog form. As to the ringing noticed in CW, there are two causes. One is due to the reduced bandwidth and the other is due to group delay distortion. These particular types of FIR filters do not add any ringing; therefore, for a given bandwidth, the FIR filter rings less than a conventional filter. Another problem can occur if the skirts are simply made steeper than necessary. There is certain splatter or bandwidth emis-

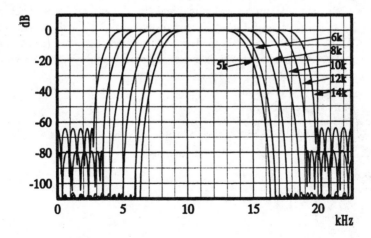

Figure 10.14 AM/FM filter width plot.

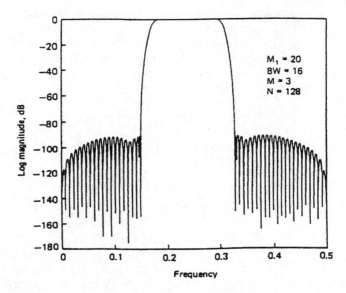

Figure 10.15 Modified presentation of the characteristic curves shown in Figures 10.13 and 10.14.

sions from all amplitude modulated signals and since we seldom have the task of removing only a constant carrier (this is done at a different place than the DSP), excessive shape factors only punch holes into noise, producing a sinusoidal voltage output much like ringing. Table 10.3 gives a selection of bandwidths and shape factors for receivers that have been found useful.

The DSP unit not only provides the selectivity, but also handles the AGC functions. For these types of systems, a dual-loop AGC is required. The analog portion of the AGC, which is wrapped around the input stage, takes over after the input signal reaches sufficiently high levels. This threshold is typically set between 1 mV and 5 mV.

10.2.3 Noise Calculations

The *digital* S/N ratio is determined as follows:

$$\frac{P_s}{N_{0q}} = -1.25 + 10 \times \log(F_s) + 602 \times b \tag{10.1}$$

where F_s = the A/D converter sample rate and b = the number of bits. This equation provides the theoretical maximum *signal to quantization noise density*

TABLE 10.3 Receiving Bandwidths and Shape Factors for DSP Receiver

3 dB (all values ±)	60 dB (all values ±)
25 Hz	75 Hz
75 Hz	150 Hz
150 Hz	225 Hz
300 Hz	430 Hz
500 Hz	770 Hz
750 Hz	990 Hz
1050 Hz	1600 Hz
1200 Hz	1760 Hz
1350 Hz	1900 Hz
1550 Hz	2100 Hz
3000 Hz	4200 Hz
4000 Hz	5200 Hz

ratio (P_s/N_{0p}). This value is somewhat similar to the S/N using sinusoidal wave-forms [10.5]. Consider the following example. For 5 V p-p to a given A/D converter:

$$N_q = \frac{V_{pp}^{\,2}}{12 \times 2^{32}} = \frac{25}{12 \times 4.3E9} \tag{10.2}$$

$N_q = 484\,E{-}12$ watts

From the equation $U^2/R = N$, we solve:

$$U = \sqrt{50 \times 23E{-}12} = 34 \ \mu V$$

or 95 μV p-p. The noise figure can be expressed as:

$$F = 1 + \frac{V_{pp}^{\,2}}{6\,K\,T\,R_s\,F_s} \tag{10.3}$$

where:

K = Boltzman's constant (1.38E–23)
T = room temperature in degrees Kelvin (300)
R_s = generator impedance of the A/D converter (15 kΩ)
F_s = sample frequency of the A/D converter (25E3)

Therefore:

$$F = 1 + \frac{(95E\text{–}6)^2}{6 \times 1.38E\text{–}23 \times 300 \times 15E3 \times 25E3} = 970$$

$$F = 10 \times \log(970) \approx 30 \text{ dB}$$

To determine the analog gain, we solve the Friis formula for VG_1:

$$F_{TOT} = F_{input} + \frac{F_2 - 1}{VG_1} \qquad\qquad (10.4)$$

where:

F_{input} = the NF of the input stage
F_2 = NF of the A/D converter
VG_1 = voltage gain of the preamplifier system

   We solve this equation for VG_1 with the understanding that we want a total NF of equal or better than 10. Assuming that the input storage NF is 3 dB, we now obtain:

$$10 = 3 + \frac{970}{x}$$

Solving for x we arrive at 138.57 or 42.8 dB. This is the required voltage gain to match the prescribed NF. Recall that we had already set the preamplifier gain to 48 dB, thereby allowing for a safety margin toward tolerance and differentiating between sinusoidal noise and quantization noise.

10.2.4 Noise Cancellation

Besides providing superb selectivity in filters, the other attractive feature of DSP is its ability to deal with correlated and uncorrelated signals. DSP processors, either at RF or IF levels or at audio frequencies, can compare signals and determine whether they are not correlated, slightly correlated, or totally correlated. Such adaptive filters were developed for echo cancellation in long-distance telephone lines [10.6]. DSP techniques can be used to detect certain classes of signals and subtract them from a wide frequency band, thereby, effectively canceling the signals. This is done in the time domain. Noncorrelated signals can also be removed, thereby enhancing the correlated signals. Modern noise reduction techniques and automatic notch filters utilize a combination of both techniques.

There are a number of types of adaptive filters. For the purposes of radio reception, we are primarily interested in the following technologies:

- Adaptive noise canceler
- Adaptive self-tuning filter
- Canceling periodic interference without an external reference source
- Adaptive line enhancer
- System modeling
- Linear combiner

Regardless of whether the noise reduction algorithm is the Widrow-Hoff *least means square* (LMS) algorithm, or the *discrete Fourier transform* (DFT) -based "spectral subtraction" algorithm, DSP noise reduction works by finding the most significant spectral lines in a signal and then forming bandpass filters around the strongest energy concentrations. The particular algorithm only determines how this is accomplished. To understand adaptive noise reduction, it helps to think about filters in a different way (References [10.7] and [10.8]).

Without going into details on how to design DSP-based noise reduction systems (which is beyond the scope of this book), we will consider some examples of what can be accomplished with these technologies. Figure 10.16 shows a spectrum from zero (0) to 4 kHz with a desired signal around 1.5 kHz and two interfering signals—one at 1 kHz and one at 3 kHz. After activating the auto-notch feature on the DSP receiver (Kenwood TS-870), the interfering signals are significantly reduced, as can be seen in Figure 10.17. The DSP-based filtering can be done at the AF (mostly) and at the IF (hopefully). The IF filtering requires significantly

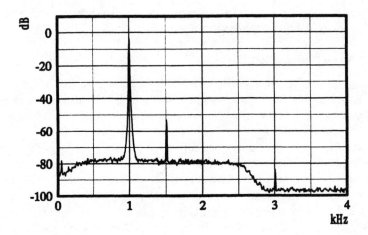

Figure 10.16 A desired signal of 1.5 kHz in the presence of 1 kHz and 3 kHz interfering signals.

more processing power. Figure 10.18 shows the automatic notch filter, rather than beat cancellation done via DSP at the IF level. Figure 10.19 shows the effect of the automatic notch filter compared with an analog version, such as the Kenwood TS-950SDX. The TS-870 was the first unit that allowed beat cancellation in the audio and auto-notch in the IF. Depending on the signal type, these systems have different effectiveness. Figure 10.20 shows the general diagram of the adaptive

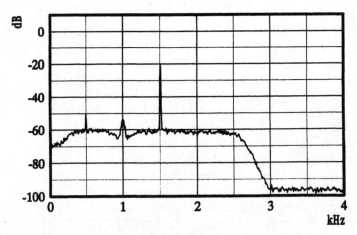

Figure 10.17 Results of DSP removal of the interfering signals at 1 kHz and 3 kHz shown in Figure 10.16.

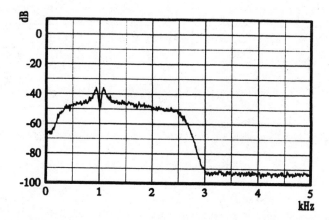

Figure 10.18 Performance trace of the TS-870 automatic notch (IF). Note the single beat rejection at 1 kHz.

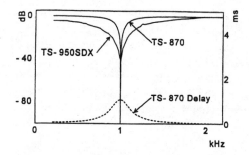

Figure 10.19 An automatic notch filter performed using DSP techniques. This filtering can be performed at IF (preferably) or at AF (typically). IF notch filters designed for multitones require significant computational power.

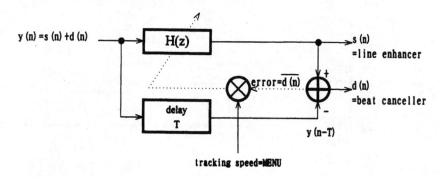

Figure 10.20 Block diagram of the adaptive filter used in the TS-870 radio.

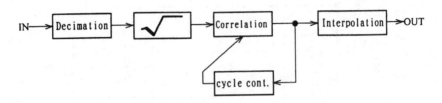

Figure 10.21 Block diagram of the correlated method used for noise improvement.

filter used in the TS 870 and Figure 10.21 shows the block diagram of the correlation method used for noise improvement. Finally, Figure 10.22 shows the S/N as a function of the various adaptive filters or correlation match system.

10.2.5 Spectral Subtraction

Spectral subtraction is another way to reduce the noise in voice signals. This technique accomplishes much the same thing as the LMS algorithm, but in a different way. Up to this point, all of the DSP algorithms we have discussed work by processing a series of numbers that represent the signal waveform as a function of time. Spectral subtraction, on the other hand, works by processing a series of numbers that represent the frequency content of the input signal. To do this, DSP devices use a relatively complex mathematical operation (transform) to change the signal representation from the time domain to the frequency domain.

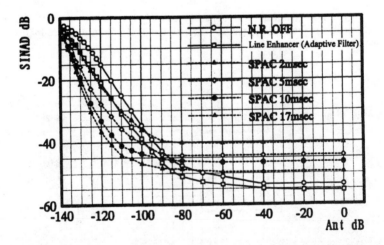

Figure 10.22 S/N as a function of the various adaptive filters and correlation match system.

For example, what comes out of an A/D converter is a series of numbers that represent the audio voltage in time increments at 0 μs, 100 μs, 200 μs, and so on. The transformation operation yields a series of numbers that indicate signal energy in *frequency* increments at 300 Hz, 320 Hz, 340 Hz, and so on, up through 3000 Hz or more. A complementary inverse transform returns the frequency data to a time-domain signal. If we perform the time-to-frequency transform and follow it immediately with the frequency-to-time (inverse) transform, we get our original signal back.

Spectral subtraction is a three-step process:

- Transform the signal to the frequency domain
- Process the frequency domain data
- Inverse-transform back to the time domain

This process repeats for successive short segments (a fraction of a second) of audio. Spectral subtraction relies on two basic assumptions:

- Voice-frequency energy is concentrated in a small number of frequencies
- Noise energy is uniformly distributed throughout the audio spectrum

Spectral subtraction algorithms attempt to determine the "noise floor" of a signal. The process assumes that any frequency-domain value at or below the noise floor is noise and sets the energy at that frequency to zero. Conversely, it considers signals above the noise floor to be voice components and allows them to pass.

The use of DSP spectral subtraction for noise reduction involves several disadvantages, including the following [10.7]:

- It takes a substantial amount of time to perform the forward and inverse transforms. The resulting delay through the DSP (a fraction of a second) can create an annoying "electrical backlash" condition. The delay makes it difficult to rapidly tune receivers, because the audio and dial position are not synchronized.

- Spurious audio "tones" result when processing noisy signals in some implementations. These appear as seemingly random beeps at random frequencies. They are caused by the algorithm's imperfect ability to distinguish between the signal and noise in the frequency domain.

- Spectral subtraction requires much more DSP computing power than the LMS algorithm.

All algorithms have their disadvantages; yet under some conditions, spectral subtraction may provide the best performance. The JPS NIR-12 DSP, for example, provides both spectral subtraction and LMS noise reduction (which JPS refers to as *adaptive peaking*). The user can choose the best noise reduction method for each situation.

10.3 Spread Spectrum

Spread spectrum techniques expand bandwidth to gain transmission advantages. These techniques were originally developed for military applications, but their properties are rapidly bringing them into more general use. Some properties that can be achieved by using spread-spectrum waveforms include the following:

- Resistance to jamming
- Reduction of the probability of intercept, location, and identification
- Multiple use of a common wide band with small interference among users, sometimes referred to as *code division multiplex* (CDM)
- Provision of accurate range information
- Resistance to multipath distortion in transmission
- Resistance to nongaussian noise and unintentional interference from other signals

A significant disadvantage of spread spectrum is the need for bandwidth expansion that is roughly proportional to the degree of performance improvement in any of these areas. Therefore, spread-spectrum modulations tend to require relatively high carrier frequencies to send moderate data rates (several kilohertz) or a substantial reduction in data rate (a hundred or more times) to be used at lower carrier frequencies. As of this writing, there are a large number of commercial, consumer, and military applications throughout the spectrum. A review of the development of spread-spectrum techniques to meet various applications is given in [10.9].

The principal modulation techniques for spread spectrum are *frequency hopping* (FH), *time hopping* (TH), *direct-sequence spreading* (DS), also referred to as *pseudonoise* (PN), and *chirp* (closely related to wide-band FM and FH). Not all of the techniques are equally useful for all applications. Where the intention is to provide resistance to jamming or interception, it is necessary to control the spectrum spreading by a process that cannot be duplicated by opponents within a period sufficiently timely to be of use to them. Thus, there must be a large number of potential spreading waveforms selected by the intended users in a manner that has no obvious rule that can be readily analyzed by an opponent. A review of the theory of FH and DS techniques is presented in [10.10], and there are a number of other publications (References [10.11] to [10.18]) on the details of spread spectrum. Only a brief overview is given here.

A conventional transmitter has a relatively narrow band centered about the carrier frequency, to which the narrowband receiver can be tuned. Any other signal in the narrow band can interfere with and possibly disrupt the communication. Because of high power density in the band, the signal is easy for others to detect and locate using direction-finding techniques. Pseudorandom spreading distributes the transmitter power over a much wider frequency range, with much lower power density. The spreading may be over contiguous channels, or it may be distributed over a wide band with gaps in its spectrum. Because the spreading is reversed at the receiver, narrowband interferers are spread before demodulation, and wideband interferers remain wideband. The interference power density in the reconstructed narrow band remains low, while the higher-power density of the desired signal is available to the receiver demodulator. Therefore, interference and disruption tend to be reduced. The lower-density transmission makes intercept and location more difficult, especially in a band that contains more than one signal.

The receiver and the transmitter must both use the same randomly or pseudorandomly generated control signal, and the receiver must synchronize it with the incoming signal. This presents a number of operating problems to the user stations, as well as to the opponent. The controlling codes, if pseudorandomly generated, must be protected while in use, and must be changed from time to time to prevent discovery. When the change is made, it must be done at the same time by all code users. If precise timing is not available, which is often the case, a synchronization recovery technique is required. When the interference is natural or unintentional, the synchronization problem is much simplified.

The character of FH modulation systems changes as the rate of hopping is lower or higher than the symbol rate of the basic digital signal (or the highest modulation rate of an analog baseband signal). Where the hopping rate is less than or equal to the data symbol rate, we refer to it as *slow FH* (SFH); where it is higher, we refer to it as *fast FH* (FFH). In either case, the degree of spreading may be the same; however, with SFH, band occupancy during any one hop is close to what it would have been without FH. In the case of FFH, each hop occupies a broader bandwidth than it would in the absence of hopping. This difference has implications in the signal design, the receiver design, and the performance of the two FH types.

Figure 10.23 shows a block diagram of a typical FH system, with the spectrum spreading indicated. Note that in the case of SFH, the individual channels are comparable in width to the original channel, but because of the short transmission duration, the average power in each channel is reduced. In FFH, the individual

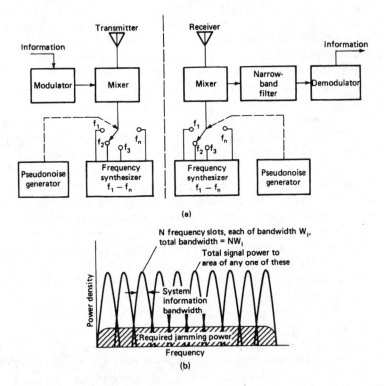

Figure 10.23 FH system: (*a*) block diagram and (*b*) spectrum. (*Adapted from [10.12].* *Reprinted with permission from* Naval Research Review.)

channel widths are broadened in addition to the hopping, so that the density may be further reduced. In the figure, the spread channels are shown as nonoverlapping, but this is not essential. Because the individual spread channels are not occupied at the same time by the signal, overlapping channels may be used, with a consequent increase in power density. SFH requires minimum modification of the receiver design. SFH can also allow sharing of many channels by a cooperative group of users without mutual interference, using sufficiently accurate clocks, as long as the transmission distances are not too great. Between hops, an allowance must be made for the maximum range between users and the clock drifts. On the other hand, the interference to and from noncooperative receivers on any of the channels results in pulses at the original power level, which can be almost as disruptive as the original signal. Even without jamming, there are likely to be noncooperative users in the spectrum that will cause interference and loss of information in some hops. Sufficient redundancy is required to operate through such interference.

FFH requires greater complexity in the receiver. During each digital symbol, the signal is hopped over a number of frequencies. At carrier frequencies that are sufficiently high that the medium remains nondispersive over a relatively broad band, phase-coherent multitone synthesizers can be built and controlled by standards with sufficient accuracy to allow coherent recombination of the hops into each symbol. In most cases this is not possible, and noncoherent recombination is required. This results in a loss of sensitivity in the demodulator (up to a maximum of 3 dB), but does provide redundancy within the symbol, which can make transmission more reliable when the transmission is dispersive. Because the individual channels are broadened in FFH, the short-term power density is reduced, resulting in lower interference and more difficulty in intercepting, locating, and jamming. However, with FFH it is difficult or impossible to maintain cooperative simultaneous use of the same spectrum without mutual interference. Consequently, some fraction of the antijamming capability is used against friendly stations when they operate simultaneously. As long as the number of friendly stations that are operating produce a power density that is lower than the possible power density from jammers, the interference can be tolerated.

FFH tends to be used for line-of-sight radio paths at high UHF and above. SFH is generally used at HF and low UHF where multipath is a problem. It is also applicable to tropospheric scatter modes at higher frequencies, although FFH with noncoherent demodulation can also be used.

In TH, the symbol is represented by one or more very short pulses, with relatively long time intervals between them. The average repetition rate must be great enough to maintain the information throughput, but the time of occurrence of each pulse is determined by the pseudorandom control process. The receiver can thus gate on when a pulse is expected and remain off otherwise, thereby eliminating interference. Although this process spreads the spectrum with a low power density, the pulses themselves must make up in power what they lose in duration in order to maintain the same average power. Because of the high peak power requirements and need for accurate timing, pure TH systems are seldom used in communications. The technique may be combined with FH to add an additional complication for jammers, or it may be used in cooperative multiple-access applications where the overall power requirements are low. If the separate users are not synchronized, but use different codes, interference occurs rarely and error control redundancy can be used to overcome it.

DS techniques spread the spectrum by modulating the carrier with a signal that varies at a sufficiently rapid rate to accomplish the spectrum expansion. For this purpose, a high-rate binary waveform, generated by either a maximum-length sequence generator or a generator using a nonlinear binary processor, is used. The former structure is obtained by the use of feedback with a shift register [10.19]. Some types of the latter are given in (References [10.20] and [10.21]).

The maximum-sequence generator can produce sequences with pseudorandom properties of considerable length ($2^N - 1$) for shift registers of N-bit length, and a much smaller number of selection parameters to set the feedback taps and choose the starting contents of the register. For example, a 31-bit shift register produces a sequence that does not repeat for 4.3×10^9 operations. At a rate of 2400 bits/s, the repetition period is 20.7 days. This type of generator is useful for applications against which hostile action is not expected. However, because the processing is linear, it is possible to analyze the sequence from a relatively short sample and determine the tap settings and the register contents at a future time. Once the sequence is known, it can be duplicated at an enemy interceptor or jammer, and the spread-spectrum advantage is lost. Nonlinear processes, such as DES, are much more difficult to analyze and break, and are therefore preferred for cases where jamming or intercept is expected.

Any standard modulation technique can be used for the spreading sequence. However, BPSK or QPSK are most often used for a pseudorandom binary-spreading sequence. The information modulation, which is much slower, can be applied to the carrier, using the same or some other modulating technique applied either before or after spreading. For digital transmission, the spreading waveform rate is made an integral multiple of the modulation rate, so that the two waveforms can be synchronized and combined at baseband prior to carrier modulation. The receiver demodulates the wave by the equivalent of generating a synchronous, identically spread carrier waveform and mixing it with the incoming wave so that the resulting difference wave is reduced to narrowband while retaining the information modulation. Figure 10.24 is a simplified block diagram of a DS system, indicating the spectra and waveforms at various points in the system. The degree of expansion may be hundreds, or as much as a few thousand in some cases. The individual bits of the DS code are referred to as *chips* to distinguish them from the data bits.

DS spreading may be employed either alone or in combination with FH to achieve a required bandwidth expansion. Because signal processing is mostly digital, DS is often preferred when a sufficiently wide contiguous bandwidth is available. At low frequencies, when the ultimate in spreading protection is required, DS may be used with the narrow bandwidth available (from a few kilohertz down to a few tens of hertz) to send much lower speed data (less than a few tens of hertz). DS has been employed in ranging, antijam, and anti-intercept applications. It is suitable for multiple-access use. However, in such applications, it is at a disadvantage in mobile surface applications because of the *near-far* problem. Such systems may have users located within ranges from a few feet to 40 or 50 mi of one another. This can result in interference on the order of 120 dB. The required spreading ratio of a trillion times is impractical, since the entire radio spectrum from 1 Hz to infrared would be needed to send a data rate less than 1 bit/s. SFH can handle the near-far problem much more readily because the sharing stations generally do not occupy the same spectrum at the same time; hence, the near transmissions can be filtered from the far ones. If a common repeater is available that is distant from all users (such as a satellite), DS code division multiple access becomes practical.

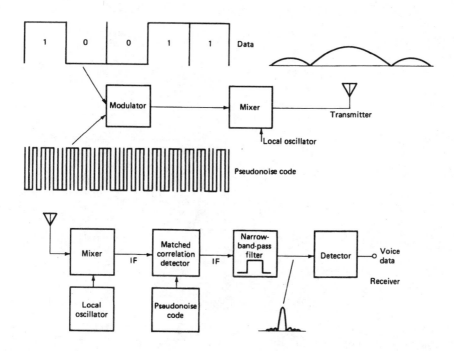

Figure 10.24 Block diagram, waveforms, and spectra for direct-sequence spreading.

The chirp waveform (References [10.22] and [10.23]) for spreading the spectrum comprises a linear swept frequency wave. When the datum is a 1, the signal is swept in one direction; when it is a 0, it is swept in the other direction. The sweeping occurs over a wide band compared to the modulating data rate, and because it is linear, it produces a spectrum spread relatively uniformly across the sweeping band. Detection of chirp may be made by generating synchronous up and down sweeps at the receiver, mixing them separately with the incoming signal, and comparing the outputs of the separate amplifier channels to determine which has the greatest energy. An alternative is to use broadband filters with parabolic phase characteristic to provide linear delay dispersion across the band. This has become practical through the use of SAW filters.

If a constant-envelope signal is swept linearly in frequency between f_1 and f_2 in a time T, its compression requires a filter whose delay varies linearly such that the lower frequencies are delayed longer than the upper frequencies, with such a slope that all components of the chirp pulse input add up at the output. In practice, a fixed delay is also added to maintain nonnegative overall delay for all positive frequencies. When the pulse is passed through the filter, the envelope of the resulting output

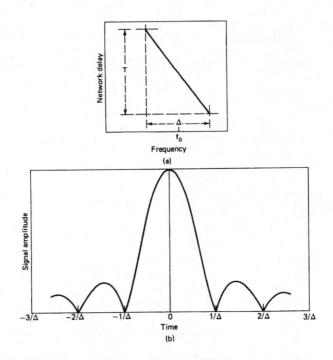

Figure 10.25 Chirp filter: (*a*) delay function for dispersive chirp filter and (*b*) output envelope of chirp signal passed through filter. (*From [10.16]. Reprinted with permission from* Bell System Technical Journal.)

is compressed by the ratio $1/D$, where $D = (f_2 - f_1)T$. The pulse power is also increased by this ratio. The dispersion factor D is a measure of the effectiveness of the filter. The delay function and the envelope of the filter output for the chirp input are indicated in Figure 10.25. Analogous results occur for the inverse chirp input (high-to-low frequency) with appropriate change in the delay function of the dispersive filter. It is possible to generate a chirp signal by exciting a chirp network with a pulse, rather than using an active frequency sweep technique. This method of generation can be advantageous in some applications.

SAW technology has made it possible to construct small filters with large D (References [10.24] and [10.25]). Figure 10.26 shows the envelope traces of the expanded and compressed pulses for a filter design with $D = 1000$.

The original use of chirp signals was in radar systems to provide pulse compression. Swept FM (chirp) is also used in aircraft altimeters and in intercept systems where compressive receivers are used for rapid spectrum analysis. Swept FM is also employed in HF ionospheric sounder systems as an alternative to pulse sounding.

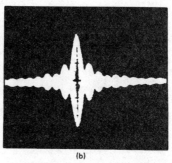

(a)   (b)

Figure 10.26 Expanded and compressed envelopes for a linear FM dispersive filter: (*a*) down-chirp linear FM expanded pulse (10-µs gate bias), 2.0 µs/division, and (*b*) recompressed pulse envelope (spectral inversion), 20 ns/division. (*From [10.20]. Reprinted with permission of IEEE.*)

The primary problem of any form of spread-spectrum receiver (which does not have a transmitted reference) is acquisition and tracking of the spreading waveform. Without this, it is not possible to demodulate the signal. It is also necessary to determine whether the data waveform shall be obtained from the spread waveform using coherent or noncoherent techniques and, finally, whether the data shall be demodulated in a coherent or noncoherent manner. A side problem in acquisition, if the receiver is to be used in a multiuser environment, is *cold entry* into the network, that is, how a new user who is not synchronized to the tracking waveform shall acquire synchronization when the other users are already synchronized and operating. The final question, which concerns all receivers, is how well does the receiver perform relative to its goals. We can only address these problems briefly.

10.3.1 Frequency Hopping

There are two approaches to demodulation of a spread-spectrum signal. The first (Figure 10.23) uses a local reference signal (synchronized with the spreading wave for the incoming signal) to eliminate the spreading and reduce the wave to a narrowband signal that may be processed by conventional techniques. The second approach changes the frequency of the incoming signal, amplifies it, and then presents it to a matched filter, or filters. The matched filter eliminates the spreading and may include signal detection. Both approaches produce equivalent performance if they can be implemented equally well. In some cases, the instability of the medium may make one process superior to the other.

Figure 10.27 is a more complete block diagram of an SFH receiver. It may also apply for fast-hopping applications where the medium is sufficiently stable. While analog modulation of the signal can be used, it is not generally desirable because of the interruptions encountered in the spreading waveform between frequency

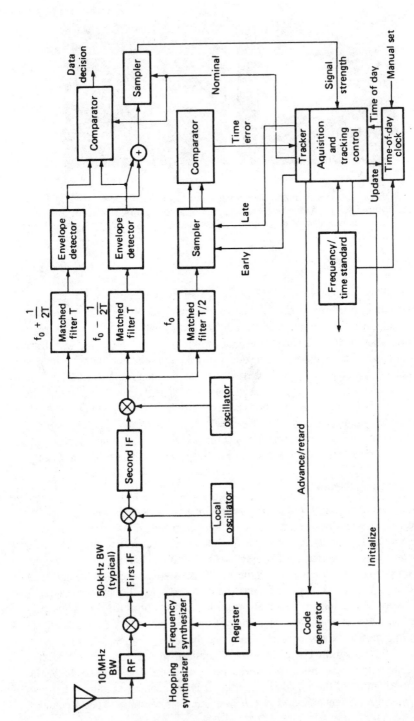

Figure 10.27 Typical FH receiver block diagram.

hops. Analog storage systems, which can speed up the modulation at the transmitter and slow it at the receiver, can be used to eliminate the interruptions, but it is generally more satisfactory to use digital signal coding and storage of the analog signals. Although voice quality suffers from interruptions, good intelligibility can be achieved at certain rates when the interruptions are short [10.24].

The receiving synthesizer hopping pattern must be synchronous with the (delayed) hopping pattern of the received signal. When the receiver is first turned on, even if the proper coding information is available, timing is likely to be out of synchronism. To acquire proper timing, it is first necessary that the operator set a local clock to the correct time as closely as possible, to reduce the range of search necessary to find exact timing at the receiver. When this *time-of-day* (TOD) estimate has been entered, the acquisition and tracking computer advances the receiver time to the earliest possible time, considering the range and clock errors. The receiver time is then retarded gradually toward the nominal clock time. Before the receiver time is retarded to the latest possible time, the code generator, driven by the receiver tracker, will overlap with the received code and the signal will be passed through the IF amplifiers to be demodulated by the envelope detectors. When full synchronization is achieved, the receiver output becomes maximum, and further retardation results in a reduction in output. At this point, the control processor switches to the tracking algorithm and passes the signal demodulator output.

The tracking algorithm for Figure 10.27 uses early and late samples of the tracking channel output. The data channel is sampled at the nominal symbol time. The tracking channel is sampled at equal early and late fractions of the hop duration. The early and late outputs vary as shown in Figure 10.28a and b as signal delay changes. The comparator subtracts the two values and provides an error signal (Figure 10.28c) to the synchronization controller to correct the delay via oscillator feedback control. The pull-in range increases with increasing early and late delays. When the receiver time has been pulled to zero error output, the controller updates the reference TOD clock, either by using an estimated range to the transmitter or by initiating a protocol for range measurement (when the link is bidirectional).

When the transmission interval is so short that a search might require most of it, another acquisition technique must be used. Such a technique is the use of a prearranged synchronizing signal at the start of each transmission. The receiver is set to the starting frequency of the synchronizing sequence, ready to hop to the succeeding frequencies once this is detected. If the other frequencies do not verify the synchronization, the receiver returns to the first frequency. This approach is easier to jam than the search approach, although particular synchronization sequences need not be used more than once for each message period. After each message period, an advanced receiver clock resets the receiver for the next sequence, until the synchronizing signal has been recognized, acquired, and tracked.

Figure 10.27 is based on orthogonal FSK data modulation. With slow FH, any data modulation technique appropriate to the medium may be used. BPSK, MSK,

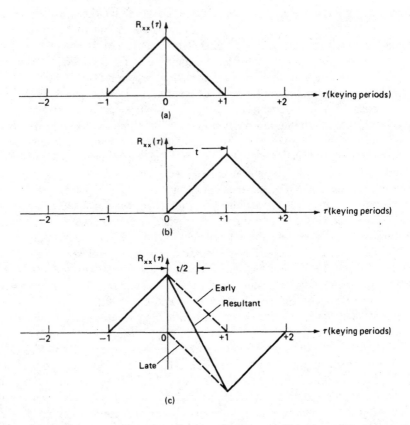

Figure 10.28 Tracking waveforms: (*a*) early correlator output, (*b*) late correlator output, and (*c*) tracking error signal (difference.)

QPSK, and *m*-ary FSK are common. FFH, however, poses different problems, both as to data and to spreading demodulation. For FFH, several frequency hops occur for each data symbol; the signal is spread over a substantial bandwidth and is not necessarily coherent. If the medium will sustain coherent transmission over a wide band (e. g., some SHF or EHF applications where multipath is not a problem), then the earlier techniques are possible, provided that the frequencies generated in the hop sequence are coherent. Coherent generation requires frequencies derived from a single oscillator and in-phase synchronism with that oscillator. Synthesizers of this type generate many frequencies from a common oscillator and subsequently generate the hop frequency by successive mixing processes.

If a more limited range of coherence is available, for example, over a single information symbol, then the output may be hopped between symbols using an oscillator that is not necessarily locked in phase to the reference. This sort of technique can be used with a small number of coherent frequencies (on the order of 10 or so), and the signal sets are usually *m*-ary symbols chosen from orthogonal or almost orthogonal groups selected from subsets of the coherent hopped frequencies.

10.3.2 Direct Sequence

The two types of spread-spectrum receivers for DS are indicated in Figure 10.29. In Figure 10.29*a*, the DS code at the receiver is used to modulate the second oscillator so that the spreading is removed and the signal to the final IF is a narrowband signal that may be demodulated appropriately. In Figure 10.29*b*, a broadband IF is used, and the second oscillator converts the signal to baseband where a matched filter correlator compresses the bandwidth for demodulation. Figure 10.30 shows a typical baseband correlator for direct sequence. The PN code for the next expected symbol is fed into the upper shift register at a very high rate and remains in the register while the signal modulated by this code is shifted into the lower shift register at the chip rate. The contents of the registers are multiplied and summed to produce the output. When the sequences of the two registers are not in alignment, the output is a relatively noisy variation about zero. However, when the two sequences line up (i. e., are correlated), the outputs from all the multipliers add up to produce an output peak whose level is determined by the original (modulated) signal before spreading.

Acquisition and tracking for the DS receiver is analogous to those processes in FH receivers. Acquisition is dependent on TOD in systems where maximum jamming protection is required, and a search "from the future" is used. For some applications, special synchronization codes may be sent, and correlators may be set in anticipation of those codes. Several tracking arrangements may also be used. One type of correlation receiver uses one channel for the signal demodulation and two channels to provide the early and late references for tracking. A drawback of this system is that the tracking channels may change independently with temperature, humidity, supply voltage, or aging, causing a shift in the locking frequency. Another early-late tracking arrangement requires only one separate channel. The code generator output to the signal channel is delayed by $\tau/2$. The early sequence has no delay, and the late sequence has τ delay. Because the modulation and demodulation processes used for processing the early and late gate signals are linear, the subtraction is made prior to modulation of the reference oscillator with the DS. Thus, the tracking channel output from a narrowband filter will be zero when the signal is in tune, and will rise on either side, having opposite phase for an early and late generation. A tracking signal is fed to a product modulator that uses the IF reference from the signal channel, limited to provide constant level. Because this channel contains the signal modulation, it serves to eliminate signal modulation from the tracking channel. Thus, when the code generator falls ahead or behind the incoming code, the tracking channel generates the voltage required to correct it.

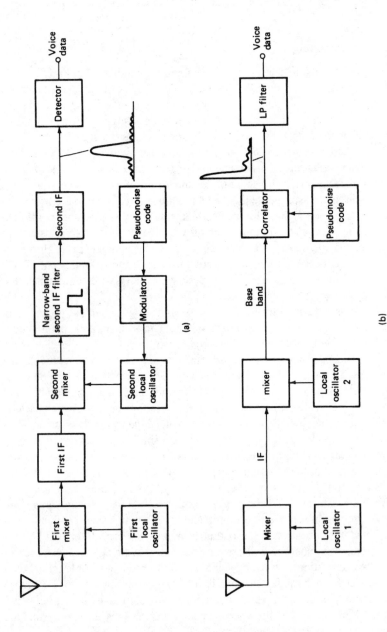

Figure 10.29 Direct-sequence spread-spectrum receivers: (*a*) correlation type and (*b*) matched-filter type.

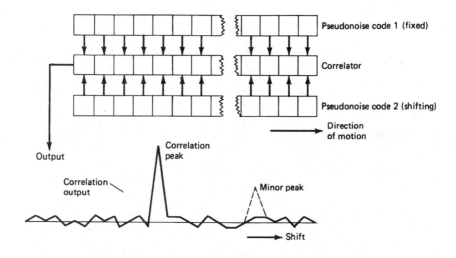

Figure 10.30 Typical baseband correlator arrangement for direct sequence.

Rather than separate-channel early-late tracking, another technique, known as *dither tracking*, may be used. In alternate data symbol intervals, the output of the DS coding generator is delayed by a time τ. An envelope or rms demodulator is used at the output of the IF channel as well as the data demodulator. The output of the envelope demodulator is switched to separate low-pass filters in alternate data symbol intervals, synchronously with the DS generator switching. One channel receives the output during early-code intervals and the other during late-code intervals. The difference of the outputs is used to control tracking, as when separate early and late channels are used (Figure 10.31).

Tracking may also be accomplished by use of baseband correlators, serving as matched filters. In this case, separate correlators are provided for I and Q channels, and the sampling rate is twice the chip rate. The PN code is fed to the reference registers, and because of the double sampling rate, each chip occupies two stages of the register. As the signal is shifted through the register, the summed output is sampled, and three successive samples that exceed a predetermined threshold are stored. They represent samples of the correlation triangle. When the center of the three samples becomes highest, the samples on either side have values in excess of the noise level in amplitude. The threshold is set to exclude the noise pulses and side lobes of the correlation. The I and Q register outputs are used to lock the VCO at either of two phases (separated by 180°), to control the code generator phase and to detect the output modulation.

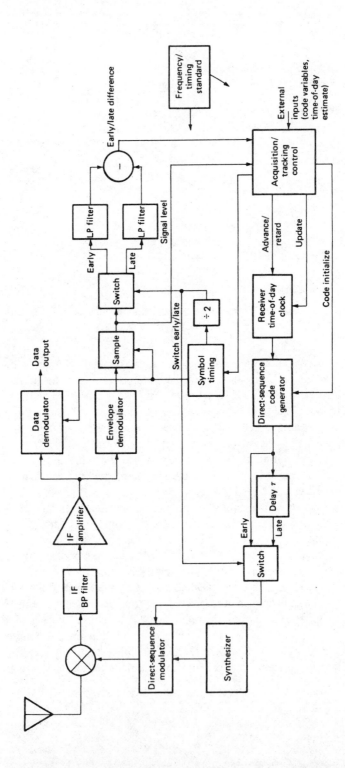

Figure 10.31 Correlation receiver with dither tracking channel for direct sequence.

10.3.3 Performance

There are many potential applications for spreadspectrum receivers and a number of different techniques for achieving spread-spectrum. Further, there are differences in transmission distortions for different frequency bands. These variations demand that many different performance criteria be considered. In almost all systems, synchronization and tracking with the spreading code is required. Thus, the time required to achieve synchronization, and the accuracy with which it can be retained, are among the most important performance criteria. Once the system is synchronized, it should accomplish the functions for which it was designed.

If spread spectrum is used as an ECCM technique, the degree of protection from all kinds of jamming is the prime performance criterion. The lack of easy detection, location, and identification by an enemy (*anti-intercept*) is important, as is the degree of protection against an enemy injection of erroneous messages (*antispoofing*). These characteristics are determined by the waveform and not the receiver. When spread spectrum is used for multiple access, the number of simultaneous users who can use the system without degrading performance for the weakest users is important. If users are mobile, the ability to protect against very strong nearby friendly users is essential (near-far performance). If spread spectrum is used for ranging, then maximum range capability, range resolution, and the time required for various degrees of resolution are the important performance criteria. If it is used to increase transmission rate or accuracy through difficult media, then these improvements contrasted with those for standard waveforms are the important performance criteria.

In all cases, the complexity and cost of the spread-spectrum receiver relative to the alternatives must be considered. In most cases, system designers base the choice of spreading technique and parameters on theoretical models, and allow margins for "implementation" losses in various parts of the system. The receiver parameters may be expressed in terms of a maximum permissible loss relative to a theoretical ideal. Many spread-spectrum receivers have multiple functions and may need to be evaluated on many of the foregoing parameters. In this book it is not possible to detail methods of prediction and measurement of the performance criteria. The references provide volumes dealing with such matters.

10.4 Simulation of System Performance

Advancements in computer technology have placed at the disposal of the radio designer a powerful engineering tool. The more complex problems in receiver design involve not only input-output relationships of simple circuits, which result in the evaluation of complex functions, but the effects of interactions of many circuits and controls. In most cases, they involve elements whose specific values can only be specified by a statistical distribution (and sometimes not even then). Computer software can be used to deal with such difficulties by simulating the situation with a series of *algorithms*, a well-defined set of rules or processes for the solution of a problem in a finite number of steps. A convenient method of displaying an algorithm is a flowchart, where the steps and their interrelationships are clearly indicated.

One clear advantage of simulation is that it provides a technique for evaluating design alternatives without first building each one. Changing parameters of a complex design in a computer run is far more economical than building, testing, and altering breadboards. Simulation of the medium permits the design to be subjected to a wide range of conditions faster and surer than nature will provide in field testing. Finally, if the design is intended to combat an unusual medium condition, the probability of finding and repeating such a condition for testing or comparison may be practically negligible. A well-designed medium simulator, however, can be made to repeat a particular set of conditions at any time, present or future. With all their advantages, however, even the best simulators are imperfect models, so that field testing of the real equipment remains necessary for the final design.

Simulations can use a sequence of analytic solutions, already known individually, by defining them in the proper sequence and taking into account any interactions (e.g., the expressions for an FM wave, a particular linear filter, a limiter, and a frequency detector). Because of the sampling theorem, when a modulating wave of finite duration is defined, the response of such a cascade can be accurately estimated by sampling at a sufficient rate to avoid aliasing. The samples are processed through each expression in sequence. The use of the z transform allows us to model circuits in sampled systems without first solving the equations analytically.

Event-driven simulations are useful in some applications, but they have only limited applicability in receiver design. Such a simulation might be used in estimating the performance of a particular receiver in conjunction with an ARQ transmission link. Mostly, however, such problems can be divided into separate simulations—in this case, first a simulation to determine the receiver's probability of message acceptance under various conditions; then the determination of the overall performance of the link by a second simulation using the message error statistics. Where there is an easy separation of problems, it is usually best to simulate each one separately.

The problems of tolerance and of noise generally do not have analytic solutions. In the former case, individual parts in a circuit may have any value in a range, with a known (or in many cases unknown) distribution of the values. Differences in overall performance can be determined easily with all parameters at the high, low, or nominal values. However, the poorest performance in a complex circuit may occur for some intermediate values of parameters. Simulating performance for all permutations of parameter increments between their upper and lower limits is usually prohibitively expensive. With only 30 parameters, each having 10 possible values, 10^{30} tests would be required. In such cases, it is best to use Monte Carlo techniques by selecting the value of each parameter for each test in a random manner, in accordance with its known or imputed distribution. Such Monte Carlo techniques are useful in problems where there are randomly distributed variables. By making enough tests with the controllable parameters kept constant, confidence can be achieved in the resulting distribution of the circuit performance.

Some examples of the use of circuit design optimization programs were given in Chapter 7. In the remainder of this chapter we describe a few examples of analytic and Monte Carlo simulations to illustrate the range of applicability. In most cases, the designer should use simulations that have already been programmed and used successfully. This is especially true at the system level. It is unwise to try to simulate full receiver designs or full link designs, simulating every individual circuit and evaluating the whole design at every level. Such a simulation is so specialized that much expensive individual program effort is usually necessary. Problems should be generalized and broken into easily solvable subproblems. For the larger of these, a program is likely to be available. The smaller may well yield to simple modification of existing software.

10.4.1 Spectrum Occupancy

In general, analytic expressions can be found for the spectrum density (and lines) resulting from common forms of modulated waves. Integral expressions for the spectrum occupancy can be given, but the integration in known form is not always easy. For that reason, numerical integration is often used. As an example, Tjhung [10.26] used Pelchat's [10.27] expression for FM spectrum density and performed a numerical integration to provide curves of spectrum occupancy that he presented two ways. It was also easy to determine the occupancy after the wave had been passed through an RF filter, by multiplying the spectral density by the filter selectivity curve (power) before integration.

For spectrum occupancy with premodulation filtering, however, rather than develop a new expression for the density, Tjhung used a heuristic approximation developed by Watt et al. [10.25] and applied it to the original density before integration. This leaves the results somewhat suspect in that case, but doubtlessly produced good approximate results more rapidly than the other process. Prabhu [10.28] also used numerical integration of spectrum density for continuous PM, treating one case of FM (MSK). Rather than deal with premodulation filtering, he used analytically defined limited time modulation functions, for which techniques for the density had been developed previously [10.29].

A related problem has been described [10.30] involving the spectrum occupancy of keyed short segments of FSK modulated waves. Time-division multiple-access systems and SFH systems often use such signals. The usual spectrum occupancy calculations have been made for a wave that is long enough to be considered infinite. It is clear that a short-keyed signal may have a wider spectrum than the infinite signal, but in most treatments this factor has been overlooked.

The technique considers the wave as an FSK wave multiplied by a pulse having a specified rise and fall shape, and a duration that can be varied. The same procedure could be used for PSK pulses or indeed any waveform that is the product of two time waveforms whose spectrum density is known or can be derived readily. If the two waveforms are $a(t)$ and $g(t)$, the spectral density of their product is determined by first identifying their autocorrelation functions and then determining the spectral density of the product of the two autocorrelation functions. This works as long as at least one of the functions is wide-sense stationary. An alternative, of course, is to use the convolution of the individual spectral densities.

In the cases examined (FSK with $h = 0.5$ and 0.7, rectangular keying, or rectangular keying with gaussian or Butterworth filter shapes having rise and fall times of one symbol), the density spectra are readily available. Spectrum occupancy in specific cases was evaluated. Final results were obtained by using the product of the autocorrelation functions and transforming to the frequency domain. FFTs were used. The analytic expression for the FSK autocorrelation function was available in the literature, so there was no need to transform it from the spectrum. The keying spectra were transformed to autocorrelations using FFT, the product autocorrelation function samples were calculated, and the resultant was retransformed to the time domain. After normalization, the occupancy was calculated by summing spectrum samples. Where needed, fractional occupancy was calculated by interpolating between the points just below and just above the desired fraction. Figure 10.32 is a flow diagram of the program.

Figure 10.33 shows some of the results. Other results are given in [10.30]. As one would expect, a square keying pulse results in a substantial increase in occupancy even for relatively large packets. The shaped pulses with single-bit rise and fall times approach the ultimate occupancy at a short packet duration (about two symbols). The difference between gaussian and Butterworth shapings (for two-, four-, and six-pole filters) is very small.

10.4.2 Network Response

Another area that has used analytic results with a computer program to arrive at a composite result not easily treated by analysis is that of network performance with modulated waves. An early example [10.31] considered the effect of limiting the transmission bandwidth of a baseband binary signal with sharp transitions between amplitudes $\pm A$. The signal was low-pass filtered at the transmitter before transmission and at the receiver after noise was added. The demodulator is a perfect integrate-and-sample-and-dump circuit, integrating over the symbol period. This is the demodulator that would be a matched filter in the absence of filtering. One of the features that make an analytic approach tractable is that until the decision process, there are no nonlinear processes and the linear processes are relatively few. The noise considered is gaussian. While the analysis is made at baseband, it applies equally to BPSK with perfect phase recovery.

To provide many patterns of intersymbol interference, 2 repetitive 40-bit sequences were used. The effect of the filters on the signal was determined by calculating the Fourier series coefficients of the periodic sequences and multiplying

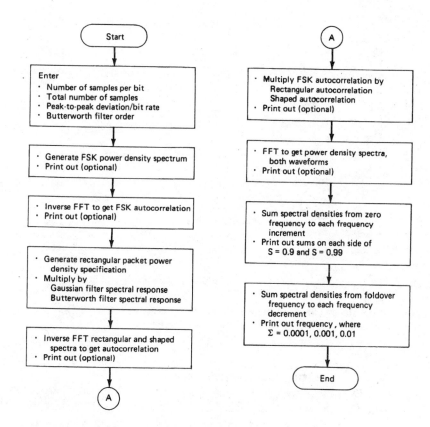

Figure 10.32 Flow diagram of spectrum occupancy computation.

them by the filter transfer functions. In the particular analysis, sharp cutoff and linear phase were used to simplify the work. Among the sequences, the different responses were integrated over the central $+A$ symbol for the 16 cases, where the 2 preceding and 2 following bits could have either sign. For multiple appearances of each case, the result was averaged to get a representative value. Because the filtered noise remains gaussian, and independent of the signal when its variance is known, the probability of error for each case can be estimated from the normal probability integral. Finally, the probability of error is averaged over the 16 cases to obtain the results.

A later result [8.53] expanded upon this technique to deal with the bit-error-rate performance with intersymbol interference from the receiver filter for an FM system. Again, a pseudorandom periodic sequence was chosen for modulation. In this case, the analytic representation of the signal allows the carrier to be separated from the modulation, which is represented by exp $[j\ m(t)]$, where $m(t)$ is the instantaneous PM resulting from the FM. Because $m(t)$ is periodic, exp $[j\ m(t)]$ can

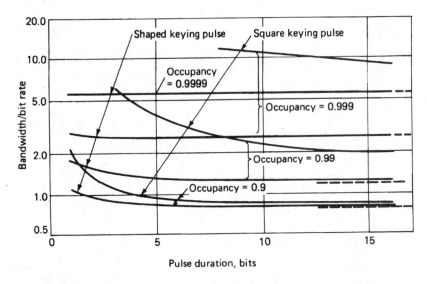

Figure 10.33 Bandwidth occupancy for pulsed FM packets (deviation 0.5; unpulsed occupancies dashed). (*From [10.21]. Reprinted with permission of IEEE.*)

be expanded into a (complex) Fourier series and the coefficients modified by the IF filter. In this case, the demodulator is assumed to be a perfect frequency demodulator, and the output is calculated using the click theory (References [8.37] to [8.39]). The action of the output low-pass filter upon the independent signal and noise components of the output is calculated, and the output decision is made at the correct sampling point.

The baseband filter was as an integrate-and-dump type with integration over the symbol period. This is a nonoptimum filter when the IF has been filtered before demodulation. Two types of IF filter were used, a rectangular and a gaussian filter, both with linear delay functions. Some typical results of the simulation are shown in Figure 10.34. For each modulation rate there is an optimum IF bandwidth, which could be different and produce different error performance if the postdemodulation filter were not fixed. The results were checked experimentally using real filters. Figure 8.68 showed the comparisons.

A somewhat more ambitious model was developed during a study of satellite communications. In this model, various filter groups were represented by their poles. Only minimum-phase filters were simulated. There were five filter groups, and the program was arbitrarily limited to handle 80 poles total. A sampled time domain simulation used the transient response of the filters. For each group of poles, the program calculated the transient complex amplitude and phase time variation. After each limiter, a convolution of its output was performed with the next filter

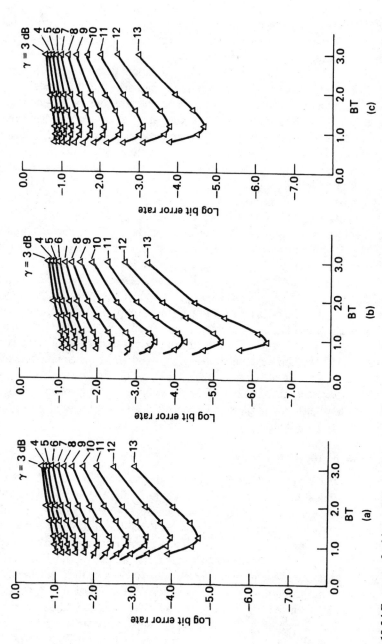

Figure 10.34 Error rate for binary FM as a function of bandwidth for a gaussian bandpass filter. $1/T$ = bit rate, B = IF bandwidth, $h = 2Tf_d a$, $-$ = $(A^2T/2)/N_0$. (a) $h = 0.5$, (b) $h = 0.7$, and (c) $h = 1.0$. (After [8.53].)

response, until the demodulator was reached. The sampling rate was chosen high enough that aliasing of the I and Q components at baseband was negligible.

Because of the convolutions, the entire output sequence at the output from each limiter was required before the next filter processing was commenced. At the output of the last filter, provision was made to run the data through an eye-plotting subroutine. The design permitted the omission of various stages to change the configuration. Noise was introduced after demodulation, based on analytic estimates of distribution, depending on the processing and demodulation type.

The first element in the program was the use of a step function in the modulator to observe the output plot. The transient response allowed the overall filter delay to be measured, so that a nominal delay in the bit position under observation could be selected. A judgment could also be made on the number of adjacent symbols giving rise to intersymbol interference. From this, the user determined how many symbols preceding and following affected the bit under observation. With the observed symbol in a reference position, the program cycled through all selected adjacent symbol permutations, processing the resulting input wave by the selected filtering, modulation, limiting, and demodulation conditions. The sampled outputs from each cycle were stored, and the error rates at selected values of E_b/N_0 were then calculated and averaged to provide an overall curve.

Figure 10.35 shows some typical plotted results for a *binary PEK* (BPEK) case with coherent demodulation. Three filter groups (a total of 80 poles) and 2 limiters were used. Noise introduced prior to the final filter group (the terminal receiver) was assumed negligible. Figure 10.35a shows curves of the calculated BER versus S/N in the receiver (last filter group) bandwidth for various sampling time and phase errors. Figure 10.35b shows eye patterns of the in-phase and quadrature components of the observed output symbol, to the left with limiting omitted and to the right with limiting operational. Figure 10.35c is a portion of the (nonlimited) step response. In the figure, the in-phase components are shown by solid lines, the quadrature components dotted.

10.4.3 Medium Prediction

Medium prediction is another area of simulation that uses determinate algorithms and tables. The most complex of these are programs that predict the operating paths between two points for propagation with ionospheric refraction. Among this class, those that predict *maximum usable frequency* (MUF) and atmospheric noise level in the HF band are the most sophisticated. A number of organizations began working in this area in the late 1950s to computerize the tables and calculation procedures that had been developed during the prior 30 years. In the United States, the *Central Radio Propagation Laboratory* (CRPL) of the National Bureau of Standards, which has since become the Institute for Telecommunication Sciences (ITS), issued a succession of programs based on ionospheric models. The CCIR has also adopted similar models [10.32]. Different programs or software modules

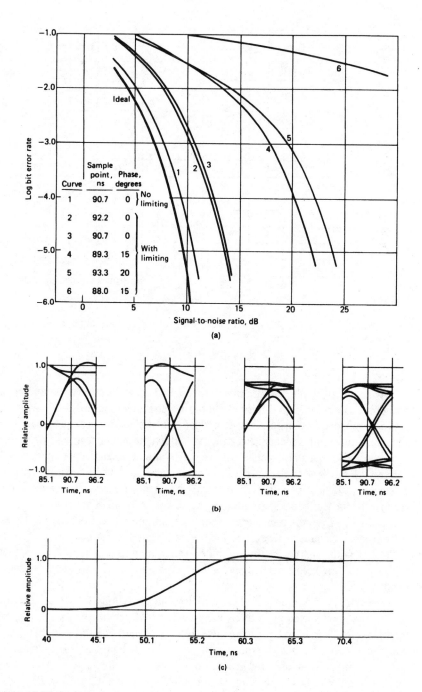

Figure 10.35 Error performance curves and eye patterns for all filters properly centered, showing the effects of various phase- and timing-recovery offsets.

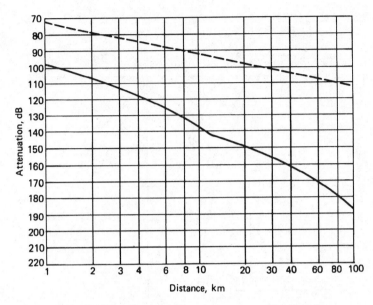

Figure 10.36 Typical plot of transmission loss from program. The dotted curve represents free space loss.

are typically available for dealing with VLF, LF, and MF, since ionospheric behavior is approximated differently in these frequency ranges.

For tropospheric transmission with clear line of sight between terminals, there are standard formulas that are easily converted to computer programs. When there is not a clear line of sight, techniques are still available when detailed path profiles are known. For land mobile service, in many cases, a clear line of sight does not exist, nor can the specific profile be predicted in advance. Based on limited empirical data, Egli [10.33] proposed a simple model for predicting the median loss at a distance d from the transmitter. There is a minimum antenna height to be used in each case, which is dependent on the frequency and the character of the soil. The variations with position at the estimated range include a general terrain factor variation assumed to be normal, in decibels, with a mean of zero and a variance of about 5.5 dB.

There were a few more implied variables, but everything can be programmed easily. This model is comparatively simple and usually produces somewhat pessimistic results. The Longley-Rice model (References [10.34] and [10.35]) is more complex. The model has two modes, one when the path profile is known and one for calculating the median for ground mobile types of application. It includes a terrain irregularity factor (which was an implied variable in Egli's model) to allow for the differences among various types of terrain, such as flat prairie land, rolling hills, and mountains. Figure 10.36 is a typical output plot.

Another set of formulas was developed [10.36] that covers a limited range of the Okamura et al. [10.37] graphic method. In general, programs that use known tables and algorithms may be programmed in a straightforward way for a wide variety of machines, if the expected usage is enough to warrant the time for programming. While bugs can be expected in any programming effort, it is usually easy to track them down in determinate programs of this sort.

10.4.4 System Simulation

The most ambitious Monte Carlo type programs endeavor to simulate the entire communications link, from the input of the transmitter to the output of the receiver. Figure 10.37 is a block diagram indicating the functions in a complete one-way system simulation. Two-way systems include feedback returning from the receiver location to the transmitter location by a simultaneous, though independent, link through the same medium to further improve the reliability of the overall transmission. The simulation functions include generations of the transmitted signal and simulating effects in the transmitter that may not be easily analyzed. The transmitted signal is coupled through its antenna to the selected transmission medium. In many cases, this will be a complex multipath medium. If multiple antenna elements are used at the receiver, there will be differences in the propagation path caused by their physical separation. Each may also have slight differences in received atmospheric and man-made noise. All separate thermal noise sources of interest introduced following the antenna element are assumed to be independent.

All the receiver circuit effects, including the linear and nonlinear processing, influence the end signal fed to the demodulators and decoders. To account for jammers or nonhostile interferers, other simulated waveforms must be impressed on the medium simulator, with the proper parameters for the simultaneous different paths involved taken into consideration, to produce outputs that add appropriately in the receiver antenna input elements. If the probability of intercept detection or interference with other receivers is to be examined, still other paths must be provided through the medium, with appropriate receivers. Such simulations are usually designed to be very flexible, by using simulation modules that may be called up for use if required, but which need not be used unless they are required. Such programs can grow slowly about an overall computer organization, providing an ever-increasing block of modules for use.

10.4.5 HF Medium Simulation

The representation of time-varying signals by multipliers and delay lines goes back a long way (References [9.26], [10.38], and [10.39]). An experimental real-time model was built and tested [10.40] using the delay-line approach shown in Figure 10.38. The $G_i(t)$ are complex numbers and the H_i are unity in the simple case. The variation in gain in each channel is changed by a band-limited random process, assumed to be gaussian, having gaussian-shaped spectrum with the mean

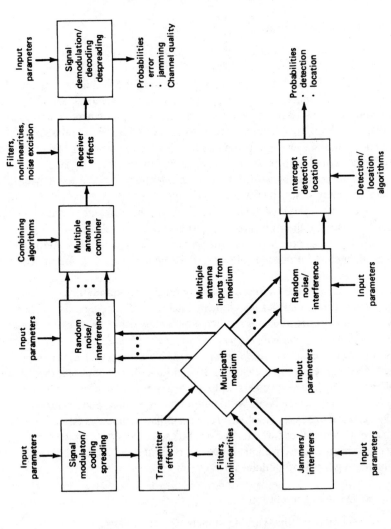

Figure 10.37 Block diagram of a complete one-way link simulation.

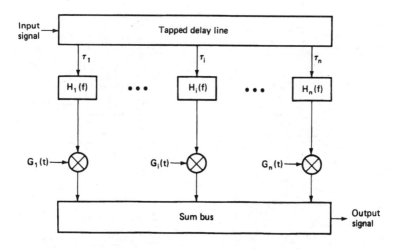

Figure 10.38 Block diagram of a tapped delay-line simulation model for wideband channels.

frequency offset to represent the average Doppler effect in the path, and the spread determined by the variance of the frequency curve. The model was found to provide a good fit for real narrowband channels common at HF. By selecting the delays, mean Dopplers, and variances of the gain selectivities, it was possible to duplicate the performance of the real channels.

Because of recent interest in the use of broadband channels, a wideband computer simulation model was developed. To ensure the capability of the model to handle wideband channels, it was necessary to allow for dispersion over the individual paths as well as gain variation. This is illustrated in the $H_i(f)$ in Figure 10.38. While the model was being constructed, we learned of another such model being developed for the Navy [10.41]. The model provided handling a total of 13 ionospheric paths, which could include ground wave, low and high, and ordinary and extraordinary waves, as appropriate for the particular prediction. Figure 10.39 shows a high-level flow diagram for the program.

To keep the processing rates to the minimum possible, the simulation is performed using the I and Q components of the modulated wave. The carrier frequency value is stored for subsequent printout and reference, if desired. It must also be used in establishing the number of paths supported, their mean attenuations, delays, and so on. The path parameters are supplied as an input to the program. The sampling rate must also be entered, set at the minimum value that will avoid aliasing of the input modulation components. Table 10.3 lists the necessary input parameters. To provide adequate records for the future, other types of data can also be entered to appear in a final printout, such as the assumptions about the date, time, and

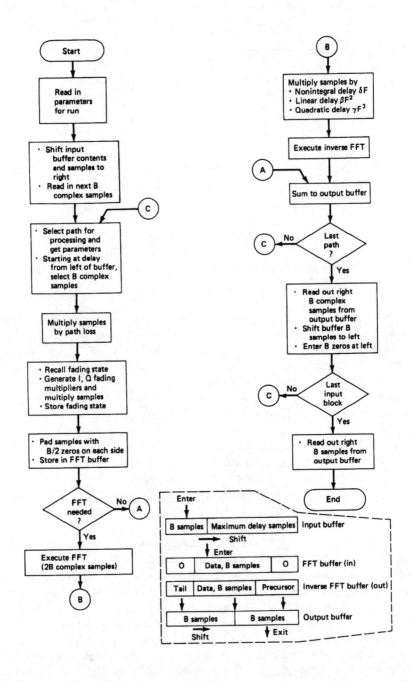

Figure 10.39 Overall flowchart of HF channel simulation model.

TABLE 10.3 List of Input Parameters for HF Medium Simulation Program

Sampling frequency, Hz
Carrier frequency, MHz
Number of active paths
For each path:
Path designation
Path median attenuation, dB
Total path delay, ms
Path dispersion
Linear delay distortion, μs/MHz
Quadratic delay distortion, μs/(MHz)^2
Path mean Doppler shift, Hz
Fading bandwidth (equal to one-half Doppler spread), Hz

geography that led to the predictions, and the type of path represented by each path number.

The program first reads in successive blocks of (complex) sample points from the transmitted waveform. After the points have been read into the input buffer, the same block is extracted from the reference (shortest delay) path for processing. The path is subjected to fading, in accordance with the input parameters. If the data parameters indicate that the path has dispersion, it is padded on each side by one-half the number of zeros in the input block and then converted by FFT to the frequency domain. The frequency samples are processed for the appropriate delay function (i. e., linear and quadratic delay) and then reconverted to the time domain by inverse FFT. The output samples include the precursor period (half-block), the processed block, and the tail (half-block), all of which are stored in the output buffer.

The second and subsequent paths are processed sequentially, each with its own parameters. The processing is analogous to that of the first path, except that the block read from the input filter includes a number n_i of samples preceding the first path block, where n_i is the largest integral number of samples in its delay. Thus, the first readout of each of the paths other than the reference has a number of zeros at the head of its block, corresponding to the delay n_i/f_m. Also, if a fractional delay remains, the block is processed in the frequency domain, whether or not there is dispersion, so that a linear phase shift may be added, corresponding to the fractional delay. The output from the inverse FFT is added to the output already in the output buffer.

When all the paths have been processed, a number of bits is read out of the output buffer, equal to the number in the original input block. This includes a half-block of zeros and the precursor the first time, plus a half-block of processed input samples. The second half-block of processed samples with the half-block of tail are shifted to the head of the output buffer, the remainder being filled with zeros, to await the next round of processing. The samples in the input buffer are then shifted by one block and the new block of input data is read into the buffer behind them. The processing repeats as before. However, because the input buffer now contains two blocks of samples, the later paths, with a delay no larger than one block, no longer have leading zeros when their input blocks are processed. Thus, the program takes in one block of samples at a time, and puts out one block at a time, delayed by a fixed period of a half-block. The block taken in must be smaller than the FFT by the amount of padding.

The Rayleigh fading process is generated by multiplying the I and Q components by samples generated by two independent gaussian processes. The two processes are identical except for the input samples, which are independent white gaussian samples. A path that fades in a Rayleigh fashion at a rate similar to that expected from an HF path (perhaps a few tenths of a hertz) and has a mean frequency offset equal to the average Doppler shift of such a path (as much as 1 or 2 Hz) is achieved by feeding the two white processes through identical bandpass filters with the center frequency offset by the average Doppler shift. A two-pole Butterworth shape was adopted. It is necessary to relate the filter parameters to the fading parameters. References [10.40] and [10.41] refer to the gaussian shape and define the Doppler spread as 2σ. This is presumably based on Bello's definition of Doppler spread [10.42]. Goldberg [10.43] gives curves and refers to the *fading rate* (FR) and *fading bandwidth* (FB), although he does not define the terms. If we define the FR as the average number of times per second that the envelope goes through its median value, using Rice's results, FB = $0.41628(b_2/\pi b_0)^{1/2} = 1.475665$FR, where b_0 and b_2 are defined by Rice from the power spectrum density of the process [8.42].

For the gaussian filter, the FB is σ, the Doppler spread is 2σ, and the fading rate is 1.476σ. For the Butterworth filter, the FB is $BW_3/2$, the Doppler spread is BW_3, and the fading rate is $0.738BW_3$, where BW_3 is the 3-dB bandwidth of the filter. None of these numbers can be derived for the single-pole filter since b_2 is infinite. For the rectangular filter, the FB is $BW/\sqrt{12}$, the Doppler spread is $BW/\sqrt{3}$, and the fading rate is $0.426BW$, where BW is the pass bandwidth.

10.4.6 Simple Simulations

In the time-gated equalizer (Chapter 9) partial autocorrelation was used to determine the location of multipath delays. This concept was evaluated using a simulation program. The purpose of the test program was to simulate a short packet transmitted in multipath and noise. At the receiver, once the timing had been recovered using a synchronization preamble, individual data packets would be processed for autocorrelation to determine whether multipath was occurring and if so, with what delays. If the packet is about twice as long as the longest expected multipath, autocorrelation of the wave truncated at the end of the packet results in a linear reduction of the length of the correlation with the delay, tending to increase the side lobes from data and noise, and thus to obscure the later multipath peaks. The program was designed to produce the partial autocorrelations from random data packets for various S/N.

Figure 10.40 is the top-level flow diagram of the program. The length of the packet was set at 32 bits, with 1-bit raised-cosine rise and fall times for its envelope.

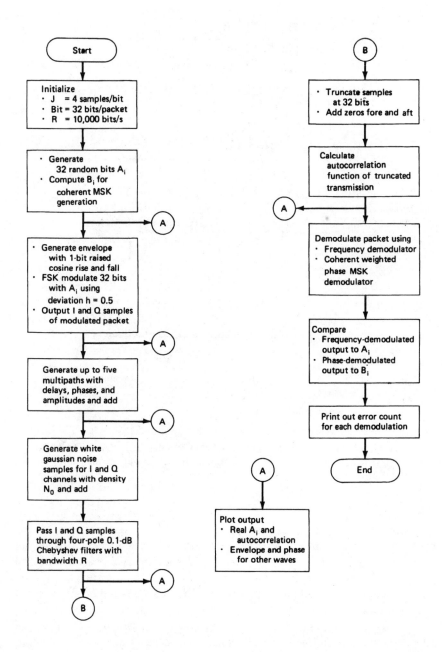

Figure 10.40 Top-level flowchart for simulation of multipath autocorrelation location.

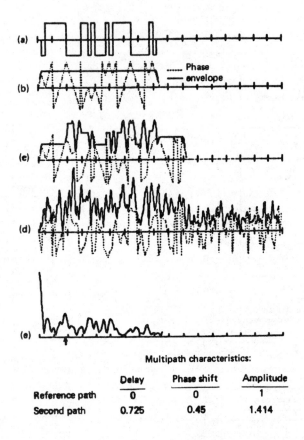

Multipath characteristics:

	Delay	Phase shift	Amplitude
Reference path	0	0	1
Second path	0.725	0.45	1.414

Figure 10.41 Graphic printout from autocorrelation program: (*a*) input 32-bit group, (*b*) modulated wave, (*c*) wave with added multipath, (*d*) multipath and gaussian noise, and (*e*) packet autocorrelation.

The number of samples per bit was set at four after initial tests showed that this produced adequate results. The packet generation is started after entry of the run parameters. The first step is the generation of 32 random bits, which are then plotted as ±1's on the screen (Figure 10.41*a*). The next step is the generation of an MSK packet, using the selected bits to produce FSK with a peak-to-peak deviation of exactly 0.5 bit rate. Also, baseband samples $B(N)$ are derived, which would be required for generation of the MSK packet using offset quadrature carriers generating the I and Q samples of the MSK modulated packet. These are used later for comparison with an MSK demodulator output. The I and Q samples $X(N)$ and $Y(N)$ of the complex envelope are then generated at J samples per bit. The amplitude and phase are plotted as shown in Figure 10.41b.

Multipath delay, phase, and amplitude may then be entered for up to a total of five multipaths. To add a particular multipath, each signal I and Q sample pair is multiplied vectorially by the multipath amplitude and phase shift to produce the multipath I and Q components. These are delayed by the number of samples equivalent to the selected multipath delay and are then combined with the composite samples from the original signal and other multipaths already added. This is plotted by another subroutine, as shown in Figure 10.41c.

The next step is to generate gaussian noise samples independently in the I and Q channels. This was done by converting a randomly generated uniform distribution to Rayleigh by using the formula $V = \sigma [-2 \ln (RND)]$, where RND is generated from a random distribution, uniform between 0 and 1, and σ is chosen to give the assigned S/N. A second random variable, θ, was generated from a similar distribution by rescaling to make the resultant distribution uniform between $\pm\pi$. The I and Q samples of the noise components $V \cos \theta$ and $V \sin \theta$ are added to the composite signal components, and the resultant is passed through a four-pole Chebyshev 0.1-dB low-pass filter with a bandwidth equal to the bit rate. The output is plotted as shown in Figure 10.41d.

Finally, the partial autocorrelation is carried out. The first 32J samples received (first packet) are padded with an equal number of zeros, and the autocorrelation is carried out by successively offsetting the wave by one sample and multiplying and adding samples of the offset wave with those of the nonoffset wave. The correlation tends to be random except where the delay corresponds to a multipath, when a correlation occurs over the number of samples in the packet less the number in the delay. Because 32 bits is rather short, in some cases data side lobes are also higher than random. Figure 10.41e displays the autocorrelation. The example is a two-multipath case, and the largest peak is seen to be at the point of multipath delay (arrow under axis). In this case, other peaks are only slightly below the multipath peak, which has a rather large amplitude. A number of runs were made with various multipath delays, amplitudes, and phases, and over a range of S/N values. In most cases, troublesome multipath levels produced the highest peaks, but in a few cases a side-lobe peak was higher. In some runs, a third path was added and was also usually distinguishable. The results of the simulation served as a guide for the design described in Chapter 9.

In addition to the correlation, the program included two demodulator routines, operating on the filtered data—a frequency demodulator and a coherent MSK demodulator. The former simply used the arcsine to determine the phase change per sample and summed over the samples in the bit. Because the wave had passed through a filter with finite delay and there was no bit timing recovery routine, the demodulator bit intervals could be delayed a finite number of samples, set by the user. The optimum was normally found by cut and try. If the resultant was positive, the frequency was increasing; if it was negative, the frequency was decreasing. If on some occasion the average envelope for the two samples dropped to zero, the

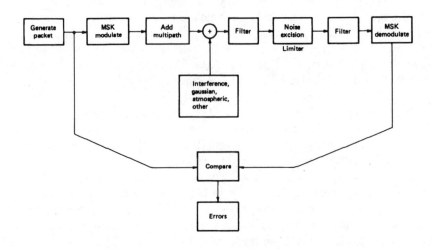

Figure 10.42 Block diagram of a link simulation model for effects of limiter on impulsive noise and interferers.

phase change was made zero. The output bits were compared with the originally selected bits to determine the number of errors.

The MSK demodulator used a reference that could be set to have a different phase from the first path. The incoming samples were projected (vectorially) on the reference I and Q channels to produce the output channels. Each channel was weighted by the half sine wave required to match the MSK modulation (one offset by a bit interval from the other). The samples in each channel after weighting were summed over the symbol ($2J$ samples), and a plus or minus decision was made. The decision is made alternately for each channel to get the output wave, which is compared to the MSK output values $B(N)$ derived before transmission. The program could be set to use either or both demodulators, and print out the number of errors in the packet from the demodulator.

This program was subsequently adapted for use primarily in error rate determination and to check the effects of filter bandwidth and hard limiting on the error rate in the presence of noise and adjacent channel interference. The size of the packet was increased and a noise program was added that was intended to better simulate the impulse character of man-made noise and atmospheric noise in the lower HF, MF, and LF portions of the spectrum. Figure 10.42 is a block diagram of the link functions that were simulated. The principal differences from the earlier model were the longer packet, the provision to add impulsive noise or sinusoidal interference in addition to gaussian noise, the addition of a provision to provide a noise limiter and an additional filter before demodulation, and deletion of the correlation and the detailed displays.

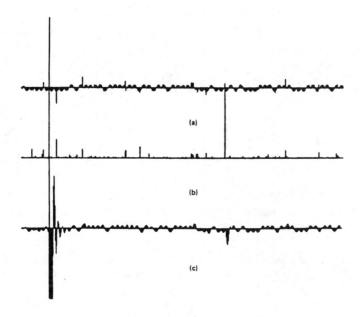

Figure 10.43 Graphic display from simulation model.

Displays were available optionally to show the noise waveform. Figure 10.43 shows a printout of one such display. In this case, Figure 10.43a is the in-phase channel of the modulation plus the in-phase portion of the noise, Figure 10.43b represents the envelope of the noise samples, and Figure 10.43c shows the effect of the roofing filter, which limits the noise bandwidth as well as the signal bandwidth. The impulsive nature of the noise is clearly evident.

The noise model chosen was a truncated Hall [10.44] model with coefficient 2. This closely matches the envelope distributions of atmospheric noise given in CCIR Report 322. The inputs required are the mean noise and V_d [10.45]. There are indications that real atmospheric noise differs from the Hall model because of its occurrence in bursts (References [10.46] and [10.47]). However, an analytic expression was not readily available for another model. The maximum envelope level is determined by the value of V_d. A truncation is required, because otherwise the total power is infinite and the mean power cannot be obtained.

The results obtained from this model (Figure 10.44) show the advantages of using a wider-band filter prior to a limiter for reducing the effects of impulse noise (a well-known rule for reducing impulse noise). It should be noted that with the parameters shown, the MSK demodulator noise bandwidth, following the filter, is about 740 Hz. Other data show the effect of a nearby CW jamming signal for the two bandwidths, as derived from the model. These show that we must be careful

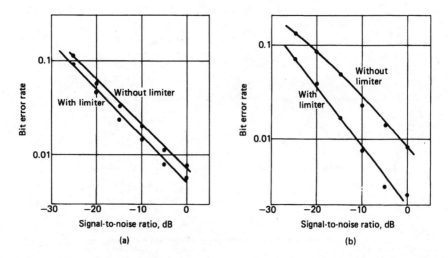

Figure 10.44 Effects of changing the prelimiter bandwidth on the error rate. Atmospheric noise only (V_d = 15.9 in 1200 Hz and MSK modulation at 1200 bits/s): (*a*) 1.5 kHz filter and (*b*) 3.0 kHz filter.

in such a selection of bandwidth to avoid widening the band too much. The relative importance of impulsive noise and nearby interfering signals must be carefully weighed. A simple model of this sort can provide initial tradeoffs in design, once the requirements are known.

10.4.7 Applications of Simulation

The importance of simulation as a tool for the receiver designer cannot be over-stated. However, some cautions are in order. Where possible, existing programs should be used. There are a variety of professional design and simulation programs available from a number of vendors. Each is tailored to meet specific needs. Many are structured as modular elements that can be added as dictated by the project at hand. Developing job-specific simulation programs in-house is possible, however, it can require a very large amount of time to develop, and—perhaps—nearly as much time to debug. In any event, simulations can become very complex and require a great deal of time from the designer and the machine. This needs to be carefully considered in any specific case but should not deter the designer from benefiting from the extremely useful tool of simulation.

References

10.1. J. B. Knowles and E. M. Olcayto, "Coefficient Accuracy and Digital Filter Response," *IEEE Trans.*, vol. CT-15, Mar. 1968.

10.2. E. Avenhaus and H. W. Schüssler, "On the Approximation Problem in the Design of Digital Filters with Limited Wordlength," *Arch. Elek. Übertragung*, vol. 24, p. 571, 1970.

10.3. U. L. Rohde, "A High-Performance Hybrid Frequency Synthesizer," *QST*, p. 30, Mar. 1995.

10.4. W. E. Sabin and E. O. Schoenike (eds.), *Single Sideband Systems and Circuits*, 2nd ed., McGraw-Hill, New York, 1995.

10.5. M. E. Frerking, *Digital Signal Processing in Communication Systems*, Van Nostrand Reinhold, New York, 1994.

10.6. S. Haykin, *Introduction to Adaptive Filters*, Macmillan Publishing Company, New York, 1984.

10.7. D. L. Hersberger, "DSP—An Intuitive Approach," *QST*, p. 39, Feb. 1996.

10.8. S. E. Reyer and D. L. Hershberger, "Using the LMS Algorithm for QRM and QRN Reduction," *QEX*, p. 3, Sept. 1992.

10.9. R. A. Scholtz, "The Origin of Spread Spectrum Communications," *IEEE Trans.*, vol. COM-30, p. 822, May 1982.

10.10. R. L. Pickholtz, D. L. Schilling, and L. B. Milstein, "Theory of Spread-Spectrum Communications—A Tutorial," *IEEE Trans.*, vol. COM-30, p. 855, May, 1982.

10.11. R. C. Dixon (ed.), *Spread Spectrum Techniques*, IEEE Press, New York, 1976.

10.12. D. Torrieri, *Principles of Military Communications Systems*, Artech House, Dedham, Mass., 1981.

10.13. J. K. Holmes, *Coherent Spread Spectrum Systems*, Wiley, New York, 1982.

10.14. R. C. Dixon, *Spread Spectrum Systems*, 2d ed., Wiley, New York, 1984.

10.15. M. K. Simon, J. K. Omura, R. A. Scholz, and B. K. Levitt, *Spread Spectrum Communications*, vols. I-III, Computer Science Press, Rockville, Md., 1985.

10.16. L. A. Gerhardt and R. C. Dixon (eds.), "Special Issue on Spread Spectrum Communications," *IEEE Trans.*, vol. COM-15, Aug. 1977.

10.17. C. E. Cook, F. W. Ellersick, L. B. Milstein, and D. L. Schilling (eds.), "Special Issue on Spread-Spectrum Communications," *IEEE Trans.*, vol. COM-30, May 1982.

10.18. A. G. Cameron, "Spread Spectrum Technology Affecting Military Communication," *Naval Research Rev.*, vol. 30, Sept. 1977.

10.19. W. W. Peterson and E. J. Weldon, *Error Correcting Codes*, M.I.T. Press, Cambridge, Mass., 1972.

10.20. "Data Encryption Standard," Federal Information Processing Standard, Pub. 46, National Bureau of Standards, Washington, D.C., Jan. 15, 1977.

10.21. "DES Modes of Operation," FIPS Pub. 81, National Bureau of Standards, Washington, D.C., Dec. 2, 1980.

10.22. J. R. Klauder, A. C. Price, S. Darlington, and W. S. Albersheim, "The Theory and Design of Chirp Radars," *Bell Sys. Tech. J.*, vol. 39, p. 745, July 1960.

10.23. C. E. Cook, "Pulse Compression, Key to More Efficient Radar Transmission," *Proc. IRE*, vol. 48, p. 310, Mar. 1960.

10.24. E. J. Nossen, "The RCA VHF Ranging System for Apollo," *RCA Eng.*, vol. 19, p. 75, Dec. 1973/Jan. 1974.

10.25. A. D. Watt, V. J. Zurick, and R. M. Coon, "Reduction of Adjacent Channel Interference Components from Frequency-Shift-Keyed Carriers," *IRE Trans.*, vol. CS-6, p. 39, Dec. 1958.

10.26. T. T. Tjhung, "Band Occupancy of Digital FM Signals," *IEEE Trans.*, vol. COM-12, p. 211, Dec. 1964.

10.27. M. G. Pelchat, "The Autocorrelation Function and Power Spectra of PCM/FM with Random Binary Modulating Waveforms," *IEEE Trans.*, vol. SET-10, p. 39, Mar. 1964.

10.28. V. K. Prabhu, "Spectral Occupancy of Digital Angle-Modulation Signals," *Bell Sys. Tech. J.*, vol. 55, p. 429, Apr. 1976.

10.29. V. K. Prabhu and H. E. Rowe, "Spectra of Digital Phase Modulation by Matrix Methods," *Bell Sys. Tech. J.*, vol. 53, p. 899, May-June 1974.

10.30. T. T. N. Bucher, "Spectrum Occupancy of Pulsed FSK," *MILCOM '82 Conf. Rec.*, vol. 2, p. 35.6-1, IEEE, New York, 1982.

10.31. H. F. Martinides and G. L. Reijns, "Influence of Bandwidth Restriction on the Signal-to-Noise Performance of PCM/NRZ Signal," *IEEE Trans.*, vol. AES-4, p. 35, Jan.1968.

10.32. "CCIR Atlas of Ionospheric Characteristics," Rep. 340, Oslo, 1966, and Suppl. 2 to Rep. 340, Geneva, 1974, ITU, Geneva, Switzerland.

10.33. J. J. Egli, "Radio Propagation above 40 MHz, over Irregular Terrain," *Proc. IRE*, vol. 45, p. 1383, Oct. 1957.

10.34. A. G. Longley and P. L. Rice, "Prediction of Tropospheric Radio Transmission Loss over Irregular Terrain—A Computer Method—1968," ESSA Tech. Rep. ERL-79-ITS-67, July 1968.

10.35. G. H. Hufford, A. G. Longley, and W. A. Kissick, "A Guide to the Use of the ITS Irregular Terrain Model in the Area Prediction Mode," NTIA Rep. 82-100, U.S. Dept. of Commerce, Boulder, Colo., Apr. 1982.

10.36. M. Hata, "Empirical Formula for Propagation Loss in Land Mobile Radio Services," *IEEE Trans.*, vol. VT-29, p. 317, Aug. 1980.

10.37. Y. E. Okamura, E. Ohmori, T. Kawano, and K. Fukudu, "Field Strength and Its Variability in VHF and UHF Land-Mobile Service," *Rev. Tokoyo Elec. Commun. Lab.*, vol. 16, p. 825, Sept./Oct. 1968.

10.38. T. Kailath, "Channel Characterization: Time-Variant Dispersive Channels," in E. J. Baghdady (ed.), *Lectures on Communication System Theory*, McGraw-Hill, New York, 1961.

10.39. P. A. Bello, "Characterization of Randomly Time-Variant Linear Channels," *IEEE Trans.*, vol. CS-11, p. 360, Dec. 1963.

10.40. C. C. Watterson, J. R. Juroshek, and W. D. Bensema, "Experimental Confirmation of an HF Channel Model," *IEEE Trans.*, vol. COM-18, p. 792, Dec. 1970.

10.41. R. Lugannani, H. G. Booker, and L. E. Hoff, "HF Channel Simulator for Wideband Signals. A Mathematical Model and Computer Program for 100-kHz Bandwidth HF Channels," Final Rep. on Contract N000123-76-C-1090, NOSC Tech. Rep. TR-208, Mar. 31,1978.

10.42 P. A. Bello, "Some Techniques for the Instantaneous Real-Time Measurement of Multipath and Doppler Spread," *IEEE Trans.*, vol. COM-13, p. 285, Sept. 1965.

10.43 B. Goldberg, "300 kHz–30 MHz MF/HF," *IEEE Trans.*, vol. COM-14, p. 767, Dec. 1966.

10.44 T. A. Schonhoff, A. A. Giordano, and Z. McC. Huntoon, "Analytic Representations of Atmospheric Noise Distributions, Constrained in V_d," *IEEE Conference Rec. ICC-'77*, vol. 1, p. 8.3-169, June 1967.

10.45 "World Distribution and Characteristics of Atmospheric Radio Noise," CCIR Rep. 322, ITU, Geneva, 1964.

10.46 R. T. Disney and A. D. Spaulding, "Amplitude and Time Statistics of Atmospheric and Man-Made Noise," ESSA Tech. Rep. TR-ERL 150-ITS98, Feb. 1970.

10.47 J. Herman, X. DeAngelis, A. Giordano, K. Marsotto, and F. Hsu, "Considerations of Atmospheric Noise Effects on Wideband MF Communications," *IEEE Commun. Mag.*, vol. 21, p. 24, Nov. 1983.

Additional Suggested Reading

"Digital HF Radio: A Sampling of Techniques," Third International Conference on HF Communication Systems and Techniques, London, UK, Feb. 26, 1985.

"Eight Ways to Better Radio Receiver Design," *Electronics*, Feb. 20, 1975.

Index

Z

Ulrich L. Rohde is president of Compact Software, Chairman of Synergy Microwave Corp., and a partner of Rohde & Schwarz, Munich, West Germany, a multinational company specializing in advanced test and radio communications systems. Previously, he was the business area director for Radio Systems of RCA, Government Systems Division, Camden, New Jersey, responsible for implementing communications approaches for military secure and adaptive communications. Having studied electrical engineering and radio communications at the universities of Munich and Darmstadt, Germany, he holds a Ph.D. in electrical engineering and an Sc.D. (hon.) in radio communications.

From 1990 to 1992, Dr. Rohde was appointed Visiting Research Professor at New Jersey Institute of Technology, Department of Electrical Engineering, Newark, NJ. Dr. Rohde is also a member of the staff at George Washington University and as an adjunct professor teaching in the Electrical Engineering and Computer Sciences departments gave numerous lectures worldwide regarding communications theory and digital frequency synthesizers. In addition, as a Professor of Electrical Engineering at the University of Florida, Gainesville, he has been teaching radio communications course.

Dr. Rohde has published more than 50 scientific papers in professional journals, as well as four other books: *Communications Receivers: Principles and Design,* McGraw-Hill, 1987; *Digital PLL Frequency Synthesizers: Theory and Design,* Prentice-Hall 1983; *Transistoren bet hoechsten Frequenzen,* in German, 1965, (on microwave transistors). His latest book, *Microwave Circuit Design Using Linear and Nonlinear Techniques,* with co-authors George Vendelin and Anthony M. Pavio was published by John Wiley & Sons, NY, January, 1990.

Dr. Rohde is a member of the following: ETA KAPPA NU Honor Society; senior member of the IEEE; Executive Association of the Graduate School of Business, Columbia University, New York; the Armed Forces Communications & Electronics Association; fellow of the Radio Club of America; and a life member of the Amateur Radio Relay League. He holds an Extra Class U.S. amateur license (call sign KA2WEU, 1991) and is an ARRL accredited VE. He is known for his German call sign DJ2LR (1956), and Swiss call sign HB9AWE (1973). He also holds an FCC General Radiotelephone Operator License, a Global Maritime Distress and Safety System (GMDSS) maintainer license, both with radar endorsements and a second class Radiotelegraph operator license.

Jerry C. Whitaker is President of Technical Press, a Beaverton (OR) based consulting company. Mr. Whitaker has been involved in various aspects of the electronics industry for over 20 years, with specialization in communications. Current book titles include the following:

- *Radio Frequency Transmission Systems: Design and Operation*, McGraw-Hill, 1990

- *Electronic Displays: Technology, Design, and Applications*, McGraw-Hill, 1993

- *Maintaining Electronic Systems*, CRC Press, 1991

- *AC Power Systems Handbook*, CRC Press, 1991

- *Power Vacuum Tubes Handbook*, Van Nostrand Reinhold, 1994

- Editor-in-chief, *The Electronics Handbook*, CRC Press, 1996

- Co-author, *Interconnecting Electronic Systems*, CRC Press, 1992

- Co-author, *Television and Audio Handbook for Engineers and Technicians*, McGraw-Hill, 1989

- Co-editor, *Television Engineering Handbook* (revised edition), McGraw-Hill, 1991

- Co-editor, *Information Age Dictionary*, Intertec Publishing/Bellcore, 1992

Mr. Whitaker has lectured extensively on the topic of electronic systems design, installation, and maintenance. He is the former editorial director and associate publisher of *Broadcast Engineering* and *Video Systems* magazines, and a former radio station chief engineer and television news producer.

Mr. Whitaker is a Fellow of the Society of Broadcast Engineers and an SBE-certified professional broadcast engineer. He is also a member of the following professional organizations: SPIE, Society of Motion Picture and Television Engineers, International Television Association, and the Institute of Electrical and Electronics Engineers.

Mr. Whitaker has twice received a Jesse H. Neal Award *Certificate of Merit* from the Association of Business Publishers for editorial excellence. He has also been recognized as *Educator of the Year* by the Society of Broadcast Engineers.